Risk and Safety Analysis of Nuclear Systems

Risk and Safety Analysis of Nuclear Systems

John C. Lee

Norman J. McCormick

WILEY

A JOHN WILEY & SONS, INC., PUBLICATION

Published by John Wiley & Sons, Inc., Hoboken, New Jersey.

Published simultaneously in Canada.

Library of Congress Cataloging-in-Publication Data:

Lee, John C., 1941– author.
Risk and Safety Analysis of Nuclear Systems / John C. Lee, Norman J. McCormick.
p. cm
ISBN 978-0-470-90756-6 (hardback)
1. Nuclear facilities—Security measures. 2. Nuclear engineering—Safety measures. 3. Nuclear engineering—Risk assessment. I. McCormick, Norman J., 1938– author. II. Title.
TK9152.L44 2011
621.48'35—dc22 2010049603

oBook ISBN: 978-1-118-4346-2 ePDF
ISBN: 978-1-118-04344-8 ePub ISBN:
978-1-118-04345-5

MWEP416408281125

CONTENTS

PREFACE

Nuclear power provides over 20% of the U. S. electricity generation and in several other countries the percentage is much higher (e.g., in France it is nearly 80%). After a multi-decade hiatus, it appears that nuclear power again may become a viable option for new electrical generation facilities in the United States. Enrollments in undergraduate and graduate nuclear science and engineering programs around the country are now increasing and recently there have been applications to the U. S. Nuclear Regulatory Commission for the licensing of proposed nuclear power plants. We hope that this book will help enhance the safety, reliability, and availability of nuclear energy systems in the coming decades and serve to remind the next generation of nuclear professionals that *a nuclear accident anywhere is a nuclear accident everywhere.* This was demonstrated with the tsunami-initiated events of March 2011 at the Fukushima Daiichi nuclear complex.

The first part of the book covers the principles of risk and reliability analysis found in courses typically offered in mechanical engineering or industrial engineering departments, as well as in nuclear engineering programs. The second part of the book covers applications of the methods for probabilistic risk assessment of complex engineered systems, together with deterministic safety analysis of nuclear power plants. A review of major accidents and incidents for nuclear power plants over the past thirty years also is presented, as well as passive safety features of advanced nuclear systems under development. The advanced systems are expected to efficiently generate electricity and process heat as well as transmute transuranics from used nuclear fuel.

The book has been developed in conjunction with a course taught every year to seniors and beginning graduate students in the Nuclear Engineering and Radiological Sciences department at the University of Michigan by the first author. A portion of that course was based on the textbook *Reliability and Risk Analysis Methods and Nuclear Power Applications* (Academic, 1981) by the second author that was used a couple of decades ago for a course in the University of Washington Nuclear Engineering department. Portions of that book have been extensively revised and additional exercises have been included to form the first part of this book.

The first author acknowledges help from Josh Hartz and Kwang Il Ahn, and a number of his current and former students, especially John Lehning, Douglas Fynan, Athi Varuttamasenni, Fariz Abdul Rahman, and Nick Touran. He also wishes to thank the late Professor Thomas H. Pigford for an introduction to the emerging field of nuclear reactor safety and the late Professor William Kerr for sustained opportunities to learn the reactor safety culture. Finally, he offers thanks to his wife Theresa and daughter Nina for all their loving care and sustained support. The second author thanks his wife Millie for her patience and not asking too frequently "Are you sure you want to be doing this when retired?"

March 2011

John C. Lee
Ann Arbor, Michigan

Norman J. McCormick
Seattle, Washington

This second printing of the book includes an updated Section 9.4 on the Fukushima Daiichi accident and corrections of typographical errors, which are also available at http://rsans.engin.umich.edu.

March 2016

JCL and NJM

PERMISSIONS AND COPYRIGHTS

Many figures and tables in this book have been reproduced from copyrighted sources. Permission from the publishers and authors for the use of the material is gratefully acknowledged. Some of the sources are directly identified in captions and footnotes, while many others are cited by alphanumeric references. Citations for these sources are listed below:

Introduction to Nuclear Power, 2nd ed., G. F. Hewitt and J. G. Collier
Copyright © 2000 by Taylor & Francis. Figures 8.13, 8.14, 8.15, 8.16, 8.17, 8.18, 8.19.

Handbook of System and Product Safety, 1st ed., pp. 242, 243, 245, W. Hammer
Copyright © 1972 by Pearson Education, Inc., Upper Saddle River, NJ. Figures 6.3, 6.4, 6.5.

Nuclear Engineering and Design
Copyright © 1987 by Elsevier Science and Technology. Figures 8.20, 8.21, 11.1, 11.5, 11.6.

Nuclear Engineering International
Copyright © 2002 by Progressive Media Group. Figure 11.9.

Nuclear News
Copyright © 1986 by the American Nuclear Society, La Grange Park, IL. Figure 9.8.

Nuclear Science and Engineering
Copyright ©1981, 1987, 2006 by the American Nuclear Society, La Grange Park, IL. Figures 13.1, 9.15, 13.4, 13.9, 13.10, 13.11, 13.12, 13.13, Table 13.2.
Nuclear Technology
Copyright © 1989 by the American Nuclear Society, La Grange Park, IL. Figures 9.1, 9.2, 9.4, 9.5, 9.6, 9.7.
Reliability Engineering and System Safety
Copyright © 1988, 1993, 2008 by Elsevier Science and Technology. Table 13.1. Figures 7.4, 9.1, 9.2, 13.2, 13.6, 13.7, 13.8.
The New York Times, K. Chang
Copyright © June 8, 2003 by *The New York Times*. All rights reserved. Used by permission and protected by the copyright laws of the United States. The printing, copying, redistribution, or retransmission of the material without express written permission is prohibited. Figure 9.11.

A number of figures and tables were also obtained from publications of various government agencies and laboratories: Tables 6.1, 6.4, 6.5, 6.7, 9.1, 9.2, 10.1, 10.2, 10.3, 10.4, 10.5. Figures 2.2, 2.4, 6.8, 7.1, 7.2, 7.3, 8.1, 8.3, 8.4, 8.6, 8.7, 8.8, 8.9, 8.12, 8.26, 8.27, 8.28, 8.29, 9.3, 9.9, 9.10, 9.12, 9.13, 9.16, 9.17, 9.18, 9.19, 10.1, 10.2, 10.3, 10.5 10.6, 10.7, 10.8, 10.10, 10.11, 10.12, 10.13, 10.14, 10.15, 10.16, 10.17, 10.18, 10.19, 11.11, 11.12, 11.13, 11.19, 11.22, 11.23, 12.1, 12.2.

List of Tables

List of Figures

CHAPTER 1

RISK AND SAFETY OF ENGINEERED SYSTEMS

1.1 RISK AND ITS PERCEPTION AND ACCEPTANCE

Risk and safety concerns for the engineering of nuclear power plants are somewhat analogous to the opposing yin and yang energies that represent the ancient Chinese understanding of how things work. The outer circle represents "everything", while the "yin" (black) and "yang" (white) shapes within the circle represent the interaction of two energies that cause everything to happen. As such, risk (yin) is the performance downside of a nuclear system and safety (yang) is what happens when the system performs its intended function. In the Chinese interpretation of yin-yang, there is a continuous movement between the two energies, just as there is when a nuclear system operates. Just as the Chinese have observed, risk and safety are intertwined, even though the engineering principles for each have a different emphasis.

Risk is the combination of the predicted frequency of an undesired initiating event and the predicted damage such an event might cause if the ensuing follow-up events were to occur. In essence, it combines the concepts of "How often?" with "How bad?"

In this book we are concerned with *probabilistic* risk assessment (PRA) and the methods used to analyze the safety of nuclear systems. For this reason we are investigating risks that *might* occur to society as a whole, rather than risks that might be incurred by an individual in society. A PRA typically models events that only very rarely occur. Hence it differs from an investigation in which there is an operating history from which to predict risks. Although most of the licensing and regulations governing the current generation of operating nuclear power plants are based on deterministic assessment of the consequences of postulated accidents and operating conditions, there is an increasing emphasis placed on implementing PRA techniques in licensing decisions. With this perspective, the terminology *probabilistic safety analysis* often is used to represent the safe assessment that combines the elements of both probabilistic and deterministic methods. Thus, the dichotomy between risk and safety has become somewhat fuzzy in recent years.

When thinking about a complex technology it is not difficult to conjecture a series of questions: What if undesired event A happened? Or if undesired event B happened? Or if undesired event C happened? ... To scientifically answer such questions requires clearly defining what the consequences of events A, B, C, ... are, but an often overlooked aspect is the frequency of occurrence of such events. Risk analysis techniques are needed to assess both the frequency and the consequence of an undesired event while safety analysis techniques are for preventing the occurrence of such events.

Perception of the risk associated with any human activity, including that associated with the utilization of man-made systems, is quite subjective. This can be illustrated by the way the news media typically report on airplane crashes involving the injury or death of even a few passengers and crew, while the annual casualties of 40,000 to 50,000 individuals due to automobile accidents in the United States do not receive special coverage. The distinction between perhaps a few hundred casualties resulting from airplane accidents and a much larger number of deaths from automobile accidents in the United States every year can be characterized in two ways: (a) voluntary versus involuntary risks and (b) distributed versus acute or catastrophic risks. We consider the risk associated with traveling in private automobiles a voluntary one that is under our personal control, in contrast to the involuntary risk involved with commercial airline flights in which we do not have control. Similarly, an automobile-related accident typically does not result in a large number of casualties so the risk is distributed, while a catastrophic airline crash can result in a large number of casualties.

Acceptability of risk is often inversely proportional to the consequences. In the risk space shown in Fig. 1.1, the abscissa represents the consequences or dreadfulness and the ordinate the observability or familiarity of the hazard. Events in the upper right quadrant, entailing significant consequences and significant unfamiliarity or limited observability, generally require strict regulations. In the case of postulated accidents in nuclear systems, the consequences could be significant although the probability of the accidents is predicted to be very small. Thus, the traditional method of risk evaluation is often subject to public skepticism, despite the extensive efforts made in implementing scientific principles in the design, construction, and operation of nuclear systems.

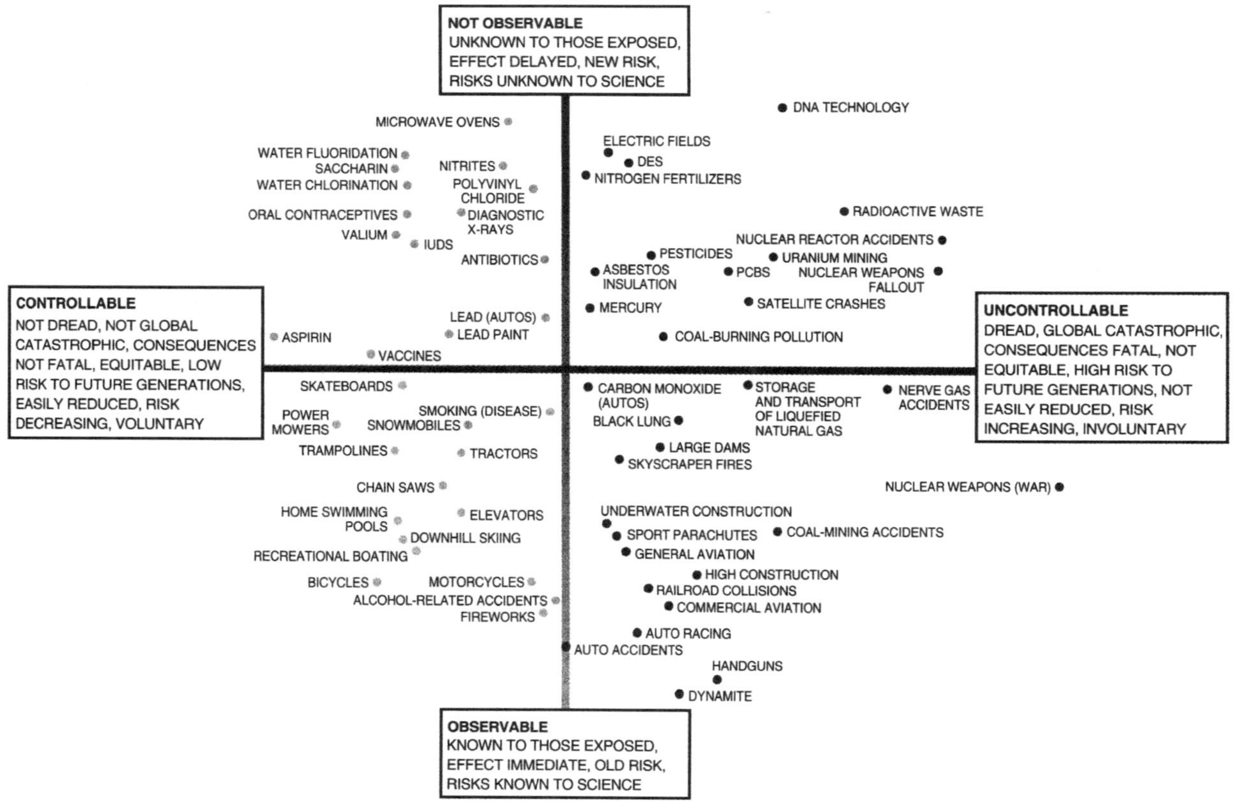

Figure 1.1 Risk space illustrating acceptability of risks as a function of consequences and observability. *Source:* Reprinted with permission from [Mor93]. Copyright © 1993 Scientific American, a division of Nature America, Inc.

Risks are incurred in everyday life by everyone, of course. So what distinguishes such risks from those from the operation of a nuclear power plant, for example? An important distinction in whether an individual accepts a risk is whether he or she has control over the risk to be incurred. Other factors are important as well and have been summarized in Table 1.1.

Table 1.1 Factors Affecting Acceptance of Risks

Effect	Opposite Effect
Assumed voluntarily	Incurred involuntarily
Consequences occur immediately	Consequences delayed
Consequences reversible	Consequences irreversible
Consequences short term	Consequences long term
No alternatives available	Many alternatives available
Small uncertainty	Large uncertainty
Common hazard	Unknown or "dreaded" hazard
Exposure is necessary	Exposure is optional
Incurred occupationally	Incurred nonoccupationally
Incurred by other people	Not incurred by other people

Source: Modified and expanded from [Low76].

The use of nuclear power for the generation of electricity has the disadvantage of many factors working against its acceptance. By its very nature, a probabilistic analysis of any system can never yield a result for a "risk known with certainty." The potential for a delay of the effects and the irreversible consequences following a catastrophic event at a radioactive waste disposal site are contributing effects to the siting of such sites, for example. Public concerns over the potential for delayed climate changes arising from the buildup of CO_2 also can be understood in the context of the Table 1.1 factors.

One might think that the response of the public to modern medical imaging methods might provide a clue for the eventual acceptance of nuclear power. Widespread acceptance of x-rays shows that a radiation technology can be tolerated once its use becomes familiar, its benefits clear, and its practitioners trusted. In spite of the three most widely publicized nuclear power accidents, at Three Mile Island Unit 2, Chernobyl and Fukushima Daiichi, the nuclear power safety record is outstanding in light of the benefits obtained from the electricity generated without CO_2 emissions. But yet several decades have passed, with countries like France generating upward of 80% of its electricity by nuclear power, and the acceptance of nuclear power in the United States has remained lower than most engineers with a nuclear background could have imagined at those earlier times.

It can be argued that unfavorable media publicity has played a role in the lack of acceptance of nuclear power by a large fraction of the U.S. population. An outstanding example of this is what transpired after the Three Mile Island nuclear reactor accident in March 1979 in which some radioactive gas was released a couple of days after the accident, but not enough to cause any dose above background levels to local residents.

Indeed, for 18 years the Pennsylvania Department of Health maintained a registry of more than 30,000 people who lived within 5 miles of Three Mile Island at the time of the accident, but that was discontinued in mid 1997 without any evidence of unusual health trends in the area. Yet an explosion at the Union Carbide India pesticide plant in Bhopal in December 1984 released toxic gas in the form of methyl isocyanate and its reaction products over the city. The estimated mortality of this accident is believed to have been between 2500 and 5000 people, with up to 200,000 injured [Meh90]. But such an accident was largely ignored by the media in comparison to the publicity surrounding the Three Mile Island accident. One reason for this disparity was that the consequences of the Bhopal accident were known within days while the effects of the Three Mile Island accident took years to assess.

Industrial facilities such as nuclear reactors and chemical plants have been studied, by the techniques presented in this book, for their risks to the public at large. But such investigations are entirely different than what people do in making their own individual decisions about risks in their everyday lives. Because ordinary citizens do not have direct control over how their electricity is generated or various products are manufactured, the operation of such industrial facilities must lead to the probability of undesired consequences much lower than the risks from everyday occurrences.

For common risks leading to unnatural human deaths incurred involuntarily by an individual, for example, the probability of occurrence loosely can be bounded between 10^{-6}/yr and 10^{-2}/yr. The lower bound is set by the risk of death from natural events, such as lightning, flood, earthquakes, insect and snake bites, etc. (about one death per year per million people) and the upper bound arises by the death rate from disease (about one death per year per 100 people). The lower bound is not, however, appropriate for a large-scale commercial facility like a nuclear power or chemical plant.

One can argue that the risks from the operation of plant A need not necessarily be as small as those from operation of plant B if one perceives the benefits of the products produced by plant A to be greater than those from plant B. An early *comparative risk assessment* of technologies for the generation of electricity was performed by Inhaber [Inh82]. He investigated the production of electricity in MWe-yr from 11 different sources: coal, oil, nuclear, natural gas, hydroelectric, wind, methanol, solar space heating, solar thermal, solar photovoltaic, and ocean thermal sources (but did not consider ocean tidal, for example). One innovative feature of the study was to put the technologies for each power source on equal footing by also assigning the percentage of risk—for energy backup during the predicted down time for maintenance, etc.— from other electric generating plants in Canada. (Thus, interruptible power sources were assigned risks not only from their own performance.) Beside the risks from activities related to electricity production, operation, maintenance, and energy backup, his risk estimates included emissions from acquisition of materials to build the plant, energy storage, transportation, and the gathering and handling of fuels and acquiring material and equipment. For nuclear systems he also included estimates of the risks of waste management along with possible catastrophic reactor accidents. The consequences per MWe-yr included public deaths and occupational deaths and also public and occupational lost person-days. Although the numerical values of an early version of the study and some of the techniques were questioned [Hol79a,Hol79b], risks from nonconventional energy sources can be as high as or even higher than that of some conventional sources and the relative rankings of the 11 systems were not strongly influenced by whether the energy backup was included in the calculations [Inh82]. Figure 1.2 shows that in such energy comparison studies that are normalized to equal amounts of uninterruptible power generation, it is important to account for the risks from producing the materials used to construct the energy production system.

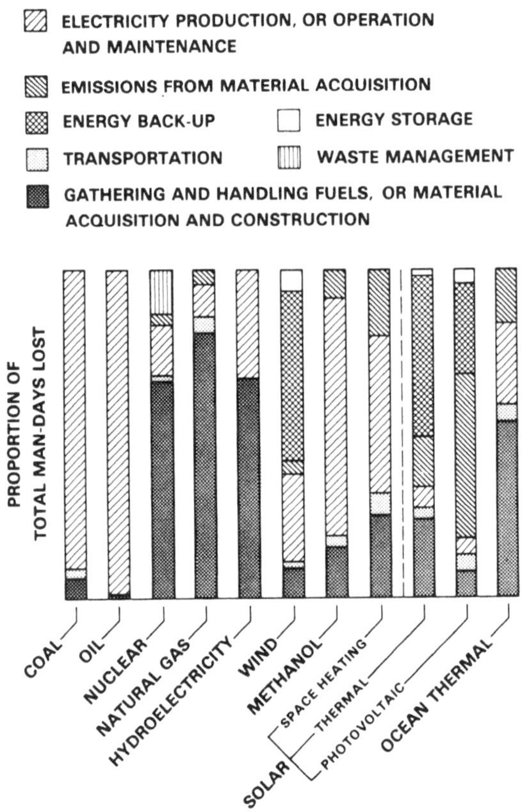

Figure 1.2 Proportions of risk by source, normalized to the sum of the occupational and public risks for each source. *Source:* Reprinted with permission from [Inh82]. Copyright © 1982 Gordon and Breach.

1.2 OVERVIEW OF RISK AND SAFETY ANALYSIS

The objective of a risk analysis is to predict what might happen, beginning with an undesired initiating event and following that event in time to predict an undesired consequence if the active and passive safety systems fail to perform their intended function. In other words, risk involves the occurrence or potential occurrence of some accident sequence involving one or more events, together with the ensuing consequences from such an accident. On the other hand, the objective of a safety analysis is to design the components of a system so that undesired initiating events do not occur or, if they do, that backup systems intervene in the progression of following events to prevent or mitigate any undesired consequences.

What types of undesired initiating events can occur? There are postulated events such as a large pipe break caused by an earthquake or an electrical short in a safety system caused by a local fire. Indeed, part of the focus in the latter part of this book is to focus on some of these initiating events. What happens after such an initiating event? Because of the inherent potential danger of an uncontrolled release of ionizing radiation, nuclear plants have backup safety systems to reduce the undesired consequences from the undesired accident sequence. The failure of such backup systems causes an initiating event to become a sequence of failure events to form the accident sequence.

What kind of consequences are of concern? The loss of human life immediately comes to mind, such as in the catastrophic Chernobyl accident with the loss of life to plant workers and citizens in the surrounding countryside. Of course there also are differences between the length of time people lived following that event: some died within hours and others from the prolonged exposure to radionuclides that affected their thyroid glands, for example.

The potential consequences from a release of the radiological source contained in a typical nuclear power plant (NPP) pose a unique safety concern. An estimate of the inventory in an operating reactor may be obtained based on a simple physical analysis with the approximation that every fission event is a binary fission yielding two fission products and that every fission product (FP) undergoes one radioactive decay in an equilibrium operating condition. With this simple but reasonable approximation, together with a recoverable energy of 200 MeV released per fission, 1 W of thermal energy generated requires 3.1×10^{10} fissions/s, which then produces approximately 2 Ci of radioactivity. Thus, a 1.0-GWe nuclear power plant with a thermal efficiency of 33% produces 3.0 GWt, which then yields an equilibrium radioactivity of 6000 MCi (6 BCi). This simple estimate compares favorably with a total radioactivity inventory of 5.6 BCi, including 3.8 BCi of FP radioactivity, in the tally of radioactivity in Appendix A for a 3.56-GWt reactor [Rah84]. This huge inventory of radionuclides accounts for about 6 to 7% of the total power in a typical operating plant, and this power must be dissipated after the chain reaction is terminated. (These two features provide distinctly different risk and safety concerns from a coal-fired plant.) For this reason Appendix A also contains an introduction to the fission product inventory and decay heat in a nuclear reactor, health effects of radiation exposure, and current regulations governing radiation exposure.

As engineers analyzing a nuclear system we have a moral obligation to develop the safest possible system. By performing a risk analysis we may obtain sufficient information to redesign it and lower the probability of the occurrence of an accident or mitigate the ensuing consequences. Alternatively, it may be possible to show that the probability of occurrence of a postulated accident is negligibly small enough that the potential accident can be neglected compared to other potential accidents.

A PRA can provide either a point estimate or an interval estimate of an event. Although the point estimate may give the best single value for the probability of occurrence, it does not give any indication of the uncertainty in the estimate. An interval estimate, on the other hand, is useful because the width of the interval conveys how well, in a probabilistic sense, the point estimate is known. Confidence limits for an estimated parameter provide a point estimate combined with functions of the standard errors. Hence both estimates are useful.

In addition to the need to calculate the risk of any technology it is necessary to represent the state of knowledge uncertainty and population variability [Kap83]. The state of knowledge uncertainty is also known as "assessment uncertainty" and covers the uncertainty that could be reduced by further research. The "population variability" for nuclear power plants accounts for variability in engineered systems, e.g., differences in engineered safety systems of individual plants. The first PRA of a family of potential system failures for a boiling water reactor (BWR) and a pressurized water reactor (PWR) was the *Reactor Safety Study* [NRC75] completed in 1975. Although that study is now dated, because it was based on nuclear plants that were operating in 1972 and designed much earlier, it is still of interest to engineers interested in risks from nuclear systems because the study established methods used in all later investigations and because it was very comprehensive.

1.3 TWO HISTORICAL REACTOR ACCIDENTS

The importance of risk and safety analysis becomes obvious when considered in the context of history. The Three Mile Island accident in 1979 in which the reactor was destroyed by a core meltdown—but which led to only a very minor release of radioactivity outside the turbogenerator building—provided an incentive to further develop techniques to predict potential events leading to system malfunctions. Follow-on reports augmenting the procedures developed in the Reactor Safety Study and used in probabilistic risk analyses were published in the early 1980s, including a guide to fault tree analysis [Ves81] and a PRA procedures guide [NRC83]. Another important report was an assessment of risks for five U.S. nuclear power plants [NRC90].

The accident at the Chernobyl nuclear power plant in 1986 also contributed to the current emphasis on the use of probabilistic techniques for the analysis of nuclear systems, even though that plant was of an entirely different type than those built outside of the former Soviet Union because the RBMK reactors had a positive void coefficient of reactivity. A power excursion was initiated when the reactor operators were testing the performance of the coolant pumps operated with electrical power from the plant's turbine generator rather than off-site power. After overheated fuel from the reactor core was ejected into the coolant, causing it to boil off, reactivity was added to the reactor core, which increased the power excursion so rapidly that the control systems could not shut the system down. A steam explosion subsequently destroyed the pressure vessel, which led to the release of massive amounts of radioactivity, causing early fatalities and subsequent long-term health consequences from radiation exposure.

These two reactor accidents, along with the 2011 Fukushima accident and other incidents of major concern, will be discussed in much more detail in Chapter 9.

1.4 DEFINITION OF RISK

To express the concept of risk in more mathematical terms, risk $\mathcal{R}_i$ combines the frequency $\mathcal{F}_i$ of an event sequence i, in events per unit time, with the corresponding damage $\mathcal{D}_i$, which is the magnitude of the expected consequence. A traditional definition of risk is

$$\mathcal{R}_i = \mathcal{F}_i \mathcal{D}_i. \tag{1.1}$$

Other definitions could be used, however, if one wished to amplify the importance of undesired events with large consequences, such as with $\mathcal{R}_k = \mathcal{F}\mathcal{D}^k$ for $k > 1$. Risk differs from *hazard*, which is a condition with the potential of causing an undesired consequence, and from *danger*, which is exposure to a hazard.

More generally, the damage from an accident sequence can be analyzed with a continuum of outcomes between x and $x + \Delta x$. Then, instead of Eq. (1.1), the *risk density* $\mathcal{R}_i(x)$ of magnitude $\mathcal{D}_i(x)$, per unit of damage, can be interpreted as

$$\mathcal{R}_i(x) = \mathcal{F}_i \mathcal{D}_i(x). \tag{1.2}$$

Usually, however, of more interest is the risk of damages $\mathcal{D}_i(x)$ exceeding the magnitude X, in which case the risk in Eq. (1.2) is replaced by

$$\mathcal{R}_i(\geq X) = \mathcal{F}_i \int_X^\infty \mathcal{D}_i(x)\,\mathrm{d}x. \tag{1.3}$$

The risk $\mathcal{R}_i(\geq X)$ is the *complementary cumulative distribution function* (CCDF) for accident sequence i.

In the case of a severe release of radioactivity, more than one type of potential damage could occur. For example, there could be early deaths, within days to weeks after the release, due to acute doses of radiation. Or latent somatic effects after lesser radiation exposures, leading to cancer fatalities, might occur typically within a few years or a few decades. In addition, loss of work time (in person-days) and property losses also are potential damages. For such cases, when a catastrophic initiating event i causes a variety of predicted consequences of type j, leading to damages with a magnitude between $\mathcal{D}_{ij}$ and $\mathcal{D}_{ij} + \Delta\mathcal{D}_{ij}$, then Eqs. (1.2) and (1.3) are replaced by, respectively,

$$\mathcal{R}_i(x) = \mathcal{F}_i \sum_j \mathcal{D}_{ij}(x) \tag{1.4}$$

and

$$\mathcal{R}_i(\geq X) = \mathcal{F}_i \sum_j \int_X^\infty \mathcal{D}_{ij}(x)\,\mathrm{d}x. \tag{1.5}$$

A cornerstone of the risk and safety assessments for nuclear systems is the principle of defense in depth (DID), originating from the various safety measures that Enrico Fermi and his colleagues incorporated in the planning and execution of the first self-sustaining chain reaction at the University of Chicago in 1942. Thus, the DID principle has been implemented at every stage of design, construction, and operation of nearly every nuclear reactor around the world, with an ultimate objective of protecting the health and life of the population at large, although some people would argue that this was not done with the Russian RBMK reactors. The principle may be accomplished through the diversity and redundancy of various equipment and safety functions. The safety principle may also be represented in terms of multiple layers of radiation barriers, including the fuel matrix, fuel cladding, reactor pressure vessel, and ultimately the reactor containment building. In terms of safety functions, three basic levels may be illustrated: (a) *prevention* of accidents via reactor shutdown, (b) *mitigation* of accidents through the actuation of an auxiliary coolant system, and (c) *protection* of the public via containment sprays minimizing the release of radionuclides to the environment. The DID principles are fully reflected in the General Design Criteria, promulgated as Appendix A to Title 10, *Code of Federal Regulations*, Part 50 [NRC71].

1.5 RELIABILITY, AVAILABILITY, MAINTAINABILITY, AND SAFETY

The risk and safety issues of a nuclear plant initially depend on the plant design and construction. Thereafter, because a plant naturally cannot operate indefinitely without intervention, the degree of risk versus safety depends on the maintenance procedures and operator actions intended to improve the plant operation.

To determine a risk $\mathcal{R}_i(\geq X)$ of an undesired event, it is necessary to predict the availability of the safety systems that should operate after the initiating event to mitigate the consequences. The availability of a safety system is analyzed with the concepts of reliability engineering used for predicting whether the system is "up" or "down." When performing an availability or reliability analysis, there are several issues related to performance that must be considered: hardware and software failures, human errors, and incorrect operating procedures as well as the interactions between these.

What are the differences between a reliable system and an available system? Or to phrase the question in a different way, can a safety system, for example, be available but not very reliable? Reliability $R(t)$ is the probability that a system can perform a specified function or mission under given conditions *for* a period of time t, while availability $A(t)$ is the probability that a system can perform a specified function or mission under given conditions *at* time t. The difference between $R(t)$ and $A(t)$ arises because reliability does not account for the possibility that a given system can be repaired after its failure. This means that $R(t)$ predicts the time of interest t until the system has undergone its first failure, whereas the system may have failed in the past but been repaired so that it is operational at time t with predicted availability $A(t)$. Reliability, also called the survival function of a system, is the complement of the failure probability $F(t)$ that defines the probability of failure after a time period t, i.e., $R(t) = 1 - F(t)$.

It is important to note that reliability refers to the first system failure, but a system with redundant subsystems can exhibit subsystem failures without system failure. For a reliability analysis, once a system has failed, any incomplete repair actions are considered to cease, whereas for an availability analysis the on-going repair actions continue. Thus, if a system can be repaired, then the *mean time between failures* (MTBF) should exceed the *mean time to failure* (MTTF).

The assumptions about the way a system degrades with age and how it responds to a failure affect the type of model that can be assumed for repair of a system. A *minimal repair* returns the system to the state the system was in immediately preceding failure, while a *perfect repair* or renewal repair returns it to the state it was in when new. A minimal repair model allows the analysis of systems that are deteriorating or improving with time, while a perfect repair model does not. A minimal repair model for which improvements with time might be appropriate, for example, is if the repair people can learn from identical previous repairs.

Maintainability, on the other hand, is the ability of a system component, during its prescribed use, to be restored to a state in which it can perform its intended function when the maintenance is performed under prescribed procedures. It involves actions typically performed according to procedures established by the manufacturer of the component. Although manufacturers may have tabulated data that prescribe regular maintenance procedures, the frequency of maintenance actions is guided by experience and depends not only on the quality of a system's components but also on the operating environment of the equipment, such as the operating temperature or pressure. Although probabilistic failure analyses can be incorporated when developing a scheduled maintenance procedure, maintenance procedures for nuclear systems tend to be developed more through operating experience, with the objective of increasing the safety of the plant and decreasing the system downtime caused by an unscheduled outage.

Reliability, availability, and maintainability (often abbreviated RAM) all contribute to improving the safe operation of a nuclear plant. A plant operated with good RAM procedures provides safety, which can be defined as eliminating those conditions, to an acceptable level of risk, conditions that can cause death, injury, occupational illness, or damage to or loss of equipment or property. Because safety is the single most overriding consideration of plant operation, one is most interested in the availability of the plant safety systems to perform their intended functions *at* the time they are needed. From the perspective of decreasing the plant downtime, on the other hand, one is interested in the reliability of the system components *for* the duration of time between routinely scheduled maintenance activities. A RAM program coupled to safety (S) enhancement of the plant leads to a RAMS structure.

The RAM program did not develop as a unique discipline, but rather it has grown out of the integration of activities previously used by engineers to achieve a reliable, safe and cost-effective system. Engineered systems have been growing more and more complex over the past decades, which now requires increased attention to maintain the performance of the systems with minimal cost. Thus, it has been a constant challenge for engineers to apply preventive maintenance on engineered systems in a cost-effective way to avoid failures, which would usually require more costly repair or maintenance procedures.

Reliability-centered maintenance (RCM) provides a framework for developing optimally scheduled maintenance programs that are cost effective. The RCM concept was first developed in the aircraft industry when the first Boeing 747 was built. There were many requirements for maintaining such a complex aircraft and there was a need to identify a maintenance strategy that could reduce unnecessary maintenance tasks. By 1978 the first full description of RCM was published [Now78], and in the 1980s the Electric Power Research Institute introduced RCM to the nuclear industry.

Maintenance activities are usually classified [Rau04] as either preventive or corrective activities. Preventive maintenance (PM) represents planned maintenance that is performed when the equipment is functioning properly to avoid future failures. It may involve inspection, adjustments, lubrication, parts replacement, calibration, and repair of items that are beginning to wear out. PM may be carried out on a regular basis, regardless of whether the functionality or performance is degraded or not. PM activities can be classified into the following categories:

(a) *Clock-based maintenance.* This is the simplest form of PM where maintenance is carried out according to a fixed maintenance schedule on a regular basis.

(b) *Age-based maintenance.* This form of PM is carried out at a specified age of the item, often according to manufacturer's specification. Aging may be measured in terms of time in operation, number of times operated, or other time concepts.

(c) *Condition-based maintenance.* This PM is based on one or more condition variables of the equipment. It requires a monitoring scheme of the variables and a set threshold to initiate maintenance. Examples of condition variables are temperature, pressure, and vibration of a component.

Corrective maintenance (CM), or in a simpler word repair, is carried out when an item has failed. The objective of CM is to quickly restore the equipment to functionality or to switch in a standby equipment to restore the system. Corrective maintenance is also called run-to-failure maintenance, which effectively represents the result of a deliberate decision to operate the system until a failure occurs.

1.6 ORGANIZATION OF THE BOOK

Chapters 2 through 5 provide an introduction to some of the more important concepts from the first several weeks of a course in reliability engineering as typically taught on most university campuses in mechanical or industrial engineering departments. Chapter 2 covers the elements of probability and reliability theory and some widely used probability distributions for a system that can be modeled as a single component. Chapter 3 presents aspects of statistics used in working with data for a reliability analysis on one component. In Chapter 4 the reliability of multiple-component systems is introduced, while Chapter 5 illustrates a way to incorporate repair of components into an analysis. The PRA discussion begins in Chapter 6 with methods for combining failure probabilities and consequences, followed by PRA computer programs in Chapter 7.

Nuclear power plant safety analysis is treated in Chapter 8 before considering major nuclear power plant accidents and incidents in Chapter 9. With this background, past PRA studies of nuclear plants are discussed in Chapter 10. Advanced nuclear power plant designs with enhanced passive safety features are considered in Chapter 11, followed by topics related to risk-informed regulations and reliability-centered maintenance in Chapter 12. Chapter 13 discusses recent developments of probabilistic techniques to accurately represent dynamic system evolutions for reliability evaluation and system diagnostics. A number of mathematical and statistical techniques as well as specific data relevant to the risk and safety analysis of nuclear systems are provided as appendices.

References

[Hol79a] J. P. Holdren, K. Anderson, P. H. Gleick, I. Mintzer, and G. Morris, "Risk of Renewable Energy Sources: A Critique of the Inhaber Report," ERG 79-3, Energy and Resources Group, Univ. of California, Berkeley (1979).

[Hol79b] J. P. Holdren, *Nucl. News* 25 (March 1979); H. Inhaber, *ibid.* 25 (March 1979); J. P. Holdren, *ibid.* 32 (April 1979); H. Inhaber, *ibid.* 26 (May 1979); see also *Nucl. News* 42 (September 1979).

[Inh82] H. Inhaber, *Energy Risk Assessment*, Fig. 7, Gordon and Breach (1982).

[Kap83] S. Kaplan, "On a 'Two-Stage' Bayesian Procedure for Determining Failure Rates from Experiential Data," *IEEE Trans. Power App. Sys.* **PAS-102**, 195 (1983).

[Low76] W. W. Lowrance, *Of Acceptable Risk*, Kaufman (1976).

[Meh90] P. Mehta *et al.*, "Bhopal Tragedy's Health Effects, A Review of Methyl Isocyanate Toxicity," *JAMA* **264**, 2781(1990).

[Mor93] M. G. Morgan, "Risk Analysis and Management," *Sci. Am.* **269**, 32 (1993).

[Now78] F. S. Nowlan and H. F. Heap, "Reliability-Centered Maintenance," A066-579, U. S. Department of Commerce (1978).

[NRC71] "General Design Criteria for Nuclear Power Plants," Title 10, *Code of Federal Regulations*, Part 50, Appendix A, U.S. Nuclear Regulatory Commission (1971).

[NRC75] "Reactor Safety Study—An Assessment of Accident Risks in U.S. Commercial Nuclear Power Plants," WASH-1400 (NUREG 75/014), U.S. Nuclear Regulatory Commission (1975).

[NRC83] "PRA Procedures Guide," NUREG/CR-2300, U.S. Nuclear Regulatory Commission (1983).

[NRC90] "Severe Accident Risks: An Assessment for Five U.S. Nuclear Power Plants," NUREG- 1150, U.S. Nuclear Regulatory Commission (1990).

[Rah84] F. J. Rahn, A. G. Adamantiades, J. E.. Kenton, and C. Braun, *A Guide To Nuclear Power Technology: A Resource for Decision Making*, Wiley (1984).

[Rau04] M. Rausand and A. Hoyland, *System Reliability Theory–Models, Statistical Methods, and Applications*, Wiley (2004).

[Ves81] W. E. Vesely, F. F.Goldberg, N. H. Roberts, and D. F. Haasl, "The Fault Tree Handbook," NUREG-0492, U.S. Nuclear Regulatory Commission (1981).

CHAPTER 2

PROBABILITIES OF EVENTS

This chapter contains an introduction to the underlying principles of probabilities and their application to the analysis of failure events.

2.1 EVENTS

In order to understand probability concepts we first need to define a sample space S with unique events E_n, $n = 1, 2, \ldots$, being members of S. For brevity of equations, in this book we write E_1E_2 for the intersection of two events E_1 and E_2, although elsewhere such an intersection may be written as $E_1 \cap E_2$. Note that we cannot "multiply" events, so "E_1 AND E_2" is *not* "E_1 times E_2." It is helpful to illustrate such an occurrence with the aid of the Venn diagram in Fig. 2.1. For a sample space with N events, the intersection of all events is $E_1E_2 \cdots E_N$.

Another concept arising with events is the union of unique events such as E_1 or E_2. This will be denoted here by $E_1 + E_2$, although elsewhere it may appear as $E_1 \cup E_2$. Either convention means "E_1 OR E_2," *not* "E_1 plus E_2." For a sample space with N events, the union of all events is $E_1 + E_2 + \cdots + E_N$. The additional symbol $\overline{E}$, "NOT E," is for the complement of E.

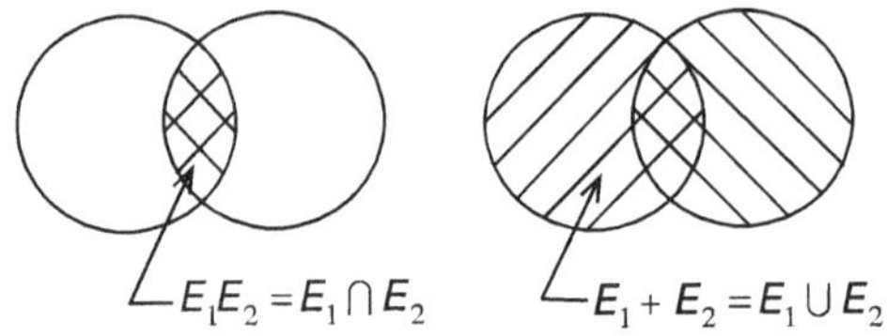

Figure 2.1 Venn diagram illustrating the intersection and union of two events E_1 and E_2.

Table 2.1 Boolean Algebra for Events

Word Description	Rules
1. Commutative law	a. $XY = YX$
	b. $X + Y = Y + X$
2. Associative law	a. $X(YZ) = (XY)Z$
	b. $X + (Y + Z) = (X + Y) + Z$
3. Idempotent law	a. $XX = X$
	b. $X + X = X$
4. Absorption law	a. $X(X + Y) = X$
	b. $X + XY = X$
5. Distributive law	a. $X(Y + Z) = XY + XZ$
	b. $(X + Y)(X + Z) = X + YZ$
6. Complementation	a. $X\overline{X} = \phi$ or 0 (null)
	b. $X + \overline{X} = \Omega$ or I (universal)
	c. $\overline{\overline{X}} = X$
7. De Morgan's theorems	a. $\overline{XY} = \overline{X} + \overline{Y}$
	b. $\overline{X + Y} = \overline{X}\,\overline{Y}$
8. Useful relationships	a. $X + \overline{X}Y = X + Y$
	b. $\overline{X}(X + \overline{Y}) = \overline{X}\,\overline{Y}$

A compound event H may consist of many events, in which case the use of parentheses may be needed to appropriately group the events. Some rules of Boolean algebra for events, given in Table 2.1, are used to simplify the writing of a compound event. The commutative and associative laws are similar to those laws for ordinary algebra. The idempotent laws enable redundancies for the same event to be eliminated. Absorption law 4a is easily justified by observing that if event X occurs then event $(X + Y)$ also has occurred so $X(X + Y) = X$; a similar argument holds for absorption law 4b. The distributive laws 5a and 5b are very useful in fault tree analysis (Chapter 6) and may be verified by using the preceding rules. De Morgan's theorems are useful if the search for a system failure event H is switched to the search for the successful operation $\overline{H}$ of that system.

For a failure analysis, a system failure event H might consist of many component failure events nested together. Boolean algebra facilitates the reduction of H to a set of single-component failure events, double-component failure events, etc. The resulting single- and multiple-component events are cut sets, i.e., combinations of events, any of which could cause failure of a system. That is, a *cut set* is defined as a set of system events that, if they all occur, will cause system failure while a *minimal cut set* of a system is a cut set that has no other cut set as a subset. Another way of saying this is that the removal of any event from a minimal cut set would cause it not to be a cut set, i.e., the system would no longer fail.

Example 2.1 Construct the minimal cut sets for a system failure event H consisting of component failure events A to E, where

$$H = A + BD(E + B) + (B + C)(D + E).$$

We observe that the first term obviously cannot be reduced, while

$$\begin{aligned} BD(E + B) &= BDE + BDB && \text{(law 5a)} \\ &= BDE + BD && \text{(law 3a)} \\ &= BD && \text{(law 4b)} \end{aligned}$$

and

$$(B + C)(D + E) = BD + BE + CD + CE \qquad \text{(law 5a)}.$$

Combining terms gives

$$H = A + BD + BE + CD + CE$$

and we conclude that a system failure could occur from a single failure event A or any of the four double failure events. In Example 2.2 we will consider the probability of occurrence of failure event H. ◇[1]

2.2 EVENT TREE ANALYSIS AND MINIMAL CUT SETS

An event tree depicts the evolution of a series of events with time. In a safety analysis of a nuclear system, for example, it provides an inductive logic method for identifying the various possible outcomes of a given (undesired) initiating event. Event trees are similar to decision trees, but they differ in that human intervention is not required to influence the outcome of the initiating event. In risk and safety analysis applications, the initiating event of an event tree can be the failure of a system itself or it can be initiated externally to the system, with the subsequent events determined by the performance of the system components. Different event trees must be constructed and evaluated to analyze a set of possible accidents.

In a given accident analysis, once an initiating event is defined, all the safety systems that possibly can be utilized after the failure event must be identified, and the set of possible failure and success states for each system must be defined. These safety systems are then structured in the form of headings for the event tree. To be conservative, usually each system is defined to have only one success state S, where

[1]The diamond symbol denotes the end of an example.

everything is working "as good as new," and a single system failure state F comprised of all possible system failures. This is illustrated with a classic tree structure in Fig. 2.2. The accident sequences that result from the tree structure are shown in the last column of the figure. Each branch of the tree yields one particular accident sequence; for example, S_1F_2 denotes the accident sequence in which the initiating event (I) occurs and system 1 is called upon and succeeds (S_1) but system 2 either is in a failed state or fails to perform upon demand. For larger event trees, this stepwise branching analysis would simply be continued.

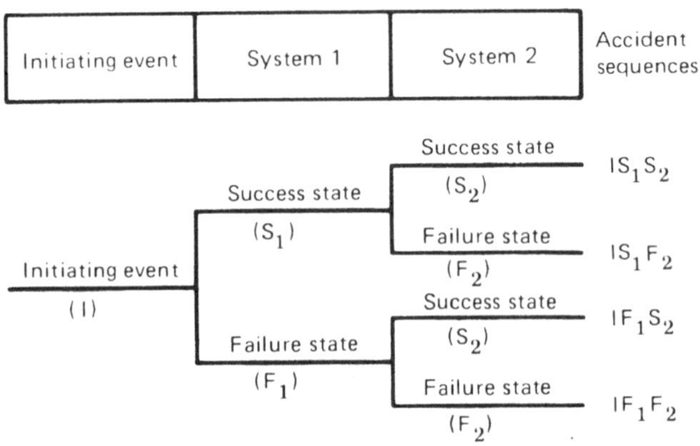

Figure 2.2 Illustration of event tree branching. *Source:* [NRC75].

It should be emphasized that the system states on a given branch of an event tree are conditional on the previous states having already occurred. In Fig. 2.2, for example, the success and failure of system 1 must be defined under the condition that the initiating event has occurred; likewise, in the upper branch of the tree corresponding to system 1 success, the success and failure of system 2 must be defined under the conditions that the initiating event has occurred and system 1 has succeeded.

A major concern in event tree construction involves accounting for the timing of the events. In some instances, the failure logic changes depending on the time at which the events take place; such a case occurs, for example, in the operation of emergency core cooling systems in nuclear plants. Then dynamic event tree analysis techniques are needed to model the system that changes during the accident, even though the safety system components remain the same [Dev92,Izq96,Lab00]. Dynamic event tree analysis will be discussed in Chapter 13.

Successful construction of an event tree provides a *qualitative* analysis of what happens after an initiating event, but if a *quantitative* analysis is desired, then each branch of the event tree must be quantitatively evaluated. This can be done by a variety of techniques, but typically the states in nuclear systems are assigned numerical values from fault tree analyses. The probabilities obtained must be *conditional* probabilities for each branch in a sequence, as schematically illustrated in Fig. 2.2. Nonconditional and conditional probabilities are the subject of the next section.

2.3 PROBABILITIES

2.3.1 Interpretations of Probability

The classic mathematical interpretation of the probability of an event E, which is the relative frequency approach, requires that if event E in sample space S occurs X number of times out of a number n of repeated experiments whose outcomes are described by S, then the probability $P(E)$ of the outcome of event E is defined by

$$P(E) = \lim_{n \to \infty} \left(\frac{X}{n} \right). \tag{2.1}$$

For a fixed n, the quantity X/n is the relative frequency of occurrence of E. Because it is impossible to actually conduct an infinite number of trials so that $n \to \infty$, usually $P(E)$ is just approximated by X/n. The strong law of large numbers and the central limit theorem [Fel68,Man74,Pap02] provide a justification that improved estimates of $P(E)$ will follow by increasing n.

The difficulty of such an interpretation for engineers interested in risk and safety analysis is that usually we do not have the option of performing n experiments because we are dealing with rarely occurring events, and sometimes it is preferable to not even perform a single experiment if the outcome would damage a system. In such instances it is necessary to resort to the axiomatic or subjective approach to the concept of probability, which we shall use from now on.

The axiomatic interpretation begins with the broad view that probability is nothing more than a measure of uncertainty about the likelihood of an event. Stated more precisely, "a probability assignment is a numerical encoding of a state of knowledge" [Tri69]. With such a broad definition it is necessary to impose some constraints before obtaining something that can be used in quantitative analysis. Examples of several kinds of knowledge are:

- *Symmetry* Sometimes a system is known to be symmetrical, as in the case of honest dice or coins. As an example, if an experiment consisting of 1000 flips of a coin gave 534 heads and 466 tails, the probability of the event that heads will appear would be assigned a probability of 0.5 because it would be believed that an insufficient number of flips had been performed to give the outcome 0.5, as expected from the relative frequency interpretation of probability.
- *Averages* Sometimes the average result of what has occurred in the past is known, such as the average annual rainfall in a given year, so this would be used as an estimate of the expected rainfall in the next year unless there were reason to believe that global warming, for example, had affected the frequency of recent occurrences.
- *Frequencies* Sometimes historical data concerning a system are known, e.g., how many years the annual rainfall in a given year exceeds the expected amount, so the frequency of occurrence in past years would be used in making predictions about rainfall in future years, again assuming no meteorological changes had occurred.

2.3.2 Axiomatic Approach to Probabilities

2.3.2.1 Probabilities for Discrete Events With the axiomatic approach, a risk and safety analysis must assign probabilities in a "coherent" manner, which requires that such probabilities obey the axioms and laws of probability. The axiomatic approach is formally developed in a deductive way from only three axioms. The *normalization axiom* states that the probability P for the outcome of any event E is a real number between zero and unity,

$$0 \le P(E) \le 1. \tag{2.2}$$

This axiom gives a way of constraining the magnitude of the probability of the outcome of an event in space S.

The second axiom deals with two mutually exclusive events, E and its complement $\overline{E}$. The *addition axiom for probabilities* states that

$$P(E) + P(\overline{E}) = 1, \tag{2.3}$$

a result that follows because either E or $\overline{E}$ is certain to occur. Here, because $P(E)$ and $P(\overline{E})$ are numbers, the plus sign between them is for addition.

The third axiom deals with the intersection of events. Earlier we introduced the intersection of events E_1 and E_2 with the notation $E_1 \cap E_2 = E_1E_2$. Now we want to determine the probability $P(E_1E_2)$. The *product axiom for probabilities*, the third and final axiom for probabilities, may be stated as

$$P(E_1E_2) = P(E_1|E_2)P(E_2) = P(E_2|E_1)P(E_1). \tag{2.4}$$

Here the "conditional probability" $P(E_1|E_2)$ is defined as the probability of event E_1 *given* event E_2 has occurred. In the special case that events E_1 and E_2 are independent, so the probability that event E_1 occurs is independent of the occurrence of event E_2, then $P(E_1|E_2) = P(E_1)$ and $P(E_1E_2) = P(E_1)P(E_2)$. A second special case occurs if events E_1 and E_2 are mutually exclusive (i.e., "disjoint"), so that $P(E_1|E_2) = 0$ and $P(E_1E_2) = 0$.

The concept of conditional probabilities is important when doing risk and safety analyses because the probability for any system is conditional on knowing—for the time of interest—the state of that system, which can change as the operating environment around the system changes with time. So keep in mind that all event probabilities $P(E)$ are really $P(E|H)$ because they are conditional in the sense that they are based on certain hypotheses or assumptions H about the system.

2.3.2.2 Probabilities for Continuous Events For a system in which the set of events E_n, $n = 1, 2, \ldots, N$, is so numerous that $N \to \infty$, a continuous variable, say x, is needed to describe the set of probabilities associated with an event. The probability for an event to occur between x and $x + \Delta x$ depends on the magnitude Δx, so we must define $p(x)\mathrm{d}x$ as the probability that the event occurs within $\mathrm{d}x$ about x. This means that $p(x)$ is a *probability density function* (PDF) or a probability *per unit* $\mathrm{d}x$ for the event to occur at x.

The probability for the continuous variable X, which is the analog to the summation of time-independent probabilities $\sum_{n=1}^{N} P(E_n)$ for mutually exclusive events E_n, is the *cumulative distribution function* (CDF),

$$P(X) = \int_{x_{\min}}^{X} p(y)\,\mathrm{d}y. \tag{2.5}$$

If $X = x_{\max}$, then

$$P(X) = 1 \tag{2.6}$$

because x is certain to occur within the range of $x_{\min}$ and $x_{\max}$. From the last two equations it is evident that

$$0 \leq p(x)\,\mathrm{d}x \leq 1. \tag{2.7}$$

2.3.3 Intersection of Events

The results for the product rule easily can be extended to cover more than two events. For discrete events $E_1, E_2, \ldots, E_N$,

$$P(E_1E_2\cdots E_N) = P(E_1)P(E_2|E_1)\cdots P(E_N|E_1E_2\cdots E_{N-1}). \tag{2.8}$$

If the events are independent, then

$$P(E_1E_2\cdots E_N) = P(E_1)P(E_2)\cdots P(E_N), \tag{2.9}$$

whereas if the events are mutually exclusive, then

$$P(E_1E_2\cdots E_N) = 0. \tag{2.10}$$

For probabilities in risk and safety analysis, often only an upper bound of $P(E_1E_2\ldots E_N)$ is needed. For $P(E_1E_2)$, for example, in view of Eq. (2.4) it follows that

$$P(E_1E_2) \leq \min[P(E_1), P(E_2)], \tag{2.11}$$

and similarly

$$P(E_1E_2\cdots E_N) \leq \min[P(E_1), P(E_2), \ldots, P(E_N)]. \tag{2.12}$$

One also can obtain an estimate for the probability of three events in terms of the probabilities of pairs of events, as in the "double-event bound"

$$P(E_1E_2E_3) \leq \min[P(E_1E_2), P(E_1E_3), P(E_2E_3)]. \tag{2.13}$$

Similarly,

$$P(E_1E_2\cdots E_N) \leq \min[\text{probabilities of all triple combinations}],\ N \geq 4. \tag{2.14}$$

2.3.4 Union of Events

Earlier we introduced the union of events E_1 and E_2 with the notation $E_1 \cup E_2 = E_1 + E_2$. Now we want to determine the probability $P(E_1 + E_2)$. From the diagram of Fig. 2.1, we can interpret the areas of events E_1 and E_2 as probabilities, in view of Eqs. (2.2) and (2.3). Thus it follows from the Venn diagram that

$$P(E_1 + E_2) = P(E_1) + P(E_2) - P(E_1 E_2), \tag{2.15}$$

where the right-hand side can be interpreted as the sum of the probabilities of the two events considered independently, with the third term to eliminate the possible double counting arising from the "overlap" caused by the intersection of the two events. Of course, if the two events are independent, then from Eq. (2.9)

$$P(E_1 + E_2) = P(E_1) + P(E_2) - P(E_1)P(E_2), \tag{2.16}$$

while if the two events are mutually exclusive, then

$$P(E_1 + E_2) = P(E_1) + P(E_2). \tag{2.17}$$

The preceding equations can be generalized to the case of more than two events, where in general

$$\begin{aligned} P(E_1 + E_2 + \cdots + E_N) &= \sum_{n=1}^{N} P(E_n) - \sum_{n=1}^{N-1} \sum_{m=n+1}^{N} P(E_n E_m) \\ &\quad + \cdots + (-1)^{N-1} P(E_1 E_2 \cdots E_N). \end{aligned} \tag{2.18}$$

The rth term on the right-hand side of Eq. (2.18) contains

$$\binom{N}{r} = \frac{N!}{r!(N-r)!} \tag{2.19}$$

probabilities for all possible combinations of the N events E_n considered r at a time. From Eq. (2.10) it follows that only the first term on the right-hand of Eq. (2.18) is nonzero if all the events are mutually exclusive. If all the events are independent, then the form of Eq. (2.18) can be improved by collecting terms and rearranging to obtain a product of factors on the right-hand side,

$$1 - P(E_1 + E_2 + \cdots + E_N) = \prod_{n=1}^{N} [1 - P(E_n)]. \tag{2.20}$$

The alternating signs in the series in Eq. (2.18) immediately suggest the bounds for $P(E_1 + E_2 + \cdots + E_N)$,

$$P(E_1 + E_2 + \cdots + E_N) \le \sum_{n=1}^{N} P(E_n), \tag{2.21}$$

$$\ge \sum_{n=1}^{N} P(E_n) - \sum_{n=1}^{N-1} \sum_{m=n+1}^{N} P(E_n E_m). \tag{2.22}$$

Equation (2.21) is important for bounding the probability of failure F_{sys} of a system for which the minimal cut sets consisting of combinations of events are known.

Example 2.2 From Example 2.1 the system minimal cut sets for H consist of $C_1 = A$, $C_2 = BD$, $C_3 = BE$, $C_4 = CD$, and $C_5 = CE$. Thus $P(H) = P(C_1 + C_2 + C_3 + C_4 + C_5) \leq \sum_{n=1}^{N} P(C_n)$. ◇

An important simplification of Eqs. (2.16) and (2.18) arises for some risk and safety analyses in instances when the events are independent and highly infrequent. In such cases, the *rare-event approximation* often is invoked so that Eq. (2.18) simplifies to

$$P(E_1 + E_2 + \cdots + E_N) \simeq \sum_{n=1}^{N} P(E_n). \tag{2.23}$$

Also, Eq. (2.9) remains applicable, so

$$P(E_1 E_2 \cdots E_N) \simeq \prod_{n=1}^{N} P(E_n). \tag{2.24}$$

Let us now examine what happens if the bounds of Eqs. (2.21) and (2.22) are not used, but instead the full expansion of the probability in Eq. (2.18) is used. To evaluate $P(H)$ for event H given in Examples 2.1 and 2.2, it is important to note that components B, C, D, and E occur in more than one minimal cut set C_n. Thus, when calculating the system failure probability $P(H)$, it is necessary to avoid "double counting" the failure probability of those components. This perhaps can be best illustrated by an example.

Example 2.3 We wish to determine $P(H)$ for the H of Examples 2.1 and 2.2 from Eq. (2.18). To simplify the notation, only in this example the symbol A will stand for $P(A)$, etc. Thus,

$$\begin{aligned} P(H) \;=\; & [A + BD + BE + CD + CE] \\ & - [ABD + ABE + ACD + ACE + BD\underline{B}E + BDC\underline{D} \\ & + BDCE + BECD + BEC\underline{E} + CD\underline{C}E] \\ & + [ABD\underline{B}E + ABDC\underline{D} + ABDCE + ABECD + ABEC\underline{E} \\ & + ACD\underline{C}E + BD\underline{B}EC\underline{D} + BD\underline{B}EC\underline{E} + BDC\underline{D}\,\underline{C}E \\ & + BECD\underline{C}\,\underline{E}] - [ABD\underline{B}EC\underline{D} + ABD\underline{B}EC\underline{E} + ABDC\underline{D}\,\underline{C}E \\ & + ABECD\underline{C}\,\underline{E} + BD\underline{B}EC\underline{D}\,\underline{C}\,\underline{E}] \\ & + [ABD\underline{B}EC\,\underline{D}\,\underline{C}\,\underline{E}]. \end{aligned}$$

This initial result shows that the initial five minimal cut sets have resulted in 31 terms in the probability expansion and that, because of lack of independence, 18 of the 31 terms contain redundant factors (denoted by the underlined letters). (For example, had the number of minimal cut sets been 10, the complete expansion would have contained 1023 terms; if the number of cuts had been 20, then 1,048,575 terms would have been needed. The general relation is $2^n - 1$ for n minimal cut sets.)

Table 2.2 Results for Example 2.4

Terms Included	Failure Probability	Error (%)
1	0.05	+9.3
1 and 2	0.0454	−0.74
1, 2, and 3	0.045842	+0.22
1, 2, 3, and 4	0.045738	−0.0022

After elimination of the redundant factors and algebraically adding the probability products for identical terms,

$$\begin{aligned} P(H) &= [A + BD + BE + CD + CE] \\ &\quad - [ABD + ABE + ACD + ACE + BCD + BCE \\ &\quad + BDE + CDE + 2\,BCDE] \\ &\quad + [ABCD + ABCE + ABDE + ACDE + 4\,BCDE + 2\,ABCDE] \\ &\quad - [BCDE + 4\,ABCDE] \\ &\quad + [ABCDE]. \diamond \end{aligned}$$

The last example illustrates how tedious the computation of a system failure probability can be when there are a lot of different minimal cut sets C_n containing many events E_n. Fortunately, for this purpose computer programs exist that will be discussed in Chapter 7. Alternatively, $P(H)$ can be bounded as in Example 2.2.

Example 2.4 To illustrate the accuracy of a failure probability as in Eq. (2.18) for the system in Example 2.3, we assume the failure probabilities in that example to be

$$\begin{aligned} P(A) &= 0.01, \\ P(B) &= P(C) = P(D) = P(E) = 0.1. \end{aligned}$$

Then the final result for $P(H)$ from Example 2.3 is

$$P(H) = 0.05 - 0.0046 + 0.000442 - 0.000104 + 0.0000001 = 0.045739.$$

If approximate answers had been obtained by truncating the series expansion in Eq. (2.18), the results would have been as in Table 2.2 [GEC74].

If the probabilities of failure of each component had been one order of magnitude smaller, then the exact result would have been

$$P(H) = 0.00139561399$$

and taking only the first term would have given 0.0014, for an error of 0.31%. This illustrates the fact that the error bounds in Eqs. (2.21) and (2.22) are closer together when the failure probabilities of the events are smaller. $\diamond$

2.3.5 Decomposition Rule for Probabilities

When analyzing systems with multiple components it is sometimes useful to break down the analysis into parts corresponding to the occurrence and nonoccurrence of one or more of the events. For the probability of occurrence of event E_1 in terms of the conditional probabilities $P(E_1|E_2)$ and $P(E_1|\overline{E}_2)$,

$$P(E_1) = P(E_1|E_2)P(E_2) + P(E_1|\overline{E}_2)P(\overline{E}_2), \tag{2.25}$$

which follows from Eqs. (2.4) and (2.17). This can be generalized, if necessary, such as

$$\begin{aligned} P(E_1) = {} & P(E_1|E_2E_3)P(E_2)P(E_3) + P(E_1|E_2\overline{E}_3)P(E_2)P(\overline{E}_3) \\ & + P(E_1|\overline{E}_2E_3)P(\overline{E}_2)P(E_3) + P(E_1|\overline{E}_2\overline{E}_3)P(\overline{E}_2)P(\overline{E}_3), \end{aligned} \tag{2.26}$$

with the number of terms for expressing $P(E_1)$ in terms of conditional probabilities $P(E_1|E_2E_3\cdots E_N)$, etc., given by 2^{N-1}. If events $E_2E_3\cdots E_N$ are all independent, then Eq. (2.10) can be used to break down the right-hand side into terms involving only $P(E_n)$ and $P(\overline{E}_n)$, $n = 2, \cdots, N$.

⇒ **Summary.** What we have learned in this section about probabilities:

1. Probabilities are a numerical encoding of states of knowledge.
2. Probabilities satisfy the normalization, addition, and product axioms.
3. Probabilities can be unconditional or conditional.
4. Probabilities can be for discrete events or a continuum of events.
5. Equations for probabilities are simplified if the events are independent and mutually exclusive or if the rare-event approximation can be assumed.
6. Equations for bounding a probability are available.

2.4 TIME-INDEPENDENT VERSUS TIME-DEPENDENT PROBABILITIES

Quantitative risk and safety analyses are performed for the failure events of systems. The probability of a system failure depends on the probabilities of failure events of the components comprising the system. A failure event for any system component occurs during a finite period of time. If the time period is short compared to the time of interest in an analysis, then the failure event can be assumed to be nearly instantaneous and hence time independent. On the other hand, if the period of time is not short, then the failure event is time dependent. Such an event can be viewed as occurring due to a degradation failure.

Simple examples of a time-independent failure are the failure of a light switch on demand or the instantaneous fracture of the filament of an incandescent lightbulb or the rupture of the casing for a set of bearings. Examples of degradation failures, on the other hand, are those due to wear from continued use. One problem associated with analyzing such failures is defining when a "failure" actually occurs; many times a component will be marginally serviceable and hence replacement or repair will not be required until later. In some cases, it is convenient to define such a failure event as occurring when unscheduled maintenance or repair actions must be initiated. In other situations, a degradation failure can be defined to occur when the component performs outside its acceptable performance limits.

Probabilities for failure events that occur in a time-independent mode and those that occur in a time-dependent manner require that the data tabulated for each failure type must be different. While time-independent failure events can be given a numerical value for the probability of failure, time-dependent failure events are analyzed with data for probabilities of failure per unit time.

2.5 TIME-INDEPENDENT PROBABILITIES

2.5.1 Introduction

For a component operated only in an "on-or-off" mode, we shall denote event E by D. The failure probability for demand event D is $F(D)$ or simply F, and the corresponding probability that the component does not fail is $\overline{F}(D)$, where for the simplest of systems there are only two outcomes of an event: either the system fails on demand, with probability F, or it functions as designed and does not fail, with probability $\overline{F}$. Thus, from Eq. (2.3), the two probabilities are related by

$$F = 1 - \overline{F}. \tag{2.27}$$

Such demand failures occur in a system component during its intermittent, possibly repetitive operation: The component either fails or does not fail at the Nth demand, event D_N. The probability $\overline{F}(W_{N-1})$ that the component works for each of $N-1$ operations is

$$\overline{F}(W_{N-1}) = \overline{F}(D_1 D_2 \cdots D_{N-1}). \tag{2.28}$$

Just because the system works for $N-1$ operations does not mean that it will operate at the Nth demand. That is, $\overline{F}(D_N|W_{N-1})$ is the conditional probability that the component will operate at the Nth demand given that it did not fail for $N-1$ demands, while $F(D_N|W_{N-1})$ is the corresponding conditional probability of failure. By Eq. (2.4), the probability that a component will fail to operate on the Nth demand after it worked for all previous demands is

$$F(D_N W_{N-1}) = F(D_N|W_{N-1})\overline{F}(W_{N-1}). \tag{2.29}$$

From Eq. (2.8), the last equation also can be written as

$$\begin{aligned} F(D_1 D_2 \cdots D_N) &= F(D_N|D_1 D_2 \cdots D_{N-1}) \\ &\quad \times \overline{F}(D_{N-1}|D_1 D_2 \cdots D_{N-2}) \cdots \overline{F}(D_2|D_1)\overline{F}(D_1). \end{aligned} \tag{2.30}$$

For demand-type failures, one ideally would like to have a complete tabulation of all the probabilities in Eq. (2.30) for every intermittently operating component in a system. Usually it is necessary, because of limitations in the experimental data available, to assume the demand events are identical and independent; then any failure is assumed to be *random* so that $F(D_N|W_{N-1}) = F(D)$ and $\overline{F}(D_N|W_{N-1}) = \overline{F}(D)$. Then Eq. (2.30) reduces to

$$F(D_1D_2\cdots D_{N-1}D_N) = F(D)[\overline{F}(D)]^{N-1} = F(D)[1-F(D)]^{N-1}. \quad (2.31)$$

Note that Eqs. (2.30) and (2.31) give the probability of failure on the Nth demand, which differs from the probability that a repairable system will undergo a failure sometime during N demands. For example, for random failures, the latter probability would be N times the former since the failure could occur on any one of the N demands.

Example 2.5 A light switch fails randomly with a demand failure probability of 10^{-4}. On the average, the switch is used 20 times per week. (a) What is the probability that the switch will fail at the end of a 6-year period? (b) What is the probability it could fail exactly once during the 6 years if it was immediately repaired after failure?

(a) Over a 6-year period, the switch could be used $20 \times 52 \times 6 = 6240$ times, so from Eq. (2.30) the probability of failure on the 6240th demand is

$$10^{-4}[1-10^{-4}]^{6239} = 5.36 \times 10^{-5}.$$

(b) The probability it could fail exactly once during the 6 years is

$$6240(5.36 \times 10^{-5}) = 0.334. \diamond$$

2.5.2 Time-Independent Probability Distributions

Two parameters of interest for any discrete probability distribution $P(r)$ of the random variable r are the mean m and the variance σ^2. For outcomes $r = 0, 1, \ldots, N$, the mean is defined as

$$m = \sum_{n=0}^{N} nP(n), \quad (2.32)$$

while the variance, which measures the deviation of values about the mean, is

$$\sigma^2 = \sum_{n=0}^{N} (n-m)^2 P(n). \quad (2.33)$$

The square root of the variance is the *standard deviation* σ.

We now consider two useful distributions that involve time-independent events which are "instantaneous" demands on the system, the binomial distribution and the Poisson distribution.

2.5.2.1 Binomial Distribution Suppose the performance of a device is not known, so that an experiment consisting of N demands is to be performed, where N is fixed. The demands are specified to be independent (or Bernoulli trials) such that F is constant for each trial. In order to describe the experiment with the binomial distribution, it is necessary that the ordering of the events not affect the result of the experiment. The probability of M failures *out of* the N demands, $P_N(M)$, is obtained by selecting the proper term from the binomial expansion of the equation

$$(F + \overline{F})^N = 1. \tag{2.34}$$

The result is

$$P_N(M) = \begin{pmatrix} N \\ M \end{pmatrix} F^M \overline{F}^{N-M} = \frac{N!}{M!(N-M)!} F^M \overline{F}^{N-M} \tag{2.35}$$

and the mean and variance for the binomial distribution are

$$m = NF, \tag{2.36}$$

$$\sigma^2 = NF\overline{F}. \tag{2.37}$$

Another probability distribution that follows from Eq. (2.35) is the probability that the device fails for M or fewer demands,

$$P_N(\leq M) = \sum_{n=0}^{M} P_N(n), \tag{2.38}$$

or for more than M demands

$$P_N(> M) = 1 - \sum_{n=0}^{M} P_N(n). \tag{2.39}$$

For a large enough N, the binomial distribution can be approximated by a normal distribution with the same mean m and variance σ^2. This approximation gives good results if NF and $N\overline{F}$ are both at least 5.

The binomial distribution is used for a single device that operates on demand—or does not—and can be repaired to an "as good as new" state immediately after it fails. Then M is the number of failures in N demands and $P_N(M)$ is the probability the device will fail on M demands. Equation (2.35) is a generalization of Eq. (2.30) for situations where the device operated on demand and can undergo instantaneous repairs.

The binomial distribution is for devices that can either fail or not fail. If there are more than two outcomes possible, then the multinomial distribution must be used.

The beta distribution is a generalization of the binomial distribution of Eq. (2.35) in which the failure on demand F is replaced by the continuous random variable θ, M by a possibly noninteger parameter $\alpha - 1$, and $N - M$ by parameter $\beta - 1$ to obtain the probability density function

$$p(\theta) = \begin{cases} \dfrac{\Gamma(\alpha+\beta)}{\Gamma(\alpha)\Gamma(\beta)} \theta^{\alpha-1}(1-\theta)^{\beta-1}, & \alpha > 0,\ \beta > 0,\ 0 \leq \theta \leq 1, \\ 0, & \text{otherwise}, \end{cases} \tag{2.40}$$

where $\Gamma(x)$ is a gamma function defined in Appendix B. The beta distribution is used when a random variable θ takes values in the interval $[0, 1]$. The probability density is properly normalized because

$$\int_0^1 p(\theta)\mathrm{d}\theta = 1. \tag{2.41}$$

Example 2.6 A switch that can be repaired has a failure rate of 10^{-4} per demand. Compute the probability that (a) in 1000 operations the switch will fail exactly two times and (b) it will fail two or more times.

(a) Because $F = 10^{-4}$ and $N = 1000$, Eq. (2.35) gives

$$P_{1000}(2) = \frac{1000!}{2!(1000-2)!}(10^{-4})^2(1-10^{-4})^{1000-2} = 0.0045.$$

(b) To evaluate the probability the switch will fail two or more times, from Eq. (2.35) we obtain $P(0) = 0.9048$ and $P(1) = 0.0905$, so we use Eq. (2.39) with $M = 1$ to obtain

$$P_{1000}(>1) = 1 - P_{1000}(0) - P_{1000}(1) = 0.0047. \diamond$$

A second interpretation of the binomial distribution for failure analyses involves the case of N identical units that initially function, each with failure probability F. Then $P_N(M)$ gives the probability that M of the N units in the system will fail.

Example 2.7 A system has four identical components that operate simultaneously and independently. Three components must remain operating or the system will fail. If the failure of each component is 0.02 over the design life of the system, compute the probability of system failure.

With $F = 0.02$ and $N = 4$, from Eq. (2.35) we compute $P_4(0) = 0.9223$ and $P_4(1) = 0.0753$, so from Eq. (2.39)

$$P_4(>1) = 1 - P_4(0) - P_4(1) = 0.0024. \diamond$$

2.5.2.2 Poisson Distribution The Poisson distribution is like the binomial distribution because it describes phenomena for which the average probability of an event is constant, independent of the number of previous events. In this case, however, the system undergoes transitions randomly from one state to another by an *irreversible* process so that the order of the events cannot be interchanged, as it can for the binomial distribution. The distribution is used when the number of possible events is large but the probability of occurrence over a given time interval is small.

The Poisson distribution can be derived from the binomial distribution of Eq. (2.35) in the limit of a small failure probability F such that, for a large number N of events, the expected number of occurrences of the event of interest remains finite and assumes a value on the order of unity,

$$NF = \mu = \mathrm{O}(1), \tag{2.42}$$

where μ is the mean value of the number of occurrences. After substitution of $F = \mu/N$ and $\overline{F} = 1 - \mu/N$ into Eq. (2.35) and taking the limit as $N \to \infty$,

$$\begin{aligned} P(n) &= \lim_{N\to\infty} P_N(n) \\ &= \lim_{N\to\infty} \frac{N(N-1)\cdots(N-n+1)}{n!}\left(\frac{\mu}{N}\right)^n\left(1-\frac{\mu}{N}\right)^{N-n} \\ &= \lim_{N\to\infty}\left(1-\frac{\mu}{N}\right)^N \frac{\mu^n}{n!}\frac{1\cdot(1-1/N)\cdots[1-(n-1)/N]}{(1-\mu/N)^n} \\ &= \lim_{N\to\infty}\left(1-\frac{\mu}{N}\right)^N \frac{\mu^n}{n!}. \end{aligned} \tag{2.43}$$

After invoking the relation

$$e = \lim_{N\to\infty}\left(1+\frac{1}{N}\right)^N, \tag{2.44}$$

the Poisson distribution is obtained,

$$P(n) = \exp(-\mu)\mu^n/n!. \tag{2.45}$$

The Poisson distribution $P(n)$ physically represents the probability of n occurrences of an event given that the expected (average) number of occurrences equals μ. In other words,

$$\begin{aligned} \exp(-\mu) &= \text{the probability that an event will not occur,} \\ \mu\exp(-\mu) &= \text{the probability that an event will occur exactly once,} \\ (\mu^2/2!)\exp(-\mu) &= \text{the probability that an event will occur exactly twice.} \end{aligned}$$

A quick check shows that the probability of obtaining any number of occurrences is (correctly) unity,

$$\sum_{n=0}^{\infty} P(n) = \exp(-\mu)\sum_{n=0}^{\infty}\frac{\mu^n}{n!} = 1. \tag{2.46}$$

The mean and the variance of the Poisson distribution follow from Eqs. (2.32) and (2.33) for Eq. (2.45) as the number of occurrences approaches infinity. From the definition of μ,

$$m = \mu, \tag{2.47}$$

although it is less obvious that

$$\sigma^2 = \mu \tag{2.48}$$

which gives a fractional standard deviation of $\sigma/m = \mu^{-1/2}$.

The Poisson probability that an event will occur k times or less is

$$P(\leq k) = \sum_{n=0}^{k} P(n). \tag{2.49}$$

The probability $P(>k)$ that the event will occur $k+1$ or more times is the complement of $P(\leq k)$.

The Poisson distribution can be used to analyze the failure of a system consisting of a large number of identical components that, upon failure, cause irreversible transitions in the system. Each component is assumed to fail independently and randomly. Then μ is the most probable number of system failures over the system lifetime.

Example 2.8 A nuclear reactor is fueled with 200 assemblies, each of which can fail if the cladding on a fuel rod fails. If each assembly fails in an independent and random manner over its lifetime in the reactor core, compute the probability of three assemblies failing if, on average, 1% of the fuel assemblies are known to fail.

The mean number of assembly failures is $\mu = 2$, so from Eq. (2.45) for $n = 3$,

$$P(3) = \exp(-2)2^3/3! = 0.1804.$$

As an approximate check, we can use the probability of a single assembly failing, $F = 0.01$, and the binomial distribution result of Eq. (2.35) with $N = 200$ to obtain

$$P(3) = \frac{200!}{3!(200-3)!}(0.01)^3(0.99)^{200-3} = 0.1814. \quad \diamond$$

⇒ **Summary.** What we have learned in this section about time-independent probabilities:

1. Time-independent probabilities are used to analyze system components that fail "on demand."
2. The binomial distribution is useful for systems with two possible outcomes of events, either failure or no failure, in cases where there is a known, finite number of (Bernoulli) trials and the ordering of the trials does not affect the outcome.
3. The Poisson distribution treats systems in which randomly occurring phenomena cause irreversible transitions from one state to another.

2.6 NORMAL DISTRIBUTION

We now consider a probability density function $p(x)$ for a random variable X and a function $g(x)$. The outcome of a set of experiments or measurements involving X is the *expectation* of g,

$$E(g) = \langle g \rangle = \int_{-\infty}^{\infty} g(x)p(x)\mathrm{d}x. \tag{2.50}$$

The *sample mean of* X,

$$E(x) = \langle x \rangle = \int_{-\infty}^{\infty} xp(x)\mathrm{d}x, \tag{2.51}$$

is the average value of a large number of measurements x if X is distributed by the probability density function $p(x)$. The *variance* associated with measurements $g(x)$ is the *expected or mean-square deviation* from the sample mean,

$$\begin{aligned} V(g) &= \int_{-\infty}^{\infty} [g(x) - \langle g \rangle]^2 p(x) \mathrm{d}x \\ &= \langle g^2 \rangle - 2\langle g \rangle \int_{-\infty}^{\infty} g(x) p(x) \mathrm{d}x + \langle g \rangle^2 = \langle g^2 \rangle - \langle g \rangle^2. \end{aligned} \quad (2.52)$$

The *standard deviation* $\sigma(g)$ is then defined as the square root of the sample variance,

$$\sigma(g) = \sqrt{V(g)}. \quad (2.53)$$

The standard deviation is a measure of the flatness or sharpness of the probability density function: The smaller σ is, the sharper the distribution, whereas the distribution spreads out as σ increases.

For a discrete random variable with the probability P_n and functional value $g_n = g(x_n)$, corresponding to measurements x_n, $n = 1, 2, \ldots, N$, the sample mean and variance are

$$E(g) = \langle g \rangle = \sum_{n=1}^{N} g_n P_n, \quad (2.54)$$

$$V(g) = \sum_{n=1}^{N} [g_n - \langle g \rangle]^2 P_n. \quad (2.55)$$

For N repeated measurements of a random variable X, for which $P_n = 1/N$ for $n = 1, 2, \ldots, N$, the sample mean and variance of the measurements are

$$E(x) = \langle x \rangle = \overline{x} = N^{-1} \sum_{n=1}^{N} x_n, \quad (2.56)$$

$$V(x) = \sigma^2 = N^{-1} \sum_{n=1}^{N} [x_n - \overline{x}]^2 = \langle x^2 \rangle - \langle x \rangle^2. \quad (2.57)$$

The *Gaussian*, or *normal, distribution* for random variables for a mean value $m = \langle x \rangle$ is

$$p(x) = \frac{1}{\sqrt{2\pi}\sigma} \exp\left[-\frac{(x-m)^2}{2\sigma^2} \right] \quad (2.58)$$

or, in *standard form*,

$$p(\widehat{x}) = \frac{1}{\sqrt{2\pi}} \exp\left[-\frac{\widehat{x}^2}{2} \right], \quad \widehat{x} = \frac{x-m}{\sigma}. \quad (2.59)$$

In a manner similar to that used in deriving the Poisson distribution in Section 2.5, the normal distribution also can be derived [Mar56] from the binomial distribution in

Table 2.3 Confidence Levels for Mean of Normal Distribution for Large Sample Size

Two-Sided Confidence Level (%)	One-Sided Confidence Level (%)	k
99.73	99.86	3.00
99	99.5	2.58
98	99	2.33
96	98	2.05
95.45	97.72	2.00
95	97.5	1.96
90	95	1.645
80	90	1.28
68.27	84.14	1.00
50	75	0.6745

the limit of a large sample size with the mean and standard deviation kept the same as those for the binomial distribution. The Gaussian distribution is an important special case of $p(x)$ because it can be used for establishing *interval estimates* or *confidence estimates* for predicting the range of a random variable. There are two types of interval estimates: two sided and one sided. A two-sided estimate for the probability that the mean m is within a specified range is given by $P(|x-m| \leq k\sigma)$ whereas either $P(x \leq m + k\sigma)$ or $P(x \geq m - k\sigma)$ gives a one-sided estimate. The values of k must be obtained from *Student's t-distribution* and depend on the sample size N. Values of k for large N are given in Table 2.3; more complete tables are available for cases where the sample size is small [Abr64,Lip73]. The Gaussian distribution represents the optimal or natural probability density function [Bar74] given two statistics, mean and variance, in the sense that it introduces the least bias or smallest possible information beyond the mean and variance.

Another distribution closely related to the normal distribution of Eq. (2.58) is the lognormal distribution (sometimes spelled out as the logarithmic-normal distribution) which is obtained with the change of variables

$$x = \ln z, \qquad \sigma = \alpha, \qquad m = \ln \beta. \tag{2.60}$$

When changing the variables of any probability density function, it is necessary to include the Jacobian of the transformation so that

$$p(y) = \left| \frac{\mathrm{d}x}{\mathrm{d}y} \right| p(x) \Big|_{x=y}. \tag{2.61}$$

Example 2.9 If the probability density function $p(x)$ for the normal distribution is to be expressed in terms of z, where $x = \ln z$, then

$$p(z) = z^{-1} p(x)|_{x=\ln z}. \ \diamond$$

Thus, with the variables in Eq. (2.60), the lognormal distribution becomes

$$p(z) = \frac{1}{\sqrt{2\pi}\alpha z} \exp\left\{-\left[\frac{\ln(z/\beta)}{\sqrt{2}\alpha}\right]^2\right\}, \ \alpha, \beta > 0, \ 0 \le z \le \infty. \tag{2.62}$$

The mean and variance of the distribution are

$$m = \beta \exp(\alpha^2/2) \tag{2.63}$$
$$\sigma^2 = m^2[\exp(\alpha^2) - 1] = \beta^2 \exp(\alpha^2)[\exp(\alpha^2) - 1]. \tag{2.64}$$

With the value of k from Table 2.3, the one-sided distributions for the 5th-percentile and 95th-percentile values are

$$z_{0.05} = \beta \exp(-1.645\alpha) \quad \text{and} \quad z_{0.95} = \beta \exp(1.645\alpha) \tag{2.65}$$

while the median is given by

$$z_{0.5} = \beta. \tag{2.66}$$

The *error factor* is then defined as

$$\frac{z_{0.95}}{z_{0.5}} = \exp(1.645\alpha), \tag{2.67}$$

so that the product of the error factor and the median yields the 95th percentile value. The lognormal distribution $p(z)$, plotted as a function of $\ln z$ and z in Fig. 2.3, clearly illustrates that $p(z)$ is a normal distribution $p(x)$ expressed in terms of $x = \ln z$.

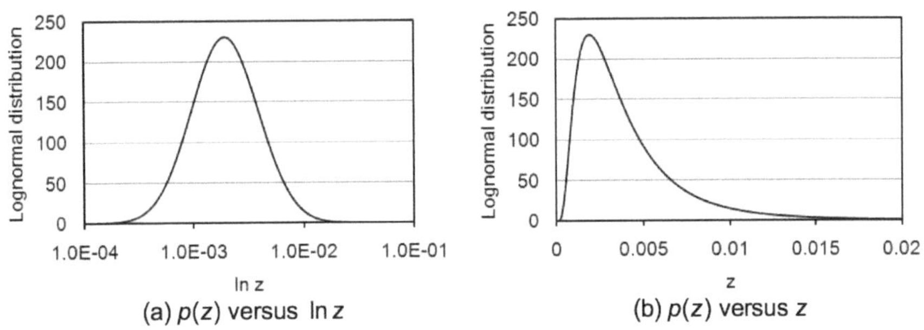

Figure 2.3 Lognormal distribution for $\alpha = 0.7$ and $\beta = 3.162 \times 10^{-3}$ plotted as a function of $\ln z$ and z.

The lognormal distribution is a nonsymmetric distribution with a long tail. The distribution is useful when factors or percentages characterize the variations under consideration, e.g., if z varies between some values $z_0 c$ and z_0/c. In particular, if $z = 10^x$, $x = \log z$, variable x normally distributed represents the order of magnitude of a random variable. Thus the lognormal distribution is often used in probabilistic risk assessments of nuclear power plants where the probability of certain events can vary over several orders of magnitude.

2.7 RELIABILITY FUNCTIONS

We turn now to investigating probability density functions $f(t)$ for failures as a function of time t. For systems that do not undergo repair, the analog to Eq. (2.27) for the failure of a system component is

$$f(t)\mathrm{d}t = \lambda(t)\,\mathrm{d}t[1 - F(t)]. \tag{2.68}$$

Here

$$\begin{aligned} f(t)\mathrm{d}t &= \text{the probability for failure in } \mathrm{d}t \text{ about } t, \\ \lambda(t)\mathrm{d}t &= \text{the probability for failure in } \mathrm{d}t \text{ about } t, \text{ given that it} \\ &\quad\ \text{survived to time } t, \\ 1 - F(t) &= \text{the probability the device did not fail prior to time } t. \end{aligned} \tag{2.69}$$

Another way of saying the same thing is

$$f(t) = \lambda(t)[1 - F(t)], \tag{2.70}$$

where $f(t)$ is the failure probability density and $\lambda(t)$ is the instantaneous failure rate (often called the *hazard rate*) in units of inverse time. It should be pointed out that $\lambda(t)$ does not obey an equation like Eq. (2.61). A slightly different way of thinking about the relationship between the $f(t)$ and $\lambda(t)$ is to compare the probability of failure Pr for two different situations:

$$f(t) = \lim_{\Delta\tau\to 0} \frac{\Pr(\tau < t \le \tau + \Delta\tau)}{\Delta\tau}, \tag{2.71}$$

$$\lambda(t) = \lim_{\Delta\tau\to 0} \frac{\Pr(\tau < t \le \tau + \Delta\tau | t > \tau)}{\Delta\tau}. \tag{2.72}$$

For a system component known to be functioning at time $t = 0$, the probability for failure between 0 and t, which is the cumulative distribution function $F(t)$, is related to the failure probability density by

$$F(t) = \int_0^t f(\tau)\,\mathrm{d}\tau. \tag{2.73}$$

This result is consistent with Eq. (2.2) because every system component is considered to be operable when new, so

$$F(0) = 0. \tag{2.74}$$

[If $f(t) = 0$ for $0 \le t \le T$, the system continues to remain "as good as new" so $F(T) = 0$.] Because every device must eventually fail,

$$F(\infty) = 1, \tag{2.75}$$

and this equation serves as the normalization condition for the failure probability density $f(t)$. [There are devices, however, that need not eventually fail in the finite time of interest for a risk and safety analysis even if they do not undergo repairs. This could occur, for example, if the device is always operated below a threshold loading required for failure; then care must be exercised in evaluating $F(t)$ as t becomes large, but such complications are beyond our interest here.] Differentiation of Eq. (2.73) shows that

$$f(t) = \mathrm{d}F(t)/\mathrm{d}t, \tag{2.76}$$

and this result can be used in Eq. (2.70) to obtain

$$\lambda(t) = \frac{\mathrm{d}F(t)/\mathrm{d}t}{1 - F(t)} = -\frac{\mathrm{d}\ln[1 - F(t)]}{\mathrm{d}t}. \tag{2.77}$$

Because nuclear systems and other well-engineered systems tend to rarely fail, the numerical values of the probabilities for failure for time-independent components, $P(E)$, and time-dependent components, $F(t)$, tend to be very small. But generally, the successful performance of such components is of more interest. This leads us to follow the lead of reliability analysts and define

$R(t)$ = the probability that a system or system component performs a specified function or mission under given conditions *for* a prescribed time.

Stated another way, $R(t)$ is the probability that a specified fault event has not occurred in a system or system component for a given period of time t under the specified operating conditions. Reliability is just the complementary probability to $F(t)$, i.e.,

$$R(t) = 1 - F(t). \tag{2.78}$$

In other words, $F(t)$ is the *un*reliability, the probability that the device will fail at some time between 0 and t, and $R(t)$ is the probability that it will *not* fail during that time period. With Eq. (2.78) the immediately preceding five equations become

$$R(t) = \int_t^\infty f(\tau)\,\mathrm{d}\tau, \tag{2.79}$$

$$R(0) = 1, \tag{2.80}$$

$$R(\infty) = 0, \tag{2.81}$$

$$f(t) = -\mathrm{d}R(t)/\mathrm{d}t, \tag{2.82}$$

$$\lambda(t) = -\frac{\mathrm{d}\ln R(t)}{\mathrm{d}t} = -\frac{1}{R(t)}\frac{\mathrm{d}R(t)}{\mathrm{d}t} = \frac{f(t)}{R(t)}. \tag{2.83}$$

Equation (2.83) can be integrated, followed by use of Eq. (2.80), to obtain the very useful equation

$$R(t) = \exp\left[-\int_0^t \lambda(\tau)\,\mathrm{d}\tau\right]. \tag{2.84}$$

Once the hazard rate $\lambda(t)$ is specified, the last equation gives the probability that a device or system will survive to the time t of interest. A combination of Eqs. (2.83)

Table 2.4 Summary of Equations for $\lambda(t),\ R(t),\ F(t)$, and $f(t)$

Quantity	Equality 1	Equality 2	Equality 3
Hazard rate $\lambda(t)$	$-[1/R(t)]\mathrm{d}R(t)/\mathrm{d}t$	$f(t)/R(t)$	$f(t)/[1-F(t)]$
Reliability $R(t)$	$\exp[-\int_0^t \lambda(\tau)\,\mathrm{d}\tau]$	$1-F(t)$	$\int_t^\infty f(\tau)\,\mathrm{d}\tau$
Failure probability $F(t)$	$1-\exp[-\int_0^t \lambda(\tau)\,\mathrm{d}\tau]$	$1-R(t)$	$\int_0^t f(\tau)\,\mathrm{d}\tau$
Failure probability density $f(t)$	$\lambda(t)R(t)$	$-\mathrm{d}R(t)/\mathrm{d}t$	$\mathrm{d}F(t)/\mathrm{d}t$

and (2.84) leads to the important result

$$f(t) = \lambda(t)\exp\left[-\int_0^t \lambda(\tau)\,\mathrm{d}\tau\right]. \tag{2.85}$$

Other equations relating $f(t),\ F(t),\ R(t)$, and $\lambda(t)$ are given in Table 2.4. Although there are four parameters, only one is independent. Generally $\lambda(t)$ is the one tabulated because it is measured experimentally and because it tends to vary less rapidly with time than the other three parameters.

For many devices, the behavior of $\lambda(t)$ follows the classic "bathtub curve" of Fig. 2.4. Early in life, $\lambda(t)$ for such a device is high because of "wear-in failures" or failures arising because of poor quality control practices. During the middle portion of useful life, failures occur at a rather uniform rate corresponding to random failures. Finally, late in life, $\lambda(t)$ begins to increase because of "wear-out failures." As one might expect, for different devices the shape of the curves and the length of time for each stage of life can be different. The bathtub curve is an example of a piecewise continuous hazard rate.

For nuclear systems, early life failures tend to be random because of high-quality control requirements, while maintenance of components helps mitigate against wear-out failures. Besides, when devices fail infrequently and are sufficiently complex and costly so that many tests cannot be performed to characterize patterns of failure, only an estimate of $\lambda(t)$ is available. Thus the usual procedure in many risk and safety analyses is to assume that failures are random so that

$$\lambda(t) = \lambda. \tag{2.86}$$

In instances where $\lambda(t)$ is piecewise continuous and given by $\lambda_n(t)$ for $t_{n-1} \leq t \leq t_n$, it is sometimes convenient to use the equation

$$R(t_N) = \exp\left[-\sum_{n=1}^{N}\int_{t_{n-1}}^{t_n} \lambda_n(t)\,\mathrm{d}t\right] = \prod_{n=1}^{N}\exp\left[-\int_{t_{n-1}}^{t_n} \lambda_n(t)\,\mathrm{d}t\right] \tag{2.87}$$

or

$$R(t_N) = R(t_{N-1})\exp\left[-\int_{t_{N-1}}^{t_N} \lambda_N(t)\,\mathrm{d}t\right]. \tag{2.88}$$

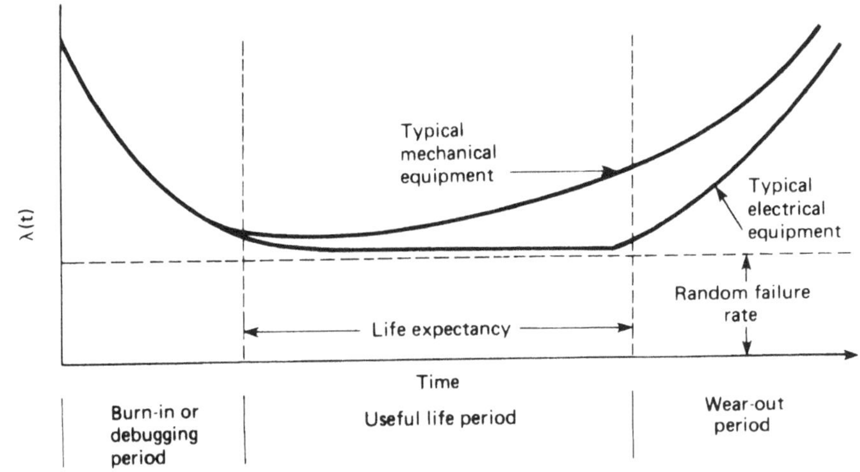

Figure 2.4 Time dependence of conditional failure (hazard) rate. *Source:* [Lam75].

Example 2.10 Calculate the reliability of a device that never fails when it is nearly new before going through a period of random failures and then, approaching its end of life, the hazard rate increases exponentially. The piecewise continuous hazard rate is

$$\lambda(t) = \begin{cases} 0, & 0 \le t \le a, \\ \lambda, & a < t \le b, \\ \lambda \exp[(t-b)/c], & b < t, \text{ with } c > 0. \end{cases}$$

Equation (2.84) gives $R(t) = 1,\ 0 \le t \le a$, and Eq. (2.88) yields

$$\begin{aligned} R(t) &= R(a) \exp\left[-\int_a^t \lambda(\tau)\mathrm{d}\tau\right] \\ &= \exp[-\lambda(t-a)], \quad a \le t \le b. \end{aligned}$$

Likewise, Eq. (2.88) also gives

$$\begin{aligned} R(t) &= R(b) \exp\left[-\int_b^t \lambda(\tau)\mathrm{d}\tau\right] \\ &= \exp\{-\lambda[b-a-c+c\exp(t-b)/c]\}, \quad t \ge b. \diamond \end{aligned}$$

Example 2.11 For Example 2.10 in which $a = 10$ hr, $b = 1000$ hr, $c = 500$ hr, and $\lambda = 10^{-4}$ per hour, calculate the minimum time T in hours before the failure probability equals 0.2.

For $F(T) = 0.2$, it is obvious that $T > 10$ hr, so we first assume $10 < T \leq 1000$ hr and use Eq. (2.88) to find

$$\begin{aligned} T &= a - \ln[1 - F(T)]/\lambda \\ &= 10 - \ln(1 - 0.2)/10^{-4} = 2241. \end{aligned}$$

Because $T > 1000$ hr, the assumption was incorrect, so Eq. (2.88) again is used,

$$\begin{aligned} T &= b + c\ln\left[1 - c^{-1}\{b - a + \ln[1 - F(T)]/\lambda\}\right] \\ &= 1000 + 500\ln\{1 - (500)^{-1}[1000 - 10 + \ln(1 - 0.2)/10^{-4}]\} = 1624 \text{ hr. } \diamond \end{aligned}$$

Example 2.12 A device fails in a random manner while in a *phased mission* mode of operation consisting of three stages: $0 \leq t < t_1, t_1 \leq t < t_2$, and $t \geq t_2$. Obtain the failure probability for the device.

Three nonsynchronous hazard rates, denoted as λ_j, $j = 1, 2, 3$, characterize the hazard rate, which can be written as

$$\lambda(t) = \lambda_1 + (\lambda_2 - \lambda_1)H(t - t_1) + (\lambda_3 - \lambda_2)H(t - t_2),$$

where the Heaviside step function is $H(x) = 1, x \geq 0$, and $H(x) = 0, x < 0$. With Eqs. (2.78) and (2.87) or (2.88), the failure probability is

$$F(t) = \begin{cases} 1 - \exp(-\lambda_1 t), & 0 \leq t < t_1, \\ 1 - \exp[-\lambda_1 t - (\lambda_2 - \lambda_1)(t - t_1)], & t_1 \leq t < t_2, \\ 1 - \exp[-\lambda_1 t - (\lambda_2 - \lambda_1)(t - t_1) - (\lambda_3 - \lambda_2)(t - t_2)], & t \geq t_2. \quad \diamond \end{cases}$$

Another way to look at the relationship between $\lambda(t)$ and $R(t)$ arises if it is desired to determine the minimum time T before the probability of failure exceeds a specified value. Then T can be determined from

$$\int_0^T \lambda(t)\mathrm{d}t = -\ln R(T) = -\ln[1 - F(T)]. \tag{2.89}$$

The first moment of the failure probability density $f(t)$ is a useful indication of the average life for a device. The *mean time to failure* (MTTF) is given by

$$\text{MTTF} = \frac{\int_0^\infty t f(t)\mathrm{d}t}{\int_0^\infty f(t)\mathrm{d}t} = \int_0^\infty t f(t)\mathrm{d}t. \tag{2.90}$$

[The equation has been simplified with Eq. (2.75).] The MTTF also can be expressed in terms of $R(t)$ if Eq. (2.82) is used in Eq. (2.90) and the result is integrated once by parts. If the integral in Eq. (2.90) is defined, which always happens if $tR(t) \to 0$ as $t \to \infty$, then

$$\text{MTTF} = \int_0^\infty R(t)\mathrm{d}t. \tag{2.91}$$

The MTTF is especially simple for random failure events where $\lambda(t) = \lambda$,

$$\text{MTTF} = 1/\lambda. \tag{2.92}$$

Another useful indicator of the performance of a device is the fraction of time a device is in a failed state, i.e., the fractional unavailability $\langle F(t)\rangle_T$, over the time period $0 \leq \tau \leq T$,

$$\langle F(t)\rangle_T = T^{-1}\int_0^T F(\tau)\mathrm{d}\tau. \tag{2.93}$$

For example, for random failure events,

$$\begin{aligned}\langle F(t)\rangle_T &= T^{-1}\int_0^T [1-\exp(-\lambda\tau)]\mathrm{d}\tau \\ &\approx 1-(\lambda T)^{-1}\{1-[1-\lambda T+(\lambda T)^2/2-(\lambda T)^3/6+\cdots]\} \\ &\approx (\lambda T/2)[1-\lambda T/3+\cdots], \quad \lambda T \ll 1. \end{aligned}\tag{2.94}$$

For a device that fails randomly and rarely so that $\lambda(t) = \lambda$ and $\lambda t \ll 1$, the "rare-event approximation" can be used to approximate $F(t)$ by

$$\begin{aligned}F(t) &= 1-\exp(-\lambda t) \\ &\approx 1-[1-\lambda t+(\lambda t)^2/2-\cdots] \\ &\approx \lambda t[1-\lambda t/2+\cdots], \quad \lambda t \ll 1. \end{aligned}\tag{2.95}$$

The rare-event approximation, *when applicable*, can be viewed as a quick way of approximately transforming a time-dependent failure probability into a demand-type failure probability.

⇒ **Summary.** What we have learned in this section about time-dependent failure probabilities:

1. For a device operated continuously, the hazard rate $\lambda(t)$ must be prescribed in order to determine the failure probability $F(t)$.
2. The reliability $R(t)$ for a device is $1 - F(t)$, the complement of the failure probability.

It is worth pointing out that the same principles used to introduce failure distributions can be employed to describe distributions for the repair of a device. Instead of Eq. (2.70), we can use the equation

$$\widetilde{r}(t) = \mu(t)[1-\widetilde{R}(t)]. \tag{2.96}$$

Here the tilde symbol is used to differentiate *repair* from *reliability* and

$$\begin{aligned}\widetilde{r}(t)\mathrm{d}t &= \text{the probability for repair in } \mathrm{d}t \text{ about } t, \\ \mu(t)\mathrm{d}t &= \text{the probability for repair in } \mathrm{d}t \text{ about } t, \text{given that it is not} \\ & \text{repaired prior to time } t, \\ 1-\widetilde{R}(t) &= \text{the probability the device is not repaired prior to time } t.\end{aligned}$$

In other words, $\widetilde{r}(t)$ is the repair probability density, $\mu(t)$ is the instantaneous repair rate in units of inverse time, and the probability of repair is

$$\widetilde{R}(t) = \int_0^t \widetilde{r}(\tau)\mathrm{d}\tau, \tag{2.97}$$

Table 2.5 Summary of Equations for $\mu(t)$, $\widetilde{R}(t)$, and $\widetilde{r}(t)$

Quantity	Equality 1	Equality 2
Instantaneous repair rate $\mu(t)$	$-\mathrm{d}\ln[1-\widetilde{R}(t)]/dt$	$\widetilde{r}(t)/[1-\widetilde{R}(t)]$
Repair probability $\widetilde{R}(t)$	$1-\exp[-\int_0^t \mu(\tau)\,\mathrm{d}\tau]$	$\int_0^t \widetilde{r}(\tau)\,\mathrm{d}\tau$
Repair probability density $\widetilde{r}(t)$	$\mu(t)[1-\widetilde{R}(t)]$	$d\widetilde{R}(t)/dt$

with $\widetilde{R}(\infty)=1$. The equations for repair are given in Table 2.5.

If it can be assumed that the device eventually can be repaired with the passage of time, the *mean time to repair* (MTTR) satisfies

$$\mathrm{MTTR}=\frac{\int_0^\infty t\widetilde{r}(t)\mathrm{d}t}{\int_0^\infty \widetilde{r}(t)\mathrm{d}t}=\int_0^\infty t\widetilde{r}(t)\mathrm{d}t. \tag{2.98}$$

In situations where a repair is completed at a random time after it is initiated, then $\mu(t)=\mu$ and $\widetilde{R}(t)=1-\exp(-\mu t)$, with $\mathrm{MTTR}=1/\mu$. On the other hand, in cases where the time for a repair is short compared to the intended operation time of the device, then a repair can be considered to be essentially instantaneous and given by a constant,

$$\widetilde{R}(t)\to\widetilde{R},\quad \mathrm{MTTR}\ll\mathrm{MTTF}\text{ or }\lambda\ll\mu. \tag{2.99}$$

Combining the MTTF of Eq. (2.90) with the MTTR of Eq. (2.98) yields the *mean time between failures* (MTBF) with due accounting for the repair time,

$$\mathrm{MTBF}\;=\;\mathrm{MTTF}\;+\;\mathrm{MTTR}. \tag{2.100}$$

Finally, it should be pointed out that there may be other time-dependent activities related to a "repair system," such as inspections after repairs have been completed or periods of time required to recertify a system. If such activities can be separated from the actual repair of a system, it is sometimes helpful to treat each activity separately in order to assess where improvements in maintenance procedures can be made.

2.8 TIME-DEPENDENT PROBABILITY DISTRIBUTIONS

For the continuous-in-time failure probability density $f(t)$, corresponding to Eqs. (2.32) and (2.33), the mean value m and the variance σ^2, in terms of the standard deviation σ, are

$$m\;=\;\mathrm{MTTF}=\int_0^\infty tf(t)\mathrm{d}t, \tag{2.101}$$

$$\sigma^2\;=\;\int_0^\infty (t-m)^2 f(t)\mathrm{d}t. \tag{2.102}$$

where Eq. (2.90) has been used. The mean and variance are the two lowest order moments of $f(t)$; higher order moments sometimes are useful for selecting the type of failure distribution that should to be used to fit failure data.

We will now consider five of the most useful probability distributions for describing failures of continuously operating devices: the Erlangian, exponential, gamma, lognormal, and Weibull distributions.

2.8.1 Erlangian and Exponential Distributions

The Erlangian distribution is the time-dependent form corresponding to the Poisson distribution for failure events of devices operated on demand. The distribution arises frequently in reliability engineering calculations involving random failures for which $\lambda(t) = \lambda$. To derive the distribution from Eq. (2.45), we recognize that the mean number of failures μ is the product of λ and time t. The probability of exactly n failures occurring in time t is then given by

$$P(n, t) = \frac{\exp(-\lambda t)(\lambda t)^n}{n!}, \tag{2.103}$$

and the probability of k or fewer failures is

$$P(\leq k, t) = \sum_{n=0}^{k} \frac{\exp(-\lambda t)(\lambda t)^n}{n!}. \tag{2.104}$$

Equation (2.103) permits calculation of the failure probability density $f(t)$ for the nth failure in dt about t, given that a device has undergone $n-1$ prior failures. Then the system is vulnerable to failing with a hazard rate λ. Thus the Erlangian distribution follows from Eq. (2.103) as

$$f(t) = \lambda P(k-1, t) = \frac{\lambda(\lambda t)^{k-1}\exp(-\lambda t)}{(k-1)!}, \quad k \geq 1. \tag{2.105}$$

The most important special case is for $k = 1$, for which the exponential distribution is obtained, with

$$f(t) = \lambda \exp(-\lambda t), \tag{2.106}$$

$$F(t) = 1 - \exp(-\lambda t). \tag{2.107}$$

The mean and variance are

$$\text{MTTF} = 1/\lambda, \tag{2.108}$$

$$\sigma^2 = 1/\lambda^2. \tag{2.109}$$

In situations where the exponential distribution is applied in the rare-event approximation, Eqs. (2.94) and (2.95) can be utilized.

The exponential distribution can be used for analyzing the (first) random failure event of a device characterized by a constant hazard rate, as was discussed in Section 2.7. Both the exponential and Erlangian distributions are special cases of the gamma distribution for which k is not restricted to integer values.

2.8.2 Gamma Distribution

The gamma failure probability density obeys the equation

$$f(t) = \frac{\lambda(\lambda t)^{\alpha-1}}{\Gamma(\alpha)} \exp(-\lambda t), \quad \alpha > 0, \tag{2.110}$$

where $\Gamma(\alpha)$ is a gamma function of Appendix B for a nonnegative and real constant α. If α is an integer, then $\Gamma(\alpha) = (\alpha - 1)!$ and Eq. (2.105) is recovered. For $\lambda = 0.5$ and $\alpha = 0.5\eta$, where η is the number of degrees of freedom associated with a set of measurements, the gamma distribution becomes the *chi-square* distribution.

The gamma distribution failure probability is

$$\begin{aligned} F(t) &= \frac{\int_0^{\lambda t} y^{\alpha-1} \exp(-y)\mathrm{d}y}{\Gamma(\alpha)}, \\ &= \frac{\gamma(\alpha, \lambda t)}{\Gamma(\alpha)}, \end{aligned} \tag{2.111}$$

where $\gamma(\alpha, \lambda t)$ is defined in Appendix B as the incomplete gamma function [Abr64]. For $t \to \infty$, $\gamma(\alpha, \lambda t) \to \Gamma(\alpha)$, which is consistent with $F(\infty) = 1$. The mean and variance are

$$\text{MTTF} = \alpha/\lambda, \tag{2.112}$$

$$\sigma^2 = \alpha/\lambda^2. \tag{2.113}$$

In situations where the gamma distribution is applied in the rare-event approximation,

$$F(t) \approx (\lambda t)^\alpha/\alpha[1 - \alpha\lambda t/(\alpha + 1) + \cdots], \quad \lambda t \ll 1, \tag{2.114}$$

and from Eqs. (2.93) and (2.111),

$$\langle F(t) \rangle_T = \frac{1}{T}\int_0^T F(\tau)\mathrm{d}\tau \approx \frac{(\lambda T)^\alpha}{\Gamma(\alpha + 2)}\left[1 - \frac{\alpha\lambda T}{\alpha + 2} + \cdots\right], \quad \lambda T \ll 1. \tag{2.115}$$

The gamma distribution is especially appropriate for systems subjected to repetitive, random shocks generated according to the Poisson distribution. The failure probability then depends on the number of shocks the device has undergone, i.e., its age. Another application for the distribution arises if the mean rate of wear of a device is a constant but the rate of wear is subject to random variations. For example, for metal devices where the onset of wear takes the form of corrosion, a time-shifted gamma failure distribution can be used with $f(t) = 0$ for $t < \tau$ by replacing t by $t - \tau$ on the right-hand side of $f(t)$ in Eq. (2.110) for $t \geq \tau$, where τ is the time delay parameter; then the MTTF of Eq. (2.112) becomes $\tau + \alpha/\lambda$.

Example 2.13 A device subject to corrosion is placed in service and does not begin to fail before 6 months, after which the failures follow the gamma distribution with $\alpha = 2$ and $\lambda = 10^{-2}$/yr. Estimate (a) the probability of failure after the device has operated for 5 years and (b) its mean time to failure.

(a) With a time delay before the onset of failures of $\tau = 0.5$ yr, first check the value of $\lambda(T-\tau) = 10^{-2}(5-0.5) = 0.045 \ll 1$, so from Eq. (2.115),

$$F(5\,\text{yr}) \approx \frac{(0.045)^2}{\Gamma(4)}\left[1 - \frac{2(0.045)}{2+2}\right] \approx 3.3 \times 10^{-4}.$$

(b) From Eq. (2.112), $\text{MTTF} = \tau + \alpha/\lambda = 0.5\,\text{yr} + 2/(10^{-2}\,/\text{yr}) = 200.5$ yr. ◇

2.8.3 Lognormal Distribution

The time-dependent lognormal distribution follows by substituting $z = t$ in Eq. (2.62), which then gives

$$\begin{aligned}
f(t) &= \frac{1}{\sqrt{2\pi}\alpha t}\exp\left[-\frac{[\ln(t/\beta)]^2}{2\alpha^2}\right], \quad \alpha,\beta > 0,\ 0 \le t \le \infty, &(2.116)\\
F(t) &= \pi^{-1/2}\int_{-\infty}^{z}\exp(-\tau^2)\mathrm{d}\tau \\
&= \begin{cases} (1/2)[1-\operatorname{erf}|z|], & z = \ln(t/\beta)/\sqrt{2}\alpha,\ t < \beta \\ (1/2)[1+\operatorname{erf} z], & z = \ln(t/\beta)/\sqrt{2}\alpha,\ t > \beta, \end{cases} &(2.117)\\
\text{MTTF} &= \beta\exp(\alpha^2/2) &(2.118)\\
\sigma^2 &= \beta^2\exp\alpha^2(\exp\alpha^2 - 1), &(2.119)
\end{aligned}$$

where erf z is the error function [Abr64]

$$\operatorname{erf} z = \frac{2}{\sqrt{\pi}}\int_0^z \exp(-u^2)\mathrm{d}u. \tag{2.120}$$

Of course, if a time-shifted lognormal failure distribution is used with $f(t) = 0$ for $t < \tau$, as obtained by replacing t by $t - \tau$ on the right-hand side of Eq. (2.116) for $t \ge \tau$, then the MTTF of Eq. (2.118) is increased by τ.

The constant shape parameter α (which is dimensionless) and the constant scale parameter or "characteristic life" β (in units of time) are sufficient to specify $f(t)$ for the two-parameter model. One should always check the units of the scale parameter because often the symbol $\widetilde{\beta} = \ln\beta$ (in units of ln-time) is used instead of β, in which case, for example,

$$\begin{aligned}
f(t) &= \frac{1}{\sqrt{2\pi}\alpha t}\exp\left[-\frac{(\ln t - \widetilde{\beta})^2}{2\alpha^2}\right], \quad \alpha,\ \widetilde{\beta} > 0, &(2.121)\\
\text{MTTF} &= \exp\left(\widetilde{\beta} + \frac{\alpha^2}{2}\right). &(2.122)
\end{aligned}$$

The lognormal distribution is skewed to longer times than the Gaussian distribution, which is symmetric about its mean value; the skewness increases with increasing values of α. In situations where the lognormal distribution is applied in the rare-event approximation,

$$F(t) \approx \frac{\exp(-z^2)}{2\sqrt{\pi}z}\left[1 - \frac{1}{2z^2} + \cdots\right], \quad z = \left|\frac{\ln(t/\beta)}{\sqrt{2}\alpha}\right| \gg 1,\ t < \beta. \qquad (2.123)$$

One use of the lognormal distribution is when a random variable at the "nth step" can change by a random proportion of the variable at the "$(n-1)$th step." For example, if the incremental growth of a crack changes by a random length under each loading event, then a lognormal distribution would be appropriate for time-dependent failures. Another feature of the lognormal distribution is that the skewness to longer times incorporates the general behavior of outlier events because the skewness accounts for the occurrence of those infrequent but large deviate values. For this reason the lognormal distribution is often used for analyzing the risk and safety of systems for which failures rarely occur, as in nuclear systems. This was discussed in Section 2.6.

Example 2.14 A testing program has shown that failures of a particular type of pump satisfy a lognormal distribution with a mean time to failure of 2.5×10^4 hr. It is also known that 40% have failed by $T = 10^4$ hr. Calculate the probability of failure for a pump intended for $T_* = 5 \times 10^4$ hr of service.

From Eq. (2.118),

$$\ln \text{MTTF} = \ln \beta + \alpha^2/2.$$

Because $F(T) < 0.5$, Eq. (2.117) gives

$$\begin{aligned} 2F(T) &= 1 - \text{erf}\left|\frac{\ln T - \ln \beta}{\sqrt{2}\alpha}\right| \\ &= 1 - \text{erf}\left|\frac{\ln T - \ln \text{MTTF} + \alpha^2/2}{\sqrt{2}\alpha}\right|, \qquad T < \beta. \end{aligned}$$

Substitution of $F(T) = 0.4$ along with the values for T and MTTF gives

$$\text{erf}\left|\frac{-0.9163 + \alpha^2/2}{\sqrt{2}\alpha}\right| = 0.2,$$

and a table of error functions yields

$$\left|\frac{-0.9163 + \alpha^2/2}{\sqrt{2}\alpha}\right| = 0.179.$$

If we remove the absolute-value sign from the left-hand side of the equation by assuming the argument is positive, we obtain the root $\alpha = 1.632$ and a negative root that does not satisfy the constraint $\alpha > 0$. The values of α and MTTF then lead to $\beta = 6620$ hr, an answer that must be incorrect because $\beta < T$. Therefore we remove the absolute-value sign by taking the negative argument and obtain $\alpha = 1.125$, which leads to $\beta = 1.330 \times 10^4$ hr. Finally, Eq. (2.117) is used with $T_* > \beta$ to obtain

$$\begin{aligned} F(T_*) &= \frac{1}{2}\left[1 + \text{erf}\left(\frac{\ln(T_*/\beta)}{\sqrt{2}\alpha}\right)\right] \\ &= 0.5[1 + \text{erf}(0.833)] = 0.78. \diamond \end{aligned}$$

2.8.4 Weibull Distribution

The Weibull distribution is a very general and popular one for analyzing the reliability of many systems. For this failure distribution, with constant shape factor α, constant scale parameter or "characteristic life" β, and a time delay τ, the three-parameter form is

$$f(t) = \begin{cases} \dfrac{\alpha}{\beta}\left(\dfrac{t-\tau}{\beta}\right)^{\alpha-1} \exp\left[-\left(\dfrac{t-\tau}{\beta}\right)^{\alpha}\right], & \tau \le t \le \infty, \\ 0, & 0 \le t \le \tau \end{cases} \tag{2.124}$$

for $\alpha > 0$ and $\beta > 0$, with

$$F(t) = 1 - \exp\{-[(t-\tau)/\beta]^{\alpha}\}, \quad 0 \le \tau \le t \le \infty, \tag{2.125}$$

$$\text{MTTF} = \tau + \beta\Gamma(1+\alpha^{-1}), \tag{2.126}$$

$$\sigma^2 = \beta^2\{\Gamma(1+2\alpha^{-1}) - [\Gamma(1+\alpha^{-1})]^2\}. \tag{2.127}$$

In situations where the Weibull distribution is applied in the rare-event approximation,

$$F(t) \approx [(t-\tau)/\beta]^{\alpha}\{1 - [(t-\tau)/\beta]^{\alpha}/2 + \cdots\}, \quad 0 \le [(t-\tau)/\beta]^{\alpha} \ll 1, \tag{2.128}$$

and from Eqs. (2.93) and (2.125),

$$\begin{aligned} \langle F(t)\rangle_T &= \frac{1}{T-\tau}\int_{\tau}^{T} F(\tilde{\tau})\,\mathrm{d}\tilde{\tau} = 1 - \frac{\beta}{\alpha(T-\tau)}\gamma\left[\frac{1}{\alpha}, \left(\frac{T-\tau}{\beta}\right)^{\alpha}\right], \\ &\approx \frac{[(T-\tau)/\beta]^{\alpha}}{\alpha+1} - \frac{[(T-\tau)/\beta]^{2\alpha}}{2(2\alpha+1)} + \cdots, \quad \left(\frac{T-\tau}{\beta}\right)^{\alpha} \ll 1 \end{aligned} \tag{2.129}$$

where the incomplete gamma distribution of Eq. (2.111) again has been used.

This distribution has the attractive attribute that the hazard rate satisfies a power law as a function of time,

$$\lambda(t) = \frac{\alpha}{\beta}\left(\frac{t-\tau}{\beta}\right)^{\alpha-1}. \tag{2.130}$$

For $\alpha = 1$, the exponential distribution is obtained, with hazard rate $\lambda = \beta^{-1}$. Furthermore, as α increases, the Weibull distribution tends to the normal distribution; indeed, for $\alpha \ge 4$, Eq. (2.124) and the normal distribution are almost indistinguishable! Another special case of the Weibull model is the *Rayleigh* distribution, for which $\alpha = 2$. In some cases it may be appropriate to combine together two or more Weibull models for either the failure probability density or the hazard rate.

Example 2.15 Failures of a given device operating continuously can be classified as either sudden (catastrophic) or delayed (wear-out). Construct a model for the failure probability of the device.

Catastrophic failures can occur as soon as the device is exposed to an operating environment outside the maximum tolerances for operation; then the Weibull distribution starting at $t = 0$ with a shape parameter $\alpha_1 < 1$ can be an appropriate model.

Wear-out failures are due to the aging of the device; a Weibull model with failures beginning for $t = \tau > 0$ and a shape parameter $\alpha_2 > 1$ can be an appropriate failure model. From Eq. (2.125) the failure probability is

$$F(t) = k\left\{1 - \exp\left[-\left(\frac{t}{\beta_1}\right)^{\alpha_1}\right]\right\} + (1-k)\left\{1 - \exp\left[-\left(\frac{t-\tau}{\beta_2}\right)^{\alpha_2}\right]\right\}$$

for $\beta_1, \beta_2 > 0$ and $0 \le k \le 1$. ◇

An important application of the Weibull distribution arises when analyzing the failure of a device that consists of a large number of identical components, each of which can fail independently according to the gamma distribution and all of which must function for the device to not fail. An example of such a device is a nuclear reactor fuel assembly consisting of a large number of (presumably identical) fuel rods; if the cladding fails for any rod, then the assembly is said to have "failed."

Example 2.16 An analysis of 39 Mark-IA driver fuel assemblies for the EBR-II reactor showed [Ols76] that the failure probability $F(u_{max})$ as a function of the maximum burnup u_{max} (in %) at normal EBR-II operating conditions is

$$F(u_{max}) = 1 - \exp\left[-\left(\frac{u_{max} - 3.0}{0.674}\right)^{5.91}\right], \qquad u_{max} \ge 3.0.$$

Calculate the mean percent burnup m at the time of failure.

Comparison of $F(u_{max})$ with Eq. (2.125) shows that $\alpha = 5.91$, $\beta = 0.674\%$, and there is a delay of 3.0% before the onset of failures. From Eq. (2.126), adjusted for a shifted burnup percent of 3.0%,

$$m = 3.0 + 0.674\Gamma(1 + 1/5.91),$$

and from a table of gamma functions [Abr64], $m = 3.62\%$. ◇

2.8.5 Generalized "Bathtub" Distribution

An example of a general distribution that is shaped like the bathtub curve in Fig. 2.4, with wear-in failures early in the life cycle and wear-out failures later, is one for which the failure probability density is [Dhi79]

$$\begin{aligned} f(t) &= [c\alpha(t/\beta)^{\alpha-1} + (1-c)\gamma(t/\theta)^{\varrho-1}\exp(t/\theta)^{\alpha}] \\ &\quad \times \exp\{-c\beta(t/\beta)^{\alpha} - (1-c)[\exp(t/\theta)^{\varrho} - 1]\} \end{aligned} \tag{2.131}$$

for $0 \le c \le 1$ and $\alpha, \beta, \varrho, \theta > 0$. The constants α and ϱ are shape parameters and the constants β and θ are scale parameters. For this distribution

$$R(t) = \exp\{-c\beta(t/\beta)^{\alpha} - (1-c)[\exp(t/\theta)^{\varrho} - 1]\} \tag{2.132}$$

and

$$\lambda(t) = c\alpha(t/\beta)^{\alpha-1} + (1-c)\varrho(t/\theta)^{\varrho-1}\exp(t/\theta)^{\alpha}. \tag{2.133}$$

Several distributions can be obtained as a special case of this one:

- Exponential, for $c = 1$, $\varrho = 1$
- Weibull, for $c = 1$
- "Bathtub" curve, e.g., for $\alpha = 1$ and $\varrho = 0.5$

2.8.6 Selection of a Time-Dependent Probability Distribution

Preferably a time-dependent probability distribution can be selected on the basis of the physical nature of the problem fitting most or all of the underlying assumptions associated with a particular distribution. If insufficient theoretical reasons are available for such a selection, it may be necessary to infer a suitable distribution by other means. One possibility is to examine the ratios of higher moments of $f(t)$ about the mean. If the higher moments about the mean are denoted by

$$\mu_n = \int_0^\infty (t - \mu_1)^n f(t)\mathrm{d}t, \quad n = 2,\ 3,\ 4, \dots\ , \tag{2.134}$$

so that $\mu_2 = \sigma^2$ is the variance, then the third moment is *skewness*, which is a measure of the asymmetry of the distribution, and $n = 4$ gives the *kurtosis*, which is related to the peakedness of $f(t)$. The *coefficient of skewness* $\beta_1^{1/2}$ measures the skewness of a distribution relative to its spread and is defined as

$$\beta_1^{1/2} = \mu_3/\mu_2^{3/2}. \tag{2.135}$$

The *coefficient of kurtosis* β_2 is

$$\beta_2 = \mu_4/\mu_2^2. \tag{2.136}$$

The dimensionless coefficients β_1 and β_2 provide the possibility in some instances of distinguishing between different distributions, as illustrated by Fig. 2.5. We see that the exponential distribution is the point for $\beta_1 = 4$ and $\beta_2 = 9$. All gamma distributions have parameters such that $\beta_2 - 1.5\beta_1 = 3$, which is the straight line connecting the point for the exponential distribution, with the point at $\beta_1 = 0$ and $\beta_2 = 3$ which is the normal distribution. The lognormal distribution in Fig. 2.5 is nearly a straight line given by $\beta_2 - [16(1+\epsilon)/9]\beta_1 = 3$, where $\epsilon \leq 0.1$.

The Weibull distribution for β_2 versus β_1 lies along two distinct curves depending on whether the shape parameter α exceeds 3.6, for which $\beta_1 = 0$. For $1 < \alpha \leq 11.5$, the curve always lies somewhat below that for the gamma distribution, while for $\alpha > 11.5$, the curve lies above that for the gamma distribution and rapidly approaches that for the lognormal distribution.

From the data available, if the (β_1, β_2) point is reasonably close to a point or curve corresponding to one of the standard distributions, then that distribution can be selected to represent the data. Because of fluctuations in sampling, the moments μ_3 and μ_4 are especially sensitive to data outliers, so selection of the distribution

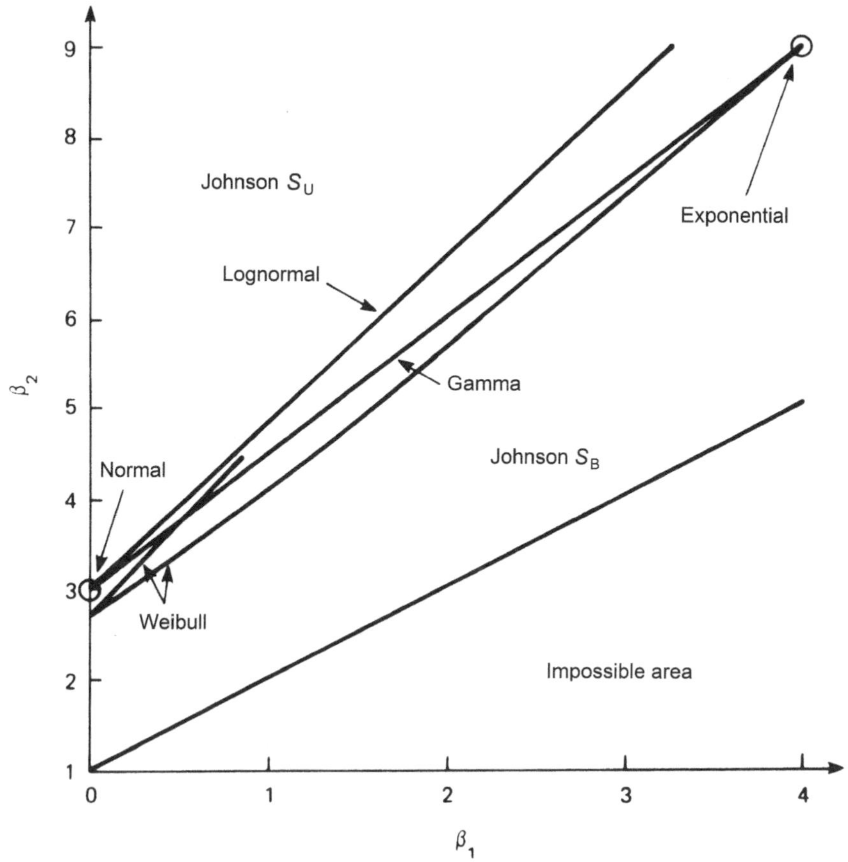

Figure 2.5 Regions of kurtosis versus skewness for various distributions.

with Eqs. (2.135) and (2.136) often does not lead to a conclusive selection of the best probability distribution.

In instances where (β_1, β_2) lies far from a point or curve corresponding to one of the standard distributions, this suggests that a more general distribution may be required. The *Johnson* distribution is a three-parameter distribution (actually a four-parameter distribution if there is a time delay) that spans the entire possible area in Fig. 2.5. There are three alternate forms of the Johnson distribution: the Johnson S_L is just the lognormal distribution, the Johnson S_U spans the area above the lognormal curve in the figure, and the Johnson S_B spans the possible area below the lognormal curve. This entire area also may be fitted to one of the forms of the *Pearson* distribution. Such distributions are beyond our scope of interest here.

⇒ **Summary.** What we have learned in this section about time-dependent probability distributions:

1. The gamma distribution is a natural extension of the Poisson discrete distribution and encompasses the Erlangian and exponential distributions as special cases; it is frequently useful for characterizing fatigue failures arising from repetitive shocks or, in the case of the exponential distribution, random failures.
2. The time-dependent lognormal distribution has the form of the normal distribution except that the independent variable is $\ln t$ rather than t; it is very widely used to describe failures for risk and safety studies and has an important role for treating possible errors in random variables, as was described in Section 2.6.
3. The Weibull distribution includes both the exponential and normal distributions as special cases; it is widely used because it encompasses all cases in which the hazard rate varies according to a power of t.
4. The parameters for these five distributions are determined from the first and second moments of $f(t)$, i.e., the mean and variance, and from the time of the beginning of possible failures.
5. The coefficients of skewness and kurtosis may be helpful in determining which distribution to use if there is not a physical basis for selecting one of them.

2.9 EXTREME-VALUE PROBABILITY DISTRIBUTIONS

Extreme-value distributions, or weakest link distributions, are for the probability of occurrence $P^*(x_*)$ of either the maximum or minimum x_* when a large number of independent events are sampled from an initial distribution $P(x)$ [Gum58]. [Here the asterisk emphasizes that the distribution $P^*(x_*)$ differs from the parent distribution $P(x)$ and x_* differs from x.] By restricting the sampling to a large number of events, only the *asymptotic* distributions for maximum or minimum values need be considered. A classification system has been developed according to the behavior of $F(\infty)$ and to whether the minimum or maximum values are selected. Results of the classification scheme are shown in Table 2.6, along with a few natural phenomena for which the extreme-value distributions apply.

There are three types of extreme-value distributions. Type III is the Weibull distribution previously described and Type II distributions are not of interest to us. We first consider the Type I distribution for maximum values for which

$$P^*(x_*) = \exp\{\exp[-\alpha(x_* - \beta)]\}, \quad -\infty < x_* < \infty,\ \alpha > 0,\ -\infty < \beta < \infty. \tag{2.137}$$

Differentiation of this equation with respect to x_* leads to the probability density for the Type I asymptotic distribution of maximum values as

$$p^*(x_*) = \alpha \exp\{-\alpha(x_* - \beta) - \exp[-\alpha(x_* - \beta)]\}. \tag{2.138}$$

The mean value of x_* and the variance are

$$m = \beta + 0.577\alpha^{-1}, \tag{2.139}$$

$$\sigma^2 = 1.645\alpha^{-2}, \tag{2.140}$$

Table 2.6 Classification Scheme and Applications for Extreme-Value Distributions

Values Sampled	Initial Distribution Sampled From $P(x)$	Extreme-Value Distribution Type $P^*(x_*)$	Some Applications
Maximum	Gamma	Type I (maximum values)	Sea wave height River level
	Lognormal	Type I (maximum values)	Flood damage magnitude Earthquake magnitude and frequency
Minimum	Normal (Gaussian)	Type I (minimum values)	Material fracture strength
Minimum	Gamma	Type III (identical to Weibull)	Drought occurrence Wind speed minimum

while the hazard function in units of x_*^{-1} is

$$\lambda^*(x_*) = \frac{\alpha \exp[-\alpha(x_* - \beta)]}{\exp\{\exp[-\alpha(x_* - \beta)]\} - 1}. \tag{2.141}$$

Floods, tornadoes, hurricanes, and earthquakes are examples of natural disasters that can be fit to a Type I asymptotic distribution of maximum values. The data for the probability of occurrence and the severity of each of these phenomena generally are quite site specific and depend on the locality under consideration. Therefore detailed meteorological studies are required to use a Type I distribution for any risk analysis.

Another quantity of interest is the return period $T(x_*)$ of an extreme of magnitude $\geq x_*$, given by

$$T(x_*) = [1 - P^*(x_*)]^{-1}, \tag{2.142}$$

where $P^*(x_*)$ is given in Eq. (2.137). As an example, if x is the rainfall in a year, then it takes an average of $T(x_*)$ years for an annual maximum rainfall of at least x_* to occur once—provided there are no modifications in rainfall patterns due to environmental changes.

The Type I asymptotic distribution $F^*(x_*)$ of minimum values x_* can be used to investigate the fracture strength of a material that has a Gaussian distribution of crack sizes. This extreme-value distribution is given by

$$P^*(x_*) = 1 - \exp\{-\exp[\alpha(x_* - \beta)]\}, \quad -\infty < x_* < \infty,\ \alpha > 0,\ -\infty < \beta < \infty, \tag{2.143}$$

with Eqs. (2.138) through (2.141) replaced by

$$\begin{aligned} p^*(x_*) &= \alpha \exp\{\alpha(x_* - \beta) - \exp[\alpha(x_* - \beta)]\}, & (2.144)\\ m &= \beta - 0.577\alpha^{-1}, & (2.145)\\ \sigma^2 &= 1.645\alpha^{-2}, & (2.146)\\ \lambda^*(x_*) &= \alpha \exp[\alpha(x_* - \beta)]. & (2.147) \end{aligned}$$

The Type I distribution of maximum values is closely related to that for minimum values except that the values of $\lambda^*(x_*)$ are dramatically different.

2.10 PROBABILITY MODELS FOR FAILURE ANALYSES

The traditional failure models introduced in Sections 2.5 and 2.6 provide a framework with which to express the failure probabilities of different *individual* components of a complicated device or system. That is, they can be used to describe only a system or device that can be modeled as a single component. The models do not take into account how the components are linked together. For example, they do not take into account how the pumps, pipes, and valves of a piping system are physically linked together, nor do they incorporate the way in which wires and switches of an electrical circuit are arranged. We defer an introduction to the principles of linking the possible failure of components of a system to the failure of the system itself until after we have considered in Chapter 3 how to incorporate experimental data into the failure models.

In some instances it may be appropriate to model the failure of a component with a combination of distributions. We have already seen examples of the application of different models during different phases of the lifetime of a device, such as the bathtub model, for which the hazard rate and the failure probability were continuous with time. But the failure of any component or system can be discontinuous with time if the operating environment undergoes a "step" change. Then, for example, the hazard rate of a system component can be discontinuous with time, as can occur if the failure of a different system component causes overheating of the component that continues to function in the different operating environment.

Yet another example in which the possible failure of one component might be linked with the failure of other "identical" components in a system arises from common cause or common mode failures. These failures occur because of a common event in the system operating environment that is external to the operation of the system itself, as in the case of an earthquake. Such common cause failures cause a decrease in the system reliability.

If one can assume that all the components can be simultaneously destroyed with a single random event, then the probability of failure of the system due to such a common cause is governed by the exponential distribution. If we define the reliability of the system against such an event as $R_c(t) = \exp(-\lambda_c t)$, where λ_c is the hazard rate for the common cause, from the product rule for probabilities in Eq. (2.4) a model for the reliability of the system $R_{sys}(t)$ then is

$$R_{sys}(t) = R_{sys|c}(t)R_{sys|\overline{c}}(t) = \exp(-\lambda_c t)R_{sys|\overline{c}}(t), \tag{2.148}$$

where $R_{sys|\overline{c}}(t)$ denotes the system reliability to time t given no common cause failures. Common cause failures will be discussed in more detail in Section 6.2.

References

[Abr64] M. Abramowitz and I. Stegun, eds., *Handbook of Mathematical Functions with Formulas, Graphs, and Mathematical Tables*, U.S. Government Printing Office (1964); reprinted by Dover (1970).

[Bar74] Y. Bard, *Nonlinear Parameter Estimation*, Academic Press (1974).

[Dev92] J. Devooght and C. Smidts, "Probabilistic Reactor Dynamics—I: The Theory of Continuous Event Trees," *Nucl. Sci. Eng.* **111**, 229 (1992).

[Dhi79] B. S. Dhillon, "A Hazard Rate Model," *IEEE Trans. Rel.* **29**, 150 (1979).

[Fel68] W. Feller, *An Introduction to Probability Theory and Its Applications*, vol. 1, 3rd ed., Wiley (1968).

[GEC74] "Reliability Manual for Liquid Metal Fast Breeder Reactors (LMFBR) Safety Programs," SRD-74-113, General Electric Company (1974).

[Gum58] E. J. Gumbel, *Statistics of Extremes*, Columbia Univ. Press (1958).

[Izq96] J. M. Izquierdo, E. Melendez, and J. Devooght, "Relationship Between Probabilistic Dynamics and Event Trees," *Reliab. Eng. Sys. Safety* **52**, 197 (1996).

[Lab00] P. E. Labeau, C. Smidts, and S. Swaminathan, "Dynamic Reliability: Towards an Integrated Platform for Probabilistic Risk Assessment," *Reliab. Eng. Sys. Safety* **68**, 219 (2000).

[Lam75] H. E. Lambert, "Fault Trees for Decision Making in Systems Analysis," UCRL-51829, Lawrence Livermore Laboratory (1975).

[Lip73] C. Lipson and N. J. Sheth, *Statistical Design and Analysis of Engineering Experiments*, McGraw-Hill (1973).

[Man74] N. R. Mann, R. E. Schafer, and N. D. Singpurwalla, *Methods for Statistical Analysis of Reliability and Life Data*, Wiley (1974).

[Mar56] H. Margenau and G. M. Murphy, *The Mathematics of Physics and Chemistry*, Van Nostrand Reinhold (1956).

[NRC75] "Reactor Safety Study—An Assessment of Accident Risks in U.S. Commercial Nuclear Power Plants," WASH-1400, U.S. Nuclear Regulatory Commission (1975).

[Ols76] N. J. Olson, C. M. Walter, and W. N. Beck, "Statistical and Metallurgical Analyses of Experimental Mark-1A Driver Fuel Element Cladding Failures in the Experimental Breeder Reactor II," *Nucl. Technol.* **28**, 134 (1976).

[Pap02] A. Papoulis and S. U. Pillai, *Probability, Random Variables, and Stochastic Processes*, 4th ed., McGraw-Hill (2002).

[Tri69] M. Tribus, *Rational Descriptions, Decisions and Designs*, Pergamon (1969).

Exercises

2.1 Construct the minimal cut sets for event H,

$$H = (A + B)(A + C)(D + B)(D + C).$$

2.2 Construct the minimal cut sets for event H,

$$H = \overline{AB + A\overline{B} + \overline{A}\,\overline{B}}.$$

2.3 Construct the minimal cut sets for event H,

$$H = (\overline{\overline{A}B\overline{C}})(\overline{A\overline{B}\,\overline{C}}).$$

2.4 Construct the minimal cut sets for event H,

$$H = (AB + CDE)[AB + (\overline{C} + \overline{D} + \overline{E})].$$

2.5 Write the equation for $P(E_1 + E_2 + E_3 + E_4)$ if the events E_n are mutually exclusive.

2.6 Write the equation for $P(E_1 + E_2 + E_3 + E_4)$ if the events E_n are *not* mutually exclusive.

2.7 An old slot machine has 3 reels, each with 10 different symbols. What is the probability of obtaining 2 lemons (a) twice in 20 trials and (b) more than twice in 20 trials?

2.8 The probability of an event occurring per trial is 0.1. What is the probability of exactly 12 events occurring out of 100 trials if it is calculated with (a) the binomial distribution and (b) the Poisson approximation to the binomial distribution?

2.9 Electric motors in a widget factory fail at the rate of 1.37/yr. Determine the probability that three or more motors fail in a single year.

2.10 During a prescribed length of time a system undergoes a mean of 2 breakdowns. Determine the probability for (a) 0, (b) 1, (c) 2, and (d) 10 breakdowns during that time.

2.11 The multinomial distribution is a generalization of the binomial distribution that accounts for more than two outcomes of any event. In N trials, if mutually exclusive outcomes O_n, $n = 1, 2, \ldots, N$, occur M_n times with probabilities P_n, respectively, then

$$P_N(M_1 M_2 \cdots M_N) = \frac{N!}{M_1! M_2! \cdots M_N!} P_1^{M_1} P_2^{M_2} \cdots P_N^{M_N}.$$

A reactor has control rods that fail to deploy on activation with a probability of 10^{-5} and that fail to fully insert with a probability of 4×10^{-5}; otherwise they deploy as designed. With 5000 activations, what is the probability that (a) a single rod does not correctly operate and (b) the probability that either one or two identical rods acting independently will not correctly operate once activated simultaneously?

2.12 If a reactor has three or four coolant loops, each of which has a failure probability of 0.05 over the lifetime of the plant, what is the probability that at least one loop eventually will fail if there are (a) three coolant loops, and (b) four coolant loops?

2.13 If a reactor of the preceding exercise with four coolant loops requires at least two loops to be operating, what is the probability that the reactor cooling system eventually will fail over the lifetime of the plant?

2.14 Some control units have failures that are given by the exponential distribution with a mean time to failure of 5000 hr. What is the probability that a unit will survive an additional 1000 hr if it has survived (a) 1000 hr, (b) 5000 hr, and (c) 10,000 hr?

2.15 The probability a system with constant hazard rate λ will fail to survive for a mission of 100λ is 0.5. What are the probabilities that (a) it survives for 500λ and (b) it fails within 1000λ?

2.16 The hazard rate for a pressure valve is given by $\lambda(t) = 1/(t+2)$. Determine (a) the probability of failure $F(t)$ and (b) the failure probability density $f(t)$.
2.17 For a system with $\lambda(t) = 0.007t^{-0.3}/\text{yr}$, determine (a) the probability the system will fail in 5 yr and (b) the mean time to failure.
2.18 A lightbulb operates continuously to illuminate a display panel. Determine the mean time to failure for the bulb if its conditional failure rate is (a) $\lambda(t) = 5 \times 10^{-5}t$ and (b) $\lambda(t) = 5 \times 10^{-7}t$ with t in days.
2.19 For the failure probability density $f(t) = kt^a \exp(-t)$, where $a \geq 0$, determine (a) the normalization constant k so that $F(\infty) = 1$, (b) the mean of $f(t)$, and (c) the variance of $f(t)$.
2.20 A system hazard rate $\lambda(t)$ is piecewise continuous,

$$\lambda(t) = \begin{cases} \lambda, & 0 \leq t < t_1, \\ \lambda + k(t - t_1), & t \geq t_1. \end{cases}$$

Determine (a) the probability the system has not failed during time t, (b) the failure probability density, and (c) the MTTF.
2.21 A device fails with a hazard rate given by

$$\lambda(t) = \begin{cases} at, & 0 \leq t < t_1, \\ at_1^2/t, & t \geq t_1, \end{cases}$$

where a is a constant and $at_1^2 > 1$. If $a = 10^{-2}$ hr^{-2} and $t_1 = 15$ hr, (a) calculate the time τ after the device is placed in service before the probability of its failure is 0.95 and (b) derive the equation for MTTF and determine its numerical value.
2.22 A device fails with a hazard rate given by

$$\lambda(t) = \begin{cases} 0, & 0 \leq t < t_1, \\ at_2^2/t, & t_1 \leq t \geq t_2, \\ at_2, & t > t_2, \end{cases}$$

where a is a constant and $0 < at_2^2 < 1$. If $a = 0.5 \times 10^{-4}$ hr^{-2}, $t_1 = 50$ hr, and $t_2 = 100$ hr, (a) derive the failure probability, (b) calculate the time after the device is placed in service before the probability of its failure is 0.99, and (c) compute the MTTF.
2.23 A device fails with a hazard rate given by

$$\lambda(t) = \begin{cases} at^2, & 0 \leq t < t_1, \\ at_1^3/t, & t \geq t_1, \end{cases}$$

where a is a constant and $at_1^3 > 1$. If $a = 2 \times 10^{-6}$ hr^{-3} and $t_1 = 100$ hr, (a) derive the failure probability, (b) calculate the time τ after the device is placed in service before the probability of its failure is 0.90, and (c) derive the equation for the MTTF in terms of the incomplete gamma function.
2.24 A device fails with a hazard rate given by

$$\lambda(t) = \begin{cases} at, & 0 \leq t < t_1, \\ at_1 \exp[b(t - t_1)], & t \geq t_1, \end{cases}$$

where a and b are constants. (a) Determine the failure probability as a function of time and (b) calculate the time τ after a device is placed in service before the failure probability will be 0.8 if $a = 10^{-6}$ hr^{-2}, $b = 10^{-4}$ hr^{-1}, and $t_1 = 10^3$ hr.

2.25 A device has a hazard rate of

$$\lambda = \begin{cases} \lambda, & 0 \le t < t_1, \\ \lambda(t_1/t)^{0.5}, & t \ge t_1, \end{cases}$$

for positive constants λ and t_1. Derive the equation for (a) $F(t)$ and (b) MTTF and (c) compute the time τ at which $F(\tau) = 0.7$ if $t_1 = 10^2$ hr and $\text{MTTF} = 10^3$ hr.

2.26 A device has a continuous hazard rate specified by the equations

$$\lambda(t) = \begin{cases} at^{-1/2}, & 0 \le t < t_1, \\ \lambda, & t_1 \le t < t_2, \\ b(t - t_1)^2, & t \ge t_2. \end{cases}$$

Obtain (a) the constants a and b and (b) the equations for the reliability of the device as a function of time.

2.27 A device fails with a hazard rate given by

$$\lambda = \begin{cases} 0, & 0 \le t < t_1, \\ \lambda t_1/t, & t_1 \le t < t_2, \\ \lambda t_1 t_2^{-2} t, & t \ge t_2, \end{cases}$$

where λ, t_1, and t_2 are constants and $\lambda t_1 > 0$ and $\lambda t_2 < 1$. (a) Derive the equation for $R(t)$ and (b) calculate the time τ in hours after the device is placed in service before the probability of its failure is 0.92 if $\lambda = 5 \times 10^{-4}$ hr^{-1}, $t_1 = 10^3$ hr, and $t_2 = 10^5$ hr.

2.28 The failure probability density for a device is given by

$$f(t) = k \exp(-at) \sinh(bt),$$

where k is a normalization constant and a and b are positive constants with $a > b > 0$. In terms of a and b, derive (a) the constant k, (b) the reliability, (c) the hazard rate, and (d) the MTTF.

2.29 The failure probability density for a device is given by a generalization of the gamma distribution (to which it reduces when $\beta = 1$),

$$f(t) = kt^{\alpha-1} \exp(-\lambda t^{\beta}),$$

where k, α, λ, and β are positive constants. Derive an equation for (a) the normalization constant k, (b) the failure probability $F(t)$ in terms of the incomplete gamma function, and (c) the MTTF.

2.30 The failure probability density for a generalized gamma distribution satisfies the equation

$$f(t) = k(t/t_c)^{(\beta/\alpha)-1} \exp[-(t/t_c)^{\beta}],$$

where k, α, and β are positive constants and t_c is a time scaling factor. Derive an equation for (a) the normalization constant k and (b) the MTTF.

2.31 The failure probability density for a device is given by $f(t) = kE_2(at)$, where $E_n(x)$ is the exponential integral function and k and a are constants. Some properties of $E_n(x)$ for $n = 1, 2, \ldots$ are

$$
\begin{aligned}
E_n(x) &= \int_1^\infty y^{-n}\exp(-xy)\mathrm{d}y = x^{n-1}\int_x^\infty y^{-n}\exp(-y)\mathrm{d}y \\
&= \int_x^\infty E_{n-1}(y)\mathrm{d}y, \\
\mathrm{d}E_n(x)/\mathrm{d}x &= -E_{n-1}(x), \\
E_n(0) &= (n-1)^{-1}, \quad n > 1.
\end{aligned}
$$

Determine (a) the normalization constant k, (b) the failure probability $F(t)$, (c) the reliability $R(t)$, (d) the hazard rate $\lambda(t)$, and (e) the MTTF.

2.32 The failure probability density for a device is given by

$$f(t) = k\int_0^\infty y^\alpha \exp(-\beta y^2 - ty^{-1})\mathrm{d}y, \quad \alpha > -2,\ \beta > 0,$$

where k, α, and β are constants. In terms of gamma functions, determine (a) the normalization constant k and (b) the MTTF.

2.33 The manager of a fleet of trucks has been told they require major overhauls according to a Weibull distribution, with $\beta = 250{,}000$ km and $\alpha = 1.7$. Determine the probability that a truck will not break down in (a) 100,000 km and (b) 500,000 km.

2.34 A large number of valves have times to failure that follow a two-parameter Weibull distribution, with $\beta = 10$ yr and $\alpha = 0.5$. Determine the probability that a valve will survive (a) 1 yr, (b) 5 yr, and (c) 10 yr without failure, and (d) the MTTF.

2.35 The failures of widgets in a batch follow the Weibull distribution with $\beta = 10^5$ hr and $\alpha = 2.3$. Determine (a) the probability that one of these widgets will not fail in 10,000 hr and (b) the mean value for the distribution.

2.36 A flood-prone region has a mean annual high water level expressed as "4 m above flood stage." The measured high water level for each year has a standard deviation of 1.5 m. (a) Select an appropriate extreme-value distribution, and (b) use it to estimate the water level x_* in meters that will be exceeded once in 100 yr.

2.37 Show that the variance is equal to the mean for the Poisson distribution as indicated in Eq. (2.48).

CHAPTER 3

RELIABILITY DATA

In this chapter we examine methods for obtaining a point estimate of a probability (i.e., a single number) and for estimating the uncertainty in that estimate. Statistical methods to systematically update probability distributions given new observations are presented. We also examine methods of obtaining the confidence level for a measured reliability or failure rate.

3.1 ESTIMATION THEORY

Statistical approaches for obtaining the sample mean m and variance σ^2 for a set of measurements represented via a probability density function were presented in Section 2.6. The equations for obtaining point estimates for the failure of a component depend on the operating environment for the component, as simulated during a *life test*. For example, if a life test is terminated at time t_s before all N items have failed, then Type I censoring of the life test has been done; on the other hand, Type II censoring occurs if a life test is terminated at the time of a prescribed failure K, $K \leq N$. In this chapter we shall illustrate only the simplest case of sampling in which all available units are tested to failure without censoring. More involved sampling entails statistical approaches beyond the scope of our interest here.

Risk and Safety Analysis of Nuclear Systems. By John C. Lee and Norman J. McCormick

Certainly the most common estimator is the least squares estimator, which is discussed elsewhere [Man74]. Here we focus on the moment, maximum likelihood, and maximum entropy estimators for components that operate continuously with time, as described by the probability densities for the exponential, gamma, lognormal, and Weibull distributions defined in Section 2.8.

3.1.1 Moment Estimators

If the numbers $t_1,\ t_2, \ldots,\ t_N$ represent a set of data for the actual failure times for a set of N identical components, then an unbiased estimator (denoted by the caret symbol) for the mean time to failure,

$$m = \int_0^\infty t f(t)\mathrm{d}t, \tag{3.1}$$

is

$$\widehat{t} = N^{-1} \sum_{n=1}^{N} t_n. \tag{3.2}$$

For the variance

$$\sigma^2 = \int_0^\infty (t-m)^2 f(t)\mathrm{d}t, \tag{3.3}$$

the estimator is

$$\widehat{\sigma}^2 = (N-1)^{-1} \sum_{n=1}^{N} (t_n - \widehat{t}\,)^2. \tag{3.4}$$

[Note that the factor $(N-1)^{-1}$ appears on the right-hand side of Eq. (3.4) instead of N^{-1} to keep the estimator unbiased.] Equations (3.2) and (3.4) give two equations for calculating two unknown parameters for the failure probability distributions.

Equations for obtaining moment estimates of the parameters for the exponential, gamma, lognormal, and two-parameter Weibull distributions are given in Table 3.1, as collected from the results of Section 2.8 for the mean and variance of each distribution.

Example 3.1 Ten identical devices are tested with failures occurring at 1.7, 3.5, 5.0, 6.5, 8.0, 9.6, 11, 13, 18, and 22 ($\times 10^2$) hr. Fit the data to a Weibull distribution using moment estimators.

The first step is to use Eqs. (3.2) and (3.4) to calculate $\widehat{t} = 983$ hr and $\widehat{\sigma}^2 = 4.114 \times 10^5$ hr^2, respectively, using the 10 failure times. From Table 3.1 it is necessary to solve for $\widehat{\alpha}$ and $\widehat{\beta}$ from

$$\begin{aligned} \widehat{\beta}\Gamma(1+1/\widehat{\alpha}) &= \widehat{t}, \\ \widehat{\beta}^2\{\Gamma(1+2/\widehat{\alpha}) - [\Gamma(1+1/\widehat{\alpha})]^2\} &= \widehat{\sigma}^2. \end{aligned}$$

To solve these two equations simultaneously by an iterative method, one simple approach is to rewrite the equations as

$$\begin{aligned} \widehat{\beta}_1 &= \widehat{t}/\Gamma(1+1/\widehat{\alpha}), \\ \widehat{\beta}_2 &= \{[\widehat{\sigma}^2 + \widehat{t}^2]/\Gamma(1+2/\widehat{\alpha})\}^{1/2}, \end{aligned}$$

Table 3.1 Moment Estimators for Failure Probability Distributions

Distribution[a]	$f(t)$	$\widehat{t} = \frac{\sum_{n=1}^{N} t_n}{N}$	$\widehat{\sigma}^2 = \frac{\sum_{n=1}^{N}(t_n - \widehat{t})^2}{N-1}$
Exponential	$\lambda \exp(-\lambda t)$	$1/\widehat{\lambda}$	Not needed: Eq. (2.109)
Gamma	$\lambda(\lambda t)^{\alpha-1} \exp(-\lambda t)/\Gamma(\alpha)$	$\widehat{\alpha}/\widehat{\lambda}$	$\widehat{\alpha}/\widehat{\lambda}^2$
Lognormal	$\frac{1}{\sqrt{2\pi}\alpha t} \exp\left[-\frac{\ln(t/\beta)^2}{2\alpha^2}\right]$	$\widehat{\beta}\exp(\widehat{\alpha}^2/2)$	$\widehat{\beta}^2 \exp \widehat{\alpha}^2(\exp \widehat{\alpha}^2 - 1)$
Weibull	$\frac{\alpha}{\beta}\left(\frac{t}{\beta}\right)^{\alpha-1} \exp\left[-\left(\frac{t}{\beta}\right)^{\alpha}\right]$	$\widehat{\beta}\Gamma(1 + 1/\widehat{\alpha})$	$\widehat{\beta}^2\{\Gamma(1 + 2/\widehat{\alpha})$ $-[\Gamma(1 + 1/\widehat{\alpha})]^2\}$

[a] All N units tested to failure.

and search for the $\widehat{\alpha}$ for which $\widehat{\beta}_1 = \widehat{\beta}_2 = \widehat{\beta}$. The result is $\widehat{\alpha} = 1.58$ and $\widehat{\beta} = 1095$ hr. ◇

It should be pointed out that the estimator of Eq. (3.2) assumes that all N units available are tested to failure. In the event that J units fail out of a total of N and the testing is stopped at $t = t_s$ (Type I censoring), Eq. (3.2) should be replaced by

$$\widehat{t} = \frac{1}{J}\sum_{n=1}^{J} t_n + \left(\frac{N-J}{J}\right) t_s. \tag{3.5}$$

In testing schemes such as *accelerated life testing* or other sophisticated approaches, equations other than Eqs. (3.2) and (3.5) should be used [Man74].

3.1.2 Maximum Likelihood Estimators

The maximum likelihood method requires the "likelihood function" L which is defined for N data points as [Man74,Sho90]

$$L(t_1,\ t_2, \ldots, t_N;\ \theta_1,\ \theta_2, \ldots, \theta_M) = \prod_{n=1}^{N} f(t_n; \theta_1,\ \theta_2, \ldots, \theta_M). \tag{3.6}$$

Here $f(t_n; \theta_1,\ \theta_2, \ldots, \theta_M)$ is the failure probability density for the time of each failure t_n, for $n = 1, \ldots, N$. The values of $\theta_1,\ \theta_2, \ldots, \theta_M$ are M parameters to be estimated to determine a probability density function. For example, for the gamma distribution in Table 3.1, the two theta parameters are α and λ.

If the set of data is incomplete with only J units tested out of the N units so that the testing procedure is terminated at t_s before the last failure, then the terminated-testing form of Eq. (3.6) for L must be used,

$$\begin{aligned} &L(t_1,\ t_2, \ldots, t_J;\ \theta_1,\ \theta_2, \ldots, \theta_M) \\ &= \frac{N!}{(N-J)!} \prod_{n=1}^{J} f(t_n; \theta_1,\ \theta_2, \ldots, \theta_M)[1 - F(t_s)]^{N-J}, \end{aligned} \tag{3.7}$$

where $F(t_s)$ is the CDF of the assumed PDF $f(t_s;\theta_1,\ \theta_2,\ldots,\theta_M)$. Henceforth we shall assume that all units are tested until failure.

To determine values for θ_m from Eq. (3.6), $\ln L$ must be maximized with respect to each parameter. Because

$$\ln L = \sum_{n=1}^{N} \ln f(t_n), \tag{3.8}$$

the conditions for obtaining $\widehat{\theta}_m$, the estimate of θ_m, follow from the simultaneous solution of the set of equations [Man74,Sho90]

$$\frac{\partial \ln L}{\partial \theta_m} = 0, \quad m = 1,\ldots,M. \tag{3.9}$$

Each estimate $\widehat{\theta}_m$ is a random variable so different samples will, in general, lead to different values. The variance of $\widehat{\theta}_m$ is obtained from

$$\operatorname{var} \widehat{\theta}_m = -\left[\frac{\partial^2 \ln L}{\partial \theta_m^2}\right]^{-1} \tag{3.10}$$

if N is large [Man74,Sho68].

Equations for obtaining maximum likelihood estimates of the parameters for the exponential, gamma, lognormal, and two-parameter Weibull distributions, given in Table 3.2, follow from the appropriate equations in Section 2.8.

Example 3.2 Derive the two equations for a maximum likelihood fit of data to a gamma distribution.

From the form of $f(t)$ in Table 3.1 and Eq. (3.8), it follows that

$$\begin{aligned}\ln L &= \sum_{n=1}^{N} \ln\left[\frac{\lambda^\alpha t_n^{\alpha-1}\exp(-\lambda t_n)}{\Gamma(\alpha)}\right]\\ &= N[\alpha\ln\lambda - \ln\Gamma(\alpha)] + (\alpha-1)\sum_{n=1}^{N}\ln t_n - \lambda\sum_{n=1}^{N} t_n.\end{aligned}$$

From

$$\frac{\partial \ln L}{\partial \lambda} = \frac{N\alpha}{\lambda} - \sum_{n=1}^{N} t_n,$$

and Eq. (3.9) we see that an estimate for α is implicitly obtained in terms of that for λ,

$$\frac{\widehat{\alpha}}{\widehat{\lambda}} = \frac{1}{N}\sum_{n=1}^{N} t_n$$

For $\alpha \equiv 1$ so that $\widehat{\alpha} = 1$ the equation for estimating the single parameter of the exponential distribution is obtained.

Table 3.2 Maximum Likelihood and Maximum Entropy Estimators for Failure Distributions Defined in Table 3.1

Distribution	Data Collected[a]	Maximum Likelihood	Maximum Entropy
Exponential	$N^{-1}\sum_{n=1}^{N} t_n$	$1/\widehat{\lambda}$	Same
Gamma[b]	$N^{-1}\sum_{n=1}^{N} t_n$	$\widehat{\alpha}/\widehat{\lambda}$	Same
	$N^{-1}\sum_{n=1}^{N} \ln t_n$	$-\ln\widehat{\lambda}+\psi(\widehat{\alpha})$	Same
Lognormal	$N^{-1}\sum_{n=1}^{N} \ln t_n$	$\ln\widehat{\beta}$	Same
	$N^{-1}\sum_{n=1}^{N} (\ln t_n)^2$	$\widehat{\alpha}^2+(\ln\widehat{\beta})^2$	Same
Weibull[c]	$N^{-1}\sum_{n=1}^{N} \ln t_n$	Not used	$\ln\widehat{\beta}-\gamma/\widehat{\alpha}$
Two-parameter	$N^{-1}\sum_{n=1}^{N} t_n^{\widehat{\alpha}}$	$\widehat{\beta}^{\widehat{\alpha}}$	Same
	$N^{-1}\sum_{n=1}^{N}(t_n/\widehat{\beta})^{\widehat{\alpha}}\ln(t_n/\widehat{\beta})$ $-N^{-1}\sum_{n=1}^{n}\ln(t_n/\widehat{\beta})$	$\widehat{\alpha}^{-1}$	Not used

[a] All N units tested to failure.
[b] $\psi(\widehat{\alpha})$ is the digamma function [Abr64]
[c] $\gamma = 0.5772$.

The second equation for obtaining $\widehat{\alpha}$ and $\widehat{\lambda}$ follows from

$$\frac{\partial \ln L}{\partial \alpha} = N[\ln\lambda - \psi(\alpha)] + \sum_{n=1}^{N} \ln t_n,$$

where $\psi(\alpha)$ is the digamma function, which is the tabulated derivative of the natural logarithm of the gamma function [Abr64]. With the help of Eq. (3.9) the last result can be written as

$$\psi(\widehat{\alpha}) - \ln\widehat{\lambda} = N^{-1}\sum_{n=1}^{N} \ln t_n.$$

This second equation for the gamma distribution differs from that for the moment method and is sufficiently complicated that the two equations must be solved iteratively to obtain $\widehat{\alpha}$ and $\widehat{\lambda}$. ◇

Example 3.3 Fit the data for the devices of Example 3.1 to a Weibull distribution using maximum likelihood estimators.

From Table 3.2 it is possible to eventually converge to the estimates $\widehat{\alpha} = 1.67$ and $\widehat{\beta} = 1103$ hr. ◇

3.1.3 Maximum Entropy Estimators

The maximum entropy estimator method differs from the maximum likelihood method and has been much less studied [Tri69]. If all N units are tested to failure, the Shannon entropy function H is defined in terms of the failure probability density $f(t)$ as

$$H = -\sum_{n=1}^{N} f(t_n) \ln f(t_n). \tag{3.11}$$

A comparison of Eqs. (3.8) and (3.11) shows that the H function weights the magnitude of each value of $\ln f(t_n)$ by the magnitude of $f(t_n)$ rather than by unity.

The method derives its name from information theory which has, as a basis for estimating desired parameters, the Jaynes principle, which states that the minimally prejudiced probability distribution is that which maximizes H subject to constraints supplied by moments of the assumed distribution [Tri69]. The algorithm for maximizing H can be conveniently stated by first rewriting the probability density f in the form

$$f(t_n) = \exp[-\theta_0 - \theta_1 T_1(t_n) - \theta_2 T_2(t_n) - \cdots], \tag{3.12}$$

where $\theta_0,\ \theta_1,\ \ldots, \theta_M$ are constants in terms of the M parameters to be estimated and the functions $T_1(t),\ T_2(t), \ldots, T_M(t)$ are functions of t. The estimates of the desired parameters follow from the set of equations [Tri69]

$$-\frac{\partial \theta_0}{\partial \theta_m} = \frac{1}{N} \sum_{n=1}^{N} T_m(t_n), \quad m = 1, \ldots, M. \tag{3.13}$$

Example 3.4 Derive the two equations for a maximum entropy fit of data to a gamma distribution.

From the form of the gamma distribution in Table 3.1 and Eq. (3.12) it is found that

$$\begin{aligned}
\theta_0 &= \ln\Gamma(\alpha) - \alpha \ln \lambda, \\
\theta_1 &= \lambda, \\
\theta_2 &= 1 - \alpha, \\
T_1(t) &= t, \\
T_2(t) &= \ln t.
\end{aligned}$$

Substitution of these parameters into Eq. (3.13) shows that

$$\begin{aligned}
-\frac{\partial \theta_0}{\partial \theta_1} &= \frac{1}{N} \sum_{n=1}^{N} t_n = -\frac{\partial}{\partial \lambda}[\ln\Gamma(\alpha) - \alpha \ln \lambda] = \frac{\alpha}{\lambda}, \\
-\frac{\partial \theta_0}{\partial \theta_2} &= \frac{1}{N} \sum_{n=1}^{N} \ln t_n = -\frac{\partial \theta_0}{\partial (1-\alpha)} - \frac{\partial [\ln\Gamma(\alpha) - \alpha \ln \lambda]}{\partial \alpha} = \psi(\alpha) - \ln \lambda.
\end{aligned}$$

After denoting the estimates of λ and α by $\widehat{\lambda}$ and $\widehat{\alpha}$, respectively, the results agree with those from the maximum likelihood method, as seen in Table 3.2. ◇

Table 3.3 Comparison of Results from Examples 3.1, 3.3, and 3.5

Estimator Method	$\widehat{\alpha}$	$\widehat{\beta}$ (hr)
Moment (from Example 3.1)	1.5	1100
Maximum likelihood (from Example 3.3)	1.67	1103
Maximum entropy (from Example 3.5)	1.53	1137

From Table 3.2 it is seen that the equations for estimating the parameters of the exponential, gamma, and lognormal distributions are the same for the maximum likelihood and maximum entropy methods but differ for the Weibull distribution. Because the maximum likelihood estimator (MLE) is more cumbersome to use than the maximum entropy estimator (MEE) for the Weibull distribution, a comparison of numerical results is of interest.

Example 3.5 For the devices described in Example 3.1, determine the parameters $\widehat{\alpha}$ and $\widehat{\beta}$ for the Weibull distribution using maximum entropy estimators.

An iteration procedure involving the equations in Table 3.2 eventually leads to the results in Table 3.3 that illustrate there are some differences in the values of $\widehat{\alpha}$ and $\widehat{\beta}$ for the three different methods. ◇

3.1.4 Comparison of Estimators

The maximum likelihood estimator is the normally accepted one because [Sho90]:

1. The MLE is a sufficient estimator if a sufficient estimator exists for the problem.
2. The MLE is efficient for large N (i.e., its variance is small).
3. The MLE possesses the property of invariance, which means that if w is an estimator for var x and if w is invariant, then $\sqrt{w}$ is an estimator for σ_x.
4. The var $\widehat{\theta}_m$ of Eq. (3.10) can be computed and its distribution described in the limit as $N \to \infty$.

3.2 BAYESIAN UPDATING OF DATA

3.2.1 Bayes Equation

A basic result for conditional probabilities follows by first rewriting Eq. (2.4) for the nth event or hypothesis E_n of N mutually exclusive events or hypotheses as

$$\begin{aligned} P(E_n B) &= P(E_n)P(B|E_n) \\ &= P(B)P(E_n|B), \end{aligned} \tag{3.14}$$

where B is some other event or hypothesis. Equating the right-hand sides of these two equations gives

$$P(E_n|B) = P(E_n)\left[\frac{P(B|E_n)}{P(B)}\right]. \tag{3.15}$$

This equation is an elementary form of the Bayes equation because the left-hand side gives the *posterior* probability of E_n when B is given, while the first factor on the right-hand side is the *prior* probability of E_n and the second factor represents the relative change in the probability of E_n when B becomes known. A short-hand way of writing Eq. (3.15) for a set of events $\mathbf{E}$ is

$$P(\mathbf{E}|B)P(B) = P(B|\mathbf{E})P(\mathbf{E}). \tag{3.16}$$

From addition axiom (2.3) it follows that for mutually exclusive events

$$\sum_{n=1}^{N} P(E_n|B) = 1. \tag{3.17}$$

If this equation is multiplied by $P(B)$, then

$$P(B) = \sum_{n=1}^{N} P(B)P(E_n|B) \tag{3.18}$$

$$= \sum_{n=1}^{N} P(E_n B), \tag{3.19}$$

where product axiom (2.4) has been used to obtain Eq. (3.19). Applying the product axiom in Eq. (3.19) results in the *extension rule* for $P(B)$,

$$P(B) = \sum_{n=1}^{N} P(B|E_n)P(E_n). \tag{3.20}$$

The extension rule allows $P(B)$ to be expressed in terms of the previously known probabilities $P(E_n)$ and all the conditional probabilities $P(B|E_n)$. Substitution of Eq. (3.20) in Eq. (3.15) gives the final form for the Bayes equation,

$$P(E_n|B) = \frac{P(E_n)P(B|E_n)}{\sum_{m=1}^{N} P(E_m)P(B|E_m)}, \qquad n = 1, \ldots, N. \tag{3.21}$$

A continuum form of the Bayes equation (3.21) also is available. It is often used in probabilistic risk assessments of nuclear systems to update the probability density function $P(x)$ for x representing the failure rate of a component or the frequency of an event of interest:

$$P(x|B) = \frac{P(x)P(B|x)}{\int P(x')P(B|x')\mathrm{d}x'}. \tag{3.22}$$

In this application, the summation in the denominator of Eq. (3.21) covering all possible events E_n is replaced by an integral over the entire range of the variable x.

3.2.2 Applications of the Bayes Equation

The Bayes equation shows that once the entire set of conditional probabilities $P(B|E_n)$ becomes known, the calculation of the posterior $P(E_n|B)$ becomes straightforward. It allows one to "reverse" the order when performing hypothesis testing in instances where it is easier to incorporate information about $P(B|E_n)$, $n = 1, \ldots, N$, instead of that for $P(E_n|B)$. Thus, given the *prior distribution* $P(E_n)$ and the *likelihood function* $P(B|E_n)$, updated probabilities for events E_n, $n = 1, \ldots, N$, are generated as the *posterior distribution* $P(E_n|B)$ subject to additional observation or information B.

Equation (3.21) also can be used to revise failure data for a set of events E_n, $n = 1, \ldots, N$. If nothing is known about the probability of the events, $P(E_n)$, in Eq. (3.21) prior to initiation of a testing program (or prior to obtaining new data from an expanded testing program), then one should use the "principle of insufficient reason." This means one should pick equal probabilities for each event according to the uniform prior distribution, $P(E_n) = 1/N$. Then from a testing program one may obtain information about $P(B|E_n)$ that will lead to a revised estimate.

Example 3.6 An elementary nuclear reactor core monitoring system (CMS) consists of an uncompensated ionization chamber (IC), a temperature sensor (TS), and a pressure sensor (PS). The CMS has failed because of the failure of one of the three components. From the manufacturer's operations manual the three components are known to have probabilities of failure of 0.02, 0.04, and 0.01, respectively, over the life of the CMS at the operating conditions. (a) Obtain an estimate that the temperature sensor is the component to cause a CMS failure. (b) Revise that estimate by using data from the manufacturer's operations manual that when the IC fails, the CMS fails with probability 0.1; when the TS fails, the CMS fails with probability 0.15; and when the PS fails, the CMS fails with probability 0.1.

(a) We wish to determine *P*(TS|CMS), the probability that the TS failure is the cause of a CMS failure, given that *P*(IC)=0.02, *P*(TS)=0.04, and *P*(PS)=0.01. Because nothing initially is known about which event could cause a CMS failure, we assume *P*(CMS|IC)=*P*(CMS|TS)=*P*(CMS|PS)=1/3. From Eq. (3.21) it follows that

$$P(\text{TS|CMS}) = (0.04/3)/[(0.02/3) + (0.04/3) + (0.01/3)] = 0.571.$$

(b) From the operations manual it is learned that *P*(CMS|IC)=0.1, *P*(CMS|TS)=0.15, and *P*(CMS|PS)=0.1. Again from Eq. (3.21),

$$P(\text{TS|CMS}) = [0.04(0.15)]/[0.02(0.1) + 0.04(0.15) + 0.01(0.1)] = 0.667. \diamond$$

Example 3.7 A nuclear fuel fabrication facility has three machines $\#1$, $\#2$, and $\#3$ producing 200, 300, and 500 pellets per day with defective pellet rates of 0.6%, 0.7%, and 0.8%, respectively. If one defective pellet X is produced at the end of a day, what is the probability that it was produced by machine $\#3$?

With $P(X|\#1) = 0.006$, $P(X|\#2) = 0.007$, and $P(X|\#3) = 0.008$ and with $P(\#1) = 0.2$, $P(\#2) = 0.3$, and $P(\#3) = 0.5$, it follows from Eq. (3.21) that

$$P(\#3|X) = \frac{(0.008)(0.5)}{(0.006)(0.2) + (0.007)(0.3) + (0.008)(0.5)} = 0.548. \diamond$$

If data have been fit to a probability distribution and new test data become available, then the procedure of Section 3.1 can be repeated to revise the distribution. But to update a probability distribution for which the initial test data are no longer available, or to update such a distribution with a *subjective belief* that a revision is needed, the Bayes equation is an appropriate way of modifying a data set.

Consider a set of data $\mathbf{D} = \{D_1, D_2, \ldots, D_N\}$ and an unknown distribution parameter θ. We desire to update the prior distribution $P(\theta)$ with the *likelihood function* $P(\mathbf{D}|\theta)$ that the new data are compatible with those used to generate the prior distribution. We first consider a case with time-independent probabilities for which the prior data satisfy the *beta distribution.*

Example 3.8 Suppose θ is the demand failure probability F for a component for which data previously have been fit to the beta prior distribution $P(\theta)$ of Eq. (2.40) with parameters α and β. For additional tests giving data $\mathbf{D}$ for N components in which M failed, such that the new data satisfy the binomial distribution of Eq. (2.35), from Eq. (3.16)

$$P(\theta|\mathbf{D}) \propto [\theta^M (1-\theta)^{N-M}][\theta^{\alpha-1}(1-\theta)^{\beta-1}],$$

so the posterior probability again is a beta distribution with parameters $\alpha' = \alpha + M$ and $\beta' = \beta + N - M$. [The posterior probability itself must be properly defined by computing the normalization factor.] When the prior distribution and the posterior distributions have the same functional form, the prior distribution is said to be a *conjugate prior.* $\diamond$

Let us now consider cases where the continuous form of the Bayes equation is needed.

Example 3.9 If nothing is initially known about the reliability R for a system component, then the the prior distribution is the uniform probability density $p(R)$ equals 1 for $0 \leq R \leq 1$ and 0 otherwise. Assume that a testing program T is conducted in which M of the N components failed after a specified time period so that the binomial distribution of Eq. (2.35) is valid. Determine the probability density for the reliability.

The testing program provides data in the form of

$$p(T|R) = \frac{N!}{M!(N-M)!}(1-R)^M R^{N-M}.$$

From Eq. (3.22) it follows that

$$p(R|T) \quad = \quad \frac{p(R)p(T|R)}{\int_0^1 p(R')p(T|R')\mathrm{d}R'}$$

$$= \begin{cases} \dfrac{(1-R)^M R^{N-M}}{\int_0^1 (1-R')^M R'^{N-M} \mathrm{d}R'}, & 0 \leq R \leq 1, \\ 0, & \text{otherwise.} \end{cases}$$

Thus if two components out of five failed during the specified time period, for example, then

$$p(R|T) = \begin{cases} \dfrac{(1-R)^2 R^3}{\int_0^1 (1-R')^2 R'^3 \mathrm{d}R'} = 60(1-R)^2 R^3, & 0 \leq R \leq 1, \\ 0, & \text{otherwise.} \diamond \end{cases}$$

It is also possible to easily update a time-dependent failure PDF $f(t)$ that is given by the (prior) gamma distribution.

Example 3.10 Failure times $0 < t_1 < t_2 < \cdots < t_N = \tau$ have been observed from a system that can be modeled by the gamma failure probability density of Eq. (2.110) with parameter λ. If the times between failures are random, so that they are exponentially distributed with hazard rate λ, determine (a) the probability distribution for λ given the data and (b) the mean and variance of that probability distribution.

(a) For the times between failures given by $x_i = t_i - t_{i-1}$, with $t_0 = 0$, the likelihood function is

$$p(\mathbf{x}|\lambda) = \prod_{i=1}^{N} \lambda \exp(-\lambda x_i) = \lambda^N \exp\left[-\lambda \sum_{i=1}^{N} x_i\right] = \lambda^N \exp(-\lambda\tau),\ \lambda > 0,$$

and the gamma prior distribution is

$$f(t) = \frac{\lambda(\lambda t)^{\alpha-1} \exp(-\lambda t)}{\Gamma(\alpha)}.$$

From Eq. (3.16) it follows that

$$p(\lambda|\mathbf{x}) \propto \lambda^N \exp(-\lambda\tau) \left[\frac{\lambda(\lambda t)^{r-1} \exp(-\lambda t)}{\Gamma(r)}\right] \propto \lambda^{N+\alpha} \exp[-\lambda(t+\tau)],$$

which means that the posterior failure probability density also is a gamma distribution with α replaced by $N + \alpha$ and t by $t + \tau$. Thus the gamma distribution is a conjugate prior for new data given by the exponential distribution for random failures.

(b) From Eqs. (2.112) and (2.113), the prior mean and prior variance are

$$\begin{aligned} m &= \alpha/\lambda, \\ \sigma^2 &= \alpha/\lambda^2, \end{aligned}$$

so the posterior mean and variance are

$$\begin{aligned} m &= (N+\alpha)/\lambda, \\ \sigma^2 &= (N+\alpha)/\lambda^2. \diamond \end{aligned}$$

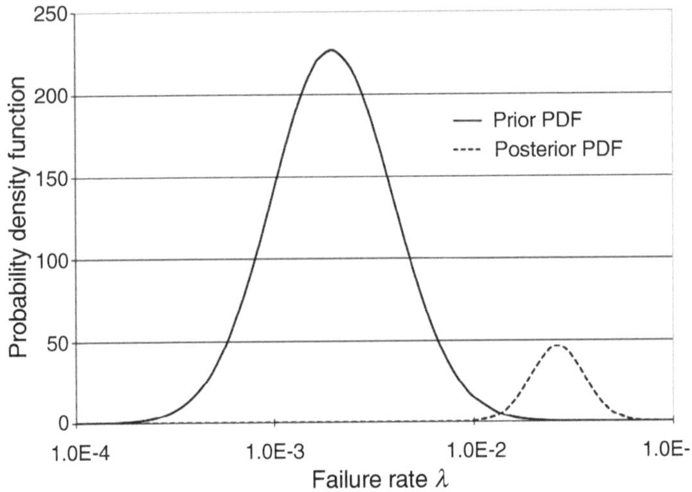

Figure 3.1 Updating a prior distribution to a posterior distribution.

Example 3.11 A manufacturer's estimate for the failure rate λ for a set of auxiliary pumps has been represented by a lognormal distribution $p(\lambda) = p(z)$ of Eq. (2.62), truncated [Atw03] to the interval $[0, 1]$, with a mean failure rate of 4.04×10^{-3} per demand and a standard deviation of 3.213×10^{-3} per demand. Plant data reveal that 12 failures to start have been observed in 250 trials. Representing the pump startup trials as a binomial distribution, obtain an updated PDF $p(\lambda|B)$, given the new observation B involving 12 failures in 250 trials, via the Bayes equation (3.22). From Eqs. (2.63) and (2.64), $\alpha = 0.7$ and $\beta = 3.162 \times 10^{-3}$ are obtained. Performing the numerical integral in the denominator of the Bayes equation

$$p(\lambda|B) = \frac{p(\lambda)\, p(B|\lambda)}{\int_0^1 p(\lambda')\, p(B|\lambda')\, d\lambda'}$$

with

$$p(B|\lambda) = \binom{250}{12} \lambda^{12} (1-\lambda)^{238}$$

yields $p(\lambda|B)$ with a mean of 3.04×10^{-2} failures per demand and standard deviation of 9.72×10^{-3} failures per demand. Figure 3.1 illustrates how the observation B updates the prior distribution $p(\lambda)$ to the posterior distribution $p(\lambda|B)$ through the Bayes equation (3.22). $\diamond$

3.3 CENTRAL LIMIT THEOREM AND HYPOTHESIS TESTING

The concept of the reliability of a component in continuous operation, such as a valve or pump in a power plant, was discussed in Section 2.7. We now turn to the task of obtaining the confidence level for a measured reliability or failure rate. Three alternate methods, two statistical techniques and one engineering approach, will be introduced for this task. We consider in this section how the *central limit theorem* (CLT) [Bru75,Spi08], which governs the statistical distribution of the sample mean of a set of measurements, can be used to test the validity of hypotheses regarding the sample mean. The CLT also can be used to establish the confidence intervals for the component reliability or failure rate. The second statistical method introduced in this section is the χ^2-distribution, which generally is used to obtain confidence intervals for the variance and test if the measurements are normally distributed. As a specialized application, the χ^2-distribution is used as a probability density function to describe component failure rates represented by the Erlangian distribution of Eq. (2.103). Finally, the reliability quantification obtained from the two statistical methods is compared with an intuitive engineering approach via the cumulative failure probability $F(t)$ of Eq. (2.107) and the Erlangian distribution of Eq. (2.105).

3.3.1 Interpretation of the Central Limit Theorem

According to the CLT, if $\overline{x}_N$ and $\sigma_{\overline{x}_N}$ are the sample mean and standard deviation for the sample mean, respectively, for a set of N measurements of a random variable X taken from a population with true mean μ and standard deviation σ, then the sample mean $\overline{x}_N$ is distributed approximately as

$$f(\overline{x}_N) = N(\mu, \sigma_{\overline{x}_N}) = \frac{1}{\sqrt{2\pi\sigma^2_{\overline{x}_N}}} \exp\left[-\frac{(\overline{x}_N - \mu)^2}{2\sigma^2_{\overline{x}_N}}\right] \text{ as } N \to \infty, \tag{3.23}$$

where $N(\mu, \sigma)$ represents the normal, or Gaussian, distribution with mean μ and standard deviation σ. Because the variance $V(\overline{x})$ of the sample mean can be obtained [Bru75,Spi08] from the population (true) $V(x)$ by the relationship

$$V(\overline{x}) = V(x)/N, \tag{3.24}$$

we rewrite Eq. (3.23) as

$$f(\overline{x}_N) = N\left(\mu, \frac{\sigma}{\sqrt{N}}\right) = \frac{1}{\sqrt{2\pi\sigma^2/N}} \exp\left[-\frac{(\overline{x}_N - \mu)^2}{2\sigma^2/N}\right] \text{ as } N \to \infty. \tag{3.25}$$

Equations (3.23) through (3.25) are valid for any random variable with an arbitrary underlying probability distribution for the population. Regardless of the PDF for the random variable itself, if multiple sets of measurements are taken, the sample mean will be normally distributed, i.e., given by Eq. (3.23) or (3.25), in the limit as $N \to \infty$. The relationship is usually satisfied approximately even for a moderate sample size. With the standard form of the Gaussian distribution $N(0, 1)$ of Eq. (2.59), for any real numbers $a < b$ Eq. (3.25) yields the result

$$\lim_{N\to\infty} P\left\{a < \frac{\overline{x}_N - \mu}{\sigma/\sqrt{N}} < b\right\} = \frac{1}{\sqrt{2\pi}} \int_a^b \exp\left(-\frac{z^2}{2}\right) \mathrm{d}z, \tag{3.26}$$

where the right-hand-side represents an area under the curve of $N(0,1)$ over the interval $[a,b]$.

With the sample mean $\overline{x} = \overline{x}_N$ determined through measurements, we can use the CLT to obtain the a priori probability, i.e., the probability you estimate before you actually take measurements, that $\overline{x}$ differs from the population (true) mean μ by less than some positive number ϵ:

$$\begin{aligned} P\{|\overline{x}-\mu| < \epsilon\} &= P\{-\epsilon < (\overline{x}-\mu) < \epsilon\} = P\left\{-\frac{\epsilon\sqrt{N}}{\sigma} < \frac{\overline{x}-\mu}{\sigma/\sqrt{N}} < \frac{\epsilon\sqrt{N}}{\sigma}\right\} \\ &= P\left\{|z| < \frac{\epsilon\sqrt{N}}{\sigma}\right\} = \frac{2}{\sqrt{2\pi}}\int_0^{\epsilon\sqrt{N}/\sigma} \exp\left(-\frac{z^2}{2}\right) dz. \quad (3.27) \end{aligned}$$

Through the proper choice of ϵ, we can interpret the a priori probability of Eq. (3.27) as the likelihood that the measurements will yield a sample mean $\overline{x}$ in the vicinity of the true mean μ. Thus, equating Eq. (3.27) to α,

$$P\{|\overline{x}-\mu| < \epsilon\} = \alpha, \quad (3.28)$$

we can say there is $100\alpha\%$ confidence that $\overline{x}$ will differ from μ by less than some value ϵ. For example, using a standard probability table such as Table 2.3, we get $\alpha = 0.95$ for $\epsilon\sqrt{N}/\sigma = 1.96$, or in terms of Eq. (3.27)

$$\begin{aligned} P\{|\overline{x}-\mu| < \epsilon\} &= P\left\{|\overline{x}-\mu| < 1.96\frac{\sigma}{\sqrt{N}}\right\} = P\{|z| < 1.96\} \\ &= \frac{2}{\sqrt{2\pi}}\int_0^{1.96} \exp\left(-\frac{z^2}{2}\right) dz = 0.95 \quad (3.29) \end{aligned}$$

and there is a 95% confidence, *before the sample is taken*, that $|\overline{x}-\mu| < \epsilon = 1.96\sigma/\sqrt{N}$. We could also say that there is 5% probability that $|\overline{x}-\mu| \geq \epsilon$ or that $\overline{x}$ will differ from the true mean μ by at least ϵ, i.e., $P\{|\overline{x}-\mu| \geq \epsilon\} = 1-\alpha$. The PDF transformation from the variable $\overline{x}$ to the normalized variable z is illustrated in Figure 3.2, where the right-hand-side PDF is $f(z) = N(0,1)$, the standard form of the Gaussian distribution, and the parameter ϵ is chosen to yield $\alpha = 0.95$. This transformation is similar to that of Eq. (2.59), except that here the standard deviation σ of Eq. (2.58) is replaced by the standard deviation $\sigma/\sqrt{N}$ of the sample mean.

For a reliability quantification of components or systems where one is primarily interested in an upper bound estimate rather than an interval estimate for the failure rate, a one-sided confidence value should be used. That is, for this purpose Eq. (3.27) should be replaced by

$$P\{(\mu-\overline{x}) < \epsilon\} = P\left\{z < \frac{\epsilon\sqrt{N}}{\sigma}\right\} = \frac{1}{\sqrt{2\pi}}\int_{-\infty}^{\epsilon\sqrt{N}/\sigma} \exp\left(-\frac{z^2}{2}\right) dz. \quad (3.30)$$

3.3.2 Hypothesis Testing with the Central Limit Theorem

The relationship of Eq. (3.27) for confidence intervals of the sample mean can be used to test a hypothesis $H\{\text{true mean } \mu \text{ is equal to some value } \mu_0\}$. For a given σ

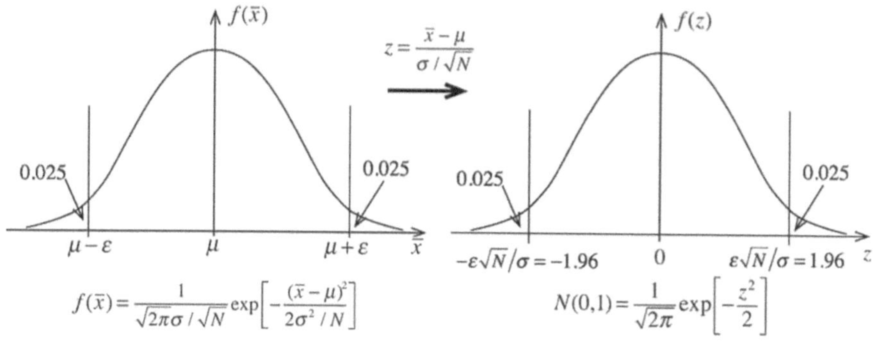

Figure 3.2 Probability density function transformation to normalized variable.

and sample size N, if we obtain a sample mean $\overline{x}$ which is different from the assumed (or claimed) value μ_0 by ϵ or more, the CLT requires that

$$P\{|\overline{x} - \mu| < \epsilon\} = \alpha, \tag{3.31}$$

or

$$\text{If } \{\mu = \mu_0\}, \text{ then } P\{|\overline{x} - \mu_0| \geq \epsilon\} = 1 - \alpha. \tag{3.32}$$

This can then be interpreted as

$$\text{If } \{|\overline{x} - \mu_0| \geq \epsilon\}, \text{ then } P\{\mu = \mu_0\} = 1 - \alpha. \tag{3.33}$$

That is, given the measurement $|\overline{x} - \mu_0| \geq \epsilon$, there is only a $100(1-\alpha)\%$ likelihood that the true mean is equal to μ_0. We can then reject the hypothesis $H\{\mu = \mu_0\}$ in favor of the hypothesis $H\{\mu \neq \mu_0\}$ at a *confidence level* of α or a *significance level* of $1-\alpha$. The significance level represents the likelihood of unnecessarily or erroneously rejecting the hypothesis.

Of course, in this hypothesis test, the true mean μ is unknown, but Eq. (3.31) provides the probability that μ lies within an interval

$$P\{\overline{x} - \epsilon < \mu < \overline{x} + \epsilon\} = \alpha. \tag{3.34}$$

Thus, if the value of μ_0 under dispute lies outside the bound of Eq. (3.34), then we conclude that there is only a slim chance, e.g., 5%, that μ_0 is equal to the true mean μ for $\alpha = 0.95$.

Example 3.12 Consider the task of determining the resistance of a precision resistor for a circuit in the reactor protection system of a nuclear power plant (NPP). After 36 measurements by a method that, based on previous experience, involves a variance of $9\,\Omega^2$, you determine a sample average of $\overline{x} = 52\,\Omega$. The manufacturer claims that the resistor has a resistance of $\mu_0 = 50\,\Omega$. Determine whether you should accept the manufacturer's claim.

For a 95% confidence level, i.e., $\alpha = 0.95$, Eq. (3.29) yields

$$\epsilon = 1.96\frac{\sigma}{\sqrt{N}} = 1.96 \times \frac{3\,\Omega}{\sqrt{36}} = 0.98\,\Omega,$$

and Eq. (3.31) suggests that there is a 95% probability that $\overline{x}$ will differ from the true mean μ by less than $0.98\,\Omega$. Given that the sample mean turns out to be different from the manufacturer's claim of $\mu_0 = 50\,\Omega$ by more than $\epsilon = 0.98\,\Omega$, Eq. (3.33) suggests that we can reject the resistor with a confidence level of 95% or significance level of 5%. Equivalently, Eq. (3.34) indicates that there is a 95% probability that the unknown true mean μ will lie between 51.02 and 52.98 Ω, and thus there is less than a 5% probability that $\mu = \mu_0 = 50\,\Omega$. In fact, in this particular example, since the sample mean $\overline{x}$ differs from the disputed population mean $\mu_0 = 50\,\Omega$ by more than 2ϵ, there is much less than a 5% chance that we are rejecting a good resistor. ◇

3.4 RELIABILITY QUANTIFICATION

For the task of determining the confidence level we may establish for the reliability of a NPP component by repetitive testing of the component, we consider a standard hypothesis testing approach via the CLT in Section 3.4.1 and compare it with a more intuitive engineering approach often used in PRA studies in Section 3.4.2. This will lead to comparison with another statistical approach via the χ^2-distribution in Section 3.4.3.

3.4.1 Central Limit Theorem for Reliability Quantification

Repetitive testing of a component can be used to determine the confidence level for the reliability of a NPP component. The component testing may now be considered a special case of hypothesis testing, with a certain number n of failures, including $n = 0$ for no failure, in N trials. With the definition of the *unreliability* p of the component,

$$p = \lim_{N\to\infty}\left(\frac{n}{N}\right), \tag{3.35}$$

and given sample mean $\bar{x} = n/N$, test hypothesis $H\,\{\text{true failure rate } p = \bar{x}\}$. For a given N, we can assume a value of p and determine the confidence level α with which to accept $H\,\{p = \bar{x}\}$ or alternately determine p corresponding to a desired confidence level α. We illustrate the latter approach. Because the outcome of each test is binary, we consider the random event X assumes value x given simply by

$$x = \begin{cases} 1: \text{ unreliability } f(x) = p\,, \\ 0: \text{ reliability } f(x) = 1 - p\,. \end{cases} \tag{3.36}$$

The expectation for X is

$$\begin{aligned} E(x) &= \langle x \rangle = \sum_x x f(x) \\ &= 1\cdot f(1) + 0\cdot f(0) = 1\cdot p + 0\cdot(1-p) = p, \end{aligned} \tag{3.37}$$

which verifies the definition of the unreliability. Similarly,

$$E\left(x^2\right) = \left\langle x^2 \right\rangle = \sum_x x^2 f(x) = 1 \cdot p + 0 \cdot (1-p) = p, \tag{3.38}$$

so that the variance is

$$V(x) = E\left(x^2\right) - E^2\left(x\right) = p - p^2, \tag{3.39}$$

giving the standard deviation $\sigma = \sqrt{p(1-p)}$.

Example 3.13 We consider three cases that illustrate the use of the CLT for repetitive testing.

(a) No failure in 60 trials, i.e., $\bar{x} = 0$, with $N = 60$. For a 95% confidence level, i.e., $\alpha = 0.95$, Eq. (3.29) yields

$$-\delta < \frac{\bar{x} - p}{\sigma/\sqrt{N}} < \delta, \quad \delta = \frac{\varepsilon\sqrt{N}}{\sigma} = 1.96. \tag{3.40}$$

Substitution of $\bar{x} = 0$ and $\sigma = \sqrt{p(1-p)}$ into the preceding equation gives

$$Np < \delta^2(1-p),$$

which finally yields, at a 95% confidence level, the desired interval estimate for the unreliability p given that no failure has been observed in 60 trials,

$$0 \le p < \frac{\delta^2}{N + \delta^2} = \frac{1.96^2}{60 + 1.96^2} = 0.060,$$

or $p \in [0, 0.060)$ per demand. Note that the lower limit $p = 0$ simply represents the nonnegative nature of the unreliability.

(b) Two failures in 169 trials, for a 99% confidence level. With $\delta = 2.58$ from Table 2.3, Eq. (3.40) results in a quadratic equation for the unknown p,

$$\frac{\bar{x} - p}{\sqrt{p(1-p)}} < \frac{\delta}{\sqrt{N}} \tag{3.41}$$

or

$$(1+\beta)\,p^2 - (2\bar{x} + \beta)p + \bar{x}^2 < 0, \quad \beta = \delta^2/N. \tag{3.42}$$

Two roots of this equation yield an interval estimate $p \in (0.002, 0.058)$ per demand.

(c) As an extension of case (b), we can also calculate a one-sided 99% confidence value,

$$P\{p - \bar{x} < \varepsilon\} = P\left\{z < \frac{\varepsilon\sqrt{N}}{\sigma}\right\} = \frac{1}{\sqrt{2\pi}}\int_{-\infty}^{\varepsilon\sqrt{N}/\sigma} \exp\left(-\frac{z^2}{2}\right) dz = 0.99, \tag{3.43}$$

which, from Table 2.3, gives $\delta = \varepsilon\sqrt{N}/\sigma = 2.33$. Solving the quadratic equation of case (b) with this new value of δ yields an upper estimate $p < 0.051$ per demand to a one-sided 99% confidence. Because we used a smaller value of δ for a one-sided confidence estimate here than in case (b), we obtained a smaller value for the upper bound of p.

Note that the upper bound estimates of 0.058 and 0.051 per demand for cases (b) and (c), respectively, are considerably larger than the sample or point estimate $\bar{x} = 0.012$ per demand. ◇

3.4.2 Engineering Approach for Reliability Quantification

For the more intuitive engineering approach for component reliability quantification, we return to Eq. (2.84) for the reliability for a component in continuous operation that can fail randomly,

$$R(t) = \exp(-\lambda t), \tag{3.44}$$

where λ is the failure or hazard rate, assumed constant. The cumulative probability of failure up to time t is likewise given by

$$F(t) = 1 - R(t) = 1 - \exp(-\lambda t) = 1 - \exp(-Np), \tag{3.45}$$

where $m = \lambda t$ is the average number of failures expected over time t or equivalently $\mu = Np$, with failure probability p per demand in N trials.

To calculate the probability of multiple failures, we extend Eq.(3.45) by using the Poisson distribution. We can then obtain the probability of n occurrences of an event of our interest, e.g., component failure, by the Poisson distribution of Eq. (2.45):

$$P(n, \mu) = \frac{\exp(-\mu)\mu^n}{n!}. \tag{3.46}$$

The reliability $R(t)$ of Eq. (3.44) is trivially given by the Poisson distribution with no failure up to time t, i.e., $R(t) = P(0, \mu)$.

The cumulative failure probability $F(t)$ of Eq. (3.45) represents the probability that any number of failures will occur in continuous operation up through time t with failure rate λ, or in N tests with unreliability p per demand. Thus, if we want to test the reliability of the component to be greater than some number, e.g., 0.95, or equivalently unreliability less than $p_0 = 0.05$ per demand, we may perform a number of tests, e.g., $N = 60$. If $p \geq p_0$, the cumulative failure probability in 60 tests will be $F(t) \geq 0.95$; that is, we expect to see at least one failure with a 95% probability. Thus, if $p \geq p_0$ and if the test is conducted repetitively, the component will pass the test without a single failure in 60 trials less than 5% of the time. In spite of this slim chance for passing the test, if no failure were to be observed in 60 trials, we then gain some confidence, say, at a 95% level, of the component unreliability $p < p_0$, or the reliability $(1 - p) > (1 - p_0) = 0.95$ per demand. The probability, calculated in advance, that the component will fail the test if it lacks the desired reliability $1 - p_0$ is usually interpreted as the confidence that the reliability is greater than $1 - p_0$ or unreliability $p < p_0$. Thus, given no failure in N trials and an assumed unreliability p_0, we set the confidence level

$$\alpha = P\{\text{observing any number of failures}\} = F(t) = 1 - \exp(-Np_0) \quad (3.47)$$

and accept the hypothesis

$$H\{0 \leq p < p_0\} \text{ at a } 100\alpha\% \text{ confidence level.} \quad (3.48)$$

The confidence level is to be calculated before the test; once the component passes the test, it is meaningless to talk about the probability of passing the test.

We may get a better feel for the engineering interpretation of the confidence level just presented if we perform another 40 tests without any failure for a total of 100 trials. In this case, we certainly should achieve a greater confidence for the component reliability $1 - p > 0.95$/demand, and indeed we calculate $F(t) > 0.99$ for $p_0 = 0.05$ and $N = 100$. That is, we now have at least a 99% confidence that the component reliability $1 - p > 0.95$/demand.

With Eq. (3.45), we have so far limited ourselves to the case with no failure in establishing the confidence level for reliability testing. This can be readily extended by considering the Poisson distribution of Eq. (3.46) for the case involving one or more failures experienced during the test. Thus, for n failures observed in N trials, calculate

$$\begin{aligned} P\{\text{observing } > n \text{ failures}\} &= 1 - \left(\sum_{m=0}^{n} \frac{\mu^m}{m!}\right) \exp(-\mu) \\ &= \alpha, \quad \mu = Np_0 = \lambda t. \end{aligned} \quad (3.49)$$

The parameter α may be interpreted now as the confidence level that, having observed n failures in N trials, the unreliability of the component is less than p_0 assumed in calculating α. This interpretation is identical to the use of Eq. (3.45) for $F(t)$ for the case of no failure, i.e., $n = 0$, but now represents the probability of observing any number of, i.e., one or more, failures. Similar to the evaluation of p_0 given no failure using $F(t)$ of Eq. (3.47), an estimate of unreliability p may be obtained by solving Eq. (3.49) for a desired confidence level α. The actual solution requires a quick iterative approach. For the case of $\alpha = 0.99$, $n = 2$, and $p_0 = 0.05$ /demand, Eq. (3.49) yields $N = 169$ trials.

3.4.3 χ^2-Distribution for Reliability Quantification

We return to the Erlangian distribution of Eq. (2.105) with a new variable $x = 2\lambda t$ representing the failure rate and $\eta = 2k$ for the kth failure at time t to obtain

$$\begin{aligned} f(x) &= \frac{1}{2^k \Gamma(k)} x^{k-1} \exp\left(-\frac{x}{2}\right) \\ &= \frac{1}{2^{\eta/2} \Gamma(\eta/2)} x^{(\eta/2)-1} \exp\left(-\frac{x}{2}\right), \quad x > 0. \end{aligned} \quad (3.50)$$

Equation (3.50) is the χ^2-distribution [Bru75,Spi08] with η *degrees of freedom* that, as noted in Section 2.8.2, can be obtained by substituting $\alpha = 0.5\eta$ and $\lambda = 0.5$ into the gamma distribution, Eq. (2.110). For a sample of η random variables normally distributed, the random variable $\chi^2 = \eta s^2/\sigma^2$ has a χ^2-distribution with $\eta - 1$ degrees of freedom, where s^2 is the sample variance and σ^2 is the population variance. The variable $u = \chi^2_{\eta,\alpha}$ for the cumulative distribution function

$$\begin{aligned} P\left\{x < u = \chi^2_{\eta,\alpha}\right\} &= F(u) = \alpha \\ &= \frac{1}{2^{\eta/2}\Gamma(\eta/2)} \int_0^u x^{(\eta/2)-1} \exp\left(-\frac{x}{2}\right) dx \quad (3.51) \end{aligned}$$

is tabulated as a function of α and η in standard statistical tables [Bru75,Spi08] and is widely used to test the goodness of samples satisfying the normal distribution.

For the purpose of reliability quantification, we follow a suggestion by Papoulis [Pap02] and integrate Eq. (3.51) by parts with $\eta = 2(n+1)$ to obtain

$$\begin{aligned} P\left\{x < u = \chi^2_{2n+2,\alpha}\right\} &= F(u) = \alpha = \frac{1}{2^{n+1}n!} \int_0^u x^n \exp\left(-\frac{x}{2}\right) dx \\ &= 1 - \sum_{m=0}^{n} \frac{u^m}{2^m m!} \exp\left(-\frac{u}{2}\right), \quad (3.52) \end{aligned}$$

which is equal to Eq. (3.49) with $u = 2Np_0$. Thus, having observed *n* failures in *N* trials, we may simply interpret that the χ^2-distribution of Eq. (3.52) with $\eta = 2(n+1)$ degrees of freedom yields the confidence level α that $x < u = \chi^2_{2n+2,\alpha}$ or $p < p_0 = \chi^2_{2n+2,\alpha}/2N$. This is the approach presented with the suggestion [Eps60] that the degrees of freedom for the χ^2-method be increased to $2(n+1)$ when the reliability test is not terminated after the *n*th failure. This method was later used in the evaluation of the anticipated transient without scram (ATWS) issues [AEC73], as discussed further in Chapter 8. The equivalence between the engineering approach of Eq. (3.49) and the χ^2-distribution of Eq. (3.52), subject to the degree-of-freedom requirement $\eta = 2(n+1)$, was recently presented [Fyn09].

Example 3.14 Consider again case (b) of Example 3.13 for the reliability quantification test resulting in two failures in 169 trials. With a one-sided confidence level of 99%, i.e., $\alpha = 0.99$, and with $\eta = 2(n+1) = 6$, we obtain $\chi^2_{6,0.99} = 16.8$ and

$$p < p_0 = \chi^2_{2n+2,\alpha}/2N = 16.8/(2 \times 169) = 0.050/\text{demand.} \diamond \quad (3.53)$$

3.4.4 Three-Way Comparison and Concluding Remarks

In this section, we presented three different methods for reliability quantification through a series of tests. To augment sample cases discussed in Sections 3.4.1 through 3.4.3, we now present [Fyn09] a bit more systematic comparison between the three methods for the case of three failures observed in a series of *N* component tests or demands. We desire to determine 95% and 99% confidence limits of the true failure rate *p* from a point estimate $\bar{x} = 3/N$. Failure rates of p_0 = [0.0001, 0.001,

Table 3.4 Upper Bound Estimates for Failure Rate Given Three Failures Observed

N	p(Eng)	p(CLT)	$p(\chi^2)$
Confidence Level = 95%			
156	5.000×10^{-2}	4.715×10^{-2}	4.971×10^{-2}
776	1.000×10^{-2}	9.634×10^{-3}	9.994×10^{-3}
1,551	5.000×10^{-3}	4.830×10^{-3}	5.000×10^{-3}
7,754	1.000×10^{-3}	9.678×10^{-4}	1.000×10^{-3}
77,537	1.000×10^{-4}	9.682×10^{-5}	1.000×10^{-4}
Confidence Level = 99%			
N	p(Eng)	p(CLT)	$p(\chi^2)$
201	5.000×10^{-2}	5.112×10^{-2}	4.998×10^{-2}
1,005	1.000×10^{-2}	1.046×10^{-2}	9.995×10^{-3}
2,010	5.000×10^{-3}	5.247×10^{-3}	4.998×10^{-3}
10,046	1.000×10^{-3}	1.052×10^{-3}	9.999×10^{-4}
100,452	1.000×10^{-4}	1.053×10^{-4}	1.000×10^{-4}

Source: [Fyn09].

0.005, 0.01, 0.05] per demand are assumed and the number N of trials required to establish that $p < p_0$ with the proper confidence level is calculated for each value of p_0 or upper bound value for p and compared in Table 3.4. The comparison begins with an iterative solution of Eq. (3.49) for the required number N of tests, which is then used to calculate p_0 by Eq. (3.43) for the CLT approach and Eq. (3.52) for the χ^2-method. The upper bound estimates p_0 for the χ^2-method agree closely with the corresponding values from the engineering approach of Eq. (3.49), confirming the mathematical equivalence of the two methods indicated in Eq. (3.52). The upper bounds for p obtained from the CLT method agree with those from the other two methods within $\pm 5\%$. The Poisson and normal distributions are both limiting forms of the binomial distribution, as discussed in Chapter 2, so the good agreement between the engineering approach and CLT method in Table 3.4 is readily understood.

Although we have discussed component reliability testing in terms of a single component, the idea can be extended to a batch of identical components from which one component will be selected at random for each test. In this context, testing a component or any widget may be also considered a Bernoulli's trial.

In principle, through repeated testing of various components of importance in NPP safety, we could establish their reliability at acceptable confidence levels. For components with high reliability, for example, those used in the reactor protection system, there is, however, a practical limit to this approach. This is because, to experimentally establish the reliability, one may have to exercise the control rod drives many more times than may actually be required during the operating life of the plant. Thus, statistical approaches via the CLT and χ^2-method, or the engineering approach, may have to be invoked to establish more meaningful, upper bound estimates for the component unreliability. This process is of considerable importance because the upper bound estimates in general are significantly larger than the sample mean or point estimates achievable based on a limited number of tests, as discussed in connection with Eqs. (3.42) and (3.43).

References

[Abr64] M. Abramowitz and I. Stegun, eds., *Handbook of Mathematical Functions with Formulas, Graphs, and Mathematical Tables*, U.S. Government Printing Office (1964); reprinted by Dover (1970).

[AEC73] "Technical Report on Anticipated Transients Without Scram for Water-Cooled Power Reactors," WASH-1270, U.S. Atomic Energy Commission (1973).

[Atw03] C. L. Atwood, J. L. LaChance, H. F. Martz, D. J. Anderson, M. Englehardt, D. Whitehead, and T. Wheeler, "Handbook of Parameter Estimation for Probabilistic Risk Assessment," NUREG/CR-6823, U.S. Nuclear Regulatory Commission (2003).

[Bru75] H. D. Brunk, *An Introduction to Mathematical Statistics*, 3rd ed., Xerox College Publishing (1975).

[Eps60] B. Epstein, "Estimation from Life Test Data," *IRE Trans. Reliab. Quality Control* **RQC-9**, 104 (1960).

[Fyn09] D. A. Fynan and J. C. Lee, "Comparison of Three Methods for Reliability Quantification," *Trans. Am. Nucl. Soc.* **100**, 435 (2009).

[Man74] N. R. Mann, R. E. Schafer, and N. D. Singpurwalla, *Methods of Statistical Analysis of Reliability and Life Data*, Wiley (1974).

[NAP05] "Health Risks from Exposure to Low Levels of Ionizing Radiation, BEIR VII-Phase 2," Biological Effects of Ionizing Radiation Committee, National Academies Press (2005).

[Pag83] T. Page, *Risk Anal.* **3**, no. 1 (March 1983).

[Pap02] A. Papoulis and S. U. Pillia, *Probability, Random Variables, and Stochastic Processes,* 4th ed., McGraw-Hill (2002).

[Sho90] M. L. Shooman, *Probabilistic Reliability: An Engineering Approach*, 2nd ed., Krieger (1990).

[Spi08] M. R. Spiegel and L. J. Stephens, *Theory and Problems of Statistics*, 4th ed., McGraw-Hill (2000).

[Tri69] M. Tribus, *Rational Descriptions, Decision and Designs*, Pergamon (1969).

[USN83] U.S. News and World Report, Dec. 19, 28 (1983).

[Wig07] C. Wiggins, "How Can Bayes' Theorem Assign a Probability to the Existence of God?," *Sci. Am.* **296**, 108 (2007).

Exercises

3.1 Table 3.5 shows the distribution of the diameters of the heads of rivets manufactured by a company. Compute (a) the mean diameter $\overline{x}$ and (b) the variance σ^2.

3.2 Determine the (a) 98%, (b) 90%, and (c) 99.73% confidence limits for the mean diameter of the rivet heads in Exercise 3.1 [Spi08].

Table 3.5 Diameters of Rivet Heads for Exercise 3.1

Diameter, cm	Frequency
0.7247–0.7249	2
0.7250–0.7252	6
0.7253–0.7255	8
0.7256–0.7258	15
0.7259–0.7261	42
0.7262–0.7264	68
0.7265–0.7267	49
0.7268–0.7270	25
0.7271–0.7273	18
0.7274–0.7276	12
0.7277–0.7279	4
0.7280–0.7282	1
Total	250

Source: Reprinted with permission from [Spi08].

3.3 Five gears were tested to failure and the following failure times were recorded: 0.5, 0.9, 1.7, 2, and 3.2×10^5 sec. For the two-parameter Weibull model, determine $\widehat{\alpha}$ and $\widehat{\beta}$ using the (a) moment, (b) maximum likelihood, and (c) maximum entropy methods.

3.4 Ten shafts were tested to failure which occurred at the following number of cycles: 3.5, 6.5, 8, 9.2, 13, 14.5, 16.8, 18, 19.5, and 24×10^5. For the two-parameter Weibull model, determine $\widehat{\alpha}$ and $\widehat{\beta}$ with the (a) maximum likelihood and (b) maximum entropy methods.

3.5 A nonreplacement life test was carried out on a sample of 10 pumps that failed at the following times after beginning the test: 4.6, 8, 10, 12, 14, 17, 20, 22, 26, and 33×10^4 sec. For the two-parameter Weibull model, determine $\widehat{\alpha}$ and $\widehat{\beta}$ with the (a) maximum likelihood and (b) maximum entropy methods.

3.6 The horn on a safety system audio alarm operates on demand 99.96% of the time. Each event is independent of all the others. How many times should the horn work with a 50% probability of not having a single failure?

3.7 Suppose 60 of 10,060 chemicals are carcinogenic and that a reliable test of any chemical correctly scores positive for any carcinogenic chemical with a probability of 95% and scores negative for a noncarcinogenic chemical 94% of the time. What is the probability that a chemical that tests positive, drawn randomly from the 10,060, is carcinogenic [Pag83]?

3.8 Two lots of bolts were used to fasten the pressure head of a nuclear reactor, with 60% from lot A_1 and 40% from lot A_2. After a bolt failure, B, testing of new bolts from both lots was performed to find that 1% of the bolts from lot A_1 and 3% of the bolts from lot A_2 could fail. Determine the probability that the failed bolt came from each lot.

3.9 If lightning L strikes a given tract of property, the probability it hits either portion A, B, or C is 0.1, 0.3, and 0.6, respectively. The probability of a fire F following a lightning strike is 0.1, 0.1, and 0.3, respectively. Following a lightning strike, what is the probability that a fire breaks out on portions A, B, and C?

3.10 Suppose that 30 of 4700 dams are highly prone to failure and that a test has been developed that correctly identifies faulty dams 90% of the time and correctly identifies dams not prone to failure 85% of the time. One dam selected randomly from the 4700 tests faulty. What is the probability that the dam is highly prone to failure?

3.11 An emergency diesel generator has undergone three series of two startup tests, each with a successful outcome D or unsuccessful outcome $\overline{D}$. Test series #1, #2, and #3 randomly result in outcomes $\{D, D\}$, $\{\overline{D}, \overline{D}\}$, and $\{D, \overline{D}\}$ or $\{\overline{D}, D\}$, respectively. On the fourth series of tests, the diesel generator starts on the first test. What is the probability that the generator will fail to start on the second test? (This is a variation of a problem known as Bertrand's Box Paradox.)

3.12 A diagnostic test for a particular disease has a 99% reliability, i.e., 99% of sick people s test positive and 99% of healthy people h test negative. Clinical data indicate that 1% of the population is sick. If a person tests positive, what is the chance he or she is suffering from the disease [Wig07]?

3.13 You are flying across the United States in a fully-loaded DC-10 airplane and have to use the restroom only to learn that only one restroom out of five on the airplane is operating. Proceeding by intuition, assume that a single restroom consists of the mechanical (M), water (W), and electrical (E) systems that function independently of one another and can fail, respectively, by a broken door latch, a stopped-up toilet, and no power to the water pump, for example. For simplicity, assume that all restrooms operate independently of one another and that on a single flight M, W, and E can fail with probabilities of 0.01, 0.02, and 0.001, respectively, and that the flight attendants are attentive enough to declare a restroom out of order with a probability of 0.8, 0.95, and 0.6, respectively. (a) For a restroom that is out of order, what are the probabilities that each system failed if the probability of failure of more than one system during a flight is negligible? (b) If all three systems must work for a restroom to be operable, what is the probability per flight for failure of a restroom? If each restroom is independent of the others, what is the probability that (c) only one in five will function by the end of a flight and (d) none or only one in five will function by the end of a flight?

3.14 During 1980 to 1983, over 175×10^6 messages were sent by the U.S. North American Aerospace Defense Command without a false message [USN83]. For the hypotheses A_1 through A_4 that the probability of sending a single false message was 10^{-8}, 10^{-9}, 10^{-10}, or 10^{-11}, test the hypotheses to estimate the probability that a false message could have been sent assuming (a) a uniform prior probability distribution and (b) nonuniform prior probabilities of 0.05, 0.15, 0.40, and 0.40, respectively.

3.15 A series of tests of a diesel-electric generator is to be performed to determine the likelihood that the diesel engine will start on demand. (a) Calculate the number of sequential trials, with one failure, required to demonstrate, to a 95% confidence, that the probability of failure to start on demand is less than 0.05. Each trial is to be considered a binary event, failure or success. (b) Using the central limit theorem, determine the one-sided 95% confidence interval for the reliability of the diesel generator given one failure in 96 consecutive trials. (c) Repeat part (b) using the χ^2-distribution. Compare the result with that of part (b).

3.16 Write a computer program that can systematically provide three-way comparisons of Exercise 3.15 for the upper bound estimates of failure rate of diesel generators. Consider confidence levels α = [90, 95, 99]%, failure rates p = [0.001, 0.005, 0.01, 0.05] per demand, and number of failures n = 0,1,2,3. Remember that, for trials with $n > 0$, the engineering approach has to use the Poisson or Erlangian distribution as indicated in Eq. (3.49). Construct your program in the following steps: (a) Start with the engineering approach for each combination of $\{\alpha, n, p\}$ to determine the number of trials N required. An iteration has to be performed in general for cases for $n > 0$. (b) When N is obtained from the engineering approach, use the central limit theorem and χ^2-distribution for each combination of $\{\alpha, n, N\}$ to determine one-sided upper-bound estimates for the unreliability p. For each value of n, i.e., the number of failures assumed, plot a family of curves comparing p for the two statistical approaches versus the values assumed in the engineering approach of part (a) as a function of confidence level α.

3.17 You have purchased a precision resistor for a circuit you are building for the reactor protection system of a nuclear power plant. You need to determine the resistance accurately, and you make 64 measurements by a method which involves, based on previous experience, a variance of 4 Ω^2. (a) Using the central limit theorem, calculate the a priori probability—the probability before you make the measurements—that you will obtain sample values for which the sample mean $\overline{x}$ will differ from the population (true) mean μ by less than 0.667 Ω. (b) If you obtain an average of 39.0 Ω for the 64 measurements you take, determine whether you should accept the manufacturer's claim that the resistor has a resistance of 40 Ω.

3.18 An epidemiological study for a sample of 3000 persons living in the vicinity of a nuclear power plant shows a cancer death rate of 21%, while the BEIR-VII report indicates a population average cancer death rate of 20%. An environmental advocacy group claims that this increase in cancer death rate is due to radiation released from the nuclear power plant. Should we accept their claim? Provide a statistical basis for your answer, listing any assumptions you make.

CHAPTER 4

RELIABILITY OF MULTIPLE-COMPONENT SYSTEMS

Chapter 2 consisted of an introduction to the analysis of failures of a single device or component of a system and introduced the associated reliability functions. This chapter contains an introduction to the reliability of a system comprised of a set of components or units n, $n = 1, \ldots, N$, each of which has a reliability $R_n(t) = 1 - F_n(t)$. In the simplest case of random failures of each component, $R_n(t) = \exp(-\lambda_n t)$, and usually this simple model will be used for purposes of illustration. But the reliability of any system component that acts in a time-independent mode, or in any time-dependent failure model more general than the exponential model, also can be substituted for the symbol R_n.

There is a strong analogy between elementary electrical circuits and the reliability of systems if one constructs a *reliability block diagram.* Such a diagram enables one to follow the flow of a "system operation" signal (i.e., reliability) from an input location to an output location and thereby effectively replace a set of system components by a single composite component. This will be illustrated in more detail after an introduction of elementary configurations of components.

4.1 SERIES AND ACTIVE-PARALLEL SYSTEMS

4.1.1 Systems with Independent Components

We first consider the simplest of systems consisting of two independent units that operate in series, as depicted in Fig. 4.1(a). For units 1 and 2 in series, both must operate for the system to function. Because the reliabilities $R_1(t)$ and $R_2(t)$ are probabilities, the product rule for probabilities of Eq. (2.4) gives the reliability for the system $R_{sys}(t)$ as

$$R_{sys}(t) = R_1(t)R_2(t). \tag{4.1}$$

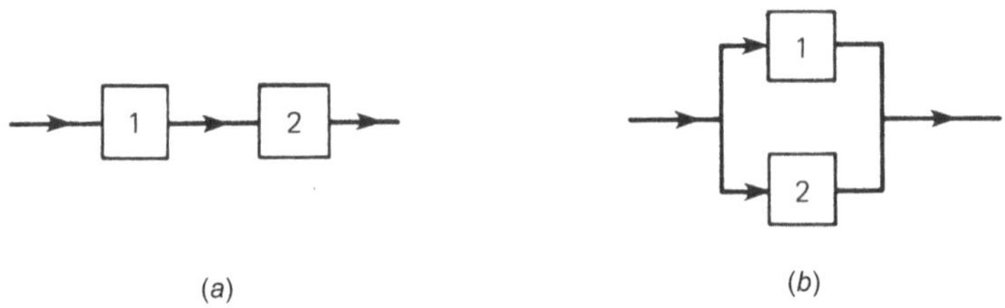

Figure 4.1 Reliability block diagram for two units in (*a*) series and (*b*) active parallel.

For two units in a traditional parallel configuration that is denoted here by an *active-parallel* configuration, either unit 1 or unit 2 must operate for the system to function, and therefore, from Eq. (2.16),

$$R_{sys}(t) = R_1(t) + R_2(t) - R_1(t)R_2(t). \tag{4.2}$$

For random failures the last two equations become, respectively,

$$R_{sys}(t) = \exp[-(\lambda_1 + \lambda_2)t], \tag{4.3}$$

$$R_{sys}(t) = \exp(-\lambda_1 t) + \exp(-\lambda_2 t) - \exp[-(\lambda_1 + \lambda_2)t]. \tag{4.4}$$

The reliability for two units in series is less than that for the less reliable of the two, while the reliability of two units in active parallel is larger than that for the more reliable of the two. In Fig. 4.2, plots of system reliability versus dimensionless time from Eqs. (4.1) and (4.2) are shown for $\lambda_1 = \lambda_2 = \lambda$ and $\lambda_1 = \lambda_2 = kt$.

Equations (2.9) and (2.16) can be employed to generalize to a system of N independent components, all in series or in active parallel, respectively,

$$\begin{aligned} R_{sys}(t) &= \prod_{n=1}^{N} R_n(t) \\ &= \exp\left[-\int_0^t \sum_{n=1}^{N} \lambda_n(\tau)\mathrm{d}\tau\right] \quad \text{(series)}, \end{aligned} \tag{4.5}$$

$$1 - R_{sys}(t) = \prod_{n=1}^{N} [1 - R_n(t)] \quad \text{(active parallel)}. \tag{4.6}$$

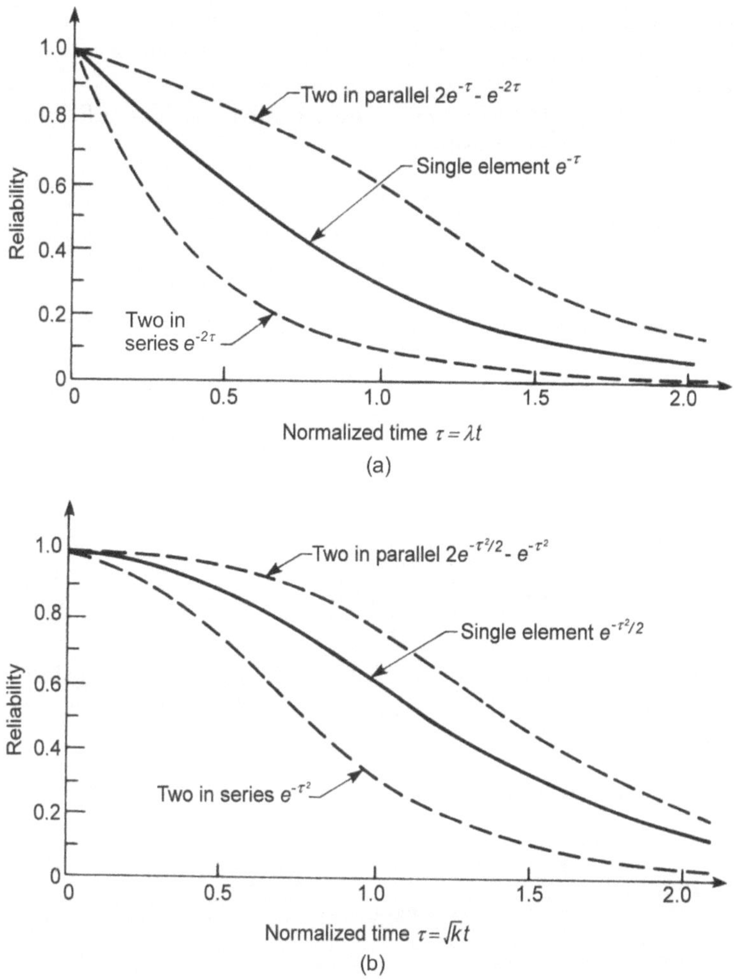

Figure 4.2 Comparison of system reliability functions for (*a*) constant hazard rate units and (*b*) linearly increasing hazard rate units with $\lambda = kt$. *Source:* Reprinted with permission from [Sho90] Copyright © 1990 Robert E. Krieger Publishing Company.

It is worth noting that the substitution of $F(t) = 1 - R(t)$ in the preceding two equations gives

$$1 - F_{sys}(t) = \prod_{n=1}^{N} [1 - F_n(t)] \quad \text{(series)} \tag{4.7}$$

$$F_{sys}(t) = \prod_{n=1}^{N} F_n(t) \quad \text{(active parallel).} \tag{4.8}$$

For this simple case, a reliability analysis on a system with active-parallel components can be transformed into a failure analysis on a system with components in series, and vice versa.

Equations (4.5) and (4.6) for the mean time to failure for N units failing randomly satisfy the relatively simple equations

$$\begin{aligned} \mathrm{MTTF}_{series} &= \int_0^\infty \exp\left[-t\sum_{n=1}^{N}\lambda_n\right] dt = \left(\sum_{n=1}^{N}\lambda_n\right)^{-1}, \qquad (4.9) \\ \mathrm{MTTF}_{active\,parallel} &= \sum_{n=1}^{N}\frac{1}{\lambda_n} - \sum_{n=1}^{N-1}\sum_{m=n+1}^{N}\frac{1}{\lambda_n+\lambda_m} \\ &\quad + \cdots + (-1)^{N-1}\left(\sum_{n=1}^{N}\lambda_n\right)^{-1}. \qquad (4.10) \end{aligned}$$

If all N units are identical, then these equations can be simplified to give the inequality

$$\mathrm{MTTF}_{series} = (N\lambda)^{-1} < \sum_{n=1}^{N}(n\lambda)^{-1} = \mathrm{MTTF}_{active\,parallel}. \qquad (4.11)$$

Example 4.1 A system consists of $2N$ identical units having a constant hazard rate λ, with N units in each of two branches in an active-parallel configuration. Determine (a) the reliability and (b) the MTTF of the system.

(a) From Eq. (4.5) the reliability of each of the two branches is $[R(t)]^N$, so from Eq. (4.2)

$$\begin{aligned} R_{sys}(t) &= [R(t)]^N + [R(t)]^N - [R(t)]^{2N} \\ &= \exp(-N\lambda t)[2 - \exp(-N\lambda t)]. \end{aligned}$$

(b) The MTTF is

$$\mathrm{MTTF} = \int_0^\infty R_{sys}(t)dt = \frac{3}{2N\lambda}. \quad \diamond$$

It may be worth pointing out that when analyzing the reliability of systems that can undergo repairs, because the counterpart to $R(t)$ is $1 - \tilde{R}(t)$, two components acting in series in a reliability analysis will act in parallel in a repairability analysis, and vice versa. For example, two repair people working independently to repair a single system component must both fail if the system is to remain failed; that is, together they act like a subsystem in a series configuration, rather than like a subsystem with two parallel components.

4.1.2 Systems with Redundant Components

Another elementary configuration is the "M-out-of-N system" consisting of N identical units of reliability $R(t)$, but only M are required for the system to operate. This system with redundant components can be analyzed with the binomial distribution of probability theory, which gives

$$\begin{aligned} R_{M/N}(t) &= \sum_{n=M}^{N} \binom{N}{n} [R(t)]^n [1 - R(t)]^{N-n} \\ &= \sum_{n=M}^{N} \frac{N!}{n!(N-n)!} [R(t)]^n [1 - R(t)]^{N-n}. \end{aligned} \tag{4.12}$$

The M-out-of-N system is a generalization of the active-parallel system of Eq. (4.6), which is a "1-out-of-N system." If the N units all fail randomly, then the MTTF corresponding to Eq. (4.12) is

$$\text{MTTF}_{M/N} = \sum_{n=M}^{N} \left(\frac{1}{n\lambda} \right). \tag{4.13}$$

The MTTF for the M-out-of-N system is between that for a series and that for an active-parallel system with N units.

Because the reliability for an M-out-of-N system in Eq. (4.12) represents the system reliability $R_{M/N} = R_{\geq M}$ with M or more identical components, each having reliability $R = 1 - F$ corresponding to a failure probability F, the failure probability $F_{M/N}$ of the system is the complement of $R_{M/N}$,

$$\begin{aligned} F_{M/N} &= 1 - R_{M/N} = R_{\leq (M-1)} = F_{\geq [N-(M-1)]} = F_{\geq (N-M+1)} \\ &= \sum_{n=0}^{M-1} \binom{N}{n} R^n (1-R)^{N-n} = \sum_{n=0}^{M-1} \binom{N}{n} F^{N-n} (1-F)^n. \end{aligned} \tag{4.14}$$

For highly reliable components, i.e., for $F \ll 1$, the failure probability for an M/N system can be approximated by

$$F_{M/N} \approx \binom{N}{N-M+1} F^{N-M+1}, \tag{4.15}$$

which indicates that the failure probability of the N-component system is essentially equal to the simultaneous failure of $N - M + 1$ components, because any additional failures contribute little to the total system failure probability.

Example 4.2 A system consists of seven components connected as in the reliability block diagram shown in Fig. 4.3. Components 2 to 4 are in active parallel with component 1, and that subsystem is in series with three components of type 5 that constitute a 2-out-of-3 subsystem. If R_n, $n = 1, \ldots, 5$, denotes the reliability of each type of component, determine the reliability of the system.

Components 2 to 4 can be replaced by an equivalent one with reliability R_{234} of

$$R_{234} = R_2 + R_3 + R_4 - R_2R_3 - R_2R_4 - R_3R_4 + R_2R_3R_4.$$

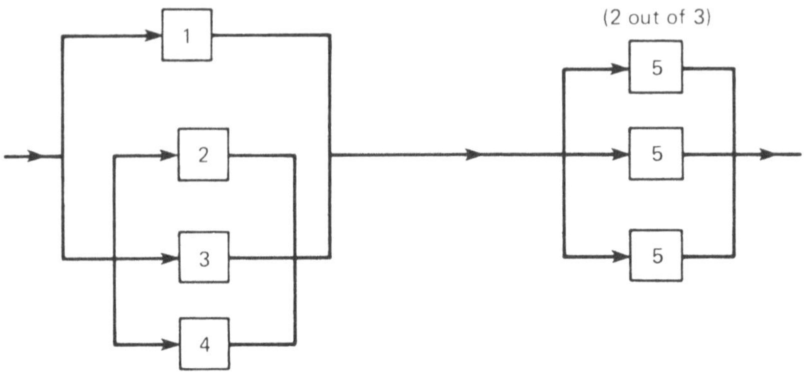

Figure 4.3 Reliability block diagram for Example 4.2.

Component 1 and subsystem 234 can be combined to give

$$R_{1234} = R_1 + R_{234} - R_1 R_{234}.$$

The 2-out-of-3 subsystem can be evaluated as an equivalent component from Eq. (4.12),

$$R_{(55)5} = 3R_5^2(1 - R_5) + R_5^3.$$

Finally, the reliability of the system is

$$R_{sys} = R_{1234} R_{(55)5}. \quad \diamond$$

4.1.3 Fail-to-Safety and Fail-to-Danger Systems

In our discussion of component reliabilities up to now, we have considered the failure of components that make the entire system incapable of performing its intended function. One example is the reactor protection system (RPS), which includes multiple scram breakers coupled to scram actuation logic modules. The scram breakers are designed to provide current to the electromagnets holding the control absorbers so that opening the breakers will automatically release the absorbers into the reactor core. For this reason, a RPS failure will occur if the scram breakers do not open when the scram actuation signal is received by the scram logic modules. Thus, the RPS failure probability should be evaluated with the component failure probability F of Eqs. (4.14) and (4.15) set equal to the probability that a scram breaker fails to open when activated.

Proper evaluation of the reliability of the RPS, however, requires evaluation of events where the scram breakers open unnecessarily when a reactor trip is not required. Such spurious reactor scrams would occur, for example, in the event of a failure in the power supply, which is configured to deenergize the circuit upon failure and send a scram signal, as a prudent measure of safety. Thus, our study of the reliability of redundant systems, such as the RPS, should consider the probability P_s that the system actuates spuriously, which is called the fail-to-safety or fail-safe probability. This is to be contrasted with the failure probability F of Eqs. (4.14) and (4.15), which is called the fail-to-danger or fail-danger probability.

Because the replacement power for a nuclear plant could cost as much as \$1 million a day and unnecessary scrams require additional maintenance and repairs, it is necessary to minimize unnecessary scrams as much as possible. This means that fail-safe probabilities should be considered, along with fail-danger probabilities, in the overall design and evaluation of systems comprising multiple redundant components such as a nuclear power plant scram system.

One way to consider such fail-danger and fail-safe systems is to examine the states of each component of the system and the effect of those components on the performance of the system itself.

Example 4.3 For a 1-out-of-2 system consisting of components 1 and 2, construct a functional state table that delineates the various combinations of component states as either functional (ok), fail-danger (fd), or fail-safe (fs), along with the probability for each component state. Also give the states of a system comprised of the two components, along with the probability for each system state.

Table 4.1 shows the functional state table for a two-component system. We can sum the probabilities for the three different operational states of the system as follows:

1. $P\{\text{ok, system functional if required}\} = (1-F-P_s)^2+2(1-F-P_s)F = (1-P_s)^2 - F^2$.

2. $P\{\text{fd, system not functional if required}\} = F^2$.

3. $P\{\text{fs, system functions unnecessarily}\} = 2(1-F-P_s)P_s + 2FP_s + P_s^2$, which reduces to $P\{\text{fs}\} = 2(1-P_s)P_s + P_s^2 = 2P_s - P_s^2 \approx 2P_s$.

Summation of the probabilities for all three system states yields unity, as it should. We also can verify that the probability for the system fail-danger state agrees with Eqs. (4.14) and (4.15). ◇

Let us now consider what happens for a fail-danger and fail-safe system if two components are combined together to form a system with redundant components. To perform such a system reliability evaluation, consider the fail-safe probability P_s as the unnecessary or spurious success probability for the system to actuate. Thus, the fail-safe probability $F_{s,M/N}$ for an M-out-of-N system is obtained as the unnecessary success probability for M or more components:

$$\begin{aligned} F_{s,M/N} &= F_s(\geq M) = \sum_{n=M}^{N} \binom{N}{n} P_s^n (1-P_s)^{N-n} \\ &\approx \binom{N}{M} P_s^M, \quad P_s \ll 1. \end{aligned} \tag{4.16}$$

It is seen that $P\{\text{fs}\}$ of Example 4.3 agrees with Eq. (4.16).

Table 4.1 Fail-Danger and Fail-Safe Functional States and Probabilities for a Two-Component System

Component 1 State	Component 1 Probability	Component 2 State	Component 2 Probability	System State	System Probability
ok	$1 - F - P_s$	ok	$1 - F - P_s$	ok	$(1 - F - P_s)^2$
ok	$1 - F - P_s$	fd	F	ok	$(1 - F - P_s)F$
ok	$1 - F - P_s$	fs	P_s	fs	$(1 - F - P_s)P_s$
fd	F	ok	$1 - F - P_s$	ok	$(1 - F - P_s)F$
fd	F	fd	F	fd	F^2
fd	F	fs	P_s	fs	FP_s
fs	P_s	ok	$1 - F - P_s$	fs	$(1 - F - P_s)P_s$
fs	P_s	fd	F	fs	FP_s
fs	P_s	fs	P_s	fs	P_s^2

Example 4.4 Consider the scram circuitry of a simple RPS that consists of two scram breakers in series or active-parallel configuration, as illustrated in Fig. 4.1. For the component probabilities $F = P_s = 10^{-3}$, determine the approximate fail-danger and fail-safe probabilities for the (a) series and (b) active-parallel configurations.

(a) For the series configuration, the M/N scram circuitry for the scram breaker system has $M = 1$ and $N = 2$ because opening either of the breakers would result in a scram and a scram failure occurs if both breakers fail to open. From Eq. (4.15), the system fail-danger probability is

$$F_{M/N} \approx F^N = F^2 = 10^{-6},$$

which is acceptably low. On the other hand, a spurious scram occurs if either breaker opens unnecessarily, which from Eq. (4.16) gives a system fail-safe probability of

$$F_{s,M/N} \approx NP_s = 2 \times 10^{-3},$$

which is unacceptably high.

(b) For the scram breakers in the active-parallel configuration, a successful scram requires opening both breakers, which results in a 2/2 system. The system fail-danger probability is

$$F_{M/N} \approx NF = 2 \times 10^{-3},$$

while a spurious scram occurs only if both breakers open, which gives the system fail-safe probability

$$F_{s,M/N} \approx P_s^2 = 10^{-6}.$$

It is seen that the acceptability of the system fail-danger and fail-safe probabilities in the active-parallel configuration is reversed from that of the series configuration.

This simple example illustrates that an actual RPS would require a higher degree of redundancy for the scram circuit breakers so that both the system fail-danger and fail-safe probabilities are acceptably low. This will be discussed further in Chapter 9. ◇

We conclude this section by highlighting one important guideline for nuclear plants in the General Design Criteria (GDC), as implemented in [AEC71]. Criterion 21 of the GDC is known as the single-failure criterion and is a key defense-in-depth concept for nuclear systems, which stipulates that the failure of any single component or subsystem should not cause the total system to fail. Thus to satisfy the single-failure criterion, if M components are needed for the system to function, an $M/(M+1)$ system would be sufficient, although the minimum redundancy should be increased to an $M/(M+2)$ system if the service of a component is allowed during operation.

4.2 SYSTEMS WITH STANDBY COMPONENTS

The purpose of standby components is to increase the reliability and the MTTF of a system from that obtained without backup components. For example, if a system has N identical independent components, with only one required for system operation and the remaining nonfailed components in standby status, then

$$(\mathrm{MTTF})_{1/N} = N/\lambda \tag{4.17}$$

if there are no failures during standby and with instantaneous switching between components.

Systems with the built-in safety of backup components can be viewed as parallel systems that are load sharing or sequential in operation in such a way that only a subsystem is in operation at a time, with the remaining components held in reserve. We begin by considering the case of a system with only two components, with component 1 operating and component 2 initially in standby and not subject to failure during standby. Also, the instantaneous switching from component 1 to component 2 is taken to be 100% reliable. The system can be depicted as in the reliability block diagram of Fig. 4.4.

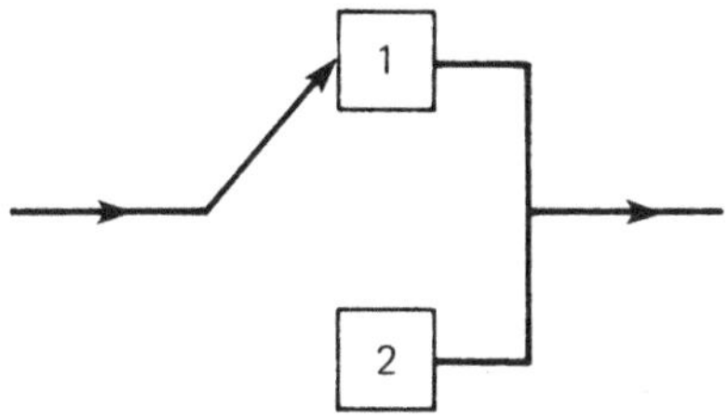

Figure 4.4 Reliability block diagram for two units with unit 2 in standby.

The system reliability can be determined from

$$R_{sys}(t) = R_1(t) + \int_0^t f_1(\tau) R_2(t-\tau)\mathrm{d}\tau. \qquad (4.18)$$

The first term on the right-hand side is the reliability of component 1, and the second term requires integration over all times to account for the reliability of component 2 after component 1 has failed at an unknown time τ with probability $f_1(\tau)\mathrm{d}\tau$. Component 2 need operate from only τ to t, which accounts for the time period $t-\tau$.

If both components fail randomly when in operation, then Eq. (4.18) gives

$$R_{sys}(t) = \exp(-\lambda_1 t) + \frac{\lambda_1}{\lambda_1 - \lambda_2}[\exp(-\lambda_2 t) - \exp(-\lambda_1 t)]. \qquad (4.19)$$

The special case with $\lambda_1 = \lambda_2 = \lambda$ yields

$$R_{sys}(t) = \exp(-\lambda t)(1 + \lambda t), \qquad (4.20)$$

which follows from Eq. (4.19) after taking the limit of $\exp[(\lambda_1 - \lambda_2)t] \approx 1 + (\lambda_1 - \lambda_2)t$ divided by $\lambda_1 - \lambda_2$, or via L'Hospital's rule. The two terms of Eq. (4.20) are those arising from the cumulative Poisson distribution of Eq. (2.49).

Equation (4.18) is not applicable if component 2 can fail while in standby or if the instantaneous switch has a reliability $R_{sw} < 100\%$. If component 2 can fail during standby with hazard rate λ_2^*, then its reliability at time τ is $R_2^*(\tau)$ and

$$R_{sys}(t) = R_1(t) + R_{sw}\int_0^t f_1(\tau) R_2^*(\tau) R_2(t-\tau)\mathrm{d}\tau. \qquad (4.21)$$

For random failures of both components, when component 2 is in standby,

$$R_{sys}(t) = \exp(-\lambda_1 t) + \frac{R_{sw}\lambda_1}{\lambda_1 + \lambda_2^* - \lambda_2}\{\exp(-\lambda_2 t) - \exp[-(\lambda_1 + \lambda_2^*)t]\}. \qquad (4.22)$$

If $R_{sw} = 1$, this equation reduces to Eq. (4.19) if $\lambda_2^* = 0$, while if $\lambda_2^* = \lambda_2$ then the reliability is just that of Eq. (4.4) for two components in active parallel.

The generalization of Eq. (4.21) to the case of three independent components operated sequentially, with perfect switching, is

$$\begin{aligned} R_{sys}(t) &= R_1(t) + \int_0^t f_1(\tau) R_2^*(\tau) R_2(t-\tau)\mathrm{d}\tau \\ &\quad + \int_0^t f_{12}(\tau) R_3^*(\tau) R_3(t-\tau)\mathrm{d}\tau, \end{aligned} \qquad (4.23)$$

where $f_{12}(\tau) = -\mathrm{d}R_{12}(\tau)/\mathrm{d}\tau$ and $R_{12}(t)$ is the sum of the first two terms on the right-hand side of this equation. Further generalizations of Eq. (4.21) can be derived in this manner. Equation (4.23) also can be applied to a system consisting of two components plus an imperfect switch that does not act in an instantaneous manner. Then the operation of component 1 is followed by the switch, now treated as component 2, before the backup component takes on the role of component 3.

Example 4.5 A reactor coolant pump has an identical pump in standby that can be successfully valved into operation 99% of the time. The reliability of a pump when operating over a year is 0.8 and during standby over two years is 0.95. What is the reliability of the pump system over a six-month time interval assuming failures occur randomly?

For $\lambda_1 = \lambda_2 = \lambda$, Eq. (4.22) reduces to

$$R_{sys}(t) = \exp(-\lambda t)\{1 + (R_{sw}\lambda/\lambda^*)[1 - \exp(-\lambda^* t)]\}.$$

The hazard rate of a pump in operation is

$$\lambda = -\ln R(1 \text{ yr})/1 \text{ yr} = -\ln 0.8/\text{yr} = 0.233/\text{yr}$$

and in standby is

$$\lambda^* = -\ln R^*(2 \text{ yr})/2 \text{ yr} = -\ln 0.95/2 \text{ yr} = 0.0256/\text{yr}.$$

Substitution of these hazard rates gives $R_{sys}(0.5 \text{ yr}) = 0.993$, which exceeds the reliability for a single pump by the factor of 1.11. ◇

The idea of switching a system component into operation following the failure of another is a general concept and can be applied to the switching of a subsystem following the failure of another. Also, the switching idea can be applied to other time-dependent activities, such as repairs or inspections following repairs.

Example 4.6 A plant component has just failed and repair is immediately begun at time $t = 0$ by a repairman who can complete the task after a random time with a mean time to repair of μ_1^{-1}. Immediately after successfully repairing the component, his boss begins to inspect the repair job and completes her inspection with a random mean completion time of μ_2^{-1}. Derive (a) the equation for the probability of successfully completing the repair and inspection as a function of time and (b) the mean time to repair and inspect the system, MTTR&I.

(a) The repairman is component 1 and the inspector is component 2 of the repair system. In this case the counterpart to the reliability $R(t)$ of a system component is the nonrepairability $1 - \widetilde{R}(t)$, where $\widetilde{R}(t)$ is defined in Eq. (2.97), because the counterpart to $F(t)$ is $\widetilde{R}(t)$. Application of the "reliability to repair transformation" shows that Eq. (4.18) can be written as

$$1 - \widetilde{R}_{sys}(t) = 1 - \widetilde{R}_1(t) + \int_0^t \widetilde{r}_1(\tau)[1 - \widetilde{R}_2(t-\tau)]\mathrm{d}\tau,$$

which simplifies to

$$\widetilde{R}_{sys}(t) = \int_0^t \widetilde{r}_1(\tau)R_2(t-\tau)]\mathrm{d}\tau$$

with Eq. (2.97). From Table 2.5, for repairs completed at random times,

$$\widetilde{R}_{sys}(t) = \int_0^t \mu_1 \exp(-\mu_1\tau)\{1 - \exp[-\mu_2(t-\tau)]\}\mathrm{d}\tau$$

gives

$$\widetilde{R}_{sys}(t) = \frac{\mu_1[1-\exp(-\mu_2 t)] - \mu_2[1-\exp(-\mu_1 t)]}{\mu_1 - \mu_2}.$$

(b) Differentiation of the last result gives

$$\widetilde{r}_{sys}(t) = \frac{\mathrm{d}\widetilde{R}_{sys}(t)}{\mathrm{d}t} = \frac{\mu_1\mu_2[\exp(-\mu_2 t) - \exp(-\mu_1 t)]}{\mu_1 - \mu_2}$$

so with $\widetilde{R}_{sys}(t) \to 1$ as $t \to \infty$, an intuitively obvious result eventually follows,

$$\text{MTTR\&I} = \int_0^\infty t\tilde{r}_{sys}(t)\mathrm{d}t = \mu_1^{-1} + \mu_2^{-1}. \quad \diamond$$

4.3 DECOMPOSITION ANALYSIS

The reliability of a time-dependent, time-independent, or hybrid system sometimes can be improved by connecting components or subsystems with a cross-link L so that they are in more than one subsystem. Such systems can be analyzed by a decomposition analysis, which is nothing more than application of a conditional probability theorem. The system reliability analyzed by decomposition relies on the selection of a "keystone component" K at either one end or the other of the link L connecting the subsystems. With Eq. (2.25) the reliability of the system can be broken down into contributions when component or subsystem K works and does not work ($\overline{K}$),

$$R_{sys} = R_K R(\text{sys}|K) + R_{\overline{K}} R(\text{sys}|\overline{K}), \tag{4.24}$$

with $R_{\overline{K}} = 1 - R_K$. (It should be noted that R can be either time dependent or time independent.)

For example, consider system A shown in Fig. 4.5 in which all components act independently and have a reliability R_n, $n = 1, \ldots, 5$. The two branches are connected by cross-link L so that a "signal" flowing from an input to output can flow along any of the paths connecting components 1–2, 1–3, 4–2, 4–3, or 4–5. If the link L in Fig. 4.5 were *not* present, then the reliability of the system could be obtained by considering first the upper paths connecting components 1–2 and 1–3. With components 2 and 3 in parallel analyzed with Eq. (4.4), combined with component 1 in series treated with Eq. (4.1), the reliability of the upper portion R_u is

$$R_u = R_1(R_2 + R_3 - R_2R_3).$$

The reliability of the path through components 4 and 5 is just R_4R_5, so the reliability R_u for system A *without* link L is

$$R_{sys\,A,\overline{L}} = R_4R_5 + R_1(R_2 + R_3 - R_2R_3)(1 - R_4R_5). \tag{4.25}$$

If the components are all identical with $R_n(t) = \exp(-\lambda t)$, then

$$R_{sys\ A,\overline{L}}(t) = 3\exp(-2\lambda t) - \exp(-3\lambda t) - 2\exp(-4\lambda t) + \exp(-5\lambda t) \quad (4.26)$$

and integration over all time yields

$$\mathrm{MTTF}_{sys,A,\overline{L}} = 13/15\lambda. \quad (4.27)$$

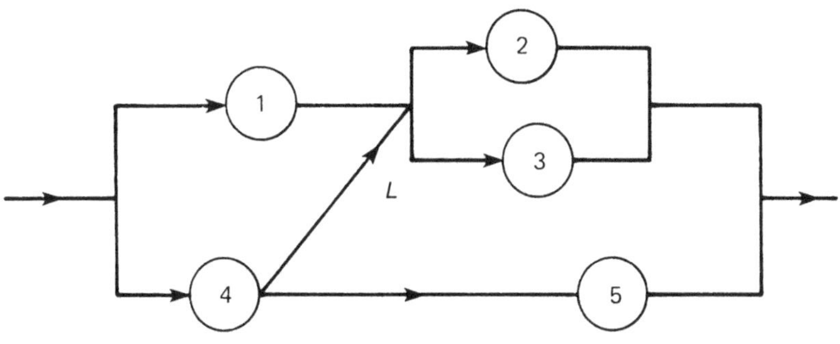

Figure 4.5 Reliability block diagram for cross-link system A.

If the link L in Fig. 4.5 *is* present, Eqs. (4.25) through (4.27) are no longer valid but a decomposition analysis can be used to obtain the reliability. For Fig. 4.5, we can select the keystone component to be component 4 or 2 or 3. We first pick component 4. Then $R(\text{sys}|4)$ is calculated for the parallel combination of components 2, 3, and 5 because the signal can bypass component 1. The reliability $R(\text{sys}|\overline{4})$ is just R_u for when cross-link L was not present, so the decomposition equation gives the reliability of system A as

$$R_{sys\ A,\overline{L}} = R_4[1-(1-R_2)(1-R_3)(1-R_5)] + (1-R_4)[R_1(R_2+R_3-R_2R_3)]. \quad (4.28)$$

In the case that all components are identical with constant hazard rate λ, then this result becomes

$$R_{sys\ A,\overline{L}}(t) = 5\exp(-2\lambda t) - 6\exp(-3\lambda t) + 2\exp(-5\lambda t) \quad (4.29)$$

and the integral over all time gives

$$\mathrm{MTTF}_{sys\ A,L} = 1/\lambda. \quad (4.30)$$

The improvement of the reliability due to the cross-link L is conveniently illustrated by comparing Eqs. (4.27) and (4.30).

Had we selected component 3 in the system to be the keystone component, then

$$R(\text{sys}|3) = R_1 + R_4 - R_1R_4 \quad (4.31)$$

because components 2 and 5 can be bypassed. A complication arises in the analysis, however, because if component 3 does not function, we are left with a reliability block diagram in which we still have not removed the coupling effects arising from cross-link L. (The same complication would have arisen had we selected component 2 instead of 3 to be the keystone component.)

To analyze the system in Fig. 4.5 after first selecting component 3 as a keystone component, we apply a *second* decomposition to calculate $R(\text{sys}|\overline{3})$. For example, if we select component 2 as the second keystone component, then we need

$$R_{sys} = R_3 R(\text{sys}|3) + R_{\overline{3}}[R_2 R(\text{sys}|2\overline{3}) + R_{\overline{2}} R(\text{sys}|\overline{2}\,\overline{3})]. \tag{4.32}$$

If component 2 works, then component 5 can be bypassed so

$$R(\text{sys}|2\overline{3}) = R_1 + R_4 - R_1 R_4, \tag{4.33}$$

whereas $R(\text{sys}|\overline{2}\,\overline{3}) = R_4 R_5$. Substitution of these results into Eq. (4.32) reproduces Eq. (4.28).

This example for system A illustrates that with the decomposition approach the final answer for the reliability does not depend on which component at the end of a link is selected as the keystone component. It also shows that one keystone component can be easier to use than another and that sometimes more than one decomposition is required.

Another subtlety of applying a decomposition is that it is important to know where the endpoints of the cross-link L are connected. This can be illustrated by considering system B in which L is connected directly between component 4 and component 3, as in Fig. 4.6, and not to the parallel combination of components 1 and 4, as it was for system A. For this system it is easiest to pick component 3 as the keystone component. Then component 5 can be bypassed so $R(\text{sys}|3)$ consists of the active-parallel combination of components 1 and 4, while $R(\text{sys}|\overline{3})$ is for the parallel combination of components 1 and 2 in series and components 4 and 5 in series. Thus the reliability of system B is

$$R_{sys\,B,\overline{L}} = R_3(R_1+R_4-R_1R_4)+(1-R_3)(R_1R_2+R_4R_5-R_1R_2R_4R_5). \tag{4.34}$$

If all five components are identical with constant hazard rate λ, then

$$R_{sys\,B,\overline{L}}(t) = 4\exp(-2\lambda t) - 3\exp(-3\lambda t) - \exp(-4\lambda t) + \exp(-5\lambda t), \tag{4.35}$$

and the integral over all time gives

$$\text{MTTF}_{sys\,B,L} = 19/20\lambda. \tag{4.36}$$

Comparison of Eq. (4.36) with Eqs. (4.27) and (4.30) shows that the cross-link for system B is not as effective as the one for system A, but it does provide an improvement in the reliability compared to the system with no cross-link.

Let us now consider system C shown in Fig. 4.7. If we pick component 4 as the keystone component, then the result for $R(\text{sys}|4)$ is just that for the parallel

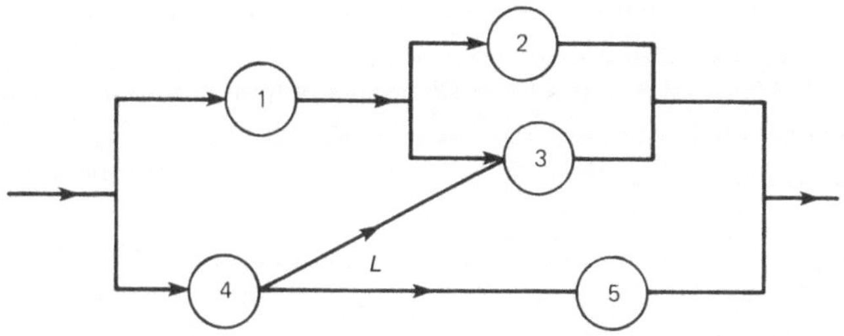

Figure 4.6 Reliability block diagram for cross-link system B.

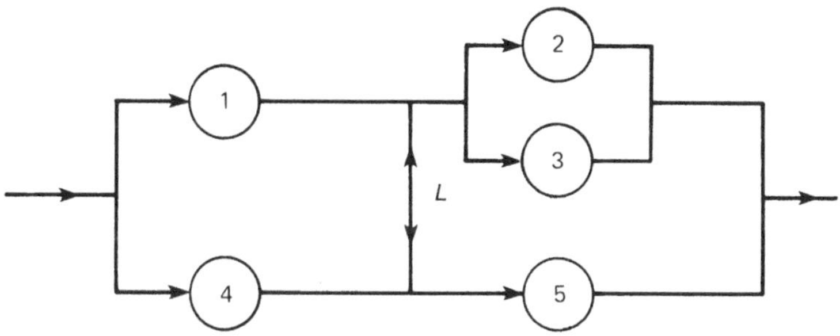

Figure 4.7 Reliability block diagram for cross-link system C.

combination of components 2, 3, and 5, but now $R(\text{sys}|\overline{4})$ must be calculated by selecting a second keystone component. With

$$R(\text{sys}|\overline{4}) = R_1 R(\text{sys}|\overline{4}1) + R_{\overline{1}} R(\text{sys}|\overline{4}\,\overline{1}) \tag{4.37}$$

and the observation that $R(\text{sys}|\overline{4}\,\overline{1}) = 0$ and $R(\text{sys}|\overline{4}\,1) = R(\text{sys}|4)$, it follows that the reliability for system C is

$$R_{sys\,C,\overline{L}} = (R_1 + R_4 - R_1 R_4)[1 - (1 - R_2)(1 - R_3)(1 - R_5)]. \tag{4.38}$$

For the case that all five components are identical with constant hazard rate λ, then

$$R_{sys\,C,\overline{L}}(t) = 6\exp(-2\lambda t) - 9\exp(-3\lambda t) + 5\exp(-4\lambda t) - \exp(-5\lambda t), \tag{4.39}$$

and the integral over all time gives

$$\text{MTTF}_{sys\,C,L} = 21/20\lambda. \tag{4.40}$$

Thus the reliability of system C is slightly better than that for system A. The alert reader will see that system C could have been analyzed without the use of the decomposition approach because the cross-link L can be shortened to a single point.

That is, system C consists of a series combination of component 1 and 4 in parallel and components 2, 3, and 5 in parallel, in agreement with the result in Eq. (4.38).

Example 4.7 A system consists of six identical components with constant hazard rate λ that are connected with two cross-links as shown in the reliability block diagram of Fig. 4.8. Determine (a) the reliability of the system and (b) the MTTF.

(a) If component B is selected as the first keystone component, then $R(\text{sys}|\overline{B}) = R_A R_C R_E$, but $R(\text{sys}|B)$ must be determined with a second keystone component such as D. Because $R(\text{sys}|BD) = R_E + R_F - R_E R_F$ and $R(\text{sys}|B\overline{D}) = R_C R_E$,

$$R_{sys} = R_B R_D(R_E+R_F-R_E R_F)+R_B(1-R_D)R_C R_E+(1-R_B)R_A R_C R_E.$$

After the substitution of $\exp(-\lambda t)$ for every R, it follows that

$$R_{sys}(t) = 4\exp(-3\lambda t) - 3\exp(-4\lambda t).$$

(b) Integration of this result over all time gives

$$\text{MTTF} = 7/12\lambda. \quad \diamond$$

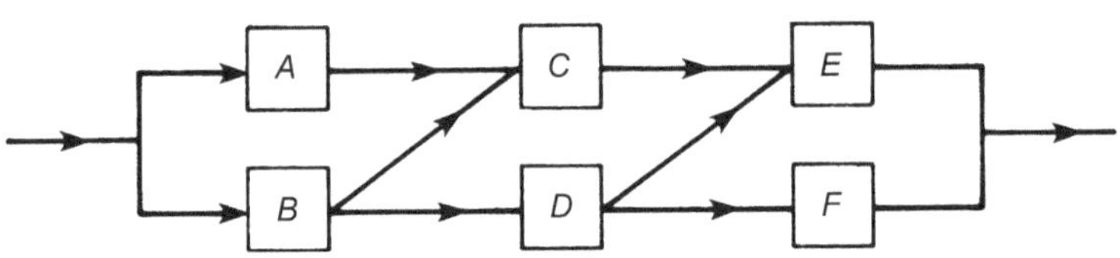

Figure 4.8 Reliability block diagram for Example 4.7.

4.4 SIGNAL FLOW GRAPH ANALYSIS

A signal flow graph is a different way of diagramming a reliability problem that, at least in one sense, is the antithesis of a reliability block diagram. In a reliability block diagram, the blocks represent components and the lines provide the linkage of the components, whereas in a signal flow graph each line or branch represents a component, with the nodes at each end of a branch representing the input and output to the component. Thus a signal flow graph has more analogy to circuits encountered in electrical engineering, for example, especially because each branch has a direction associated with it. This can be seen by comparing Fig. 4.9 for a signal flow graph and Fig. 4.1 for a reliability block diagram. The signal flow graph also is introduced here because it can be used in Section 4.5 to illustrate graphically the idea of cut sets that was introduced in Section 2.1.

The graph for a signal flow analysis of a system consists of nodes $x_1, x_2, \ldots, x_N$ and a collection of branches joining the nodes together. A path is a sequence of

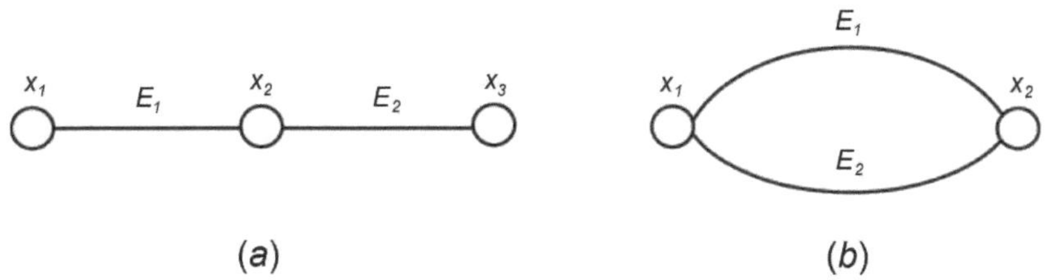

Figure 4.9 Signal flow graph for two units in (*a*) series and (*b*) active parallel.

branches in which the output node of each branch serves as the input to one or more successive branches, i.e., system components. The "weight" of a branch represents the probability of event E_n of transmitting a signal through component n. As an example, the line between nodes x_1 and x_2 labeled E_1 in Fig. 4.9(*a*) indicates the event of transmitting a signal from x_1 to x_2.

The reliability of two components in series can be written as

$$R_{sys} = P(x_1 \to x_3) = P(E_1 E_2) = R_1 R_2 \tag{4.41}$$

as seen by replacing the path between the nodes x_1 and x_3 by a single branch equal to the AND operation of both branches or components in the path. The reliability of two units in active parallel is written in the notation of Fig. 4.9(*b*) as

$$R_{sys} = P(x_1 \to x_3) = P(E_1 + E_2) = R_1 + R_2 - R_1 R_2 \tag{4.42}$$

and corresponds to the OR operation of both branches between x_1 and x_3. These equations are the direct analog of Eqs. (4.1) and (4.2) and can be extended as in Section 4.1 to systems with more than two components.

Example 4.8 Obtain the reliability for the system whose signal flow graph is shown in Fig. 4.10. The first step is to employ the AND operation to obtain the reduced graph in Fig. 4.11. Here the product rule for probabilities gives $B_1 = E_1 E_2 E_3 E_4$ and $B_2 = E_5 E_6 E_7$. The OR operation then gives the subsystem event at $(B_1 + B_2)$, and finally another AND operation yields

$$x_1 \to x_3 = E_8(B_1 + B_2).$$

The reliability of the system then follows as

$$\begin{aligned} R_{sys} = P(x_1 \to x_3) &= P[E_8(E_1 E_2 E_3 E_4 + E_5 E_6 E_7)] \\ &= R_8[1 - (1 - R_1 R_2 R_3 R_4)(1 - R_5 R_6 R_7)]. \quad \diamond \end{aligned}$$

4.5 CUT SET ANALYSIS

In Section 2.1 we defined a cut set as a set of system events that, if they all occur, will cause system failure while a minimal cut set of a system is a cut set that has no

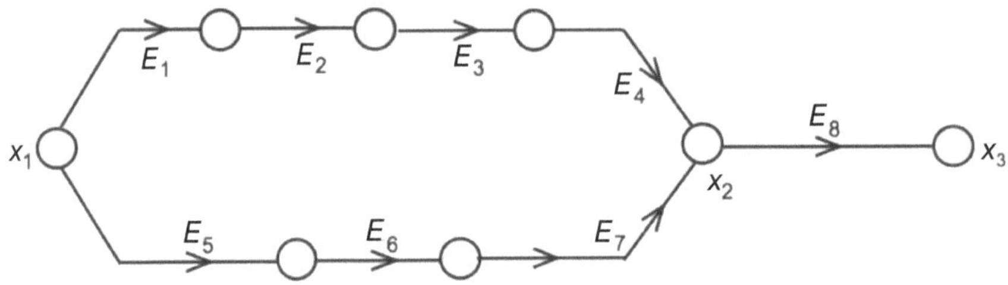

Figure 4.10 Signal flow graph for Example 4.8.

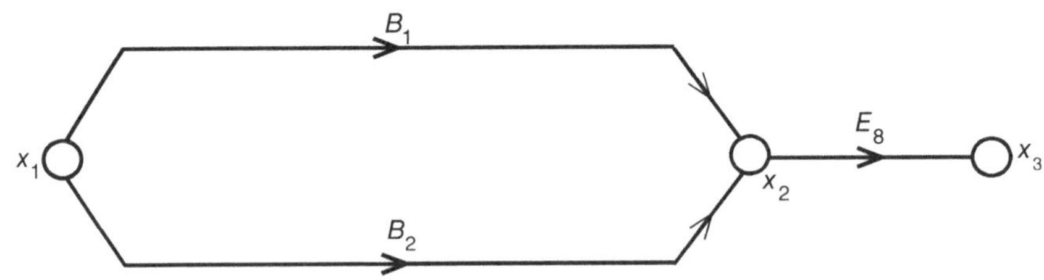

Figure 4.11 Reduced signal flow graph for Example 4.8.

Table 4.2 Other Cut Sets for Example 4.9

$C_2 = E_1E_3$,	$C_3 = E_2E_4$,	
$C_4 = E_1E_5E_3$,	$C_5 = E_2E_6E_4$,	$C_6 = E_3E_5E_6E_2$,
$C_7 = E_1E_5E_4$,	$C_8 = E_3E_6E_2$, etc.	

other cut set as a subset. Here we want to follow up on that concept by obtaining the reliability of a more complex system using minimal cut sets that follow from the ideas behind signal flow graphs.

Example 4.9 For the signal flow graph in Fig. 4.12, obtain the minimal cut sets.

One cut set, denoted as C_1, which corresponds to the wavy line L in the figure, is the event $E_1E_5E_6E_4$. Other cut sets, denoted by C_n, $n = 2,\ 3, \ldots, 8$, are in Table 4.2. If the signal flow graph is constructed such that the system input and output nodes are horizontal, the cut sets can be obtained by cutting the graph from top to bottom. The cut sets C_4 and C_5 are not minimal cut sets because C_2 and C_3 are subsets of C_4 and C_5, respectively. Cut set C_1 is not minimal because component E_6 cannot operate if E_1 and E_5 do not; likewise, C_6 is not a minimal cut set. Cut sets C_7 and C_8 are minimal, however, as well as C_2 and C_3. ◇

Now consider a general system for which all *minimal* cut sets are denoted by C_n, $n = 1, \ldots, N$. The system failure probability F_{sys} can be written as

$$F_{sys} = F(C_1 + C_2 + \cdots C_N) \tag{4.43}$$

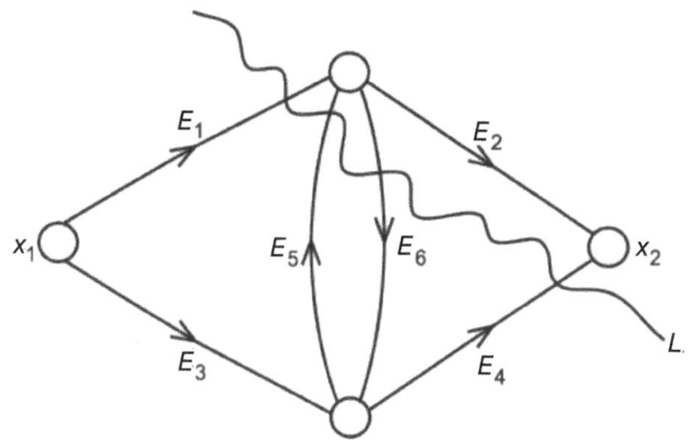

Figure 4.12 Signal flow graph for Example 4.9.

because a failure of all components in any one minimal cut set will lead to system failure. The F_{sys} can be bounded by use of Eq. (2.21),

$$F_{sys} \leq \sum_{n=1}^{N} F(C_n), \tag{4.44}$$

so the lower bound of the system reliability R_{sys} is

$$R_{sys} = 1 - F_{sys} \geq 1 - \sum_{n=1}^{N} F(C_n). \tag{4.45}$$

Likewise, use of Eq. (2.22) gives the upper bound of R_{sys} as

$$R_{sys} \leq 1 - \sum_{n=1}^{N} F(C_n) + \sum_{n=1}^{N-1} \sum_{m=n+1}^{N} F(C_n C_m). \tag{4.46}$$

Example 4.10 For the system of Example 4.9, determine (a) a minimum value of the reliability and (b) the reliability of part (a) if all components are independent and subject to random failures with hazard rates λ_n.

(a) From Eq. (4.45),

$$R_{sys} \geq 1 - [F(E_1E_3) + F(E_2E_4) + F(E_1E_5E_4) + F(E_3E_6E_2)].$$

(b) The system reliability for random failures is

$$\begin{aligned} R_{sys}(t) \;\; \geq \;\; & 1 - \{[1 - \exp(-\lambda_1 t)][1 - \exp(-\lambda_3 t)] \\ & + [1 - \exp(-\lambda_2 t)][1 - \exp(-\lambda_4 t)] \\ & + [1 - \exp(-\lambda_1 t)][(1 - \exp(-\lambda_5 t)][(1 - \exp(-\lambda_4 t)] \\ & + [1 - \exp(-\lambda_3 t][1 - \exp(-\lambda_6 t)][1 - \exp(-\lambda_2 t)]\}. \end{aligned}$$

◇

References

[AEC71] "General Design Criteria for Nuclear Power Plants," Title 10, *Code of Federal Regulations*, Part 50, Appendix A, U.S. Atomic Energy Commission (1971).
[Bou72] A. J. Bourne and A. E. Green, *Reliability Technology*, Wiley (1972).
[Sho90] M. L. Shooman, *Probabilistic Reliability: An Engineering Approach*, 2nd ed., Krieger (1990).

Exercises

4.1 If the reliability of each component is R, determine the reliability of (a) system 1 consisting of a series of N subsystems of two components in active parallel and (b) system 2 consisting of an active-parallel configuration of two subsystems, each with N components in series.

4.2 Determine the time-dependent reliability of a system that consists of one component of type 1 in series with an active-parallel combination of two components of type 2. The constant hazard rates of the components are λ_1 and λ_2, respectively.

4.3 Determine the time-dependent reliability of a system that consists of one component of type 1 in series with an active-parallel combination of three components of type 2. The constant hazard rates of the components are λ_1 and λ_2, respectively.

4.4 Determine the time-dependent reliability of a system that consists of one component of type 1 in series with a 2-out-of-3 combination of three components of type 2. The constant hazard rates of the components are λ_1 and λ_2, respectively.

4.5 The probabilities of successful operation of components A, B, and C are 0.9, 0.8, and 0.7, respectively. Determine the probability of success for the system if (a) all components must be functioning and (b) any two of the components must be functioning [Bou72].

4.6 Six identical components each have a reliability R. System 1 has two components in active parallel with each other and connected in series with the remaining four in active parallel with each other. System 2 has two identical subsystems in active parallel with each subsystem consisting of one component connected in series with two components in active parallel with each other. (a) Determine the system reliabilities $R_{sys,1}$ and $R_{sys,2}$ and (b) plot $R_{sys,1}$ and $R_{sys,2}$ versus R.

4.7 A reactor has four identical, independent coolant loops. Each loop consists of three subsystems: the pump S_1, the electrical power S_2 to the pump, and the piping S_3 that circulates the coolant. Over the mission time for operation, the probability of a pump failure is 0.2, of electrical power failure is 0.05, and of pipe failure is 0.001. Assume there is no repair during the operating period of interest. (a) Given that a coolant loop has failed, calculate the probabilities that each subsystem failed if it is known that with an electrical power failure the loop will certainly fail, but only 60% of the pump failures cause a loop failure, and only 40% of the pipe failures cause a loop failure. (b) If all three subsystems function independently, determine the probability of a loop failure over the mission time of operation. (c) If each of the loops operates independently from the others, calculate the probability that at least three of the four will function for the entire mission time.

4.8 For an aircraft with four engines that can land using only two of them, if each engine has a reliability $R = 0.95$ for completing a mission, calculate (a) the reliability of the four-engine system for a mission and (b) the reliability of the system for a mission if the airplane must have at least one active engine on each wing.

4.9 A reactor coolant system has four identical coolant loops, each with two identical pumps connected in active parallel. The reliability of each pump over the life of the plant is 0.8. At least one pump must operate for a loop to be functional, and at least three loops must operate for the coolant system to be functional. Calculate the reliability of the coolant system over the life of the plant.

4.10 A nuclear reactor has three identical coolant loops, each with two identical pumps connected in active parallel. The reliability of each pump over the life of the plant is 0.6. At least one pump must operate for a loop to be functional, and at least two loops must operate for the coolant system to be functional. Calculate the reliability of the coolant system over the life of the plant.

4.11 A nuclear reactor has three identical coolant loops, each with four identical pumps connected in active parallel. The reliability of each pump over the life of the plant is 0.4. At least two pumps must operate for a loop to be functional, and at least two loops must operate for the coolant system to be functional. Calculate the reliability of the coolant system over the life of the plant.

4.12 A reactor coolant system has three identical coolant loops, each with two identical pumps connected in active parallel. At least one pump must operate for a loop to be functional, and at least two loops must operate for the coolant system to be functional. The hazard rate of each pump is λ. For the system, determine (a) $R_{sys}(t)$ and (b) MTTF.

4.13 For the system shown in Fig. 4.13 that consists of components having a constant hazard rate of 10^{-6} per hour, (a) determine $R_{sys}(t)$ and (b) calculate the MTTF.

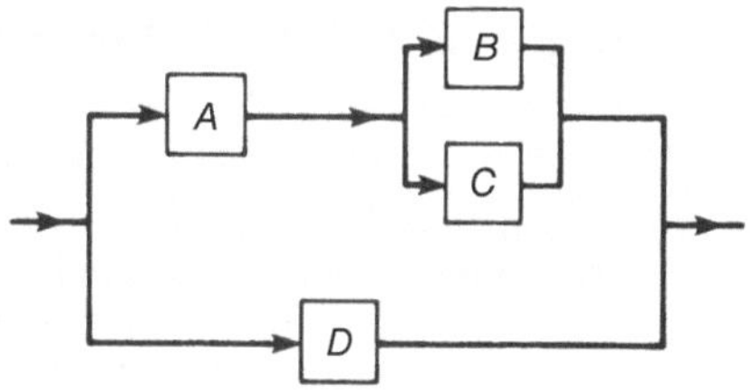

Figure 4.13 Reliability block diagram for Exercise 4.13.

4.14 Three identical units fail randomly with rate λ. If the units for the systems are (1) in active parallel, (2) in active parallel with only two of the three required for successful operation, and (3) with one active unit and two in standby, not subject to standby or switching failures, then (a) derive $R_{sys}(t)$ for each system and (b) calculate the MTTF for each system.

4.15 A reactor has two identical coolant pumps of type 1 that operate in active parallel and fail randomly. If both pumps fail, an emergency pump of type 2 can be valved into operation. The pumps of type 1 each have a MTTF of λ_1^{-1} and the pump of type 2 has a hazard rate of λ_2 during operation and λ_2^* during standby. The failure probability of the valve operation is 0.01. Determine the system reliability as a function of time.

4.16 A main coolant pump in a light water reactor has a hazard rate of 10^{-6}/hr. A backup pump that automatically starts upon failure of the main pump possibly can be used. The hazard rate for the backup pump is 10^{-5}/hr during operation and 10^{-7}/hr during standby. Calculate the reliability of the system after 2000 hr (a) with the backup pump and (b) without the backup pump.

4.17 There are two independent units in a system, with unit 2 in standby. The units fail randomly with hazard rates λ_1 and λ_2, and there are no failures during standby or switching. Derive (a) $R_{sys}(t)$ and (b) MTTF for the system and then (c) compare your result from part (b) for $\lambda_1 = \lambda_2 = \lambda$ to those obtained if the two units would have been connected either in series or in parallel.

4.18 A system consists of a main unit with constant hazard rate λ, a first standby unit with constant hazard rate λ_1, and a second standby unit with constant hazard rate λ_2. If idle standby units undergo no failures, switching will take place when required, and there is no repair of failed units, derive the reliability of the system.

4.19 A system consists of three identical components of type 1 connected in a 2-out-of-3 subsystem and two identical components of type 2 connected in a 1-out-of-2 subsystem used as a backup. All components fail randomly with hazard rate λ_n for type n. The 2-out-of-3 subsystem initially is in operation, and once that subsystem fails, a switch for the 1-out-of-2 subsystem is quickly used in an attempt to activate the backup subsystem. The switching system works on each demand with a reliability of 0.8, but activation can be quickly attempted only three times if necessary or the switching system itself fails. Determine (a) the reliabilities $R_n(t)$ for subsystems $n = 1, 2$ and the switching system reliability R_{sw} and (b) the system reliability $R_{sys}(t)$.

4.20 A system D is like system B shown in Fig. 4.6 except that component 2 is in standby and cross-link L always can be immediately switched over to component 2 whenever component 3 fails. Each component fails randomly with hazard rate λ_n, $n = 1, \ldots, 5$, and the switch is perfectly reliable. (a) Determine the reliability of system D, $R_{sys\,D}$. (b) Determine the MTTF for system D for the case that $\lambda_n = \lambda$, $n = 1, \ldots, 5$. (c) Compare the results of system D to those for systems A, B, and C of Section 4.3.

4.21 Use the decomposition method of Section 4.3 to verify Eq. (4.2) for two units in active parallel.

4.22 In case of loss of offsite power at a nuclear power plant, the peak electrical load to be supplied by the emergency electrical system is 100 kW, with an average load of 70 kW, while the vital functions alone may be run with only 50 kW. To provide high reliability of the emergency power system, three different diesel generator units are proposed: (1) one 100-kW generator, (2) two 50-kW generators, and (3) three 35-kW generators. Assuming that the reliability of the diesel generators is identical, calculate the reliability for each of the three arrangements of the generators for each of the three modes of operation: (a) peak load, (b) average load, and (c) minimum load. Plot the results as a function of the reliability R of each generator.

4.23 Repeat the fail-danger and fail-safe probability calculations of Example 4.3 and Table 4.1 for a 2-out-of-2 system consisting of components 1 and 2. Show that the fail-danger and fail-safe probabilities agree with Eqs. (4.14) and (4.16), respectively, and that probabilities for the three system states sum to unity.

4.24 Repeat Exercise 4.23 for a 2-out-of-3 system.

CHAPTER 5

AVAILABILITY AND RELIABILITY OF SYSTEMS WITH REPAIR

Chapter 4 provided an introduction to the analysis of failures of elementary systems in which the components were connected in a variety of elementary ways. Also discussed were models of the repair of very elementary systems. This chapter gives an introduction to the reliability of somewhat more complicated systems in which repairs and other time-dependent actions can occur during times in which the system operates or does not operate as it did originally.

5.1 INTRODUCTION

The capability of making inspections and repairs of a system is essential for extending the system operating lifetime. Where possible, a good system maintenance procedure will incorporate specific information about the failure rates of the system components in order to schedule periodic inspections. Such procedures are beyond the scope of the focus here, however, so only a general model requiring minimal data input will be developed.

In order to keep the analysis as simple as possible, all time-dependent events such as failures, repairs, inspections, maintenance, etc., will be assumed to be completed at a random time after they were initiated. Thus the hazard rate is $\lambda(t) = \lambda$ and the repair rate is $\mu(t) = \mu$, with the mean time to failure and mean time to repair given by λ^{-1} and μ^{-1}, respectively. Any action that occurs relatively instantaneously can be approximately incorporated into the analysis as long as that action is assigned a very rapid time-dependent rate.

The assumption of a time-independent repair rate of a system component means that every completed repair restores the component to a like-new state, i.e., the repair is perfect. This excludes the possibility of incorporating repairs on a system component that has undergone some aging and is only minimally repaired to its state existing just prior to failure. Such minimal repairs typically are not done in nuclear power systems.

To analyze more general systems in which repairs of subsystems can be done while the system is operating without that subsystem, the definition of reliability is replaced by the definition of availability

$$A(t) = \text{the probability a system or system component performs a specified function or mission under given conditions } \textit{at} \text{ a prescribed time } t.$$

Stated another way, $A(t)$ is the instantaneous probability of the system performing its intended function or mission *at* time t, whereas $R(t)$ is the probability the system performs its intended function or mission *for all times up to* time t. The system unavailability is just the complementary probability to $A(t)$, i.e., $\overline{A}(t) = 1 - A(t)$.

The determination of $A(t)$ and/or $R(t)$ for a system with repairable subsystems is more complicated than that for $R(t)$ for a system without repair because it is necessary to distinguish between the different operating and nonoperating states of the system. In risk and safety analyses, detailed calculations to obtain the instantaneous availability may not be necessary because the *steady-state availability* or asymptotic availability, $A(t) \to A(\infty)$ for $t \to \infty$, may exist and be a satisfactory solution. The value of $A(\infty)$ is a conservative estimate of the probability that the system will be operating at any random time.

Another characteristic of the system is the *interval availability*, which is the fraction of time the system operates over a mission time $(t_2 - t_1)$,

$$\langle A(t_1, t_2) \rangle = (t_2 - t_1)^{-1} \int_{t_1}^{t_2} A(t) \mathrm{d}t. \tag{5.1}$$

The limiting interval availability or *equilibrium availability* is defined by

$$\langle A(0, \infty) \rangle = \lim_{T \to \infty} T^{-1} \int_{0}^{T} A(t) \mathrm{d}t. \tag{5.2}$$

It is the fraction of time the system operates over a very long mission time. For the systems considered here that are assumed to have, for example, time-independent hazard and repair rates, the equilibrium availability equals the steady-state availability,

$$\langle A(0, \infty) \rangle = A(\infty). \tag{5.3}$$

Systems that can be repaired typically require maintenance. If the maintenance does not contribute in a negative way to the system performance (e.g., as with error-free maintenance), then $R(t) \leq A(t) \leq 1$.

In this chapter we will explore a Markov method for incorporating the effects of repair and maintenance activities on the system $A(t)$, $R(t)$, $A(\infty)$, and MTTF.

5.2 MARKOV METHOD

5.2.1 Markov Governing Equations

The mathematics of the Markov chain model for reliability and availability analyses introduced here is relatively elementary given modern computer programs. A discrete-state, continuous-time model is assumed for which the system can be in one of N unique states, $n = 1, \ldots, N$, so for this reason the method employing the Markov method is sometimes called the *system state* method. Each component of the system also is assumed to be in its own unique state. Some examples of the states of a component are: it could be (i) operating, (ii) undergoing repair, maintenance, or inspection, or (iii) in standby awaiting the beginning of a repair, maintenance, or inspection activity. The duration of any repair, maintenance, or inspection activity on a system component is taken to be random, as is failure of the component to continue to operate, so the termination of any component state is assumed to occur randomly after it was initiated.

Because all time-dependent transitions between the different operating and non-operating states are assumed to occur at random times in the Markov method of interest here, the governing equation for the probability that the system is in each possible system state n, either operational or not, is a first-order differential equation with time-independent coefficients. An important manifestation of the random-time model is that the probability for transitioning out of system state m into another system state n, or from state n into state m, depends only on system states m and n and is completely independent of all earlier states the system may have been in. An element of the transition rate matrix for the transition *from* state m and *to* state n will be denoted by M_{nm}.

The fundamental equation for the change in the probability the system is in state n, $P_n(t)$, during the following time interval Δt is

$$\begin{aligned} P_n(t+\Delta t) - P_n(t) &= \sum_{\substack{m=1 \\ m \neq n}}^{N} M_{nm} \Delta t P_m(t) \\ &- \sum_{\substack{m=1 \\ m \neq n}}^{N} M_{mn} \Delta t P_n(t), \quad n = 1, \ldots, N. \end{aligned} \tag{5.4}$$

The first term on the right-hand-side denotes the transitions into state n and the second term represents the transitions out of state n. Division by Δt, followed by the limit as $t \to 0$, gives

$$\frac{\mathrm{d}P_n(t)}{\mathrm{d}t} = \sum_{\substack{m=1\\m\neq n}}^{N} M_{nm}P_m(t) - \sum_{\substack{m=1\\m\neq n}}^{N} M_{mn}P_n(t), \quad n = 1, \ldots, N. \tag{5.5}$$

A quick check on the consistency of the set of equations (5.5) follows by summing them over n to obtain

$$\frac{\mathrm{d}}{\mathrm{d}t}\sum_{n=1}^{N} P_n(t) = \sum_{n=1}^{N}\sum_{\substack{m=1\\m\neq n}}^{N} M_{nm}P_m(t) - \sum_{n=1}^{N}\sum_{\substack{m=1\\m\neq n}}^{N} M_{mn}P_n(t) = 0 \tag{5.6}$$

following an interchange of indices n and m. Thus, integration over time gives

$$\sum_{n=1}^{N} P_n(t) = \text{constant} = 1, \tag{5.7}$$

with the constant selected to be consistent with the addition axiom for probabilities in Section 2.3. Another consistency check needed to preserve the state of a system that undergoes no change requires that

$$M_{nn} = -\sum_{\substack{m=1\\m\neq n}}^{N} M_{mn}, \tag{5.8}$$

so that

$$\sum_{m=1}^{N} M_{mn} = 0. \tag{5.9}$$

With this notation the set of equations for the probability the system is in state n can be succinctly written as

$$\frac{\mathrm{d}P_n(t)}{\mathrm{d}t} = \sum_{m=1}^{N} M_{nm}P_m(t). \quad n = 1, \ldots, N. \tag{5.10}$$

In matrix form the equations are

$$\mathrm{d}\mathbf{P}(t)/\mathrm{d}t = \mathbf{MP}(t), \tag{5.11}$$

where

$$\mathbf{P}(t) = [P_1(t), P_2(t), \ldots, P_N(t)]^T \tag{5.12}$$

is the vector of the state probabilities, with superscript T denoting the transpose. The transition rate matrix $\mathbf{M}$ is a square matrix of elements M_{nm} and size $N \times N$,

$$\mathbf{M} = \begin{bmatrix} M_{11} & M_{12} & M_{13} & \cdots & M_{1N} \\ M_{21} & M_{22} & M_{23} & \cdots & M_{2N} \\ M_{31} & M_{32} & M_{33} & \cdots & M_{3N} \\ \vdots & \vdots & \vdots & \vdots & \vdots \\ M_{N1} & M_{N2} & M_{N3} & \cdots & M_{NN} \end{bmatrix}. \tag{5.13}$$

To solve the first-order-in-time matrix equation (5.11), an initial condition must be imposed. *For consistency in working with all transition rate matrices in this chapter:* The labeling of system states adopted here is that state 1 will be the original "as-good-as-new" state in which all components are operating or operable, followed by the system state(s) in which the system is operating while undergoing some repair or maintenance procedure, and followed by the system state(s) in which the system is not operating while undergoing some repair or maintenance procedure. A typical Markov analysis assumes that system state 1 is the initial operating state, with the initial condition

$$\mathbf{P}(0) = [1, 0, \ldots, 0]^T. \tag{5.14}$$

The solution of Eq. (5.11) subject to the initial condition of Eq. (5.14) for the system availability $A(t)$ or the system reliability $R(t)$ is

$$A(t) = \sum_{n=1}^{N_u} P_n(t) \qquad \text{or} \qquad R(t) = \sum_{n=1}^{N_u} P_n(t), \tag{5.15}$$

where N_u, $N_u < N$, is the number of states in which the system is operating. The corresponding system unavailability $\overline{A}(t)$ and system unreliability $\overline{R}(t)$ are

$$\overline{A}(t) = 1 - A(t) = \sum_{N_u+1}^{N} P_n(t) \quad \text{and} \quad \overline{R}(t) = 1 - R(t) = \sum_{N_u+1}^{N} P_n(t). \tag{5.16}$$

The values of $P_n(t)$ in Eqs. (5.15) and (5.16) for an availability analysis differ from those for a reliability analysis because the transition rate matrices $\mathbf{M}$ are different, so we will denote the transition rate matrices for an availability analysis and a reliability analysis by $\mathbf{M}_A$ and $\mathbf{M}_R$, respectively.

For safety systems that typically have more operating states than failed states, the system availability or reliability may be easier to evaluate with Eq. (5.16) than with Eq. (5.15).

5.2.2 Solution of Markov Governing Equations

The general solution of Eq. (5.11) subject to the initial condition of Eq. (5.14) can be written as

$$\mathbf{P}(t) = \exp(\mathbf{M}t)\mathbf{P}(0), \tag{5.17}$$

where the matrix exponential function is a $N \times N$ matrix defined by

$$\exp(\mathbf{M}t) = \mathbf{I} + \mathbf{M}t + \frac{\mathbf{M}^2t^2}{2!} + \frac{\mathbf{M}^3t^3}{3!} + \cdots = \sum_{n=0}^{\infty} \frac{\mathbf{M}^n t^n}{n!}, \tag{5.18}$$

with $\mathbf{I}$ the unit diagonal matrix. Here, for example, $\mathbf{M}^2$ is the matrix product $\mathbf{MM}$ which is a matrix with elements $[\mathbf{M}^2]_{nm} = \sum_{j=1}^{N} M_{nj}M_{jm}$. Because $\exp(\mathbf{M}t) = \lim_{n\to\infty}(\mathbf{I} + \mathbf{M}t/n)^n$, it follows that an approximate $\mathbf{P}(t)$ is given by

$$\mathbf{P}(t) \approx \left(\mathbf{I} + \frac{\mathbf{M}t}{n}\right)^n \mathbf{P}(0) \tag{5.19}$$

for sufficiently large n.

The matrix exponential method is used in some computer programs to evaluate a vector such as $\mathbf{P}(t)$, but it also can be used to obtain a quick, approximate estimate of the availability or reliability as a function of time if the system is known to initially be in state 1 so that Eq. (5.14) is valid. Then the elements of $\mathbf{P}(t)$ needed for evaluating the availability or reliability from Eq. (5.15) are

$$\begin{aligned} P_n(t) &= \delta_{n1} + M_{n1}t + \sum_{j=1}^{N} \frac{M_{nj}M_{j1}t^2}{2!} + \sum_{j=1}^{N}\sum_{k=1}^{N} \frac{M_{nj}M_{jk}M_{k1}t^3}{3!} \\ &\quad + \sum_{j=1}^{N}\sum_{k=1}^{N}\sum_{\ell=1}^{N} \frac{M_{nj}M_{jk}M_{k\ell}M_{\ell 1}t^4}{4!} + \cdots, \quad n = 1, \ldots, N, \end{aligned} \tag{5.20}$$

where $\delta_{nm} = 1$, $n = m$, and $\delta_{nm} = 0$ otherwise. Thus, for example, from Eqs. (5.15) and (5.20), the system availability is

$$\begin{aligned} A(t) &= \sum_{n=1}^{N_u} \left[\delta_{n1} + M_{n1}t + \sum_{j=1}^{N} \frac{M_{nj}M_{j1}t^2}{2!} + \sum_{j=1}^{N}\sum_{k=1}^{N} \frac{M_{nj}M_{jk}M_{k1}t^3}{3!} \right. \\ &\quad \left. + \sum_{j=1}^{N}\sum_{k=1}^{N}\sum_{\ell=1}^{N} \frac{M_{nj}M_{jk}M_{k\ell}M_{\ell 1}t^4}{4!} + \cdots \right]. \end{aligned} \tag{5.21}$$

We turn now to the use of Laplace transforms to solve Eq. (5.11) subject to the initial condition of Eq. (5.14). Computer programs are available for this purpose, but an analytical approach works when the size of the matrix is not large. For our purposes we need only the definition of the Laplace transform,

$$\mathcal{L}[g(t)] = g(s) = \int_0^\infty g(t)\exp(-st)\mathrm{d}t, \tag{5.22}$$

and the property

$$\mathcal{L}[\mathrm{d}g(t)/\mathrm{d}t] = sg(s) - g(0). \tag{5.23}$$

The inverse Laplace transform for constant a is

$$\mathcal{L}^{-1}[(s-a)^{-k}] = t^{k-1}\exp(at)/(k-1)!, \quad k = 1,\ 2,\ \ldots\ . \tag{5.24}$$

Equation (5.23) enables the first-order differential equation (5.11) to be converted to

$$(s\mathbf{I} - \mathbf{M})\mathbf{P}(s) = \mathbf{P}(0) = [1, 0, \ldots, 0]^T \tag{5.25}$$

after use of Eq. (5.14). This matrix equation can be inverted to give

$$\mathbf{P}(s) = (s\mathbf{I} - \mathbf{M})^{-1}\mathbf{P}(0) \tag{5.26}$$

with matrix elements

$$P_n(s) = [\mathrm{cof}(s\mathbf{I} - \mathbf{M})^T]_{n1}/\Delta, \quad n = 1, \ldots, N. \tag{5.27}$$

The Δ in Eq. (5.27) is

$$\Delta = (s - s_1)(s - s_2)\cdots(s - s_N), \tag{5.28}$$

where the eigenvalues s_j, $j = 1, \ldots, N$, of Eq. (5.26) are obtained from

$$|s\mathbf{I} - \mathbf{M}| = 0. \tag{5.29}$$

For an availability analysis, one eigenvalue always will be zero and leads to the steady-state availability.

As a reminder, the elements of any $N \times N$ matrix $\mathbf{M}^T$ are related to those of $\mathbf{M}$ by $M^T_{nm} = M_{mn}$ and

$$|\mathbf{M}| = \sum_{j=1}^{N} M_{nj}(\text{cof }\mathbf{M})_{nj} = \sum_{j=1}^{N} M_{jm}(\text{cof }\mathbf{M})_{jm}. \tag{5.30}$$

The cofactor $(\text{cof }\mathbf{M})_{ij}$ for element M_{ij} is obtained by deleting the ith row and jth column, calculating the determinant of the remaining array, and multiplying by $(-1)^{i+j}$. For example, the cofactor $(\text{cof }\mathbf{M}_A)_{23}$ for $\mathbf{M}_A$ of Eq. (5.13) is

$$(\text{cof }\mathbf{M}_A)_{23} = (-1)^{2+3} \begin{bmatrix} M_{11} & M_{12} & M_{14} & \cdots & M_{1N} \\ M_{31} & M_{32} & M_{34} & \cdots & M_{3N} \\ M_{41} & M_{42} & M_{44} & \cdots & M_{4N} \\ \vdots & \vdots & \vdots & \vdots & \vdots \\ M_{N1} & M_{N2} & M_{N4} & \cdots & M_{NN} \end{bmatrix}. \tag{5.31}$$

Once the $P_n(s)$ of Eq. (5.27) have been determined in terms of the eigenvalues s_j, $j = 1, \ldots, N$, of Eq. (5.28), a partial fraction decomposition needs to be performed before Eq. (5.24) can be used. If the eigenvalues are all different, the result is a series of the form

$$P_n(s) = \sum_{j=1}^{N} b_{nj}(s - s_j)^{-1}, \tag{5.32}$$

where the b_{nj} are constants. The coefficients b_{nj} may be obtained by simple use of the residue theorem of complex variables

$$b_{nj} = \lim_{s \to s_j} (s - s_j)\, P_n(s). \tag{5.33}$$

Application of the inverse Laplace transform of Eq. (5.24) to $P_n(s)$ then gives the probability the system is in the nth state as

$$P_n(t) = \sum_{j=1}^{N} b_{nj} \exp(s_j t), \quad n = 1, \ldots, N. \tag{5.34}$$

The eigenvalue $s_1 = 0$ leads to the asymptotic solution

$$P_n(\infty) = \lim_{s \to 0+} sP_n(s) = b_{n1}. \tag{5.35}$$

5.2.3 An Elementary Example

Example 5.1 A system consists of a single component that can be repaired, as illustrated by the state transition diagram of Fig. 5.1. (a) Define the system states, (b) construct the transition rate matrix $\mathbf{M}_A$ for an availability analysis, (c) obtain an approximation for the time-dependent availability to $O(t^2)$ using the matrix exponential approach, (d) obtain the time-dependent availability using the Laplace transform approach, and (e) repeat parts (a) to (d) for a reliability analysis.

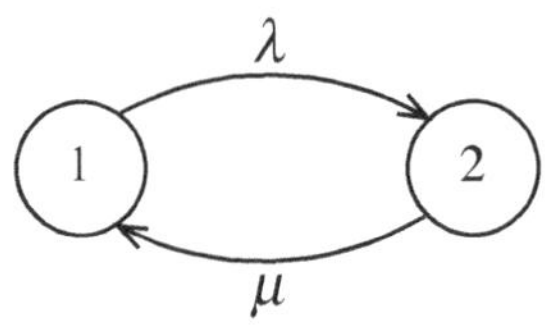

Figure 5.1 State transition diagram illustrating transitions between operational state 1 and repair state 2 of a single component.

(a) The system states are:

State	System	Components
1	Operating	Component operating
2	Not operating	Component in repair

(b) The transition rate matrix is

$$\mathbf{M}_A = \begin{bmatrix} -\lambda & \mu \\ \lambda & -\mu \end{bmatrix}.$$

(c) With

$$\mathbf{M}_A\mathbf{M}_A = \begin{bmatrix} \lambda(\mu+\lambda) & -\mu(\mu+\lambda) \\ -\lambda(\mu+\lambda) & \mu(\mu+\lambda) \end{bmatrix} = -(\mu+\lambda)\mathbf{M}_A,$$

it follows from Eq. (5.21) that

$$\begin{aligned} A(t) &\approx \sum_{n=1}^{1}\left[\delta_{n1} + M_{n1}t + \sum_{j=1}^{2}\frac{M_{nj}M_{j1}t^2}{2!}\right] \\ &\approx 1 - \lambda t + \lambda(\mu+\lambda)t^2/2. \end{aligned}$$

(d) Equation (5.29),

$$\begin{vmatrix} s+\lambda & -\mu \\ -\lambda & s+\mu \end{vmatrix} = 0,$$

gives the eigenvalues $s_1 = 0$ and $s_2 = -(\lambda + \mu)$. From Eqs. (5.27), (5.28), and (5.15) and the fact that there is only one upstate, $A(s)$ is given by

$$\begin{aligned} A(s) &= P_1(s) = [\text{cof}(s\mathbf{I} - \mathbf{M})^T]_{11}/\Delta \\ &= \frac{s+\mu}{s(s+\lambda+\mu)} \\ &= (\lambda+\mu)^{-1}[\mu/s + \lambda/(s+\lambda+\mu)] \end{aligned}$$

so Eqs. (5.32) and (5.34) give

$$A(t) = (\lambda+\mu)^{-1}\{\mu + \lambda \exp[-(\lambda+\mu)t]\}. \tag{5.36}$$

(e) For a reliability analysis,

(a′) State 2 corresponds to failure of the system

(b′) $\mathbf{M}_R = \begin{bmatrix} -\lambda & 0 \\ \lambda & 0 \end{bmatrix}.$

Again from Eq. (5.21), or by setting $\mu = 0$ in the corresponding results for $A(t)$, it is not surprising that

(c′) $R(t) \approx 1 - \lambda t + (\lambda t)^2/2,$

(d′) $R(t) = \exp(-\lambda t).$

A comparison of the time-dependent availability and reliability is shown in Fig. 5.2, where, for long times,

$$A(t) \to A(\infty) = \mu/(\lambda+\mu). \quad \diamond$$

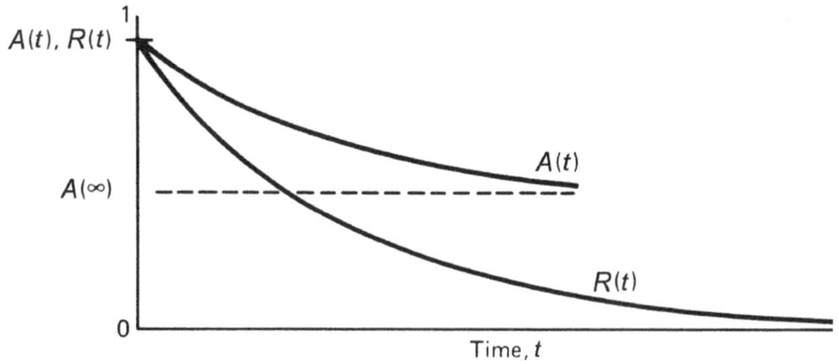

Figure 5.2 Time-dependent availability and reliability of a single unit (schematic).

For the preceding example it is also instructive to evaluate the time-dependent unavailability

$$\bar{A}(t) = 1 - A(t) = \lambda\,(\lambda+\mu)^{-1}\,[1 - \exp[-\,(\lambda+\mu)\,t] \tag{5.37}$$

and the steady-state unavailability

$$\bar{A}(\infty) = \lambda (\lambda + \mu)^{-1} \simeq \lambda \cdot \mathrm{MTTR}. \tag{5.38}$$

The last expression in Eq. (5.38) applies when $\mathrm{MTTR} = 1/\mu$ is much smaller than $\mathrm{MTTF} = 1/\lambda$, which is usually the case for well-maintained engineered systems. Equations (5.37) and (5.38) are used in fault tree analyses for system unavailability, as discussed further in Chapter 7. The result is also intuitive because the fraction of the time that the system is unavailable may effectively be calculated as

$$\begin{aligned} \text{unavailability} \simeq &\{\text{frequency of system failure}\} \\ &\times \{\text{average time required for repair}\}. \end{aligned}$$

Example 5.2 Obtain the solution for the system availability of Example 5.1 alternatively by the integral approach of Eq. (4.18) combining reliability $R(t) = e^{-\lambda t}$ of Eq. (2.84) with repairability $\tilde{R}(t) = 1 - e^{-\mu t}$ from Eqs. (2.96) and (2.97).

In this approach, the probability $P_1(t) = A(t)$ that the system is in operation at time t may be written as the sum of the probability that the system remains in operation up to time t and the probability that the system is restored from the repair state in $\mathrm{d}\tau$ about τ and remains in the operating state, after repair, for interval $t - \tau$

$$P_1(t) = R(t) + \int_0^t \mu P_2(\tau) R(t - \tau)\, \mathrm{d}\tau. \tag{5.39}$$

The convolution theorem of Laplace transform yields

$$P_1(s) = \frac{1}{s + \lambda} + \frac{\mu}{s + \lambda} P_2(s). \tag{5.40}$$

A similar balance equation for the repair state probability $P_2(t)$ may be obtained via $\tilde{R}(t) = 1 - e^{-\mu t}$ and substituted into Eq. (5.39) or (5.40) to yield Eq. (5.36). The normalization condition of Eq. (5.7) may be used equivalently in Eq. (5.39) to obtain the same result. ◇

5.3 AVAILABILITY ANALYSES

We first illustrate the construction of matrices $\mathbf{M}_A$ for availability analyses before considering solutions of examples for $A(t)$.

5.3.1 Rules for Constructing Transition Rate Matrices

Some rules for constructing transition rate matrices for time-dependent Markov analyses are:

- All matrix elements must be transition rates with dimensions of inverse time.
- Matrix element M_{nm} is the transition rate from state m to state n if $m \neq n$ and M_{nn} is the transition rate out of state n, so every diagonal element must be negative and every off-diagonal element must be positive.

- The matrix element 2λ, for example, means there are two independent components simultaneously in operation.
- All matrix elements in every column of $\mathbf{M}$ must sum to zero.

5.3.2 Availability Transition Rate Matrices

Example 5.3 A system consists of components 1 and 2 that are connected in active parallel. Each component is either in operation or under repair, with hazard rates λ_1 and λ_2 and repair rates μ_1 and μ_2, respectively. (a) Define the system states and (b) construct the transition rate matrix for an availability analysis, $\mathbf{M}_A$.

(a) The system states are:

State	System	Components
1	Operating	Components 1 and 2 operating
2	Operating	Component 2 operating, component 1 in repair
3	Operating	Component 1 operating, component 2 in repair
4	Not operating	Components 1 and 2 in repair

(b) The transition rate matrix is

$$\mathbf{M}_A = \begin{bmatrix} -(\lambda_1+\lambda_2) & \mu_1 & \mu_2 & 0 \\ \lambda_1 & -(\lambda_2+\mu_1) & 0 & \mu_2 \\ \lambda_2 & 0 & -(\lambda_1+\mu_2) & \mu_1 \\ 0 & \lambda_2 & \lambda_1 & -(\mu_1+\mu_2) \end{bmatrix}$$

where matrix element $M_{11} = -(\lambda_1 + \lambda_2)$ because the probability of component 1 failing in time interval Δt is $\lambda_1 \Delta t$, and similarly for component 2, but from Eq. (2.15) the probability that both fail in Δt is $\lambda_1 \Delta t + \lambda_2 \Delta t - \lambda_1 \lambda_2 (\Delta t)^2$ and the second-order term in Δt vanishes in the limit taken in Eq. (5.5). The same argument can be used to justify the matrix elements M_{nn}, $n = 2, 3, 4$. ◇

Example 5.4 The active-parallel system of Example 5.3 consists of two identical components, each with a hazard rate λ during operation and a repair rate μ, as illustrated in the state transition diagram of Fig. 5.3. (a) Define the system states, (b) construct the transition rate matrix $\mathbf{M}_A$ for an availability analysis, and (c) consider the changes that would occur in the analysis if only one component can be repaired at any time.

(a) The system states are:

State	System	Components
1	Operating	Both operating
2	Operating	One operating, one in repair
3	Not operating	Both in repair

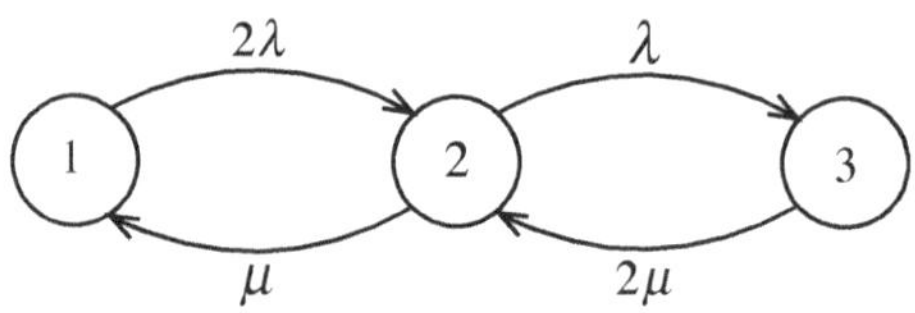

Figure 5.3 State transition diagram for the three-state system of Example 5.4.

Because the components are identical the number of system states is smaller than for Example 5.3.

(b) The transition rate matrix is

$$\mathbf{M}_A = \begin{bmatrix} -2\lambda & \mu & 0 \\ 2\lambda & -(\lambda+\mu) & 2\mu \\ 0 & \lambda & -2\mu \end{bmatrix}.$$

(c) System state 3 in part (a) becomes "One component in repair, the other waiting for repair" so

$$\mathbf{M}_A = \begin{bmatrix} -2\lambda & \mu & 0 \\ 2\lambda & -(\lambda+\mu) & \mu \\ 0 & \lambda & -\mu \end{bmatrix}. \quad \diamond$$

For a system consisting of two identical components, it is educational to examine the result from operating them with one component in standby instead of both in active parallel.

Example 5.5 A system consists of two identical components, with one component in standby until needed. Each component has a hazard rate of λ during operation, a hazard rate of λ^* during standby, and a repair rate of μ. (a) Define the system states and (b) construct the transition rate matrix $\mathbf{M}_A$ for an availability analysis.

(a) The system states are:

State	System	Components
1	Operating	One operating, one in standby
2	Operating	One operating, one in repair
3	Not operating	Both components in repair

(b) The transition rate matrix is

$$\mathbf{M}_A = \begin{bmatrix} -(\lambda+\lambda^*) & \mu & 0 \\ (\lambda+\lambda^*) & -(\lambda+\mu) & 2\mu \\ 0 & \lambda & -2\mu \end{bmatrix}.$$

If λ^* is smaller than λ, as is usually the case, this system will perform better than one with both components in active parallel. Comparison of Examples

5.4 and 5.5 shows that the state transition matrix of the system of Example 5.5 with a standby component can be converted to that of Example 5.4 with both components in active parallel if $\lambda^* = \lambda$. $\diamond$

Because of a common manufacturing defect or another flaw, sometimes components possibly can fail because their performance is coupled to other components in the system. For example, two components with failure rates λ in active operation might fail in time Δt with a probability $(2\lambda + \lambda_c)\Delta t$, where λ_c is the hazard rate arising from the coupling between them.

Example 5.6 A system consists of two identical components, with one in standby. Each component has a hazard rate of λ during operation and λ^* during standby, and when both are operable they can fail with hazard rate λ_c because of a common flaw. Only one component can be repaired at a time, with a repair rate of μ that is independent of the cause of failure. (a) Define the system states and (b) construct the transition rate matrix $\mathbf{M}_A$ for an availability analysis.

(a) The system states are the same as in Example 5.4.

(b) The transition rate matrix is

$$\mathbf{M}_A = \begin{bmatrix} -(\lambda + \lambda^* + \lambda_c) & \mu & 0 \\ (\lambda + \lambda^* + \lambda_c) & -(\lambda + \mu) & \mu \\ 0 & \lambda & -\mu \end{bmatrix}. \quad \diamond$$

Example 5.7 A system consists of two identical components, with one component in standby until needed. Following the repair of a component, the component must be recertified by testing before it can be placed back in operation or in standby. The MTTF for each component in operation is λ^{-1} and in standby is $(\lambda^*)^{-1}$, the MTTR is μ^{-1}, and the mean time for testing is τ^{-1}, as illustrated in the state transition diagram of Fig. 5.4. Both components can undergo repair simultaneously if necessary. For an availability analysis, (a) define the system states and (b) construct the transition rate matrix $\mathbf{M}_A$ for an availability analysis.

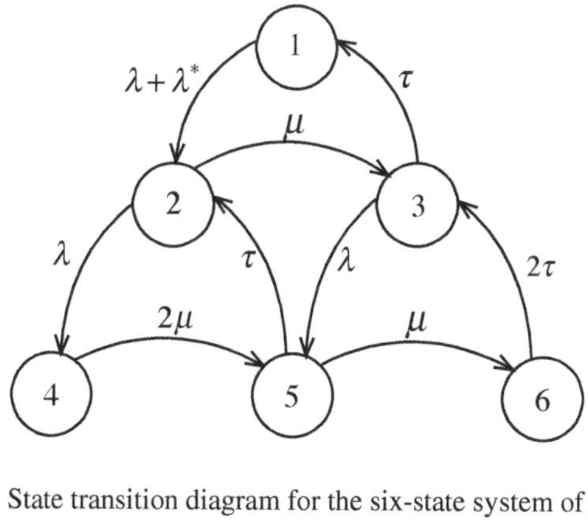

Figure 5.4 State transition diagram for the six-state system of Example 5.7.

(a) The system states are:

State	System	Components
1	Operating	One in operation, one in standby
2	Operating	One in operation, one in repair
3	Operating	One in operation, one in testing
4	Not operating	Both in repair
5	Not operating	One in repair, one in testing
6	Not operating	Both in testing

Now $N_u = 3$ and $N = 6$ for Eqs. (5.15) and (5.16).

(b) The transition rate matrix is:

$$\mathbf{M}_A = \begin{bmatrix} -(\lambda+\lambda^*) & 0 & \tau & 0 & 0 & 0 \\ (\lambda+\lambda^*) & -(\lambda+\mu) & 0 & 0 & \tau & 0 \\ 0 & \mu & -(\lambda+\tau) & 0 & 0 & 2\tau \\ 0 & \lambda & 0 & -2\mu & 0 & 0 \\ 0 & 0 & \lambda & 2\mu & -(\mu+\tau) & 0 \\ 0 & 0 & 0 & 0 & \mu & -2\tau \end{bmatrix}.$$

With a larger number of system states, it is helpful to construct a state transition diagram in order to obtain $\mathbf{M}_A$. ◇

Example 5.8 A system consists of two identical components, with one component in standby. Following the failure of a component, a repair facility must be located before the repair can begin, but after the repair is completed the component can be immediately placed back in operation or in standby. The MTTF for each component in operation is λ^{-1} and in standby is $(\lambda^*)^{-1}$, the mean time for locating a repair facility is σ^{-1}, and the MTTR is μ^{-1}. For an availability analysis, (a) define the system states and (b) construct the transition rate matrix $\mathbf{M}_A$ for an availability analysis.

(a) The system states are:

State	System	Components
1	Operating	One in operation, one in standby
2	Operating	One in operation, one awaiting repair
3	Operating	One in operation, one in repair
4	Not operating	Both awaiting repair
5	Not operating	One in repair, one awaiting repair
6	Not operating	Both in repair

Again $N_u = 3$ and $N = 6$ for Eqs. (5.15) and (5.16).

(b) Transition rate matrix $\mathbf{M}_A$ follows immediately from that of Example 5.7 after changing μ to σ and τ to μ. ◇

Example 5.9 A system consists of three identical components connected in active parallel. The components fail differently, depending on the load on each component, which, in turn, depends on how many are in operation. The instantaneous failure rate of a component is λ_i, $i = 1, 2, 3$, if $i - 1$ components have failed when there are $4 - i$ components in operation. The repair rate of the components depends on the load when they failed, with repair rate μ_i, $i = 1, 2, 3$, for components that failed under load i. For an availability analysis, (a) define the system states and (b) construct the transition rate matrix $\mathbf{M}_A$.

(a) The system states are:

State	System	Components
1	Operating	Three in operation at load 1
2	Operating	Two in operation at load 2, one in repair after load 1 failure
3	Operating	One in operation at load 3, one in repair after load 1 failure, one in repair after load 2 failure
4	Not operating	One in repair after load 1 failure, one in repair after load 2 failure, one in repair after load 3 failure

Now $N_u = 3$ and $N = 4$ for Eqs. (5.15) and (5.16).

(b) The transition rate matrix is:

$$\mathbf{M}_A = \begin{bmatrix} -3\lambda_1 & \mu_1 & 0 & 0 \\ 3\lambda_1 & -(2\lambda_2 + \mu_1) & \mu_1 + \mu_2 & 0 \\ 0 & 2\lambda_2 & -(\lambda_3 + \mu_1 + \mu_2) & \mu_1 + \mu_2 + \mu_3 \\ 0 & 0 & \lambda_3 & -(\mu_1 + \mu_2 + \mu_3) \end{bmatrix}. \quad \diamond$$

Example 5.10 A system consists of component 1 in series with components 2 and 3 that are in active parallel. The instantaneous component failure rates and repair rates are λ_n and μ_n, $n = 1,\ 2,\ 3$. Construct a state transition diagram. See Fig. 5.5. $\diamond$

5.3.3 Time-Dependent Availability Examples

Examples of $A(t)$ obtained by the Laplace transform procedure of Eqs. (5.25) to (5.34) are in Table 5.1, where $A(\infty)$ is the time-independent term in $A(t)$.

For the case of a one-component (1:) system with no standby component (0s) and one repairman (1r), the steady-state availability from Table 5.1 is

$$A_{1:0s,1r}(\infty) = \frac{\mu}{\lambda + \mu} = \frac{\text{MTTR}^{-1}}{\text{MTTF}^{-1} + \text{MTTR}^{-1}} = \frac{\text{MTTF}}{\text{MTTF} + \text{MTTR}}, \quad (5.41)$$

Table 5.1 Availability of Systems Consisting of Identical Components with $\mathrm{MTTF} = \lambda^{-1}$ and $\mathrm{MTTR} = \mu^{-1}$ with No Failures During Standby

Number of Units	System Type	Availability $A(t)$	Nonzero Eigenvalues
1	1 repairman	$\dfrac{\mu}{\lambda+\mu} + \dfrac{\lambda \exp(s_1 t)}{\lambda+\mu}$	$s_1 = -(\lambda+\mu)$
2	1 standby, 1 repairman	$\dfrac{\mu^2+\mu\lambda}{\mu^2+\mu\lambda+\lambda^2} - \dfrac{\lambda^2[s_2\exp(s_1 t) - s_1\exp(s_2 t)]}{s_1 s_2(s_1-s_2)}$	$s_1 = -(\lambda+\mu+\sqrt{\lambda\mu})$ $s_2 = -(\lambda+\mu-\sqrt{\lambda\mu})$
2	1 standby, 2 repairmen	$\dfrac{2\mu^2+2\mu\lambda}{2\mu^2+2\mu\lambda+\lambda^2} - \dfrac{\lambda^2[s_2\exp(s_1 t) - s_1\exp(s_2 t)]}{s_1 s_2(s_1-s_2)}$	$s_1 = -0.5(2\lambda+3\mu+\sqrt{4\lambda\mu+\mu^2})$ $s_2 = -0.5(2\lambda+3\mu-\sqrt{4\lambda\mu+\mu^2})$
2	2 active parallel, 1 repairman	$\dfrac{\mu^2+2\mu\lambda}{\mu^2+2\mu\lambda+2\lambda^2} - \dfrac{2\lambda^2[s_2\exp(s_1 t) - s_1\exp(s_2 t)]}{s_1 s_2(s_1-s_2)}$	$s_1 = -0.5(3\lambda+2\mu+\sqrt{4\lambda\mu+\lambda^2})$ $s_2 = -0.5(3\lambda+2\mu-\sqrt{4\lambda\mu+\lambda^2})$
2	2 active parallel, 2 repairmen	$\dfrac{\mu^2+2\mu\lambda}{\mu^2+2\mu\lambda+\lambda^2} - \dfrac{2\lambda^2[s_2\exp(s_1 t) - s_1\exp(s_2 t)]}{s_1 s_2(s_1-s_2)}$	$s_1 = -2(\mu+\lambda)$ $s_2 = -(\mu+\lambda)$

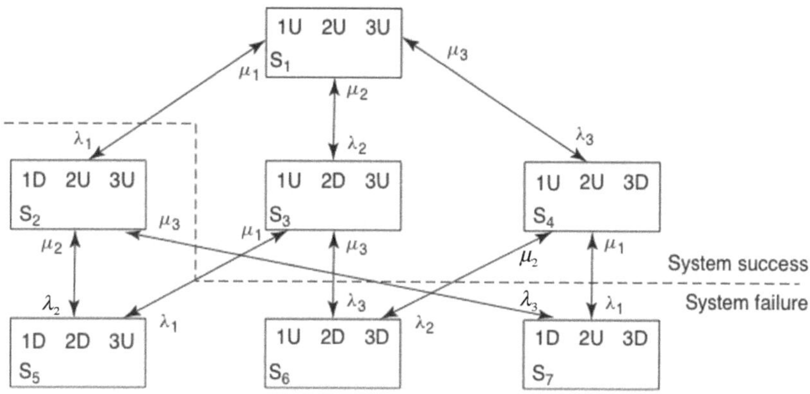

Figure 5.5 State transition diagram for Example 5.10.

Figure 5.5 State transition diagram for Example 5.10. *Source:* Reprinted with permission from [IEE98]. Copyright © 1998 The Institute of Electrical and Electronics Engineers.

which is just the long-time average fraction of time the component is available. Also of interest is an interval estimate from Eq. (5.1),

$$\begin{aligned}\langle A_{1:0s,1r}(0,T)\rangle &= T^{-1}\int_0^T A_{1:0s,1r}(t)\mathrm{d}t = \frac{\mu + T^{-1}[1-\exp(-\lambda t)]}{\lambda+\mu} \\ &\approx 1-[\lambda/(\lambda+\mu)](\lambda T/2), \quad \lambda T \ll 1, \end{aligned} \tag{5.42}$$

which can be viewed as the equilibrium availability of a single component device that is inspected after every time period T. By comparison, the interval reliability is

$$\begin{aligned}\langle A_{1:0s,0r}(0,T)\rangle &= \langle R_{1:0s,0r}(0,T)\rangle = T^{-1}\int_0^T R_{1:0s,0r}(t)\mathrm{d}t \\ &\approx 1-(\lambda T/2), \quad \lambda T \ll 1, \end{aligned} \tag{5.43}$$

which shows how repairs help to increase the interval availability.

Comparisons of steady-state availabilities for different systems also can be obtained from Table 5.1.

Example 5.11 Compare the steady-state availabilities of:

(a) Two components with one in standby and with one repairman versus one component with none in standby and with one repairman
(b) Two components with one in standby and with two repairmen versus two components with one in standby and with one repairman
(c) Two components with one in standby and with one repairman versus two components in active parallel and none in standby and with one repairman
(d) Two components with one in standby and with two repairmen versus two components in active parallel and none in standby and with two repairmen

From the leading terms in the $A(t)$ of Table 5.1, algebra gives the results

$$\begin{aligned}
(a)\quad & \frac{A_{2:1s,1r}(\infty)}{A_{1:0s,1r}(\infty)} && = 1 + \frac{\lambda\mu}{\lambda^2 + \lambda\mu + \mu^2}, \\
(b)\quad & \frac{A_{2:1s,2r}(\infty)}{A_{2:1s,1r}(\infty)} && = 1 + \frac{\lambda^2}{\lambda^2 + 2\lambda\mu + 2\mu^2}, \\
(c)\quad & \frac{A_{2:1s,1r}(\infty)}{A_{2:0s,1r}(\infty)} && = 1 + \frac{\lambda^2\mu}{2\lambda^3 + 3\lambda^2\mu + 3\lambda\mu^2 + \mu^3}, \\
(d)\quad & \frac{A_{2:1s,2r}(\infty)}{A_{2:0s,2r}(\infty)} && = 1 + \frac{\lambda^2\mu}{2\lambda^3 + 5\lambda^2\mu + 6\lambda\mu^2 + 2\mu^3}. \quad \diamond
\end{aligned}$$

Example 5.12 The active-parallel system of Example 5.4 consists of identical components, each with a hazard rate λ during operation and a repair rate μ. If the system is initially "as-good-as-new" at time $t = 0$, obtain an approximate estimate of $A(t)$ with (a) Eq. (5.15) and (b) Eq. (5.16) for times such that terms of $O(t^n) \approx 0$ for $n \geq 4$.

(a) Because there are only two operating states, $N_u = 2$ in Eq. (5.20) so after evaluating the elements needed in Eq. (5.21),

$$\begin{aligned}
\sum_{n=1}^{2} M_{n1} &= 0, \\
\sum_{n=1}^{2}\sum_{j=1}^{3} M_{nj}M_{j1} &= -2\lambda^2, \\
\sum_{n=1}^{2}\sum_{j=1}^{3}\sum_{k=1}^{3} M_{nj}M_{jk}M_{k1} &= 6\lambda^2(\lambda+\mu).
\end{aligned}$$

From Eq. (5.15) we obtain

$$A(t) = P_1(t) + P_2(t) \approx 1 - (\lambda t)^2 + \lambda^2(\lambda + \mu)t^3.$$

This result confirms our intuition that, to $O(t^2)$, the system is available until both components of the active-parallel system have failed and that the first contribution to availability from a repair is of $O(t^3)$.

(b) State 3 is the only state for which the system has failed, so $\overline{A}(t) = P_3(t)$ and $A(t) = 1 - P_3(t)$. After evaluating the elements needed for $P_3(t)$ in Eq. (5.20),

$$\begin{aligned}
M_{31} &= 0, \\
\sum_{j=1}^{3} M_{3j}M_{j1} &= 2\lambda^2, \\
\sum_{j=1}^{3}\sum_{k=1}^{3} M_{3j}M_{jk}M_{k1} &= -6\lambda^2(\lambda+\mu),
\end{aligned}$$

from Eq. (5.16), we obtain

$$A(t) = 1 - P_3(t) \approx 1 - (\lambda t)^2 + \lambda^2(\lambda + \mu)t^3,$$

in agreement with the result of part (a) and the observation that Eq. (5.16) is easier to apply than Eq. (5.15) when $N_u > N - N_u$ because of one less summation. ◇

5.3.4 Steady-State Availability

The operation of a system with repair for very long times is expected to asymptotically approach a steady-state availability $A(\infty)$, $0 < A(\infty) < 1$. To obtain that value, from Eq. (5.11) it follows that

$$\mathbf{M}_A \mathbf{P}(\infty) = 0 \tag{5.44}$$

if $\mathrm{d}\mathbf{P}(t)/\mathrm{d}t = 0$. This set of equations is ill-posed until we incorporate the normalization condition of Eq. (5.7) to replace the first row of Eq. (5.44)

$$\mathbf{M}_{A\infty}\mathbf{P}(\infty) = \mathbf{Q}, \tag{5.45}$$

where $\mathbf{Q} = [1, 0, \ldots, 0]^T$ and

$$\mathbf{M}_{A\infty} = \begin{bmatrix} 1 & 1 & 1 & \cdots & 1 \\ M_{21} & M_{22} & M_{23} & \cdots & M_{2N} \\ M_{31} & M_{32} & M_{33} & \cdots & M_{3N} \\ \vdots & \vdots & \vdots & \vdots & \vdots \\ M_{N1} & M_{N2} & M_{N3} & \cdots & M_{NN} \end{bmatrix}. \tag{5.46}$$

Thus, another rule for constructing transition rate matrices is:

- $\mathbf{M}_{A\infty}$, the matrix to obtain the steady-state availability, is created from $\mathbf{M}_A$ by setting elements $M_{1n} = 1$, $n = 1, \ldots, N$.

The solution of Eq. (5.45) for $\mathbf{P}(\infty)$ follows by matrix inversion,

$$\mathbf{P}(\infty) = \mathbf{M}_{A\infty}^{-1}\mathbf{Q}, \tag{5.47}$$

and is independent of the initial state vector $\mathbf{P}(0)$ given as Eq. (5.14) or in a more general form. In component form, the contribution of state n to the steady-state availability is given by the cofactor of the $n1$ matrix element of $\mathbf{M}_{A\infty}^T$ divided by the determinant of $\mathbf{M}_{A\infty}$,

$$P_n(\infty) = \frac{(\mathrm{cof}\,\mathbf{M}_{A\infty}^T)_{n1}}{|\mathbf{M}_{A\infty}|}. \tag{5.48}$$

Once the $P_n(\infty)$ are obtained, either $n = 1, \ldots, N_u$ or $n = N_{u+1}, \ldots, N$, the steady-state availability follows from knowledge of either the system operating states or the failed states,

$$A(\infty) = \sum_{n=1}^{N_u} P_n(\infty) \quad \text{or} \quad A(\infty) = 1 - \sum_{n=(N_u+1)}^{N} P_n(\infty). \tag{5.49}$$

Example 5.13 The active-parallel system of Example 5.4 consists of two identical components. Each component has a hazard rate of λ during operation and a repair rate of μ. For an analysis of the steady-state availability of the system, (a) construct the transition rate matrix $\mathbf{M}_{A\infty}$ if both components can be repaired at any time and (b) determine the steady-state availability of the system.

(a) The system states are the same as in Example 5.4. By setting the elements $M_{1n} = 1, n = 1, 2, 3$, in $\mathbf{M}_A$ from part (b) of Example 5.4,

$$\mathbf{M}_{A\infty} = \begin{bmatrix} 1 & 1 & 1 \\ 2\lambda & -(\lambda+\mu) & 2\mu \\ 0 & \lambda & -2\mu \end{bmatrix}.$$

(b) From part (a),

$$\mathbf{M}_{A\infty}^T = \begin{bmatrix} 1 & 2\lambda & 0 \\ 1 & -(\lambda+\mu) & \lambda \\ 1 & 2\mu & -2\mu \end{bmatrix}.$$

With state 3 the only failed state, Eq. (5.49) gives $A(\infty) = 1 - P_3(\infty)$, and from Eq. (5.48) with $(\text{cof } \mathbf{M}_{A\infty}^T)_{31} = 2\lambda^2$ and $|\mathbf{M}_{A\infty}| = 2(\lambda+\mu)^2$, it follows that $A(\infty) = (2\lambda\mu + \mu^2)/(\lambda+\mu)^2$. ◇

Example 5.14 A system consists of two identical components, with one component in standby until needed. Each component has a hazard rate of λ during operation, a hazard rate of λ^* during standby, and a repair rate of μ. Determine the steady-state availability of the system.

From Example 5.5 and Eq. (5.46) the matrix $\mathbf{M}_{A\infty}$ is given by

$$\mathbf{M}_{A\infty} = \begin{bmatrix} 1 & 1 & 1 \\ (\lambda+\lambda^*) & -(\lambda+\mu) & 2\mu \\ 0 & \lambda & -2\mu \end{bmatrix}$$

so

$$\mathbf{M}_{A\infty}^T = \begin{bmatrix} 1 & (\lambda+\lambda^*) & 0 \\ 1 & -(\lambda+\mu) & \lambda \\ 1 & 2\mu & -2\mu \end{bmatrix}.$$

With state 3 the only failed state, Eq. (5.49) gives $A(\infty) = 1 - P_3(\infty)$, and from Eq. (5.48) with $(\text{cof } \mathbf{M}_{A\infty}^T)_{31} = \lambda^2 + \lambda\lambda^*$ and $|\mathbf{M}_{A\infty}| = \lambda^2 + \lambda\lambda^* + 2\lambda\mu + 2\lambda^*\mu + 2\mu^2$, it follows that $A(\infty) = 2\mu(\lambda + \lambda^* + \mu)/[2\mu(\lambda + \lambda^* + \mu) + \lambda(\lambda + \lambda^*)]$. ◇

5.4 RELIABILITY ANALYSES

The type of analyses done for $R(t)$ are the same as for $A(t)$, but the reliability transition rate matrix $\mathbf{M}_R$ is different than the availability transition rate matrix $\mathbf{M}_A$. Even though repairs on system components can be performed unless the system has failed, once it has failed, no repairs are possible and the system is assumed to remain failed.

5.4.1 Reliability Transition Rate Matrices

For a system with only one failed system state, the $\mathbf{M}_R$ can be obtained from $\mathbf{M}_A$ by setting the repair rates in the last column to zero, i.e., $M_{nN} = 0$, $n = 1, \ldots, N$.
Example 5.15 A system consists of components 1 and 2 that are connected in active-parallel, as in Example 5.3. Each component is either in operation or under repair, with hazard rates λ_1 and λ_2 and repair rates μ_1 and μ_2, respectively. Construct $\mathbf{M}_R$.

From Example 5.3,

$$\mathbf{M}_R = \begin{bmatrix} -(\lambda_1+\lambda_2) & \mu_1 & \mu_2 & 0 \\ \lambda_1 & -(\lambda_2+\mu_1) & 0 & 0 \\ \lambda_2 & 0 & -(\lambda_1+\mu_2) & 0 \\ 0 & \lambda_2 & \lambda_1 & 0 \end{bmatrix}. \quad \diamond$$

For a system with more than one state where the system has failed, as for example if a repaired component needed for system operation must be recertified before it can be placed back in service, then there is more than one failed system state. Again, $\mathbf{M}_R$ can be constructed either by modifying transition rate matrix $\mathbf{M}_A$ or it can be constructed directly from the appropriate set of system states. From $\mathbf{M}_A$ this is done by adding all the rows (and columns) of $\mathbf{M}_A$ where the system does not operate in order to make a single row (and column) in a modified transition rate matrix and then finally setting the elements in the column for that single failed state all equal to zero (but not the row or otherwise the sum of the elements in each column would not vanish). Thus another rule for constructing transition rate matrices is:

- There is only a single failed state N for determining the reliability with $\mathbf{M}_R$, so there is only one row of failed-state elements M_{Nm} and one column of failed-state elements M_{nN}, with elements in the last column given by $M_{nN} = 0$, $n = 1, \ldots, N$.

Example 5.16 A system consists of two identical components, with one component in standby. Following the repair of a component, the component must be recertified by testing before it can be placed back in operation or in standby. The MTTF for each component in operation is λ^{-1} and in standby is $(\lambda^*)^{-1}$, the MTTR is μ^{-1}, and the mean time for recertification testing is τ^{-1}. For a reliability analysis, (a) define the system states and (b) construct the transition rate matrix $\mathbf{M}_R$.

(a) The system states are:

State	System	Components
1	Operating	One in operation, one in standby
2	Operating	One in operation, one in repair
3	Operating	One in operation, one in testing
4	Not operating	Each failed, in repair, or in testing

Now $N_u = 3$ and $N = 4$.

(b) From Example 5.7, by adding all the matrix elements in $\mathbf{M}_A$ in the columns for states 4, 5, and 6 and in the rows for states 4, 5, and 6, we obtain the matrix for the single failure state 4,

$$\begin{bmatrix} -(\lambda+\lambda^*) & 0 & \tau & 0 \\ (\lambda+\lambda^*) & -(\lambda+\mu) & 0 & \tau \\ 0 & \mu & -(\lambda+\tau) & 2\tau \\ 0 & \lambda & \lambda & -3\tau \end{bmatrix}.$$

After setting to zero all the elements in the column for the failed state, we obtain

$$\mathbf{M}_R = \begin{bmatrix} -(\lambda+\lambda^*) & 0 & \tau & 0 \\ (\lambda+\lambda^*) & -(\lambda+\mu) & 0 & 0 \\ 0 & \mu & -(\lambda+\tau) & 0 \\ 0 & \lambda & \lambda & 0 \end{bmatrix},$$

which is the transition rate matrix that could have been constructed directly from the table of system states in part (a). ◇

5.4.2 Time-Dependent Reliability Examples

After replacing $\mathbf{M}_A$ by $\mathbf{M}_R$, the solution for the time-dependent reliability follows from the solution of Eq. (5.11) obtained either from Eqs. (5.25) and (5.34) or from Eqs. (5.17) and (5.20). Examples of $R(t)$ obtained by the procedure in Eqs. (5.25) and (5.34) are in Table 5.2.

5.4.3 Mean Time to Failure

We turn now to a general consideration of the mean time to failure of a system in which components can be repaired prior to a system failure. Equation (2.91),

$$\text{MTTF} = \int_0^\infty R(t)\text{d}t, \tag{5.50}$$

can be used if $R(t)$ has been determined. But a direct way to calculate the MTTF is to form a modified transition rate matrix $\mathbf{M}_{Ru}$, which consists of *only the operating states* of either $\mathbf{M}_R$ or $\mathbf{M}_A$ (which are identical). Thus another rule for constructing transition rate matrices is:

- $\mathbf{M}_{Ru}$, the matrix for determining the MTTF, is just the submatrix formed from the first N_u rows and N_u columns of either $\mathbf{M}_R$ or $\mathbf{M}_A$.

The probability that the system is in its nth upstate is $P_n^u(t)$, so in analogy with Eq. (5.15) the reliability $R(t)$ is

$$R(t) = \sum_{n=1}^{N_u} P_n^u(t). \tag{5.51}$$

Table 5.2 Reliability of Systems Consisting of Identical Components with $\mathrm{MTTF} = \lambda^{-1}$ and $\mathrm{MTTR} = \mu^{-1}$ with No Failures During Standby

Number of Units	System Type	Reliability $R(t)$	Nonzero Eigenvalues
1	1 repairman	$\exp(-\lambda t)$	$s_1 = -\lambda$
2	1 standby, 1 repairman	$\dfrac{s_1 \exp(s_2 t) - s_2 \exp(s_1 t)}{s_1 - s_2}$	$s_1 = -0.5(2\lambda + \mu + \sqrt{4\lambda\mu + \mu^2})$ $s_2 = -0.5(2\lambda + \mu - \sqrt{4\lambda\mu + \mu^2})$
2	2 active parallel, 1 repairman	$\dfrac{s_1 \exp(s_2 t) - s_2 \exp(s_1 t)}{s_1 - s_2}$	$s_1 = -0.5(3\lambda + \mu + \sqrt{\lambda^2 + 6\lambda\mu + \mu^2})$ $s_2 = -0.5(3\lambda + \mu - \sqrt{\lambda^2 + 6\lambda\mu + \mu^2})$

The system of equations for $\mathbf{P}^u(t)$ can be written in the form

$$\mathrm{d}\mathbf{P}^u(t)/\mathrm{d}t = \mathbf{M}_{Ru}\mathbf{P}^u(t), \tag{5.52}$$

where

$$\begin{aligned} \mathbf{P}^u(t) &= [P_1^u(t),\ P_2^u(t),\ \ldots, P_{N_u}^u(t)]^T \\ \mathbf{P}^u(0) &= [1,\ 0,\ \ldots, 0]^T. \end{aligned} \tag{5.53}$$

The new feature about $\mathbf{P}^u(t)$ is that it must satisfy the *final* condition

$$\mathbf{P}^u(\infty) = \mathbf{0} \tag{5.54}$$

because there is no mechanism for recovery from the single failed state $(N_u + 1)$.

To calculate the MTTF, Eqs. (5.50) and (5.51) show that we need to calculate

$$\mathrm{MTTF} = \sum_{n=1}^{N_u} K_n, \tag{5.55}$$

where the constants K_n are the contributions of each upstate to the MTTF,

$$K_n = \int_0^\infty P_n^u(t)\mathrm{d}t, \quad n = 1, \ldots, N_u. \tag{5.56}$$

The K_n can be arranged in the form of the vector $\mathbf{K}$ that is determined by integrating Eq. (5.52) over time to obtain

$$\mathbf{M}_{Ru}\mathbf{K} = -\mathbf{P}^u(0) \tag{5.57}$$

after use of Eq. (5.54). The solution of this equation is

$$\mathbf{K} = -\mathbf{M}_{Ru}^{-1}\mathbf{P}^u(0) \tag{5.58}$$

with elements

$$K_n = -\frac{(\mathrm{cof}\ \mathbf{M}_{Ru}^T)_{n1}}{|\mathbf{M}_{Ru}|}\ . \tag{5.59}$$

Examples of the MTTF obtained from Eqs. (5.55) and (5.59) are in Table 5.3. In the table, the second term of the MTTF is the increase due to the random completion of repairs following the random failure of the components, as compared to the values in Eqs. (4.9) and (4.10) for the case of no repairs.

Example 5.17 Compare the MTTF values of:

(a) Two components (2:) with one in standby (1s) and with one repairman (1r) versus one component (1:) with none in standby (0s) and with one repairman (1r)

(b) Two components (2:) with one in standby (1s) and with one repairman (1r) versus one component (1:) with none in standby (0s) and with one repairman (1r).

Table 5.3 MTTF of Systems Consisting of Identical Components with $\mathrm{MTTF} = \lambda^{-1}$ and $\mathrm{MTTR} = \mu^{-1}$ with No Failures During Standby

Total Number	System Type	MTTF
1	1 repairman	$1/\lambda$
2	1 standby, 1 repairman	$(2/\lambda) + (\mu/\lambda^2)$
2	Parallel, 1 repairman	$(3/2\lambda) + (\mu/2\lambda^2)$

From Table 5.3, algebra gives the results

$$(a) \quad \frac{\mathrm{MTTF}_{2:1s,1r}}{\mathrm{MTTF}_{1:0s,1r}} = 2 + \frac{\mu}{\lambda},$$

$$(b) \quad \frac{\mathrm{MTTF}_{2:1s,1r}}{\mathrm{MTTF}_{2:0s,1r}} = 1 + \frac{\lambda}{3\lambda + \mu}. \quad \diamond$$

Example 5.18 A system consists of two identical components, with one kept in standby until needed. Each component has a hazard rate of λ during operation, a hazard rate of λ^* during standby, and a repair rate of μ. Determine the MTTF.

From Example 5.5 there are only two system operating states, so

$$\mathbf{M}_{Ru} = \begin{bmatrix} -(\lambda + \lambda^*) & \mu \\ (\lambda + \lambda^*) & -(\lambda + \mu) \end{bmatrix}.$$

From Eq. (5.59) it follows that

$$K_1 = \frac{\lambda + \mu}{\lambda(\lambda + \mu)}, \qquad K_2 = \frac{\lambda + \lambda^*}{\lambda(\lambda + \mu)},$$

so therefore

$$\mathrm{MTTF} = K_1 + K_2 = \frac{2\lambda + \lambda^* + \mu}{\lambda(\lambda + \mu)} = \frac{1}{\lambda} + \frac{1 + (\lambda^*/\lambda)}{\lambda + \mu}.$$

The second term on the right-hand side of MTTF gives the increase in the MTTF due to the standby component as compared to the MTTF in the first term for a single component. $\diamond$

5.5 ADDITIONAL CAPABILITIES OF MARKOV MODELS

In the previous section all transitions between different states were assumed to be characterized by a constant value for every hazard rate $\lambda = \mathrm{MTTF}^{-1}$ and a constant value for every repair rate $\mu = \mathrm{MTTR}^{-1}$. Additional time-dependent transition rates also were introduced to accommodate additional system states into the analysis. The switching between each system state was assumed to be instantaneous and perfect.

In this section instantaneous but imperfect switching between different system states, as with a switch with reliability $R_{sw} \leq 1$, is analyzed. Consideration also will be given here to the treatment of systems with time-dependent hazard and repair rates $\lambda(t)$ and $\mu(t)$, respectively.

5.5.1 Imperfect Switching Between System States

For such transitions it is necessary to modify the appropriate element(s) in a transition rate matrix to include the effects of demand-type failures. The probability for successful operation of a switch is just the reliability R_{sw}, and the probability for unsuccessful switching will be denoted by $\overline{R}_{sw} = (1 - R_{sw})$. The trick is to recognize that a transition rate M_{ij} in matrix $\mathbf{M}$ can be viewed as a rate that is conditional on the success or failure of a particular switching action, which suggests that the transition rate should be multiplied by R_{sw} or $\overline{R}_{sw}$, as appropriate. It is important to emphasize, however, that the switch reliability gives only the average probability of success of a particular switching action and does not yield the actual outcome of any single action. It is for this reason that the effects of a switch reliability *cannot* be included in a time-dependent availability or reliability analysis. On the other hand, if only a time-averaged result is needed such as the steady-state availability or the MTTF, then the appropriate transition rates *can* be multiplied by R_{sw} or $\overline{R}_{sw}$ in order to include the effects of switch failures.

Example 5.19 A system of two nonidentical components can operate with one unit in standby. Initially unit 1 is in operation. The units fail in active operation with constant hazard rates λ_i, $i = 1, 2$, and unit 2 can fail in standby at the rate λ_2^*. The switching unit operates instantly and has a reliability R_{sw}. There are no repairs possible. Derive the equation for the mean time to failure for the system.

The system states are:

State	System	Components
1	Operating	Unit 1 in operation, unit 2 in standby
2	Operating	Unit 2 in operation, unit 1 failed
3	Operating	Unit 1 in operation, unit 2 failed
4	Not operating	Both units failed or switch failed

The transition rate matrix for the system upstates is

$$\mathbf{M}_{Ru} = \begin{bmatrix} -(\lambda_1 + \lambda_2^*) & 0 & 0 \\ R_{sw}\lambda_1 & -\lambda_2 & 0 \\ \lambda_2^* & 0 & -\lambda_1 \end{bmatrix}.$$

From Eq. (5.59) it follows that

$$K_1 = \frac{\lambda_1\lambda_2}{\lambda_1\lambda_2(\lambda_1 + \lambda_2^*)}, \qquad K_2 = \frac{R_{sw}\lambda_1^2}{\lambda_1\lambda_2(\lambda_1 + \lambda_2^*)}, \qquad K_3 = \frac{\lambda_1\lambda_2^*}{\lambda_1\lambda_2(\lambda_1 + \lambda_2^*)},$$

so Eq. (5.55) gives

$$\text{MTTF} = \frac{1}{\lambda_1} + \frac{R_{sw}\lambda_1}{\lambda_2(\lambda_1 + \lambda_2^*)},$$

where the second term illustrates the benefit of the standby unit. This result checks with that obtained by integrating over all time the reliability of Eq. (4.22). ◇

Example 5.20 A system can operate with either a main unit or a standby unit. Each unit fails in active operation with a MTTF of $1/\lambda$ and the standby unit fails during standby with a MTTF of $1/\lambda^*$. The standby unit can fail to start with a probability of $\overline{R}_{sw} = 1 - R_{sw}$. Either unit can be repaired at a rate μ if it undergoes either a hardware or switch failure. Also, either of two repairmen will respond instantly to begin repairing a failed unit. Derive (a) the steady-state availability and (b) the MTTF of the system.

(a) Because the switch is a part of the standby unit, there is only one system down-state. The system states are:

State	System	Components
1	Operating	Main unit in operation, standby unit available
2	Operating	Main unit in repair, standby unit in operation
3	Operating	Main unit in operation, standby unit in repair
4	Not operating	Both units in repair

The transition rate matrix for an availability analysis is

$$\mathbf{M}_A = \begin{bmatrix} -(\lambda+\lambda^*) & \mu & \mu & 0 \\ R_{sw}\lambda & -(\lambda+\mu) & 0 & \mu \\ \lambda^* & 0 & -(\lambda+\mu) & \mu \\ \overline{R}_{sw}\lambda & \lambda & \lambda & -2\mu \end{bmatrix},$$

so from Eq. (5.46)

$$\mathbf{M}_{A\infty}^T = \begin{bmatrix} 1 & R_{sw}\lambda & \lambda^* & \overline{R}_{sw}\lambda \\ 1 & -(\lambda+\mu) & 0 & \lambda \\ 1 & 0 & -(\lambda+\mu) & \lambda \\ 1 & \mu & \mu & -2\mu \end{bmatrix}.$$

With state 4 the only failed state, Eq. (5.49) gives $A(\infty) = 1 - P_4(\infty)$, and with

$$\begin{aligned} (\text{cof } \mathbf{M}_{A\infty}^T)_{41} &= -(\lambda+\mu)\lambda(\lambda+\lambda^*+\overline{R}_{sw}\mu), \\ |\mathbf{M}_{A\infty}| &= -(\lambda+\mu)[\lambda(\lambda+\lambda^*+\overline{R}_{sw}\mu) + 2\mu(\lambda+\lambda^*+\mu)], \end{aligned}$$

it follows from Eq. (5.48) that

$$A(\infty) = \frac{2\mu(\lambda+\lambda^*+\mu)}{\lambda(\lambda+\lambda^*+\overline{R}_{sw}\mu) + 2\mu(\lambda+\lambda^*+\mu)}.$$

Table 5.4 MTTF Versus R_{sw} for $\lambda = 2.5 \times 10^{-4}$ hr^{-1} and $\mu = 0.25$ hr^{-1}

R_{sw}	MTTF (hr) for $\lambda^* = \lambda$	MTTF (hr) for $\lambda^* = 0$
1	2.006×10^6	4.008×10^6
0.99	3.343×10^5	3.644×10^5
0.98	1.824×10^5	1.909×10^5
0.95	7.715×10^4	7.858×10^4

Source: Reprinted with permission from [Dhi81].

(b) To derive the MTTF we use the matrix $\mathbf{M}_{Ru}$ obtained from the first three rows and columns of $\mathbf{M}_A$ and find

$$(\text{cof } \mathbf{M}_{Ru}^T)_{11} = (\lambda{+}\mu)^2, \; (\text{cof } \mathbf{M}_{Ru}^T)_{21} = R_{sw}\lambda(\lambda{+}\mu), \; (\text{cof } \mathbf{M}_{Ru}^T)_{31} = \lambda^*(\lambda{+}\mu),$$

and

$$|\mathbf{M}_{Ru}| = -\lambda(\lambda + \mu)(\lambda + \lambda^* + \overline{R}_{sw}\mu),$$

so from Eqs. (5.55) and (5.59) it follows that

$$\text{MTTF} = \frac{\lambda(1 + R_{sw}) + \lambda^* + \mu}{\lambda(\lambda + \lambda^* + \overline{R}_{sw}\mu)}.$$

Numerical values in Table 5.4 illustrate the effects of the reliability of the switch and the effects of failures of the standby unit during standby. ◇

5.5.2 Systems with Nonconstant Hazard Rates

The Markov method developed so far in this chapter is valid only for systems that have constant transition rates between different system states. The assumption of constant instantaneous repair rates μ does not usually introduce serious limitations because, from Eq. (2.99), repairs normally are completed quickly compared to the typical times between failures. Often, failure data may not be available or not good enough to merit using anything other than constant hazard rates λ, but that assumption can be restrictive. This is because only the exponential failure model is strictly valid, so failures described by the gamma, lognormal, or Weibull distributions then are only approximated. Another way of viewing the restrictive nature of the constant-hazard-rate model is to observe that it excludes the possibility of analyzing age-dependent failures.

One approach for using a Markov approach to treat systems with nonconstant hazard rates is to assume that

$$\lambda(t) = \lambda_j \;\text{ for }\; T_{j-1} \le t < T_j, \; j = 1, \ldots, J, \tag{5.60}$$

where T_{j-1} and T_j are the partition times for the jth time interval, with $T_0 = 0$ and $T_J = \infty$. From Eq. (5.11) it follows that

$$\mathrm{d}\mathbf{P}_j(t)/\mathrm{d}t = \mathbf{M}_j\mathbf{P}_j(t) \text{ for } T_{j-1} \leq t < T_j,\ j = 1, \ldots, J, \tag{5.61}$$

where $\mathbf{M}_j$ is the transition rate matrix for either an availability or reliability analysis and $\mathbf{P}_j(t)$ is the vector of state probabilities for time interval j with elements $\mathbf{P}_{jn}(t)$ for the states $n = 1, \ldots, N_j$, with $\sum_{n=1}^{N_j} P_{jn}(t) = 1$. Now, however, the inherent simplicity of Eq. (5.11) with the initial condition of Eq. (5.14) can be lost if the initial condition for solving Eq. (5.61) is $\mathbf{P}_j(T_{j-1}) \neq [1,\ 0,\ 0,\ \ldots, 0]^T$.

A major concern is that the manner in which the system functions may change with time, in which case the system states could change with time as well as the number of system states N_j. If the number of states increases from the $(j-1)$th time interval to the jth, then for time $T_{j-1}^+ = T_{j-1} + \epsilon$ the initial probabilities $P_{jn}(T_{j-1}^+)$ for all newly added states must vanish for ϵ vanishingly small. On the other hand, if two states, say those labeled k and ℓ, combine to form a single state m in the jth time interval, then the newly created state must have an initial probability of $P_{jm}(T_{j-1}^+) = P_{j-1,k}(T_{j-1}^-) + P_{j-1,\ell}(T_{j-1}^-)$ for $T_{j-1}^- = T_{j-1} - \epsilon$ and $\epsilon \to 0^+$.

If the system state probabilities at the beginning of a time interval can be obtained from the corresponding state probabilities at the end of the preceding time interval, then the formal solution of Eq. (5.61) can be written as

$$\mathbf{P}_j(t) = \exp[\mathbf{M}_j(t - T_{j-1}^+)]\mathbf{P}_{j-1}(T_{j-1}^+), \quad T_{j-1}^+ \leq t < T_j. \tag{5.62}$$

As can be seen, the assumption of constant hazard and repair rates simplifies everything immensely. What happens if the hazard and repair rates are not constant? In such cases one approach is to use a dynamic event tree analysis considered in Chapter 13.

References

[Dhi81] B. S. Dhillon and C. Singh, *Engineering Reliability New Techniques and Applications*, Wiley (1981).

[IEE98] "IEEE Recommended Practice for the Design of Reliable Industrial and Commercial Power Systems," IEEE Std 493-1997, IEEE Industry Applications Society (1998).

Exercises

5.1 A system consists of two identical units in active-parallel operation, with the possibility that only one unit can be repaired at any time. The mean time to failure of each unit is 1000 hr and the mean time to repair is 10 hr. Determine (a) the MTTF if repairs can be made and (b) the MTTF if repairs cannot be made.

5.2 A system with four sequentially operated units (i.e., initially three cold standby units) is operated with no repair and no switch failures. The four units fail randomly with constant hazard rates λ_n, $n = 1, \ldots, 4$. (a) Define the five system states and construct the transition rate matrix $\mathbf{M}_R$ and (b) obtain the MTTF for the system and reduce it to the simplest form.

5.3 A single-unit system can fail by either of two independent failure modes at a constant rate λ_n for mode n, $n = 1, 2$. The repair of the system after a failure occurs with the constant rate μ_n, $n = 1, 2$, depending on the mode of system failure. (a) Define the three states for an availability analysis of the system and construct the transition rate matrix $\mathbf{M}_A$, (b) derive the time-dependent availability of the system, (c) derive the steady-state availability, and (d) obtain the MTTF.

5.4 A reactor cooling system has three identical loops in normal operation. Assume the reactor can be briefly operated while one loop is under repair, but not if two loops have failed. If each coolant loop has a MTTF of λ^{-1} and if the MTTR of any loop is μ^{-1}, (a) define the three states of the system for a reliability analysis and construct the transition rate matrix $\mathbf{M}_R$, (b) derive the time-dependent reliability, and (c) obtain the MTTF.

5.5 A reactor cooling system has three identical loops, each of which has a hazard rate of λ and a repair rate of μ. Two loops are required for operation and it is possible for n loops, $n = 1, 2$, to be under repair at any time. (a) Define the three states of the system for an availability analysis and construct the transition rate matrix $\mathbf{M}_A$, (b) derive the time-dependent unavailability for rare-event failures, and (c) obtain the steady-state availability.

5.6 A system consists of three identical pumps, but only two are in use at any time with one in standby. The hazard rate of a pump in operation is λ and in standby is λ^* and the repair rate is μ no matter the cause of failure. If necessary, two pumps can undergo repair simultaneously. (a) Define the system states for an availability analysis and construct the transition rate matrix $\mathbf{M}_A$, (b) derive the steady-state availability, and (c) obtain the MTTF.

5.7 A system consists of two identical components in an active-parallel configuration. Each component has a hazard rate of λ and when both are operating a failure can occur with hazard rate λ_c because of a common flaw. If necessary the two components can undergo repair simultaneously, each with a repair rate of μ that is independent of the cause of failure. (a) Define the system states and construct the transition rate matrix $\mathbf{M}_A$, (b) derive the steady-state availability, and (c) obtain the MTTF.

5.8 A system consists of two identical pumps with only one in operation at a time. Each pump has a hazard rate of λ and a common flaw that can cause both to fail with hazard rate λ_c. If necessary the two pumps can undergo repair simultaneously, each with a repair rate of μ that is independent of the cause of failure. (a) Define the system states and construct the transition rate matrix $\mathbf{M}_A$, (b) derive the steady-state availability, and (c) obtain the MTTF.

5.9 A system consists of two identical units, with one in active standby until needed. After a unit fails, there is a mean time delay of σ^{-1} before a repair in initiated. The MTTF for each unit in operation is λ^{-1} and in standby is $(\lambda^*)^{-1}$ and the MTTR is μ^{-1}. Both units can undergo repair simultaneously if necessary and any repaired unit is immediately placed back into operation or in standby if not needed. For an availability analysis, (a) define the system states and (b) construct the transition rate matrix $\mathbf{M}_A$.

5.10 A system consists of two identical components, a system operator, and two repair people. The mean time to failure of a component in operation is λ^{-1}, the mean time for the operator to notice a failure is σ^{-1}, and the mean time to complete the repair of a failed component is μ^{-1}. Only one component operates when the system operates, and there are no failures of a component in standby. The operator must first notice that a component has failed before switching on a standby component if available and notifying a repair person to begin work. Any repaired component is immediately placed back into operation or in standby if not needed. (a) Define the system states for an availability analysis and (b) construct the transition rate matrix $\mathbf{M}_A$.

5.11 A nuclear steam supply system has two turbogenerator units, with unit 1 in operation and unit 2 in standby whenever both are operable. During operation the units have a MTTF of λ_n^{-1}, $n = 1, 2$, and during standby unit 2 has a MTTF of $(\lambda_2^*)^{-1}$. Repairs to either unit can occur with a MTTR of μ_n^{-1}, $n = 1, 2$, but only unit 1 can be repaired if both have failed. (a) Define the system states for an availability analysis, (b) construct the transition rate matrix $\mathbf{M}_A$, and (c) construct the transition rate matrix $\mathbf{M}_R$.

5.12 A system has three identical pumps, only one of which operates at a time. During operation each fails randomly at a rate λ. There are no failures during standby. The repair of any pump is completed with a mean time of μ^{-1} once work has begun, but there is a mean time delay σ^{-1} before beginning repairs of each unit after it fails. Two pumps can be repaired simultaneously. (a) Define the system states for a reliability analysis and (b) construct the transition rate matrix $\mathbf{M}_R$.

5.13 A reactor has two identical coolant loops, each with two identical pumps connected in active parallel. Only one loop is operated at a time, with at least one pump required to be functional for a loop to be operable, and the loop with the most operable pumps is always in operation. All pumps fail randomly when in active operation with a MTTF of λ^{-1}. There are no failures during standby and no failures in switching from one loop to the other. It is possible to do repairs on a maximum of two pumps, with a mean time to repair of μ^{-1}, but no repairs are performed on an operating loop with a failed pump. (a) Define the system states for an availability analysis and (b) construct the transition rate matrix $\mathbf{M}_A$.

5.14 Complete the derivation of $P_2(s)$ by setting up a balance equation for $P_2(t)$ equivalent to Eq. (5.39) and substitute it into Eq. (5.40) to obtain Eq. (5.36).

CHAPTER 6

PROBABILISTIC RISK ASSESSMENT

In this chapter the essential principles are developed for predicting the potential for system failures to occur and the consequences if they do. Probabilistic risk assessment is a systematic technique for investigating the transformation of an undesired initiating event into a set of possible outcomes and their consequences. For each initiating event, a semipictorial event tree (ET) is constructed to follow a sequence of events through multiple stages of safety systems to be activated, as discussed in Section 2.2. For a safety system comprising multiple components, a fault tree (FT) is constructed to represent the combinations of component failures that could result in the system failure, treated as the top event of the tree. The top event probability of the FT is then obtained by combining, through Boolean algebra, failure probabilities of the components treated as basic events. The combinations of basic events resulting in the top event are represented through cut sets discussed in Sections 2.1 and 4.5. Because a probabilistic analysis is normally done for a highly reliable system, the probability of the events occurring usually can be determined with the rare-event approximation. Thus such a technique is best at identifying weaknesses or vulnerabilities in a system, rather than determining highly accurate estimates of the failure frequency for undesirable or failed states.

Table 6.1 Failure Modes Used in Reactor Safety Study

Failure mode	Code
Closed	C
Disengaged	G
Does not close	K
Does not open	D
Does not start	A
Engaged	E
Exceeds limit	M
Leakage	L
Loss of function	F
Maintenance fault	Y
No input	N
Open	O
Open circuit	B
Operational fault	X
Overload	H
Plugged	P
Rupture	R
Short circuit	Q
Short to ground	S
Fault transfer	T

Source: [NRC75].

Although a PRA can be applied to any engineered system, the emphasis and applications studied here are those focusing on nuclear systems. After considering the modes of failure of system components and some inductive methods for performing a PRA, the fault tree and event tree techniques will be discussed.

6.1 FAILURE MODES

To determine what faults to include in a failure analysis, it is necessary to determine what combinations of failure events lead to the undesired event. This is done by breaking the analysis of a system down into subsystems and their components while keeping in mind how the components interact.

Table 6.1 gives an example list of some different hardware failure modes for components in a system. The short-hand codes listed were those used to keep track of failure modes in the 1975 Reactor Safety Study [NRC75]. A more extensive list of failure modes is in Table 6.2.

A general observation is that components that function in a dynamic manner usually fail much more frequently than static system components. System components such as switches and valves undergo active failures when they fail, for example,

Table 6.2 Some Generic Failure Modes

No.	Mechanical Component Mode	No.	Electrical Component Mode
1	Structural failure (rupture)	17	Structural failure
2	Fails to open	18	Fails to start
3	Fails to close	19	Fails to stop
4	Fails to remain in position	20	Fails to switch
5	Fails open	21	Premature operation
6	Fails closed	22	Delayed operation
7	Inadvertent operation	23	Erroneous input (high)
8	Intermittent operation	24	Erroneous input (low)
9	Erratic operation	25	Erroneous output (high)
10	Physical binding or jamming	26	Erroneous output (low)
11	Vibration failure	27	Loss of input
12	Internal leakage	28	Loss of output
13	External leakage	29	Shorted
14	Fails out of tolerance (high)	30	Shorted (electrical)
15	Fails out of tolerance (low)	31	Open (electrical)
16	Restricted flow	32	Leakage (electrical)

Source: Reprinted with permission from [Vil92].

but wires, pipes, and stationary structural elements that operate in a static manner undergo passive failures.

6.2 CLASSIFICATION OF FAILURE EVENTS

There are several useful ways to categorize failure events. Three ways of interest here are primary/secondary/command failures, common cause failures, and human errors.

6.2.1 Primary, Secondary, and Command Failures

Primary faults are those that occur when a system component performs its intended function but fails because of a basic mode such as a structural fault. In a system with a high performance standard, primary faults typically occur with excessive or unanticipated wear due to improper component maintenance or replacement of parts. In a system that has a lower performance requirement, primary faults typically arise as "wear-in" failures, which are those occurring because of poor design or fabrication.

Secondary faults arise when a system component is subjected to a load or an operating environment for which it was not designed. As an example, if a pressure vessel fails due to a system overpressure caused by the failure of the pressure relief valve, it fails via a secondary fault.

Command faults occur if a component operates correctly but at the wrong place or time. As an example, if a pressure vessel fails because of an operator who incorrectly closed a valve, then a command fault has occurred because of the human error.

6.2.2 Common Cause Failures

The possibility of failures of multiple components caused by a common means was briefly discussed in Section 2.10 and an elementary example of the impact of such failures was treated in Section 5.3. We now consider common cause and common mode failures in more detail. In the early years of PRA investigations, a common cause failure (CCF) was viewed as an event in which two or more components failed from the same inherent flaw in the design, manufacture, operation, or maintenance of the components. Common mode failures (CMFs), on the other hand, were those failures of two or more components arising from the same failure mode, e.g., the same inherent flaw in the operation of the components. The distinction between the two types of common failures now largely has been abandoned in more recent schemes for characterizing CCFs. Examples of reasons for CCFs are given in Table 6.3.

6.2.2.1 Elementary Analysis of CCFs When multiple system components have a common failure cause, the overall system failure probability can be significantly affected. There are several ways of analyzing CCF events for time-dependent systems, although no approach is necessarily suitable for all situations. The approach used in [NRC75], which is not unreasonable but lacks a physical basis, is to take the geometric mean of two bounds on the instantaneous failure rate. The lower bound λ_i for the system is obtained by assuming there is no common failure mode between system components so they all fail independently, and the upper bound λ_{dep} of the system failure rate is that of the system component having the largest failure rate. Thus the common mode failure rate is

$$\lambda_c = \sqrt{\lambda_i \lambda_{dep}}. \tag{6.1}$$

The *beta factor* approach differs from that based on the geometric mean because the failure rate for each component is assumed to be the sum of independent and dependent (common cause) failure contributions,

$$\lambda = \lambda_i + \lambda_c. \tag{6.2}$$

With the definition

$$\beta = \lambda_c/\lambda, \tag{6.3}$$

it follows that the CCF contribution is

$$\lambda_c = \frac{\beta}{1-\beta}\lambda_i. \tag{6.4}$$

This approach was presented in [NRC90], even though there also is no physical basis for it. Sample beta values are given in Table 6.4.

Table 6.3 Examples of Contributing Events to Common Cause Failures

Source	Category	Examples
Environment (system, components, subsystems)	Impact	Pipe whip, water hammer, missiles, earthquake, structural failure
	Vibration	Machinery in motion, earthquake
	Pressure	Explosion, out-of-tolerance system changes (pump overspeed, flow blockage)
	Grit	Airborne dust, metal fragments generated by moving parts with inadequate tolerances
	Stress	Thermal stress at welds of dissimilar metals, thermal stresses and bending moments caused by high conductivity and density
	Temperature	Fire, lightning, weld equipment, cooling-system fault, electrical short circuits
	Loss of energy	Common drive shaft, same power supply
	Calibration	Misprinted calibration instruction
	Manufacturer	Repeated fabrication error, such as neglect to properly coat relay contacts. Poor workmanship. Damage during transportation
Plant	Installation contractor	Same subcontractor or crew
	Maintenance	Incorrect procedure, inadequately trained personnel
	Operation	Operator disabled or overstressed, faulty operating procedures
	Test	Faulty test procedures that may affect all components normally tested together
Aging	Aging	Components of same materials

Source: Reprinted with permission from [Wag77]. Copyright © 1977 Society of Industrial and Applied Mathematics.

More general common cause multiparameter failure models have been developed for systems with complex redundancy [NRC90] and are discussed elsewhere [Ful88].
Example 6.1 A system consists of two identical units in active parallel that can fail independently with a constant hazard rate λ. The units have a design defect that causes the failure of one unit to overheat the second one, leading to system failure governed by λ_c. Determine the system reliability (a) without and (b) with CCF and then (c) compare the MTTF values for no CCF and with the CCF.

(a) From Eq. (4.4), the reliability with both components independent is

$$R_{sys,i}(t) = 2\exp(-\lambda_i t) - \exp(-2\lambda_i t).$$

Table 6.4 Some Generic Beta Factors for Various Reactor Components

Type of Component	Upper Bound	Mean Value
Reactor trip breakers	0.19	0.0792
Diesel generators	0.05	0.0208
Motor-operated valves	0.08	0.0333
Safety/relief valves (PWR)	0.07	0.0292
Safety/relief valves (BWR)	0.22	0.0917
Safety injection pumps	0.17	0.0708
Residual heat removal pumps	0.11	0.0458
Containment spray pumps	0.05	0.0208
Auxiliary feedwater pumps	0.03	0.0125
Service water pumps	0.03	0.0125
Batteries	0.10	0.0400

Source: [NRC90].

(b) The CCF mode acts like a pseudocomponent in series with the active-parallel one, so from Eq. (4.3),

$$\begin{aligned} R_{sys,i\&c}(t) &= R_{sys,i}(t)R_{sys,c}(t) = [2\exp(-\lambda_i t) - \exp(-2\lambda_i t)]\exp(-\lambda_c t) \\ &= 2\exp(-\lambda t) - \exp[-(2-\beta)\lambda t)] \end{aligned}$$

with the help of Eqs. (6.2) and (6.3).

(c) The system reliability decreases as β increases, as indicated by the MTTF values,

$$\text{MTTF} = \begin{cases} \dfrac{3}{2\lambda}, & \text{without CCF,} \\ \dfrac{3}{2\lambda}\left[1 - \dfrac{\beta}{3(2-\beta)}\right], & \text{with CCF.} \end{cases} \quad \diamond$$

6.2.2.2 General Approach to CCFs

Four criteria must be met for an event to be classified as resulting from a common cause [NRC03b]:

1. Two or more individual components must fail or be degraded, including failures during demand, inservice testing, or from deficiencies that would have resulted in a failure if a demand signal had been received.
2. Two or more individual components must fail or be degraded in a select period of time such that the PRA mission would not be certain.
3. The component failures or degradations must result from a single shared cause and coupling mechanism.
4. The component failures are not due to the failure of equipment outside the established component boundary.

CCFs result from the simultaneous existence of two factors: the susceptibility for components to fail or become unavailable due to a particular cause of failure and a coupling factor (or coupling mechanism) that creates the condition for multiple components to be affected by the same cause. For example, two pressure relief valves could fail to open at the required pressure due to an incorrect setpoint. Because of personnel error, the two valves fail together due to the coupling factors of a common calibration procedure and common maintenance personnel.

Coupling factors are characteristic of a group of components that are susceptible to the same causal mechanisms of failure. Such factors include similarity in design, location, environment, mission, and operational, maintenance, and test procedures. Coupling factors can be categorized into the following five groups for analysis purposes [NRC03a]:

- Hardware quality
- Hardware design
- Maintenance
- Operations
- Environment

Coupling factors are different than root and proximate causes of CCFs. A root cause is the most basic, and often the most obvious, cause of such a failure that, if corrected, could prevent a recurrence. A proximate cause, on the other hand, is a symptom of the failure which does not necessarily provide an understanding of what led to the failure. Major categories of proximate causes for CCFs, which also are proximate causes of single-component failures, are [NRC03b]:

- Design, construction, installation, and manufacture inadequacy causes
- Operational and human-related causes (e.g., procedural errors, maintenance errors)
- Internal to the component, including hardware-related causes and internal environmental causes
- External environmental causes
- State of other component
- Other causes

Proximate causes and the coupling factors sometimes appear to overlap, even though they are different. Consider maintenance, for example. As a proximate cause the term refers to errors and mistakes made during maintenance activities, but as a coupling factor maintenance refers to the similarity of maintenance among the components (e.g., the use of the same personnel and procedures).

Given the rarity of common cause events and the difficulty of building a database of prior occurrences relevant to a particular nuclear facility, mitigating the problem of CCFs with a search for defenses may be the most effective and practical approach. Three ways to defend against a CCF event are to defend against the failure proximate cause, defend against the coupling factor, or defend against both the proximate cause and the coupling factor. For example, consider two redundant components in the same room as a steam line. A barrier that separates the steam line from the components is an example of defending against the proximate cause. A barrier that separates the two components is an example of defending against the coupling factor (same location). Installing barriers around each component is an example of defending against both the cause and the coupling factor [NRC03b].

A defense based on a coupling factor is easier to assess because the coupling mechanism among failures is more readily apparent and therefore easier to interrupt. A search for defenses against coupling primarily involves looking for dissimilarities among components, in contrast to a search for coupling factors which involves looking for similarities among components. Such dissimilarities include differences in the components themselves (diversity), differences in the way they are installed, operated, and maintained, and differences in their environment and location.

Any comprehensive examination for CCFs should include identification of the root causes, coupling factors, and defenses in place against them.

6.2.3 Human Errors

Just as there are elementary approaches for incorporating CCFs into a PRA, the Markov method of Chapter 5 can be used for treating human reliability in systems with only a few components. Such an approach typically requires that a constant human error rate be assumed [Dhi86] and hence does not adequately address the complexity or the operating conditions of a nuclear plant.

There are two types of human/nuclear plant interactions that need to be considered: those involving routine plant operation, testing, and maintenance and those involving plant safety issues. The two categories differ in the amount of stress an individual might be under. Because nuclear power safety systems are designed to operate automatically during the initial stages of an accident sequence, human intervention normally would not be required. In instances, however, where it would be required, the human response undoubtedly would be done under stress.

There also are two general categories of errors involved in human reliability analysis (HRA): errors of omission and errors of commission. Errors of omission involve actions in which operators take no action or where a set of actions taken leads to no significant difference between no action and the actions taken. Errors of commission involve actions taken that can significantly increase the severity of an undesired incident.

Three objectives of a human reliability analysis are [Ger94]:

- Identify sources of human error and human failure modes to be included as human failure events (HFEs) in a PRA framework or model.
- Develop models in the PRA representing the specific HFEs of interest.
- Quantify the human error probability (HEP) associated with each HFE, including understanding the factors that may most influence the HEP estimate.

From [NRC06] we learn that HRA methods developed for NPPs use a common categorization scheme to distinguish between:

- those HFEs postulated in the PRA as contributing to the unavailability of equipment by leaving a system or individual component in a faulty undetected state due to errors during testing and maintenance,

- those HFEs contributing to an initiating event (i.e., to an abnormal event that can challenge plant safety), and

- those HFEs contributing to the failure of a safety function, system, or component modeled in the PRA in response to an initiating event.

As a result, HFEs in a HRA are classified as (1) preinitiator HFEs, (2) initiator-related HFEs, and (3) postinitiator HFEs. This categorization scheme helps distinguish the conditions under which a task is being performed and, therefore, identifies the conditions affecting human performance that could be quite different for the different tasks modeled in a PRA.

Preinitiator actions involving normal operations, such as testing and maintenance, generally are not time sensitive, and hence time is typically not an important influencing factor. But preinitiator HFEs may be related to short-cutting test and validation practices due to causes such as tedious repetition of restoration activities, tool availability or suitability, and accessibility of the component being maintained. Therefore, those types of influencing factors may be more important to take into consideration when modeling and assessing preinitiator HFEs in a PRA.

Initiator-related HFEs involve human failures that can induce or otherwise contribute to the occurrence of an initiating event (e.g., an operator inadvertently causing shutdown of a feedwater pump, which in turn causes an automatic shutdown of the plant). It is not a common practice to model these types of HFEs in PRAs. The occurrence and the frequency of such events are captured in PRAs by the use of available statistical data on initiating event occurrences.

Postinitiator HFEs, associated with actions taken in response to an initiating event and subsequent plant transient, are modeled and analyzed in a PRA/HRA. Studies of human performance under abnormal or accident conditions have identified many influencing factors. For example, in some situations, the time available to respond can be an important factor. Other factors can also be important, such as how well procedures will direct the appropriate actions to take, given the postulated accident scenario, and to what extent the operators have been trained on the type of scenario being addressed. As a result, the performance shaping factors (PSFs) for postinitiator human events are handled differently from the PSFs for preinitiator human events.

An important part of HRA is the use of expert judgment: "Expert judgment is an integral part of HRA, and it is difficult to proceed in the process of human error quantification without some recourse to its use. Expert judgment is not a model but a process of capturing information about human actions based upon the use of the knowledge and understanding of persons who are either directly involved or observe the actions of others in the operation of a plant" [Spu10].

6.3 FAILURE DATA

6.3.1 Hardware Failures

Obtaining good failure data for evaluating the potential risks of nuclear systems is not an easy task. For electrical systems primary sources are [IEE98] and [IEE07]. An example of such data is given in Appendix C. Reliability data for electrical, electronic, sensing equipment, and mechanical equipment applicable to nuclear power plants (NPPs) are presented in [IEE84], which also provides guidelines for the collection and presentation of the reliability data. The systems covered in the database include heaters, valve operators and actuators, instruments, controls, sensors, and energy transport and exchange equipment.

Besides the references at the end of the chapter, some data on CCFs are available at www.nea.fr from the intergovernmental Nuclear Energy Agency in Paris, France under the international common cause data exchange (ICDE) project. The objectives of the ICDE project are to:

- Collect and analyze CCF events over the long term so as to better understand such events, their causes, and their prevention
- Generate qualitative insights into the root causes of CCF events which can then be used to derive approaches or mechanisms for their prevention or for mitigating their consequences
- Establish a mechanism for the efficient feedback of experience gained in connection with CCF phenomena, including the development of defenses against their occurrence, such as indicators for risk-based inspections
- Generate quantitative insights and record event attributes to facilitate quantification of CCF frequencies in member countries
- Use the ICDE data to estimate CCF parameters

NEA/ICDE reports have been issued over the Internet, for example, on centrifugal pumps, diesel generators, motor-operated valves, safety and relief valves, check valves, and switching devices and circuit breakers.

6.3.2 Human Errors

Obtaining good estimates of failure rates for possible human errors is even more difficult than acquiring hardware failure data. Incorporating human performance into risk and safety analyses is complicated because an extensive database cannot be compiled that is applicable to all the situations that can be envisioned to arise during the different events in an accident sequence. Statistical data can be acquired, though, by monitoring the performance of plant personnel when they operate simulators, but there is always the concern that the human stress response is different when simulated versus during the events in a real accident. Thus a human reliability analysis is usually done by modeling whenever expert opinion is not available.

A variety of HRA models have been developed with which to simulate human behavior in various situations. The first model to gain widespread use for nuclear plant analyses was the technique for human error rate prediction (THERP) [Swa83], which investigated how internal and external factors can influence the reliability of human performance for both pre- and postinitiators. The approach uses performance shaping factors that account for interactions or dependencies between persons. A list of 50 potential factors that could affect performance under different circumstances was developed in tables that are mainly job or environment related [Swa83].

According to [NRC06]: "There are currently over 20 methods available for characterizing and predicting states of human failure. In all of these methods, human failure is characterized by humans either not performing the desired action or doing something other than the desired action. This often implies a time frame (i.e., if an action is not performed before a certain time, it can be considered a failure or error). Each of these methods provides explicit consideration of human factors and other influences that affect performance, and these methods encourage analysts to apply them to account for situational factors that, together with operator, crew, or organizational factors, may affect the likelihood of human failure."

Some other HRA methods include [NRC06]:

- The cause-based decision tree (CBDT) approach, which considers human failure modes as predominantly arising from failures at the plant information-operator interface or at the procedure-crew interface [Sin93].
- The standardized plant analysis risk-human reliability analysis (SPAR-H) method, which estimates human error probabilities (HEPs) associated with diagnosis and action [NRC05]. It involves adjusting a nominal HEP of 0.01 for diagnosis and 0.001 for actions, with multipliers that represent the strength of the effect for each PSF on the success/failure of the task analyzed, as with [Bor05]

$$\text{HEP} = \text{Nominal}_{\text{HEP}} \prod_{n=1}^{N} \text{PSF}_n. \tag{6.5}$$

- ATHEANA, a technique for human error analysis, that requires accounting for the ways that unsafe acts (UAs) can occur for different error forcing contexts (EFCs) that might arise in a given accident sequence [For04]. The method [NRC96] produces estimates of the human reliability probability $P(\text{HFE}|\text{S})$ for a human failure event (HFE) of interest, given a postulated accident scenario S, by incorporating the conditional probabilities $P(\text{EFC}_\text{i}|\text{S})$ and $P(\text{UA}|\text{EFC}_\text{i}, \text{S})$ that an unsafe act will occur,

$$P(\text{HFE}|\text{S}) = \sum_i P(\text{EFC}_\text{i}|\text{S}) P(\text{UA}|\text{EFC}_\text{i}, \text{S}). \tag{6.6}$$

Another way of categorizing the different approaches to a HRA is to consider the three main types of models depending on whether the HRA involves task-related, time-related, or context-related actions [Spu10]. The task-related model subdivides a task into subtasks that involve decisions and actions, while time-related models utilize time as the principal feature in the determination of human reliability probabilities.

THERP and SPAR-H are task-related models and ATHEANA is a context-related model, for example.

The basis for estimating human reliability values (e.g., human factors and plant conditions for the scenario of interest) that are considered by the current methods and the probability values used in the methods include both actual data and judgment. Actual data that are applicable and in a form usable for a nuclear power plant (NPP) risk assessment are sparse, however. Also, because there is significant judgment required in performing a HRA, there is considerable uncertainty in the HRA results that can lead to skepticism about the credibility of the results. Thus, HRA is among the most uncertain portions of a NPP risk assessment.

6.4 COMBINATION OF FAILURES AND CONSEQUENCES

A variety of techniques can be employed depending on the purpose of the analysis and the resources available with which to carry out risk evaluations combining failures and consequences. The techniques loosely can be categorized as either inductive or deductive techniques, depending on how estimates for the frequencies of the failure events are obtained. Inductive methods are characterized by the use of spreadsheets with different columns for summarizing the analysis whereas deductive methods typically are done with event tree and fault tree analysis. Sometimes an inductive analysis will be performed as a preliminary step in developing a deductive analysis.

6.4.1 Inductive Methods

One example of an inductive method is a preliminary hazard analysis (PHA) that emphasizes the potential hazards posed to plant personnel and other people following an undesired initiating event of a system. A PHA typically includes consideration of corrective measures to be taken following an undesired initiating event.

A failure modes and effects analysis (FMEA) is another example of an inductive approach. Different spreadsheet formats can be followed for a FMEA, but all require that the failure malfunction of each component be considered, including the mode of failure. A failure mode effects and criticality analysis (FMECA) is similar to a FMEA except that the criticality of the failure is analyzed. A criticality analysis consists of ranking each potential failure mode according to the combined influence on the system of the severity of occurrence and an intuitive estimate of the probability of occurrence of an event obtained either from tabulated sources or from an analyst's best estimate. One possible severity classification for failure modes of a criticality analysis is in Table 6.5.

The effects of each failure are traced through the system in order to assess the ultimate effect of any failure mode on the overall system performance according to the severity class. Typical column headings for a FMECA spreadsheet are in Table 6.6. Care must be taken when interpreting results from different FMECAs, however, because sometimes alternative severity classification schemes are used, as illustrated in Table 6.7.

Table 6.5 Severity Classification Scheme for Failure Modes

Severity Class	Effect of Failure Mode
Category I: Catastrophic	May cause death or complete mission loss
Category II: Critical	May cause severe injury or major system degradation, damage, or reduction in mission performance
Category III: Marginal	May cause minor injury or degradation in system or mission performance
Category IV: Minor	Does not cause injury or system degradation but may result in system failure and unscheduled maintenance or repair

Source: [MIL80].

Table 6.6 Sample Column Headings for FMECA Spreadsheet

System component
Failure mode
Effects on other system components
Effects on entire system:
Failure frequency
Severity class
Remarks

Table 6.7 Sample Classification System for FMECA

Consequence Category	Effect of Failure
Class I: Benign	Negligible effect on system
Class II: Marginal	Degraded system but not causing major damage or personal injuries
Class III: Critical	Significantly degraded system that without immediate action will cause loss of the system or serious personal injuries and/or deaths
Class IV: Catastrophic	Severely degraded system causing immediate loss of the system and personal injuries and/or deaths

Source: [Lam73].

A hazard and operability study (HAZOPS) is yet another inductive approach, like a FMEA or FMECA, for searching for initiating failure events that can cause undesired system failures. A HAZOPS approach consists of employing a set of guide words, such as those in Table 6.8, to suggest ways a system might not perform as

Table 6.8 Sample Guide Words for HAZOPS or Other Analysis Methods

Process Parameter	Failure Event
Flow	None
	Too little
	Too much
	Reverse
	To wrong place
	Erratic
Temperature	Too low
	Too high
	Erratic
Pressure	Too low
	Too high
	Erratic
Volume	Too small
	Too large
	Erratic
pH	Too low
	Too high
	Erratic
Viscosity	Too low
	Too high
	Erratic

designed. The analysis is carried out, in a manner analogous to following a piping and instrumentation diagram, by repeatedly asking what would happen if the fluid or electrons in that portion of the diagram were to not fulfill the intended design function. During the analysis, maintenance and testing issues are addressed, as well as the system performance during startup and shutdown.

The advantages of such inductive analyses are that they are simple to apply and they encourage an orderly examination of the hazard conditions of a system. The principal disadvantage of an inductive analysis is that it is difficult to consider more than one failure event at a time, and not multiple or common cause failures, so such inductive methods tend to not be widely employed for nuclear systems.

6.4.2 Event Tree Analysis

For complicated nuclear systems, event trees as introduced in Section 2.2 provide a step-by-step risk analysis technique with which to evaluate the progression of system failure events that follow an undesired initiating event E_i and lead to a predicted undesired consequence. That is, one follows a sequence of events from the initiating failure through the states of safety systems that should be activated. The outcome of

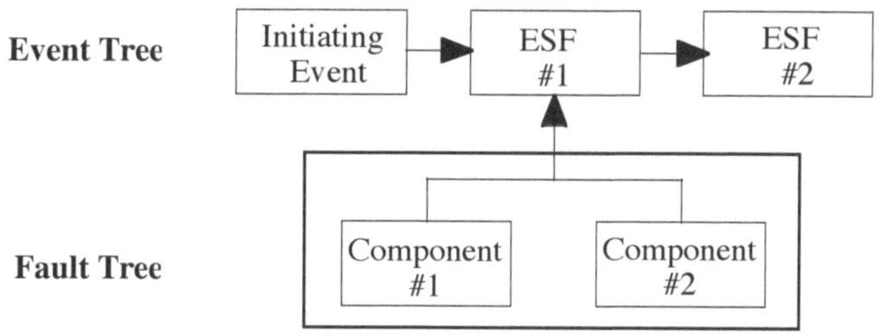

Figure 6.1 Illustrative PRA block diagram linking a fault tree to an event tree featuring two engineered safety features (ESFs).

each subsequent event leads to the question "What happens next?" that is answered by defining the following event.

One advantage of an event tree approach is that minimal cut sets can be used to give a qualitative understanding of the importance of different failure sequences. Boolean algebra techniques help eliminate redundancies arising in an event tree analysis in which an accident sequence is presumed to involve a given safety subsystem that is to be activated at different times.

Once an event tree has been constructed so that the results associated with each accident sequence have been defined, the final task is to compute the probabilities of system failure. For nuclear systems, the *conditional* probabilities of each of the system failure events in an event tree typically are obtained by fault tree analysis, as illustrated with the PRA block diagram in Fig. 6.1.

Consider the elementary event tree for a loss-of-coolant accident (LOCA), as shown in Fig. 6.2. Following the initiating event of a pipe break, with a predicted frequency F_0, electric power must be available for the emergency core cooling system to function. The probability P_1 is for the failure of electric power and subsequent *conditional* failure probabilities for system events 2, 3, and 4 follow. Multiplication of the conditional probabilities for the branches involved in a sequence then gives the probability of that sequence. The undesired outcome of loss of containment integrity occurs either by the system failure sequence $F_1 = F_0(1 - P_1)P_2P_3P_4$ or by $F_2 = F_0P_1$, so the total frequency of system failures resulting in unacceptable consequences is $F_1 + F_2$.

In nuclear power applications the type of PRA is often referred to by one of three levels:

- Level 1: A system analysis to determine the core damage frequency based on system and human factor evaluations.

- Level 2: A containment analysis to determine the core performance and fission product release to the environment.

- Level 3: A consequence analysis to determine off-site transport and health effects of fission product releases.

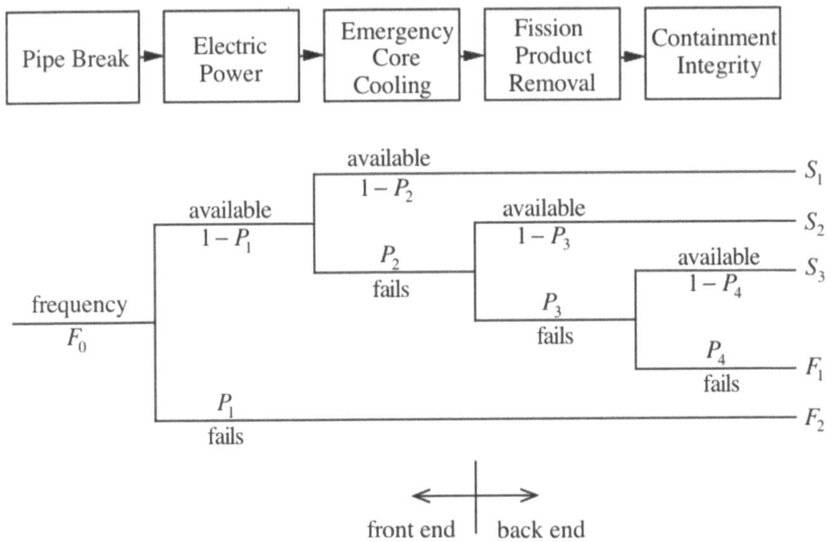

Figure 6.2 Simplified event tree for a loss-of-coolant accident.

Thus, the simplified event tree of Fig. 6.2 represents a LOCA analysis comprising Levels 1 and 2 of a PRA. The Level 1 analysis involving failures of the systems leading to the core damage is also referred to as a front-end PRA, while the Level 2 analysis representing events following the core damage is called a back-end PRA. A detailed discussion of event trees is presented with a PRA example for five nuclear plants in Chapter 10.

6.5 FAULT TREE ANALYSIS

6.5.1 Introduction

A fault tree is constructed by first defining the *top event* that must delineate the precise nature of the undesired failure event. The occurrence of the top event can be analyzed by answering the question "What caused that?" One works downward in the tree to determine the causes of subordinate failure events, each of which is described by its own fault tree of successively subordinate failure events. All subordinate fault trees are connected by logical connective functions, called *gates*, such as AND and OR. Thus the final fault tree consists of successive layers of events, connected by gates, that are possible contributors to the top failure event.

The construction of each branch of a fault tree is terminated when a contributing failure event can be divided no further or when it is decided to limit further analysis of the possible causes of a possible failure. Such terminal events are *basic events*. A basic failure event can occur because of either a structural fault or failure to open or close, or to start or stop, etc. Other basic failure events can arise because a system is out of tolerance so that it fails because of excessive operational or environmental stress placed on the system element.

As a general rule, a system is well designed if there are many AND gates near the top of the fault tree because, from Eqs. (2.4) and (2.15), the probability of an undesired fault event is decreased when two or more subordinate events are linked to another with an AND gate, whereas that probability is increased when two or more subordinate events are linked via an OR gate.

6.5.2 Fault Tree Construction

A very important first step when constructing a fault tree is to clearly define the undesired top event. This is critical because every following event must be considered in terms of its effect upon that top event. The next step is to identify contributing events that can directly cause the top event to occur. At least five possibilities exist:

1. Failure of input to the device, such as a signal to operate
2. Failure of the device itself, by each of the possible modes of failure, such as those in Table 6.2
3. Failure due to operator error, such as failure to actuate a switch or to properly install the device
4. Failure due to a common failure mode with other components, such as the same manufacturing defect in the components
5. Failure due to an external event not considered a part of the system, such as an earthquake

If it is decided that a given contributing failure event is a primary failure mode, that branch of the tree is stopped and the basic event is shown graphically by a circle, as illustrated in Table 6.9.

Sometimes it is necessary to stop the further development of a branch of the tree because of time constraints or if sufficient information is unavailable about causes of the basic failure event. These underdeveloped primary events customarily are identified by a diamond symbol rather than a circle symbol.

It may be that an event occurs only under certain conditions that cause the system component to be out of tolerance. The failure of this secondary event is denoted by an INHIBIT gate. In other cases the termination of a tree branch may be an expected event caused by the environment in which the system is placed. Such a normally occurring basic event is not a fault event so it is denoted by a house symbol.

A few general guidelines for constructing fault trees are given in Table 6.10.

6.5.3 Qualitative Fault Tree Analysis

For a fault tree analysis (FTA), once a fault tree has been constructed, it can be utilized in different ways. The purpose of a *qualitative analysis* is to reduce the tree to a logically equivalent form in terms of the specific combinations of basic events sufficient to cause the undesired top failure event to occur. Each combination of

Table 6.9 Fault Tree Symbols Commonly Used

Symbol	Name	Description
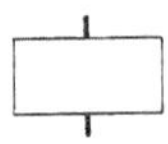	Rectangle	Fault event; it is usually the result of the logical combination of other events
	Circle	Independent primary fault event
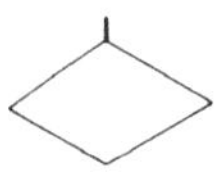	Diamond	Fault event not fully developed as to its causes; it is only an assumed primary fault event
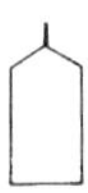	House	Normally occurring basic event; it is not a fault event
	OR gate	The union operation of events; i.e., the output event occurs if one or more of the inputs occur
	AND gate	The intersection operation of events; i.e., the output event occurs if and only if all the inputs occur
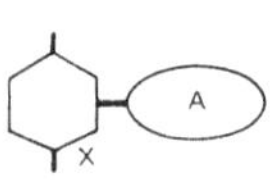	INHIBIT gate	Output exists when X exists and condition A is present; this gate functions somewhat like an AND gate and is used for a secondary fault event X
	Triangle in	Triangle symbols provide a tool to avoid repeating sections of a fault tree or to transfer the tree construction from one sheet to the next. The triangle in appears at the bottom of a tree and represents that branch of the tree (in this case "A") shown someplace else. The triangle out appears at the top of a tree and denotes that the tree "A" is a subtree to one shown someplace else.
	Triangle out	

failure modes contributing to the top failure event is a minimal cut set of failure modes for the tree. In many situations, a qualitative analysis may be sufficient to alert a system designer that changes need to be made to increase the reliability of a system. Such an approach also may be the only practical use of a fault tree if numerical data for the subordinate basic failure events are not available.

A qualitative evaluation of a fault tree provides valuable information about the different event combinations required for failure of the system defined by the top event. Such an evaluation can be carried out as follows:

Table 6.10 Fault Tree Construction Guidelines

1. A fault event description should include
   (a) the fault state of that system or component,
   (b) when that system or component is in the fault state, and
   (c) a label for performing minimal cut set evaluations.
2. Provide a label for each gate for performing minimal cut set evaluations.
3. If the fault event involves the state of the system, then an AND, OR, or INHIBIT gate is used to connect the immediate fault events and the minimum necessary and sufficient subordinate fault events.
4. If the fault event involves the state of a component, then look for
   (a) primary failure events when the component is operating within its intended design limits and environment,
   (b) secondary failure events when the component is operating outside its intended design limits or environment,
   (c) command failure events when the component is improperly actuated either by automatic or manual means, and
   (d) other possible human errors that can cause failure events.
5. Always put an event statement between any two gates to explain the event.

1. Label the primary events.
2. Label the gates and list the gate type and inputs.
3. Write a Boolean equation for each gate.
4. Use Boolean algebra to solve for the top event in terms of the cut sets.
5. Use Boolean algebra to eliminate the cut set redundancies to obtain the minimal cut sets.

Example 6.2 The input to a top event T is an OR gate labeled g_1 with inputs consisting of a primary failure event A and fault event B, which is the output of an OR gate g_2 that has inputs consisting of primary event C or the output of gate g_3. The input to gate g_3 is primary events A and D. Obtain the Boolean equation for the top event.

With $T = g_1$, $g_1 = A + B$, $B = g_2$, $g_2 = C + g_3$, and $g_3 = AD$, the resulting top event is $T = A + C + AD$ and the set of minimal cut sets is $T = A + C + D$, as determined with the aid of Boolean algebra described in Section 2.2. ◇

In most instances, a fault tree that is constructed by following through each system component in a logical manner will have redundancies that can be eliminated by the use of Boolean algebra to construct a reduced fault tree that will have fewer gates and events.

Example 6.3 Consider the electrical circuit in Fig. 6.3. (a) Construct a fault tree and (b) use Boolean algebra to construct a reduced fault tree.

(a) See the fault tree in Fig. 6.4, where transfer-in and transfer-out symbols were used to avoid needless repetition of portions of the fault tree. In this circuit, secondary failure events exist for the motor and the relay contacts, so INHIBIT

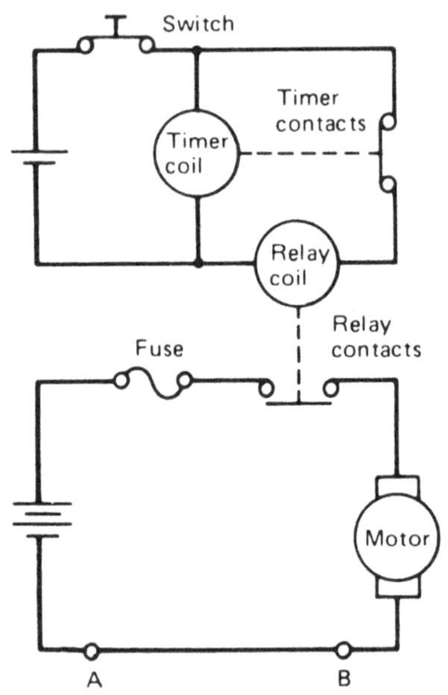

Figure 6.3 An electrical circuit. *Source:* [Ham72].

gates were needed; for example, the output of INHIBIT gate G_1 is the event represented by the output of OR gate A_3 and event Y_1.

(b) From Fig. 6.4 it follows that

$$\begin{aligned} B_1 &= (X_1 + X_2)[X_4 + X_3Y_1 + (X_7 + X_8)(X_9 + X_{10})Y_1] \\ &\quad \times [(X_1 + X_2)X_5Y_2 + X_6 + (X_7 + X_8)(X_9 + X_{10})]. \end{aligned}$$

Because some of the events are identical, i.e.,

$$\begin{aligned} X_4 = X_5 = Y_1, \\ X_3 = X_6 = Y_2, \end{aligned}$$

it follows from the idempotent law of Boolean algebra in Table 2.1 that

$$B_1 = (X_1 + X_2)X_4[X_3 + (X_7 + X_8)(X_9 + X_{10})].$$

Figure 6.5 illustrates the fault tree corresponding to this Boolean expression. ◇

6.5.4 Quantitative Fault Tree Analysis

The failure probabilities for basic events can be of two types, corresponding to "instantaneous" failures or failures that can occur after the passage of time. The

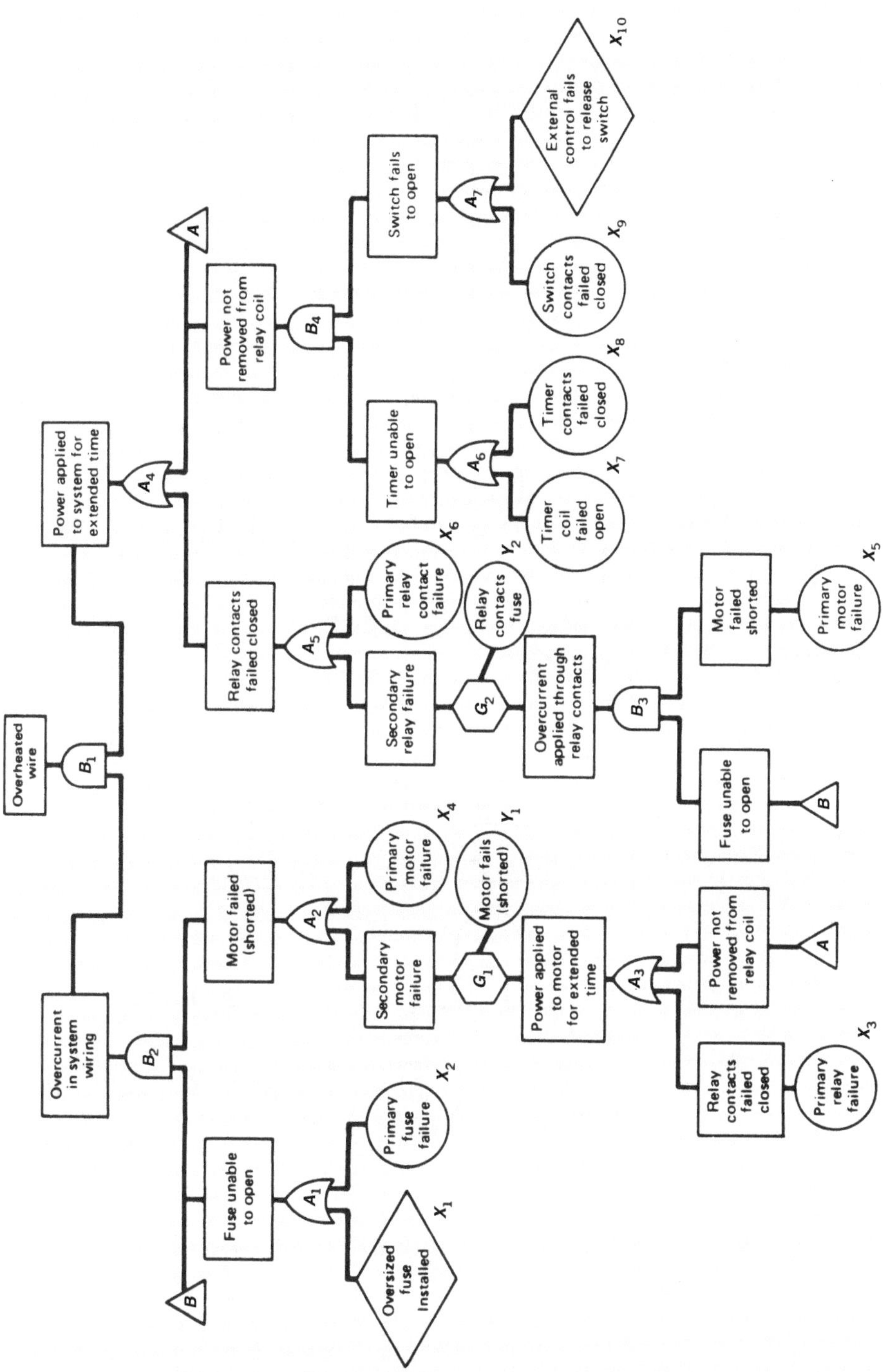

Figure 6.4 Fault tree for electrical circuit in Figure 6.3. *Source:* [Ham72].

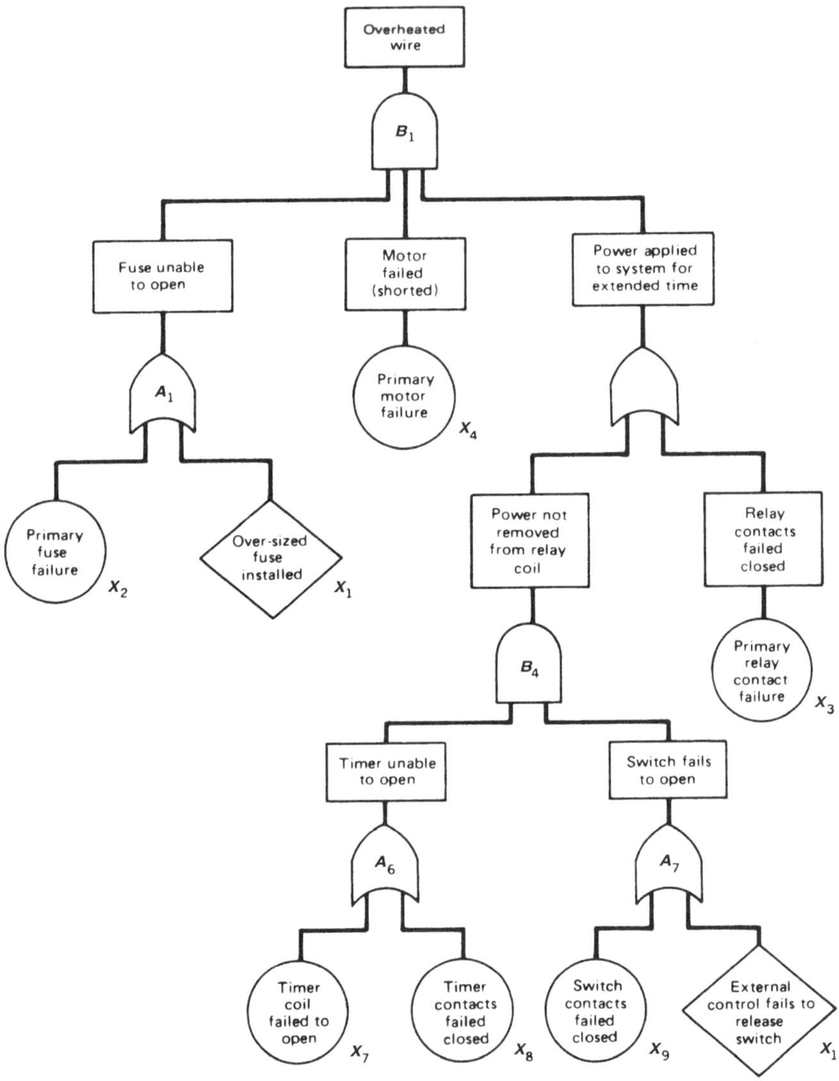

Figure 6.5 Reduced fault tree for electrical circuit in Figure 6.4. *Source:* [Ham72].

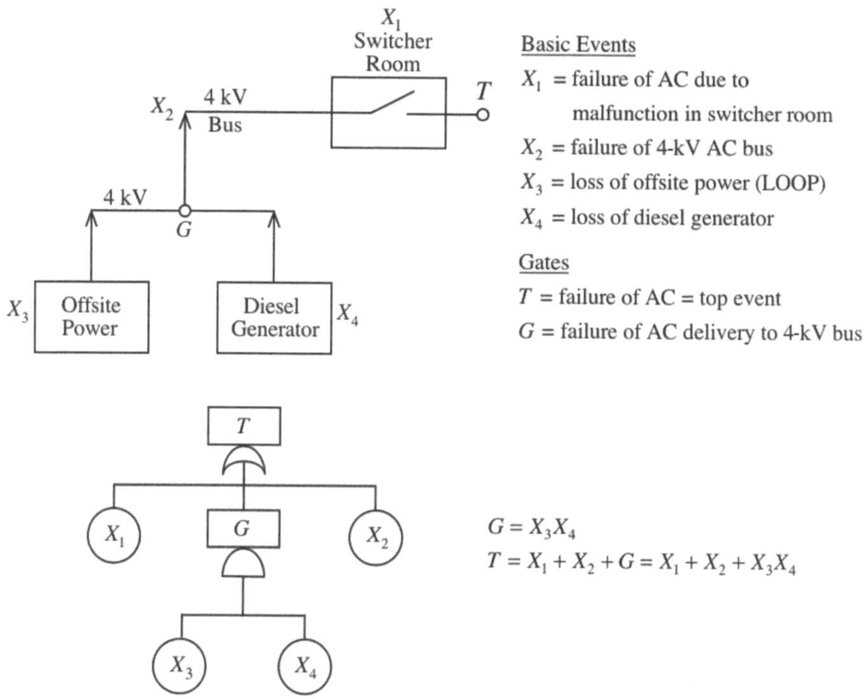

Figure 6.6 Simplified electrical system and its fault tree.

latter failures are probabilities per unit time and hence require that a mission time be defined for a component's operation so that all basic event failures are characterized by numerical values.

For a *quantitative* evaluation of a fault tree, the probability of occurrence is the probability of the union of the minimal cut sets.

Example 6.4 Consider the simple circuit in Fig. 6.6 that depicts the electric power event in the LOCA event tree of Fig. 6.2. (a) Construct a fault tree for the failure of emergency electric power as the top event and (b) from the fault tree determine the probability of a station blackout arising from failure of AC power, which is P_1 of Fig. 6.2.

(a) See Fig. 6.6 for the basic events X_1 through X_4, the gates T and G, and the resulting fault tree.
(b) From the fault tree it is seen that the probability of loss of electric power is

$$\begin{aligned} P_1 &= P(T) = P(X_1 + X_2 + X_3X_4) \\ &= P(X_1) + P(X_2) + P(X_3X_4) - P(X_1)P(X_2) - P(X_1)P(X_3X_4) \\ &\quad -P(X_2)P(X_3X_4) + P(X_1X_2X_3X_4) \\ &= P(X_1) + P(X_2) + P(X_3)P(X_4) - P(X_1)P(X_2) \end{aligned}$$

$$-P(X_1)P(X_3)P(X_4) - P(X_2)P(X_3)P(X_4)$$
$$+P(X_1)P(X_2)P(X_3)P(X_4) \quad \diamond$$

In cases where the basic event probabilities are all small so that the rare-event approximation can be used, such as in Eq. (2.23), then a quick quantitative estimate of the top-event probability can be obtained by marching through the tree from bottom to top while adding the probabilities for each input to an OR gate and multiplying them for each input to an AND gate.

Example 6.5 A system of three components labeled A, B, and C is designed to fail when any two components fail. The failure probabilities of the components are 10^{-3}, 2×10^{-3}, and 3×10^{-3}, respectively. Calculate the failure probability of the system using the rare-event approximation for (a) a fault tree not in reduced form and (b) the reduced fault tree.

(a) A fault tree with a top event T of "system failure" *could* be constructed with an OR gate g_1 connected to three secondary events consisting of the AND gates $g_2 = g_5 C$, $g_3 = g_6 B$, and $g_4 = g_7 A$, with OR gates $g_5 = A + B$, $g_6 = A + C$, and $g_7 = B + C$ in terms of the basic failure events A, B, and C. If one were to proceed directly to a quantitative evaluation of the resulting tree,

$$P(g_2) = P(A+B)P(C) \approx [P(A)+P(B)]P(C) = [(1+2)3]10^{-6} = 9 \times 10^{-6},$$
$$P(g_3) = P(A+C)P(B) \approx [P(A)+P(C)]P(B) = [(1+3)2]10^{-6} = 8 \times 10^{-6},$$
$$P(g_4) = P(B+C)P(A) \approx [P(B)+P(C)]P(A) = [(2+3)1]10^{-6} = 5 \times 10^{-6},$$
$$P(T) = P(g_2 + g_3 + g_4) \approx P(g_2) + P(g_3) + P(g_4) = 2.2 \times 10^{-5}.$$

(b) To check whether the fault tree of part (a) is in reduced form, Boolean algebra can be used to find

$$\begin{aligned} T &= (A+B)C + (A+C)B + (B+C)A \\ &= AC + BC + AB + BC + AB + AC \\ &= AC + BC + AB + \cancel{BC} + \cancel{AB} + \cancel{AC} \\ &= AB + AC + BC \end{aligned} \tag{6.7}$$

which gives a fault tree for top event T constructed with an OR gate g_1' connected to three secondary events consisting of the AND gates $g_2' = AB$, $g_3' = AC$, and $g_4' = BC$. If we now calculate the probability of the top event, we obtain

$$\begin{aligned} P(T) &= P(g_2' + g_3' + g_4') \approx P(AB) + P(AC) + P(BC) \\ &\approx P(A)P(B) + P(A)P(C) + P(B)P(C) \\ &= [1(2) + 1(3) + 2(3)]10^{-6} = 1.1 \times 10^{-5}. \end{aligned}$$

The reason the numerical answer in part (a) differs from that in part (b) is that in part (a) numerical values were included for the terms $\cancel{BC} + \cancel{AB} + \cancel{AC}$ eliminated with an absorption rule of Boolean algebra.

We conclude that it is necessary to obtain the minimal cut sets *before* quantitatively evaluating a fault tree to find the failure probability of the top event. $\diamond$

6.5.5 Common Cause Failures and Fault Tree Analysis

Common cause failures, discussed in Section 6.2.2, add complications to fault tree analysis. The commonality can involve a failure in one system that causes a failure in another system or an adverse environment causing simultaneous failures in more than one system. CCFs are frequently due to errors in the design of a system. Overlooked CCFs potentially can lead to errors of orders of magnitude in the computed probability of a top event if failure of one branch of a tree can cause a failure of another branch.

One way around the dilemma of a possible CCF is to use the Boolean NOT function and the irreducible building block shown in Fig. 6.7. The building block for dependent event B involves the conditional event $B|A$, event B given event A, and conditional event $B|\overline{A}$, event B given "not A." The building block causes the dependency condition to occur between the disjoint events A AND $B|A$ and $\overline{A}$ AND $B|\overline{A}$. To analyze a CCF, consider event A to be the possible common cause initiating event and event B to be the dependent event. If events A and B are connected with the AND gate of Fig. 6.7, then the probability of the output event AB is

$$\begin{aligned} P(AB) &= P(A\{A[B|A]+\overline{A}[B|\overline{A}]\}) \\ &= P(A[B|A]) \\ &= P(A)P(B|A). \end{aligned} \tag{6.8}$$

In a similar way, if events A and B are connected with an OR gate, then the irreducible building block is

$$\begin{aligned} P(A+B) &= P(A+\{A[B|A]+\overline{A}[B|\overline{A}]\}) \\ &= P(A)+P(\overline{A})P(B|\overline{A}) \end{aligned} \tag{6.9}$$

provided $P(AB)=0$.

Example 6.6 To illustrate the importance of properly treating a dependent event B, assume $P(A)=0.2$, $P(B|A)=0.3$, and $P(B|\overline{A})=0.8$. From Eqs. (6.8) and (6.9), $P(AB)=0.06$ and $P(A+B)=0.84$.

If it were *assumed* that events A and B were independent, then from Eq. (2.25)

$$P(B)=P(A)P(B|A)+P(\overline{A})P(B|\overline{A})=0.7,$$

from which it follows that

$$\begin{aligned} P(AB) &= P(A)P(B)=0.14, \\ P(A+B) &= P(A)+P(B)-P(A)P(B)=0.76. \end{aligned}$$

Comparison of the results shows that significant errors would arise if this common cause failure were analyzed incorrectly. ◇

6.6 MASTER LOGIC DIAGRAM

A master logic diagram (MLD) is a semigraphical top-down model in a FT structure that represents the functional relationship for a complex system consisting of multiple

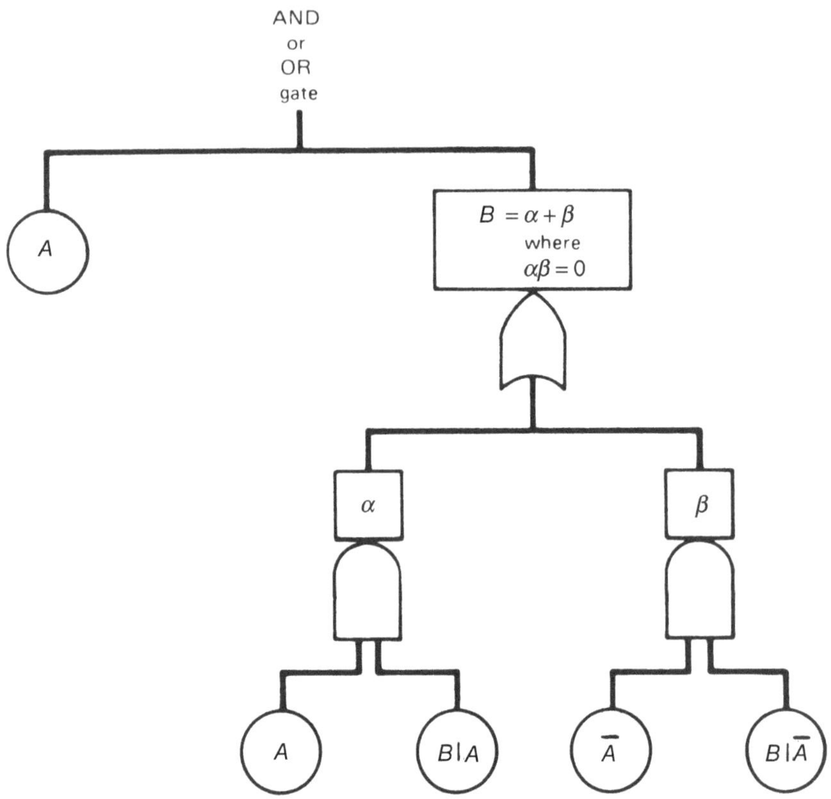

Figure 6.7 Irreducible building block for the probability of event B dependent on the occurrence of initiating event A.

components with different functions. A MLD could be used to systematically identify and group accident initiating events (IEs) that may cause the top event to occur in the system. As a deductive model, it may be considered a summary FT [NRC83] that could represent multiple levels of functionally separate subsystems.

A five-level MLD presented in Fig. 6.8 represents the excessive offsite release of radionuclides in a nuclear power plant as the top event or level 1 event. The top event could result from either an excessive direct release or excessive indirect release representing two possible pathways for the radionuclide release at level 2. Hence, the two pathways are connected through an OR gate to level 1. An excess direct release from the spent-fuel pool or similar facilities typically presents insignificant contributions to the overall NPP risk and the event is not expanded further. On the other hand, a significant radionuclide release may result from events causing an extensive core damage followed by the failure of the reactor coolant system (RCS) pressure boundary and the containment system. These three events are presented

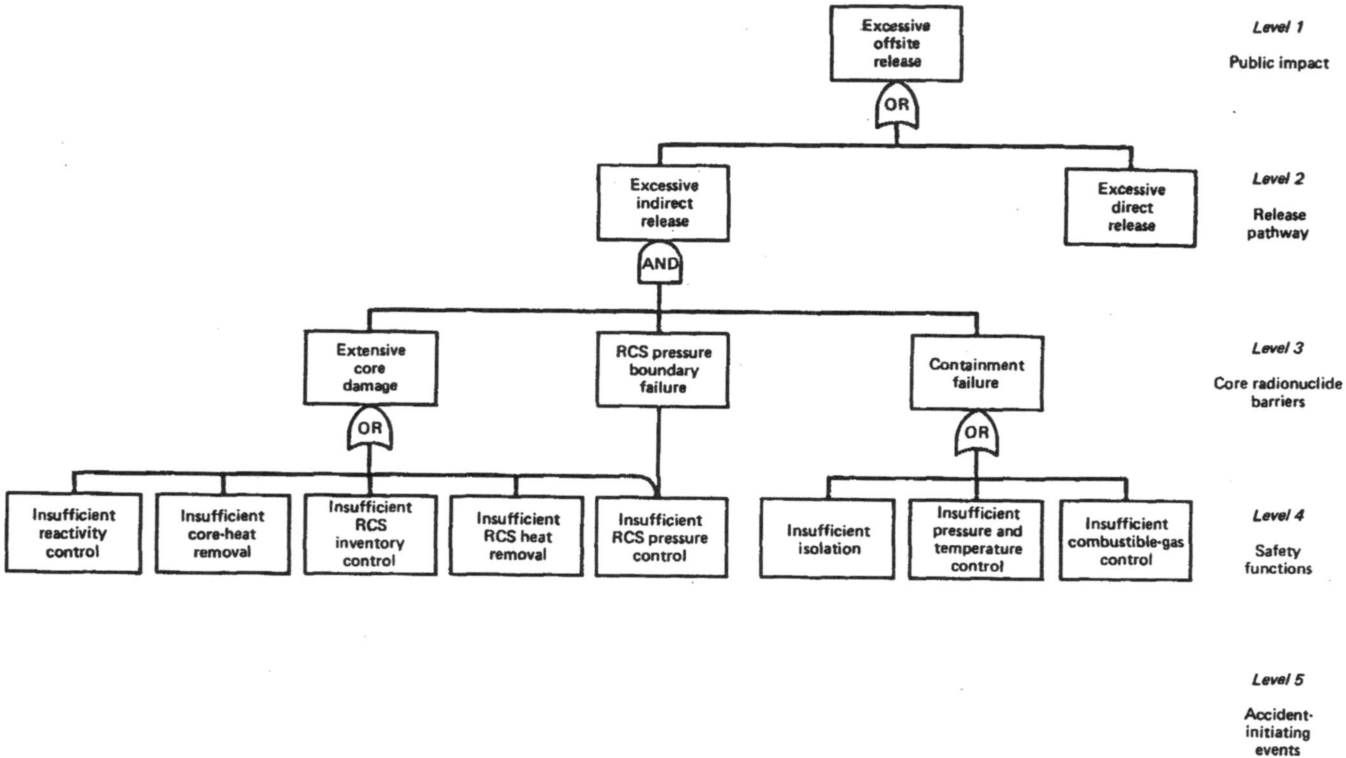

Figure 6.8 Five-level master logic diagram for events leading to excessive offsite release. *Source:* [NRC83].

at level 3 and linked to the excessive direct release pathway by an AND gate. The OR gate for the excessive core damage event illustrates that the event occurs due to the failure of some of the level 4 safety functions associated with (a) reactivity control, (b) core heat removal, (c) RCS inventory control, (d) RCS heat removal, and (e) RCS pressure control. Similar causal relationships between the other two events at level 3 and safety functions at level 4 are illustrated in Fig. 6.8. A further breakdown of the safety functions [NRC83] is usually required at level 5 to determine equipment failures or operational errors that initiate the failure of safety functions. These IEs, together with the actuation of mitigating engineered safety features (ESFs), represented at level 5 would complete the MLD. In an actual application of the MLD, the representation of system details may require additional levels, as illustrated by an eight-level MLD for the Zion PRA study in NUREG-1150, Vol. 2 [NRC90].

By constructing a top-down diagram representing system interactions, the MLD structure has been used to systematically identify IEs in chemical plants [Pap03] as well as in nuclear plants. Once the IEs are identified, the events and ESF actuations may be grouped in an ET structure illustrated in Fig. 6.2. Thus, an MLD may be considered a high-level fault tree structure that could be used systematically to construct ETs for complex systems. Since an MLD is in the structure of a FT, it may also be considered a functional block diagram [Mod99,Kum96] that could be expanded to develop a full-blown FT. A trivial example is the block diagram in Fig. 6.6, which is converted into a FT with the proper gate representations.

The MLD can also be compared with inductive methods, in particular, FMEA and HAZOPS discussed in Section 6.4.1. Both FMEA and HAZOPS may be used to search for initiating failure events in a bottom-up approach by examining deviations from normal operation, the effects of these deviations, and their effects at the functional level of a system representation. Thus, an IE can be identified as a cause of a deviation or a result of a deviation [Pap03], in a manner equivalent to the MLD approach. The MLD approach may, however, offer an advantage over the HAZOPS approach. This is because the success of a HAZOPS in identifying IEs may depend largely on the system details initially represented, while a MLD is constructed to filter down through the levels of system interactions, thereby naturally arriving at the basic functional levels and IEs.

6.7 UNCERTAINTY AND IMPORTANCE ANALYSIS

6.7.1 Types of Uncertainty in PRAs

The results of any PRA for engineered systems, including nuclear energy systems, are inherently uncertain because the FTs and ETs model processes and phenomena that occur infrequently. It is thus highly desirable that the top-event (TE) frequency of a FT or end-state probabilities of an ET be accompanied by a quantification of uncertainties in the results obtained. Uncertainties associated with PRA results are usually grouped [NRC94,Par96] into two general types: aleatory or stochastic uncertainty and epistemic or state-of-knowledge uncertainty:

- Stochastic uncertainty represents inherent variability in any measurable physical quantity and hence cannot be reduced by enlarging the database. Enlarging the database can, however, provide an improved representation of the probability distribution of the physical quantity.
- State-of-knowledge uncertainty is due to a lack of complete knowledge about systems, processes, and phenomena and may be reduced by additional measurement, testing, or analysis. Epistemic uncertainties are usually grouped into uncertainties resulting from a lack of knowledge about parameters used in the system model and those associated with the model itself, including those due to an incomplete representation of possible accident scenarios. The uncertainty in the failure rate of a component is generally considered [NRC94] epistemic, not stochastic.

Treatment of PRA uncertainties in risk-informed regulations and licensing is discussed further in Section 12.1.

6.7.2 Stochastic Uncertainty Analysis

The uncertainty analysis for a PRA typically accounts for stochastic uncertainties in numerical data for basic events in a FT. This is accomplished through propagating the uncertainty for each basic event through AND and OR gates linking the events.

As will be discussed further in Chapter 7 in connection with the SAPHIRE code [NRC08], estimates of stochastic uncertainties in the TE probabilities in FT or combined FT/ET PRA studies can be obtained by Monte Carlo sampling of the probability density functions involved. For relatively simple systems where all basic events can be assumed independent, we can also determine the mean and variance for the TE probability using analytical equations for both AND and OR gates. For AND gate $G = \prod_{n=1}^{N} G_n$,

$$m(G) = \prod_{n=1}^{N} m(G_n), \tag{6.10}$$

$$V(G) = \prod_{n=1}^{N} [V(G_n) + m^2(G_n)] - m^2(G), \tag{6.11}$$

and for OR gate $G = \sum_{n=1}^{N} G_n$,

$$m(G) = \sum_{n=1}^{N} m(G_n), \tag{6.12}$$

$$V(G) = \sum_{n=1}^{N} V(G_n). \tag{6.13}$$

In a direct Monte Carlo sampling of complex gates, the moment estimators of Section 3.1.1 can be used to determine the mean and variance, together with other probability measures, for the PDF of the TE, which may then be used in a subsequent sensitivity analysis.

6.7.3 Sensitivity and Importance Analysis

In addition to representing the inherent stochastic uncertainties via Monte Carlo sampling of the associated PDFs, sensitivity analyses can be performed to assess the impact of changes in the FT structure and numerical values of the basic event probabilities. One obvious analysis is the breakdown of the contributions from individual cut sets to the TE, illustrated as part of the standard SAPHIRE output in Chapter 7. Additional importance measures include the risk reduction measure known as the Fussell-Vesely importance measure [NRC94], often written as

$$I_{FV} = \frac{P(\mathrm{TE}|E_i = \langle E_i \rangle) - P(\mathrm{TE}|E_i = 0)}{P(\mathrm{TE}|E_i = \langle E_i \rangle)}, \tag{6.14}$$

where $P(\mathrm{TE}|E_i = \langle E_i \rangle)$ and $P(\mathrm{TE}|E_i = 0)$ are the upper bounds of a minimal cut set or TE failure rate evaluated with the frequency for basic event E_i set equal to its mean value and equal to zero, respectively. Thus I_{FV} indicates a fractional reduction in the risk associated with the decrease in the frequency of event E_i. The sensitivity of the accident sequence frequency to event E_i also can be written in terms of the Birnbaum importance measure as

$$I_B = P(\mathrm{TE}|E_i = 1) - P(\mathrm{TE}|E_i = 0). \tag{6.15}$$

The evaluation of I_B through Eq. (6.15) can be considered a partial derivative of the accident sequence frequency with respect to the frequency of event E_i and is known to often overestimate the importance of events with small frequencies. Other importance measures include [Han88]

$$\text{Risk reduction ratio} = \frac{P(\mathrm{TE}|E_i = \langle E_i \rangle)}{P(\mathrm{TE}|E_i = 0)}, \tag{6.16}$$

$$\text{Risk increase ratio} = \frac{P(\mathrm{TE}|E_i = 1)}{P(\mathrm{TE}|E_i = \langle E_i \rangle)}. \tag{6.17}$$

References

[Bor05] R. L. Boring and D. I. Gertman, "Atomistic and Holistic Approaches to Human Reliability Analysis in the U.S. Nuclear Power Industry," *Safety Reliab.* **25**, 21 (2005).

[Dhi86] B. S. Dhillon, *Human Reliability with Human Factors*, Pergamon (1986).

[For04] J. Forester et al., "Expert Elicitation Approach for Performing ATHEANA Quantification," *Reliab. Eng. Sys. Safety* **83**, 207 (2004).

[Ful88] R. R. Fullwood and R. E. Hall, *Probabilistic Risk Assessment in the Nuclear Power Industry: Fundamentals & Applications*, Pergamon (1988).

[Ger94] D. I. Gertman and H. S. Blackman, *Human Reliability & Safety Analysis Data Handbook*, Wiley Interscience (1994).

[Ham72] W. Hammer, *Handbook of System and Product Safety*, Prentice-Hall (1972).

[Han88] S. H. Han, T. W. Kim, and K. J. Yoo, "Development of an Integrated Fault Tree Analysis Computer Code MODULE by Modularization Technique," *Reliab. Eng. Sys. Safety* **21**, 145 (1988).

[IEE84] "IEEE Guide to the Collection and Presentation of Electrical, Electronic, Sensing Component, and Mechanical Equipment Reliability Data for Nuclear-Power Generating Stations," IEEE Std 500-1984, IEEE Power Engineering Society (1984).

[IEE98] "IEEE Recommended Practice for the Design of Reliable Industrial and Commercial Power Systems," IEEE Std 493-1997, IEEE Industry Applications Society (1998).

[IEE07] "IEEE Recommended Practice for the Design of Reliable Industrial and Commercial Power Systems," IEEE Std 493-2007, IEEE Industry Applications Society (2007).

[Kum96] H. Kumamoto and E. J. Henley, *Probabilistic Risk Assessment and Management for Engineers and Scientists*, 2nd ed., IEEE Press (1996).

[Lam73] H. E. Lambert, "System Safety Analysis and Fault Tree Analysis," UCID-16238, Lawrence Livermore Laboratory (1973).

[MIL80] "Procedure for Performing a Failure Mode, Effects, and Criticality Analysis," MIL-STD-1629A, U.S. Department of Defense (1980).

[Mod99] M. Modarres, M. Kaminskiy, and V. Krivtsov, *Reliability Engineering and Risk Analysis: A Practical Guide*, CRC Press (1999).

[NRC75] "Reactor Safety Study: An Assessment of Accident Risks in U.S. Commercial Nuclear Power Plants," WASH-1400 or NUREG-75/014, U.S. Nuclear Regulatory Commission (1975).

[NRC83] "PRA Procedures Guide," NUREG/CR-2300, U.S. Nuclear Regulatory Commission (1983).

[NRC90] "Severe Accident Risks: An Assessment for Five U.S. Nuclear Power Plants," NUREG-1150, U.S. Nuclear Regulatory Commission (1990).

[NRC94] "A Review of NRC Staff Uses of Probabilistic Risk Assessment," NUREG-1489, U.S. Nuclear Regulatory Commission (1994).

[NRC96] "A Technique for Human Error Analysis (A THEANA)—Technical Basis and Methodology Description," NUREG/CR-6350, U.S. Nuclear Regulatory Commission (1996).

[NRC03a] "Issues and Recommendations for Advancement of PRA Technology in Risk-Informed Decision Making," NUREG/CR-6813, U.S. Nuclear Regulatory Commission (2003).

[NRC03b] "Common-Cause Failure Event Insights Emergency Diesel Generators," NUREG/CR-6819, vol. 1, U.S. Nuclear Regulatory Commission (2003).

[NRC05] "The SPAR-H Method," NUREG/CR-6883, U.S. Nuclear Regulatory Commission (2005).

[NRC06] "Human Event Repository and Analysis (HERA) System, Overview," NUREG/CR-6903, vol. 1, U.S. Nuclear Regulatory Commission (2006).

[NRC08] "Systems Analysis Program for Hands-On Integrated Reliability Evaluations (SAPHIRE), Technical Reference," NUREG/CR-6952, vol. 2, U.S. Nuclear Regulatory Commission (2008).
[Pap03] I. A. Papazoglou and O. N. Aneziris, "Master Logic Diagram: Method for Hazard and Initiating Event Identification in Process Plants," *J. Hazard. Mater.* **A97**, 11 (2003).
[Par96] "The Characterization of Uncertainty in Probabilistic Risk Assessment of Complex Systems," *Reliab. Eng. Sys. Safety* **54**, 119 (1996).
[Sin93] A. Singh, G. W. Parry, and A. Beare, "An Approach to the Analysis of Operating Crew Responses for Use in PSAs," *Proc. Probabilistic Safety Assessment Conference*, Clearwater Beach, FL (1993).
[Spu10] A. J. Spurgin, *Human Reliability Assessment Theory and Practice*, CRC Press (2010).
[Swa83] A. D. Swain and H. E. Guttman, "The Technique for Human Error Rate Prediction," NUREG/CR-1278, U.S. Nuclear Regulatory Commission (1983).
[Vil92] A. Villemeur, *Reliability, Availability, Maintainability and Safety Assessment*, vol. 1, Wiley (1992).
[Wag77] D. P. Wagner, C. L. Cate, and J. B. Fussell, "Common Cause Failure Analysis for Complex Systems," in *Nuclear Systems Reliability Engineering and Risk Assessment*, J. B. Fussell and G. R. Burdick, eds., 289, Soc. Industrial and Applied Mathematics (1977).

Exercises

6.1 Construct a fault tree for the failure of a two-tube fluorescent fixture to emit light. The events that should be considered are human error H (the switch is "off" rather than "on"), E (external electrical power outage), S (switch failure in the "off" position), and T_i (failure of tube i). Also include primary events for failure of the wiring between the switch and tubes (W_{st}) and between the power supply and the switch (W_{es}).

6.2 Construct fault trees for failure of systems (a) through (d) of Fig. 6.9 to operate and obtain the minimal cut sets.

6.3 A system of switches is connected as shown in Fig. 6.10. The probability per demand that a switch fails in the closed position is 10^{-4} and in the open position is 10^{-3}. There are no other causes of failure. (a) Construct a fault tree for the top event "T_c, the circuit fails closed," (b) identify the minimal cut sets, (c) evaluate the probability of the top event T_c, (d) construct a second fault tree for the top event "T_o, the circuit fails open," (e) identify the minimal cut sets, and (f) evaluate the probability of the top event T_o.

6.4 For the flow system shown in Fig. 6.11 that has components A through P, (a) construct a fault tree for failure of flow, (b) determine the minimal cut sets, and (c) calculate the failure probability of the system if each component has a failure probability of 10^{-3}.

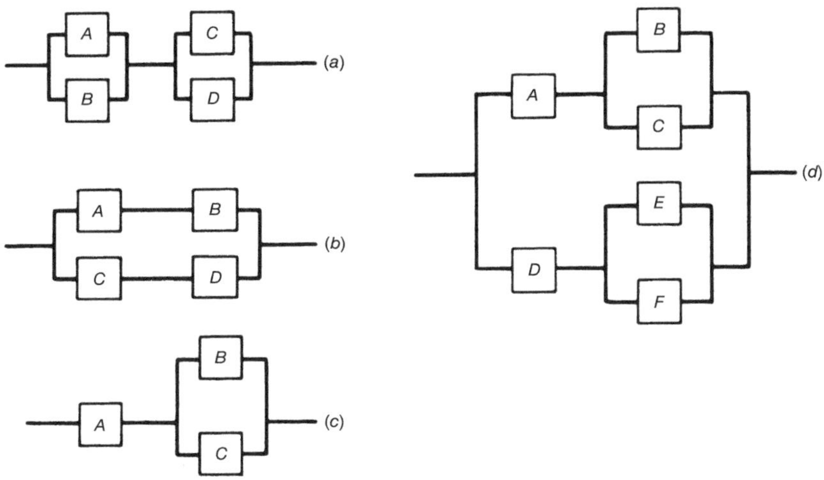

Figure 6.9 Figure for Exercise 6.2.

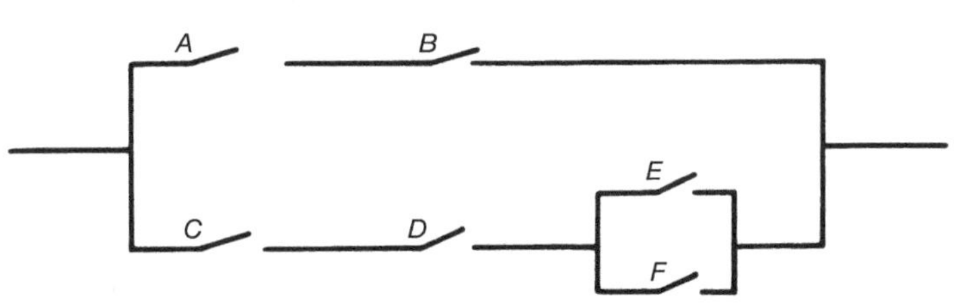

Figure 6.10 Figure for Exercise 6.3.

6.5 Construct any simple fault tree for top event T in terms of AND gates G_n for $n = 1, \ldots, N$ and OR gates for $n = (N+1), \ldots, M$ using basic fault events $E_1,\ E_2,\ \ldots$. Use your tree and De Morgan's theorems in Table 2.1 to show that the corresponding *success* tree for top event $\overline{T}$ can be constructed from the same tree structure provided the basic events are success events $\overline{E}_1,\ \overline{E}_2, \ldots$ and the gates for $n = 1, \ldots, N$ are changed to OR gates and those for $n = N+1, \ldots, M$ are changed to AND gates.

6.6 A nuclear reactor has three coolant loops, each with two pumps in active parallel. Only one loop is needed to cool the core and a loop can fail if one pump fails or the pipe between the two pumps and core ruptures. Failure of the external power supply or the plant internal power supply to the loops also will cause system failure. Construct a fault tree for failure to cool the core.

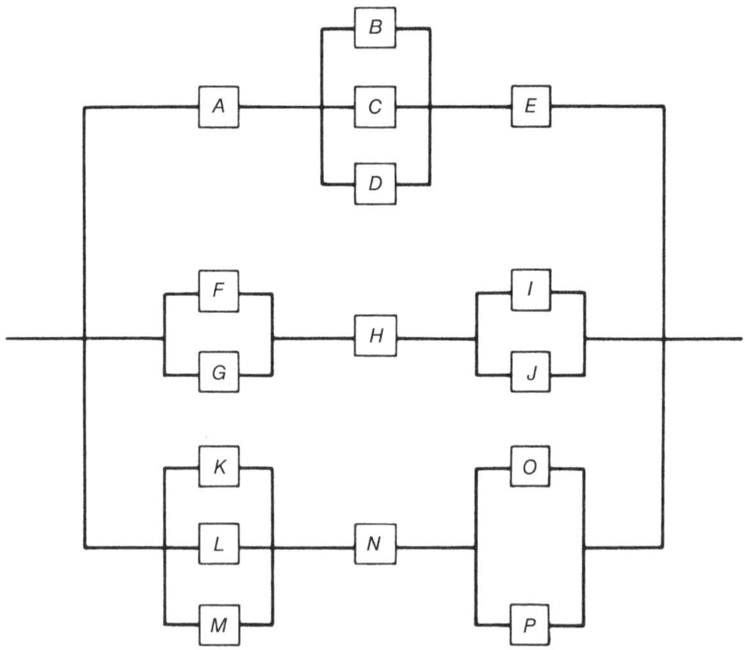

Figure 6.11 Figure for Exercise 6.4.

6.7 For a submarine reactor safety circuit there are three sensor systems, each containing two period meters. Each sensor system is wired by connecting the period meters in active parallel and it is activated whenever the reactor power level is increasing with a period of less than 10 sec. A sensor system can fail either by failure of both meters or the wiring or by failure of its own power source. An automatic scram of the reactor is initiated if two of the three sensor systems are activated and powered. (a) Construct a fault tree to compute the failure of actuation of the automatic scram system when the reactor period is less than 10 sec, (b) calculate the unavailability of the scram system for a 10-month period using the rare-event approximation if the hazard rate for every period meter is 10^{-2}/month, for the wiring is 10^{-6}/month, and for every external power source is 10^{-4}/month, and (c) repeat part (b) without using the rare-event approximation.

6.8 For the domestic hot-water system of Fig. 6.12, (a) construct a fault tree for the top event "rupture of water tank when system is in operation and faucet is closed." In your analysis, consider the following:

1. Tank failure due to rupture or damage during shipping and installation
2. Pressure relief valve failures due to its being jammed closed or to piping to and/or from the valve being blocked, or to the use of an incorrect size of valve or piping or to an incorrect installation (human error), or to an improper setting of the relief valve (human error)

3. Excess inlet water supply pressure due to the normal pressure of the water supply exceeding the design pressure limit of the tank (human error), or to the gas valve being jammed open or to the gas valve being damaged during shipping and installation, or to the controller failing open, or to the temperature measuring and comparing device failing, or to the temperature measuring and comparing device and controller being disconnected from the tank (human error)

(You may wish to compare your tree to that in [Lam73].) (b) For your fault tree, evaluate the probability of the top event occurring during the first year of operation if

1. all human errors have a probability of 10^{-3}/demand,
2. all damage during shipping and installation has a probability of 2×10^{-3}/demand,
3. blockage of pipe has a probability of 3×10^{-4}/year,
4. rupture of tank has a probability of 3×10^{-2}/year,
5. relief valve fails closed has a probability of 10^{-5}/day,
6. controller failure has a probability of 2×10^{-2}/year,
7. gas valve fails open has a probability of 10^{-3}/year, and
8. temperature measuring and comparing device fails has a probability of 10^{-2}/year.

(c) What failure probability most strongly affects your result from part (b)?

6.9 Consider the loss-of-offsite power (LOOP) for a PWR as an initiating event that could produce core melt given a sequence of subsequent failures. Draw an event tree for the possible sequence of events following LOOP that describes the success paths (adequate core cooling) and failure paths (core melt), and calculate the core melt frequency. Assume that (1) the main feedwater is unavailable without offsite power and (2) auxiliary feedwater may be supplied by either one of electric or steam-turbine-driven pumps, each of which is capable of providing enough cooling water to remove decay heat, (3) the electrically driven pump is connected to one of the diesel generators, and (4) the steam generators have sufficient water inventory to remove decay heat for 1 hour following the loss of feedwater. We also have the following estimates of system characteristics and reliability: (1) the frequency of LOOP is 0.05 per year, (2) probability that, having been lost, offsite power is not restored in 1 hour is 0.6, (3) probability that the steam-driven auxiliary pump fails to start is 0.1/demand, (4) probability of failure to start each diesel generator is 0.01/demand, and (5) probability that the electric auxiliary feedwater pump fails to start is 0.05/demand.

6.10 A station blackout (SBO) event occurred in August 1991 at the Nine Mile Point Unit 2 (NMP-2) Plant, apparently starting with a power surge in one of the station transformers. The key components of the NMP-2 emergency electrical power system involved are five uninterruptible power supplies (UPSs), a bank of 125-V DC batteries, and a UPS logic board that controls the routing of offsite power and DC current. Emergency power was not supplied to the control room because the transformer power surge resulted in the loss of current to the UPS logic circuitry and a low-voltage backup battery on the UPS logic board was dead. Assume that the unavailability of each of the UPS lines is 0.05/year and that of the 125-V DC

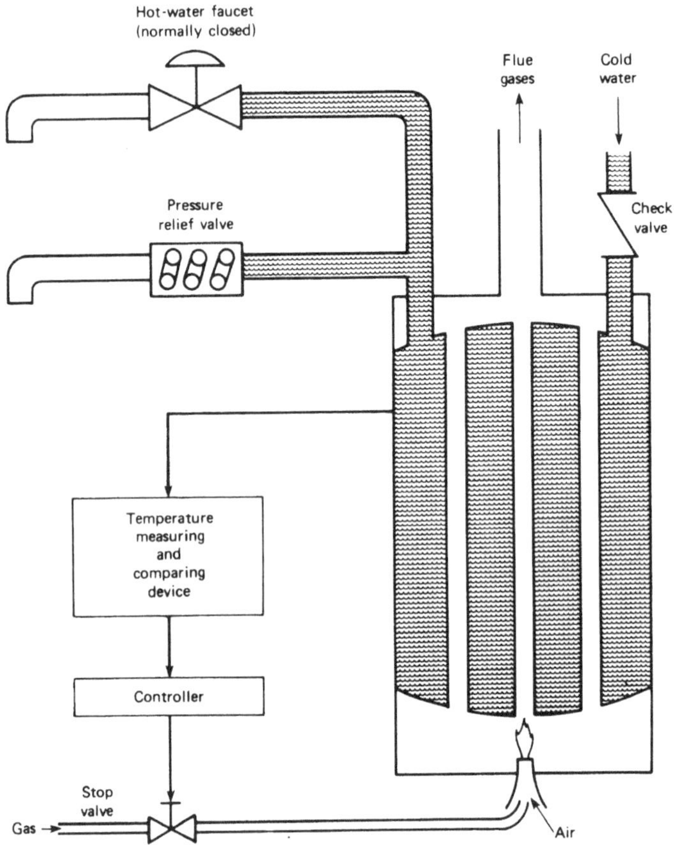

Figure 6.12 Figure for Exercise 6.8.

batteries is 0.001/demand. The failure probability of the UPS logic circuitry is 1.0×10^{-5}/demand, while the unavailability of the low-voltage UPS backup battery is estimated as 0.02/year. Draw a fault tree for the top event, Failure of Emergency Electrical Power, and calculate the expected frequency of the top event per year. Consider a mission time $T = 1.0$ year to convert any failure frequency to probability as necessary. List any assumptions you make in your analysis and make any suggestions to decrease the likelihood of SBO events at the NMP-2 Plant.

6.11 The ECCS for a PWR consists of three identical trains, each of which is sufficient to provide core cooling in case of LOCAs. The unavailability of each train is estimated as 0.02/demand. It is required that each ECCS train be tested and any required maintenance be performed once each year. Testing and maintenance of a train requires 20 hours and LOCAs requiring operation of the ECCS are expected to occur once every 100 years. Assume that each train is tested and maintained at a different time and that common mode failures can be ignored. Construct an event tree representing a LOCA with ECCS failure and calculate the expected frequency of failure modes leading to core damage.

6.12 In the emergency electrical system considered in Example 6.4, (a) construct a modified fault tree for the failure of the emergency power if the switcher room failure is due to the failure of a cooling fan for the room and the offsite AC drives the fan and (b) obtain minimal cut sets and frequency of the unavailability of the emergency power. You may assume the following with a mission time of one year: (1) frequency of 4-kV AC bus failure = 0.01/year, (2) frequency of loss of offsite power = 0.05/year, (3) probability of diesel generator failure to start = 0.02/demand, and (4) frequency of cooling fan failure = 0.02/year.

6.13 Repeat Exercise 6.12 with the probability of switcher room failure of 0.02/demand to determine the frequency of the unavailability of the emergency power for the mission time of (a) one year and (b) two years and discuss the results.

6.14 The emergency electrical system of a BWR plant consists of three trains each consisting of a UPS and a diesel generator. Each train can deliver up to 60% of electrical power required for vital control room functions. The unavailability of each UPS is 0.01/year while the diesel generators fail to start properly once every 50 times. Construct a fault tree and calculate the expected frequency for the loss of emergency electrical power. List any assumptions you make.

CHAPTER 7

COMPUTER PROGRAMS FOR PROBABILISTIC RISK ASSESSMENT

Chapter 6 presented essential principles for constructing fault and event trees used in combination for the probabilistic risk assessment of complex engineered systems. In this chapter, we discuss techniques to perform actual PRA calculations for systems involving a large sequence of events, each of which may be represented by a top event of a complex fault tree. It is not unusual to evaluate event and fault trees comprising several millions of cut sets for PRA studies of nuclear power plants. We illustrate the PRA techniques primarily via the methodology implemented in one popular computer program, SAPHIRE, developed by the Idaho National Laboratory under the auspices of the U.S. Nuclear Regulatory Commission. Only a brief review of other PRA programs is presented. An alternate cut set evaluation algorithm based on binary decision diagrams is presented in the second half of the chapter.

7.1 FAULT TREE METHODOLOGY OF THE SAPHIRE CODE

The SAPHIRE (Systems Analysis Program for Hands-on Integrated Reliability Evaluations) code evolved from the Integrated Reliability and Risk Analysis System (IRRAS) and the System Analysis and Risk Assessment (SARA) system developed to provide PRA capabilities on microcomputers in the 1980s. The IRRAS/SARA

Risk and Safety Analysis of Nuclear Systems. By John C. Lee and Norman J. McCormick

system initially used ASCII-based input/output structures, but an efficient graphic user interface (GUI) was implemented together with significant improvements in the accuracy and efficiency of the PRA algorithms in later versions of the system repackaged as SAPHIRE. With detailed documentation available [NRC08] for the methodology, implementation, and user interface, SAPHIRE version 7 is a powerful PRA tool available through a nondisclosure agreement with the U.S. Nuclear Regulatory Commission.

From the early days of the development for the fault tree methodology, the need to efficiently represent a large number of basic events and gates has been recognized. The SAPHIRE system has successfully implemented a number of algorithms that can analyze and evaluate large, complex fault trees on personal computers without sacrificing the accuracy. The FT algorithm follows the general procedure outlined in Section 6.5 and utilizes various gate structures, including AND, OR, NAND, NOR, and M/N gates. In addition, transfer gates are introduced so that pages of gates and events making up a large FT are connected efficiently in graphical displays. SAPHIRE uses both top-down and bottom-up approaches in simplifying the FT and evaluating the probability of the top event (TE) or top gate as it is synonymously called. The bottom-up algorithm is used to determine cut sets all the way up to the TE. Once the TE is identified, the entire tree is expanded from top to bottom to eliminate nonminimal cut sets and simplify and trim the tree. The FT algorithm may be represented in three major steps.

7.1.1 Gate Conversion and Tree Restructuring

To initiate the FT algorithm, it is necessary to restructure the tree so that various gates in the user-constructed FT are represented in terms of the basic AND or OR gates. The conversion of three particular gates, NAND, NOR, and $\mathrm{M/N}$ gates, are illustrated in Figs. 7.1 through 7.3. In Fig. 7.1, the gate $\mathrm{NAND}(XYZ)$ is converted to an OR gate representing $\bar{X}+\bar{Y}+\bar{Z} = /X + /Y + /Z$, with $\bar{X}$ or $/X$ representing the complement of event X. Similarly, the gate $\mathrm{NOR}(X+Y+Z)$ is converted to an AND gate $\bar{X}\bar{Y}\bar{Z}$, as illustrated in Fig. 7.2. Figure 7.3 illustrates the conversion of a 2/3 gate into an OR gate involving three binary events. The next step in the restructuring process involves merging subtrees represented by the transfer gates into one consolidated tree. In this regard, the transfer gates merely provide a convenient way for the user to check the FT logic.

Once the essential restructuring of the tree is completed, the code determines the TE, if it is not specified by the user, by searching for a gate that is not input to any other gates. It then performs checks to identify any logical loop in the tree, where a gate references itself either directly or indirectly. For example, if gate G_1 is set up as input to gate G_2, which in turn is represented as input to G_1, the code generates a fatal error so that the user may revise the tree structure.

7.1.2 Simplification of the Tree

Once the overall tree starting with the TE is restructured in terms of only AND and OR gates connecting basic events, and logical errors are eliminated, SAPHIRE

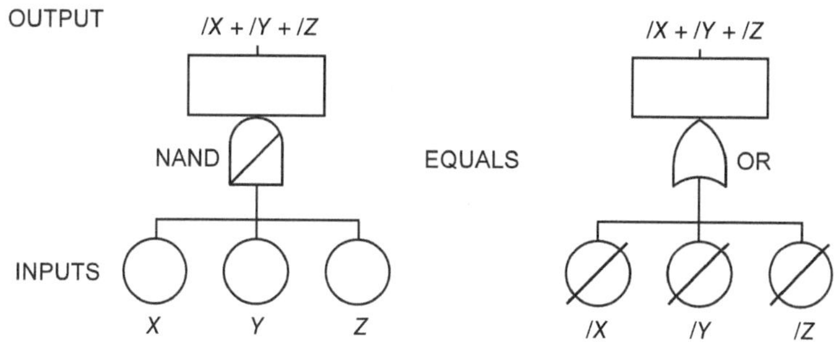

Figure 7.1 Conversion of a NAND gate to an OR gate. *Source:* [NRC08].

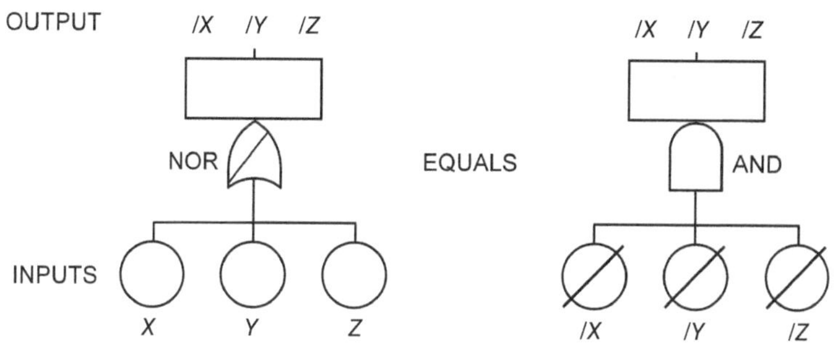

Figure 7.2 Conversion of a NOR gate to an AND gate. *Source:* [NRC08].

undertakes the task of simplifying the tree first by pruning house events. Since an event tree represents a sequence of system failures each of which is connected to a FT, an event that will not occur in the ET is defined as a house event and pruned from the tree. This step is followed by coalescing or combining gates that are input to other gates of the same type, i.e., AND gates are condensed with AND gates, and OR gates with OR gates, starting from the TE. Through this process, the number of gates is reduced and the number of inputs to a gate is increased. One simplifying assumption made throughout the FT processing is to treat all basic events as independent.

A major simplification in a FT is accomplished next through identifying independent subtrees and modules [Han88,Kum96], with the following definitions:

- An *independent event* is an event that is input to only one gate, although the gate may appear more than once in the FT.

- An *independent gate* is a gate that consists of only independent events and is input to only one other gate.

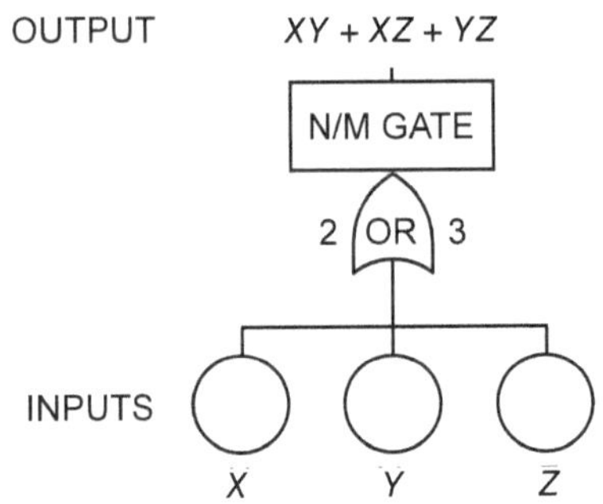

Figure 7.3 Conversion of a 2/3 gate to an OR gate. *Source:* [NRC08].

- An *independent subtree* is a gate that has only independent gates or events as input, or equivalently is a gate that consists only of single-occurrence events or gates that do not appear in any other construct in the FT. An independent subtree may, however, appear more than once in the FT.

SAPHIRE searches for independent subtrees and replaces each of them by a *module*. If a gate consists of single- and multioccurrence events or gates, then only single-occurrence events or gates are lumped into a module. The modularization process is illustrated in Fig. 7.4, where gates G_3 and G_4 consist of only single-occurrence events and hence are independent subtrees and may be designated as modules. On the other hand, G_1 has a single-occurrence event E and a single-occurrence gate G_3, together with a multioccurrence event A. In this case, only E and G_3 are lumped into module M_1. Since a module may appear more than once in a FT, it is usually not an independent gate. A gate that has a module as an input is an independent subtree if the module is an independent gate. The modularization process reduces the size of the FT but does not alter the logic of the tree. SAPHIRE uses a bit-vector approach to search for independent subtrees. A detailed search algorithm for modules is described in [Koh89].

7.1.3 Fault Tree Expansion and Reduction

Once the FT is restructured, a number of steps are taken to expand the FT from the TE and build a tree logic table capitalizing on independent gates and modules. In this process, gates are replaced by input events and reduced by using Boolean algebra summarized in Table 2.1, in particular, the idempotent, absorption, and complementation laws. The expansion process truncates events with probabilities below a user-defined limit and generates minimal cut sets. Separate algorithms are used to expand nonindependent and independent subtrees, since independent subtrees form cut sets that are minimal.

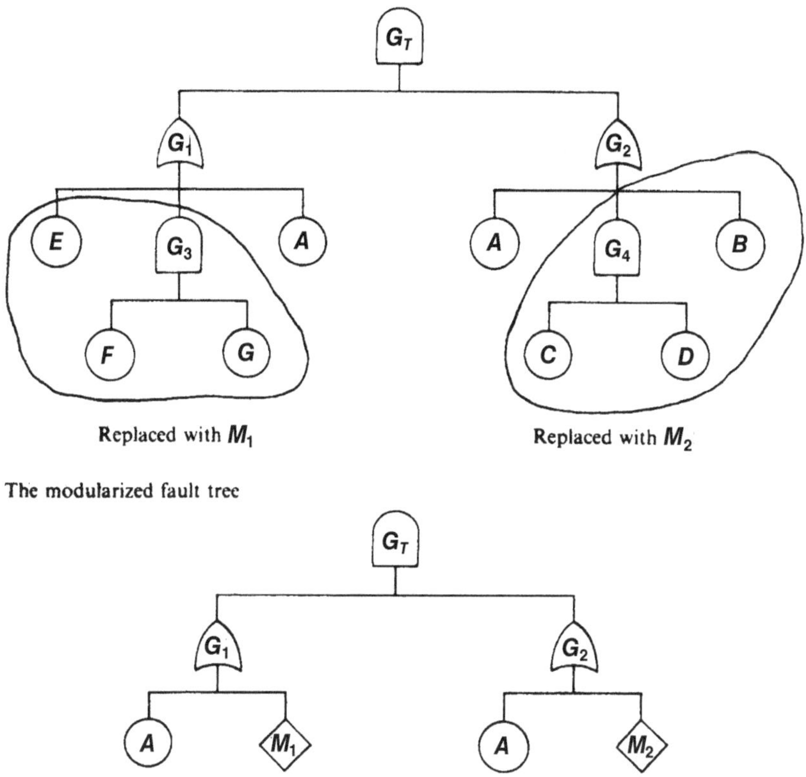

Figure 7.4 Illustration of the modularization process. *Source:* [Han88].

7.2 FAULT AND EVENT TREE EVALUATION WITH THE SAPHIRE CODE

The SAPHIRE code offers coupled ET-FT risk calculations as well as stand-alone FT calculations. The only additional requirement for the full PRA calculations is the proper linkage of every step in the sequence of events in the ET to a FT. With or without an underlying ET, SAPHIRE accepts a number of standard PDFs representing failure rates for components, including the normal, lognormal, beta, Dirichlet, gamma, chi-squared, exponential, uniform, and maximum entropy distributions. In addition, the user may invoke any arbitrary PDF by representing the desired PDF as histograms.

The code also allows the user to perform reliability calculations by specifying a mission time T for a single-component two-state Markov model for the system unavailability of any basic event. From the availability analysis of Example 5.1, the unavailability is modeled as

$$\bar{A}(t) = 1 - A(t) = \frac{\lambda\left[1 - \exp\{-(\lambda+\mu)T\}\right]}{\lambda+\mu}, \tag{7.1}$$

in terms of the failure rate λ and repair rate μ. Since for engineered systems the repair of malfunctioning systems is done promptly, i.e., the mean time to repair is short and $\mu = 1/\text{MTTR} \gg \lambda$, Eq. (7.1) may be approximated as

$$\bar{A}(t) \simeq \frac{\lambda\left[1 - \exp\left(-\mu T\right)\right]}{\lambda+\mu}. \tag{7.2}$$

For a sufficiently long mission time T, the asymptotic unavailability may be represented as in Eq. (5.38)

SAPHIRE calculates first a point estimate for the TE probability for each FT and for the probability of each end state in the ET. Because a full accounting of all possible combinations of individual events in cut sets as well as of minimal cuts themselves would be time consuming for large fault trees, the code offers three different quantification options, illustrated for a TE comprising *N* minimal cut sets, $C_n, n = 1, \ldots, N$.

1. *Min-max method* allows for the use of the exact relationship of Eq. (2.18) so that the probability of an OR gate consisting of *N* events may be calculated by

$$\begin{aligned} P\left(\sum_{n=1}^{N} C_n\right) &= \sum_{n=1}^{N} P\left(C_n\right) - \sum_{n=1}^{N-1}\sum_{m=n+1}^{N} P(C_n C_m) + \cdots \\ &\quad + (-1)^{N-1} P\left(\prod_{n=1}^{N} C_n\right). \end{aligned} \tag{7.3}$$

   Subject to the order of accuracy specified by the user, the code performs multiple traverses of the gate, beginning with the simple addition of probabilities for the individual events, moving to the probabilities associated with pairs of events, triplets of events, and eventually up to the 10th order.
2. *Minimal cut set upper bound* calculations allow for the next level of accuracy, with the assumption that all the events are independent, which allows for the use of Eq. (2.20),

$$1 - P\left(\sum_{n=1}^{N} C_n\right) = \prod_{n=1}^{N}\left[1 - P\left(C_n\right)\right]. \tag{7.4}$$

   Equation (7.4) is known to overestimate the TE probability when some of the cut set probabilities are very small [NRC08]. In such cases, it may be better to use the next method.
3. The simplest method invokes the rare-event approximation of Eq. (2.23),

$$P\left(\sum_{n=1}^{N} C_n\right) \simeq \sum_{n=1}^{N} P\left(C_n\right). \tag{7.5}$$

As a means of propagating uncertainties in the failure probabilities of components and systems represented in a FT, the code also performs sampling of the PDFs for basic events via either standard Monte Carlo or Latin hypercube sampling [Ima84] and generates various moments of the PDF for the TE. A typical summary table for the TE includes the coefficients of skewness (third moment) and kurtosis (fourth moment), defined in Eqs. (2.135) and (2.136), as well as the point estimate, mean, median, standard deviation, and 5th and 95th percentile values. The code also generates a tree logic diagram and importance measures discussed in Section 6.7.3 as well as graphical displays of the FT and ET structures.

7.3 OTHER FEATURES OF THE SAPHIRE CODE

As summarized in Sections 7.1 and 7.2, the SAPHIRE code offers the capability for stand-alone FT or combined ET-FT analyses for large, complex systems on personal computers with user-friendly GUI input/output structures. The code provides a convenient interface with other PRA codes or databases through the ASCII-based input/output capability of the models and results database (MAR-D) module. In addition to the standard two-branch ET structure, SAPHIRE allows for a three-branch ET structure, providing added flexibility to represent complex system evolutions.

Another important feature of the SAPHIRE code is a separate module called the graphical evaluation module (GEM). GEM provides a specialized user interface with SAPHIRE that allows efficient evaluation of the effects of changes in a set of basic event probabilities relative to a reference risk or reliability calculation. GEM calculations augment the importance measures discussed in Section 6.7.3 and allow an analyst to estimate the risk associated with operational maneuvers for nuclear plants in an expeditious manner.

Efficient editors allow the SAPHIRE user to modify ETs and FTs and to construct and revise piping and instrumentation diagrams (P&IDs). Finally, the SAPHIRE code offers specific models to perform seismic PRAs. Given the user's input for peak ground acceleration g associated with a postulated earthquake, the code generates fragility curves representing the probability of component failures as a function of g and interfaces with internal event PRA models for a complete seismic risk evaluation.

7.4 OTHER PRA CODES

A number of sophisticated computer programs have been developed over the years to perform stand-alone FT or combined PRA calculations for nuclear power plants and other complex engineered systems. Several of them appear to have features similar to those of the SAPHIRE code. We provide a brief review of four well-established PRA programs for the risk assessment of nuclear power plants and other complex systems. A comprehensive summary of FT analysis programs and associated references is presented in [Eri99].

1. The CAFTA (Computer Assisted Fault Tree Analysis) code [Gae89] was developed by Science Applications International Corporation for the Electric Power

Research Institute. The code offers a FT editor, basic event database, FT quantification module, and cut set editor. The FT quantification module is based on and interfaces with the FTAP code [Wil78]. The code was developed as part of the PRA study for the Crystal River Unit 3 nuclear power plant and offers the user a convenient way to utilize and revise the existing database for other nuclear power plants under study.

2. The RISKMAN code [Ful00] was developed by Pickard, Lowe and Garrick Inc. and performs integrated risk calculations with interfaces to other PRA codes, including SETS [Wor74] and CAFTA. Together with modules to perform FT and ET evaluations, the code offers the capability to account for multibranch event trees, external events, and dependencies between systems. It offers an option to efficiently perform sensitivity calculations for the whole system under study and allows for extensive error checking and configuration management for quality control.
3. The Relex software [Rel08] is a risk analysis system developed in collaboration with the National Aeronautical and Space Administration (NASA), with a focus on human factors analysis (HFA). It performs top-down evaluation of FTs invoking an underlying Markov solver and accounts for functional dependencies and common cause failures using beta factors and other models discussed in Section 6.2.2. Relex represents various performance shaping factors discussed in Sections 6.2.3 and 6.3.2, error categories, control inputs, and barriers for human performance. The code provides a simultaneous display of the FT in a compact tabular view as well as a standard graphical view and supports an interface to Microsoft Excel, Access, SQL Server and Oracle databases.
4. The PARAGON software [Par08] is a PRA system developed by ERIN Engineering Research Inc. with specific applications for configuration risk management (CRM) programs. PARAGON may combine plant-specific PRA models with other deterministic and qualitative information for CRM decisions in both online and outage modes. The software may be efficiently interfaced with other PRA tools, including CAFTA, so that importance measures of Section 6.7.3 may be evaluated for both internal and external risk-significant events.
5. The Reliability Workbench software [Iso09] is an integrated package developed by Isograph Inc. for performing a multitude of risk and reliability evaluation tasks for various disciplines including aerospace, automotive, and chemical industries. The software offers interconnected modules for (a) reliability analysis and maintainability prediction discussed in Chapters 4 and 5, (b) failure mode and effects analysis (FMEA) presented in Section 6.4, (c) reliability block diagram analysis of Sections 4.2 and 4.3, (d) FT/ET analysis, discussed in Chapter 6 and so far in this chapter, and (e) Markov chain analysis of Chapter 5.

Reliability Workbench allows users to perform risk and reliability calculations for electrical and mechanical components using a number of databases including the MIL-HDBK-217 [DoD95] and IEC TR 62380 [IEC04] standards. The maintainability prediction module calculates the MTTR for each block of components accounting for replaceable items in the block, while the FMEA module identifies potential failure modes in a system and classifies them according to their importance. The FT/ET module performs common cause failure analysis using beta factors of Section 6.2.2 and the Multiple Greek Letter (MGL) model [Bar09] representing component-specific effects. The module also provides uncertainty and importance analyses of the type discussed in Section 6.7. The Markov module of Reliability Workbench performs numerical integration of state transition equations with time-dependent transition rates and offers the capability to represent multiple stages of continuous- or discrete-system transitions or degradations. The software also offers a versatile input/output capability to construct customized reports and graphs and to import data directly from Microsoft Access databases and Excel spreadsheets.

7.5 BINARY DECISION DIAGRAM ALGORITHM

The binary decision diagram (BDD) algorithm has been developed as an alternate to traditional FT algorithms described for the SAPHIRE code in Sections 7.1 through 7.3. The algorithm is structured in the framework of a BDD introduced for digital circuit analysis and logic programming languages [Rau93]. The BDD structure when applied to FT analysis provides an efficient way of calculating the probability of the TE of a FT without the need to analyze cut sets and enumerate minimal cut sets. A basic BDD framework is first described in Section 7.5.1, followed by a few simple examples and general formulas for Boolean operations involving BDDs in Section 7.5.2. A brief discussion of more recent implementations of the BDD algorithms in actual PRA codes is presented in Section 7.5.3.

7.5.1 Basic Formulation of the BDD Algorithm

The BDD algorithm begins with an if-then-else (*ite*) formulation for a basic event (BE) or gate so that the desired probability is represented as

$$ite\,(x, e, f) = if\,\{\text{event } x \text{ occurs}\}\ then\ \{e \text{ results}\}\ else\,\{f \text{ results}\}\,, \quad (7.6)$$

which may be represented in a Boolean notation

$$ite\,(x, e, f) = (x \cap e) \cup (\bar{x} \cap f) = xe + \bar{x}f. \quad (7.7)$$

For BE x in a FT, the probability for the BE representing a failure is written as

$$P(x) = ite\,(x, 1, 0) = x^*, \quad (7.8)$$

with a value of unity associated with the path representing the event of our interest, i.e., failure, and zero with the no-failure or success path, as illustrated in Fig. 7.5. Thus, we encode $P(x)$ with the 1-state (system failure) branch leading to the 1-register and $P(\bar{x})$ with the 0-state (system success) branch leading to the 0-register. Here, the symbol x^* is introduced as a shorthand notation for the *ite* representation of BE x serving as a bottom or terminal node of a FT.

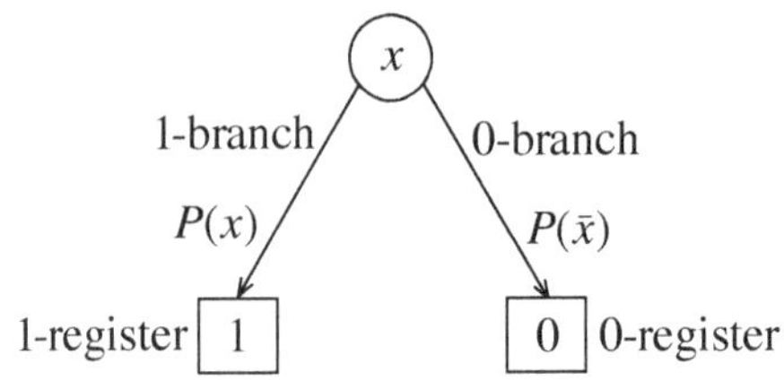

Figure 7.5 The *ite* structure of basic event x.

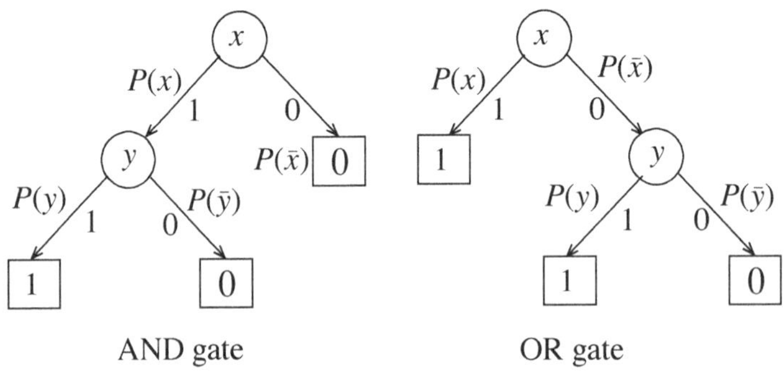

Figure 7.6 BDD representations of AND and OR gates.

The BDD representations of two fundamental FT structures, AND and OR gates, are now considered. For an AND gate comprising two basic events x and y, we seek a compound BDD expression for the probability $P(xy)$ representing the simultaneous occurrence of events x and y,

$$P(xy) = P(x)P(y) = ite\,(x, 1, 0) \cdot ite\,(y, 1, 0) = x^* \cdot y^*. \tag{7.9}$$

To encode $P(xy)$ at the terminal 1-register, the 1-branch for node x, representing the probability $P(x)$, is connected to node y, whose 1-branch representing $P(y)$ leads to the 1-register, so that a combined probability $P(x)P(y)$ is properly input to the 1-register. This logic is illustrated in the left-hand BDD of Fig. 7.6. Thus, the AND gate is encoded in a compound BDD formula

$$P(x)P(y) = xy = ite\,(x, ite\,(y, 1, 0)\,, 0) = ite\,(x, y^*, 0)\,. \tag{7.10}$$

From this point on, *we will use the simple Boolean notation xy, instead of the explicit probability notation*, whenever there is no possibility for confusion. Together with the illustration in Fig. 7.6, it should be emphasized that an AND gate encodes events along the 1-branch.

For an OR gate comprising two basic events x and y, our objective is to encode

$$P(x + y) = P(x) + [1 - P(x)]\,P(y) = P(x) + P(\bar{x})P(y). \tag{7.11}$$

Equation (7.11) suggests adding $P(x)$ of the 1-branch for node x to $P(\bar{x})$ of the 0-branch multiplied by $P(y)$ of the 1-branch for node y to determine $P(x + y)$, requiring a connection between nodes x and y via the 0-branch from node x, which is illustrated in the right-hand BDD of Fig. 7.6. Equation (7.11) is thus encoded as

$$\begin{aligned} P(x+y) &= x+y = ite\,(x,1,0) + ite\,(y,1,0) = ite\,(x,1,ite\,(y,1,0)) \\ &= ite\,(x,1,y^*)\,. \end{aligned} \tag{7.12}$$

Note that an OR gate is encoded along the 0-branch, in symmetry to the encoding of an AND gate along the 1-branch in Eq. (7.10). Note also that, in both Eqs. (7.10) and (7.12), the introduction of the shorthand symbol of Eq. (7.8) for the terminal node simplifies the visualization of the encoding logic for the two basic gate structures of FT representations. One important observation from the BDD implementation of the OR gate in Eq. (7.12) is that the overlapping probability correction $-P(x)P(y)$ inherent in the standard joint probability representation of Eq. (7.11) is simply and elegantly accounted for automatically via the BDD and *ite* structure. As illustrated further in Section 7.5.2, this is perhaps the first significant feature of the BDD algorithm for FT applications.

7.5.2 Generalization of the BDD Formulation

We now consider generalization of the two basic BDD formulations of Eqs. (7.10) and (7.12) through an FT example [Rau93] illustrated in Fig. 7.7. The top event g is an intersection of two unions and may be represented via Boolean distributive law 5.b of Table 2.1,

$$g = (x+z)(y+z) = (z+x)(z+y) = z+xy = xy+z, \tag{7.13}$$

which indicates that the sets xy and z are minimal cut sets for the FT. The TE probability is then obtained in terms of the minimal cut sets

$$\begin{aligned} P(g) &= P(xy+z) = P(xy) + P\left(\overline{xy}\right)P(z) \\ &= P(x)P(y) + \left[P\left(\bar{x}\right) + P\left(\bar{y}\right)\right]P(z), \end{aligned} \tag{7.14}$$

with application of Boolean law 7.a of Table 2.1 in the last expression. Representing Eq. (7.14) in a BDD formulation requires first a simple use of the AND gate representation of Eq. (7.10),

$$xy+z = ite(x,y^*,0) + ite(z,1,0) = ite\,(x, ite(y,1,0), 0) + ite(z,1,0). \tag{7.15}$$

Because the OR gate is encoded via the 0-branch in Eq. (7.12), Eq. (7.15) may be written as a compound *ite* expression

$$xy+z = ite\,(x, ite(y,1,z^*), z^*) = ite\,(x, ite\,(y,1,ite\,(z,1,0))\,, ite\,(z,1,0))\,, \tag{7.16}$$

where z^* replaces the value 0 both in the outer *ite* expression for node x and in the inner *ite* expression for node y. Here, we also note the usefulness of the shorthand notation z^* for the terminal node. The compound *ite* expression now clearly suggests that the 1-branch from node x should connect to node y, with the 0-branch connected

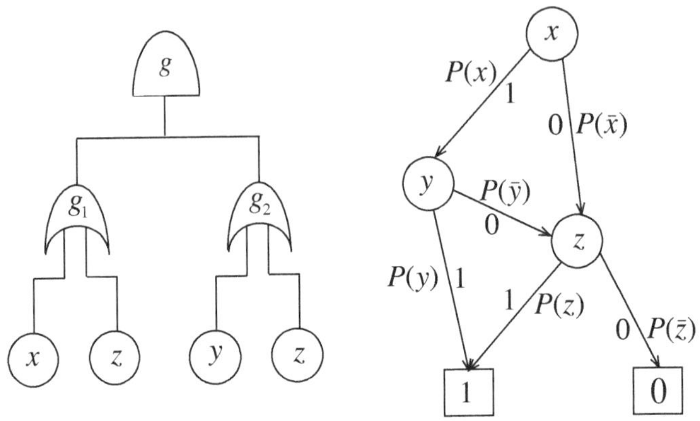

Figure 7.7 Fault tree and corresponding BDD for gate $g = (x + z)(y + z)$.

to node z. The inner *ite* expression for node y further indicates that the 0-branch from node y should also connect to node z. The BDD of Fig. 7.7 shows that the total probability input to node z is equal to the sum $[P(\bar{x}) + P(\bar{y})]$, which is multiplied by $P(z)$ via the 1-branch and added to the probability $P(x)P(y)$ exiting from the 1-branch of node y so that the total probability input to the 1-register is the desired probability $P(xy + z)$ of Eq. (7.14). This derives a key result of the original *ite* and BDD formulation [Rau93], presented only as a BDD of Fig. 7.7.

Example 7.1 Derive Eq. (7.16) alternatively by starting directly from the original gate structure

$$g = (x + z)(y + z) = ite(x, 1, z^*) \cdot ite(y, 1, z^*), \quad z^* = ite(z, 1, 0). \tag{7.17}$$

Invoking the AND gate relationship of Eq. (7.10) that the multiplication of two *ite* expressions evolves via the 1-branch allows the conversion of Eq. (7.17) to Eq. (7.16),

$$\begin{aligned} (x + z)(y + z) &= ite(x, ite(y, 1, z^*), z^*) \\ &= ite(x, ite(y, 1, ite(z, 1, 0)), ite(z, 1, 0)). \quad \diamond \end{aligned} \tag{7.18}$$

This simple derivation of Eq. (7.18) also highlights the second significant feature of the BDD formulation, which suggests the possibility to obtain the probability for gates in a FT without the need to determine the minimal cut sets via laborious processes discussed in Sections 4.5, 6.5, and 7.1.

Equation (7.18) also provides a way to obtain general formulas for combining *ite* expressions. For this purpose in mind, reuse Eq. (7.13) in a yet alternate way,

$$g = (z + x)(z + y) = z + xy = ite(z, 1, 0) + ite(x, y^*, 0), \tag{7.19}$$

where Eq. (7.10) is used for the *ite* representation of the AND gate xy. Equation (7.19) is converted, via Eq. (7.12), to a compound *ite* expression for gate g,

$$\begin{aligned} (z + x)(z + y) &= ite(z, 1, ite(x, y^*, 0)) = ite(z, 1, ite(x, ite(y, 1, 0), 0)) \\ &= ite(z, 1, x^*) \cdot ite(z, 1, y^*). \end{aligned} \tag{7.20}$$

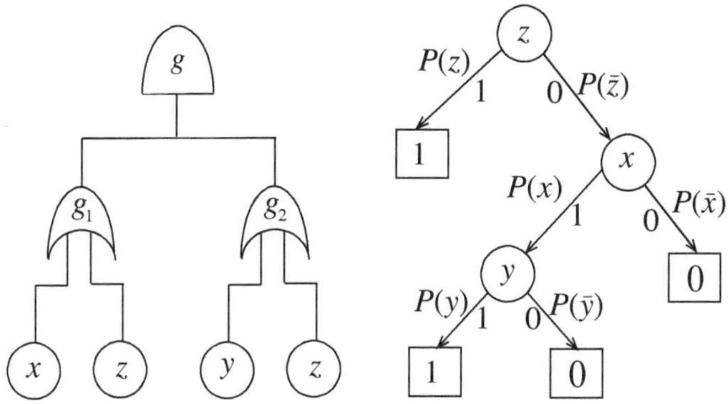

Figure 7.8 BDD for gate $g = (z + x)(z + y)$.

The *ite* expression of Eq. (7.20) is illustrated via a BDD in Fig. 7.8. The diagram clearly shows that Eq. (7.20) correctly encodes the probability for the combined gate,

$$P(g) = P(z + xy) = P(z) + P(\bar{z})P(x)P(y). \tag{7.21}$$

This example also shows that there will generally be more than one way to construct a BDD and the corresponding *ite* expression for a given gate or FT.

Using the three different *ite* representations of $g = (x + z)(y + z)$, Eqs. (7.16), (7.18), and (7.20), we now show the validity of a general formula for combining *ite* expressions proposed in the 1993 paper by Rauzy [Rau93] and subsequently used in other papers [Rem08,Jun09] without justification or proof. For the multiplication of two *ite* expressions, generalize Eq. (7.20) into a formula

$$ite(x, e_1, f_1) \cdot ite(x, e_2, f_2) = ite(x, e_1e_2, f_1f_2), \tag{7.22}$$

where $e_1 = e_2 = 1 = e_1e_2$ and $f_1f_2 = ite(x, 1, 0) \cdot ite(y, 1, 0) = ite(x, y^*, 0)$ for Eq. (7.20). Likewise, equating Eqs. (7.17) and (7.18) generalizes to

$$ite(x, e_1, f_1) \cdot ite(y, e_2, f_2) = ite(x, e_1\tilde{y}, f_1\tilde{y}), \quad \tilde{y} = ite(y, e_2, f_2), \tag{7.23}$$

where, for Eq. (7.17), $e_1 = e_2 = 1$, $f_1 = f_2 = z^*$, and $\tilde{y} = ite\,(y, 1, z^*)$, so that

$$f_1\tilde{y} = ite(z, 1, 0) \cdot ite\,(y, 1, z^*) = P(z)P(y + z) = P(z) = ite(z, 1, 0). \tag{7.24}$$

Similarly, equating Eqs. (7.15) and (7.16) generalizes to

$$ite(x, e_1, f_1) + ite(y, e_2, f_2) = ite(x, e_1 + \tilde{y}, f_1 + \tilde{y}), \quad \tilde{y} = ite(y, e_2, f_2), \tag{7.25}$$

where, for Eq. (7.15), $e_1 = y^*, f_1 = 0, e_2 = 1, \tilde{y} = z^*$. so that

$$e_1 + \tilde{y} = ite(y, 1, 0) + ite(z, 1, 0) = ite(y, 1, z^*). \tag{7.26}$$

Finally, generalizing

$$\begin{aligned} xy + xz &= ite(x, y, 0) + ite(x, z, 0) = x(y + z) = ite(x, 1, 0) \cdot ite(y, 1, z^*) \\ &= ite(x, ite(y, 1, z^*), 0) \end{aligned} \quad (7.27)$$

leads to

$$ite(x, e_1, f_1) + ite(x, e_2, f_2) = ite(x, e_1 + e_2, f_1 + f_2). \quad (7.28)$$

The four BDD rules, Eqs. (7.22), (7.23), (7.25), and (7.28), may finally be combined for either AND or OR operation $\otimes$,

$$ite(x, e_1, f_1) \otimes ite(x, e_2, f_2) = ite(x, e_1 \otimes e_2, f_1 \otimes f_2), \quad (7.29)$$

$$ite(x, e_1, f_1) \otimes ite(y, e_2, f_2) = ite(x, e_1 \otimes \tilde{y}, f_1 \otimes \tilde{y}), \quad \tilde{y} = ite(y, e_2, f_2). \quad (7.30)$$

For Eqs. (7.29) and (7.30), the following Boolean algebra may be used to simplify the operations:

$$\begin{aligned} x + y &= x + \bar{x}y = 1 && \text{if } P(x) = 1, \\ xy &= y && \text{if } P(x) = 1, \\ x + y &= y && \text{if } P(x) = 0, \\ xy &= 0 && \text{if } P(x) = 0. \end{aligned} \quad (7.31)$$

Example 7.2 Obtain an *ite* expression for the FT representing an electrical system in Example 6.4 and compare with the TE probability obtained there.

The TE is represented in terms of basic events,

$$T = x_1 + x_2 + G = x_1 + x_2 + x_3x_4,$$

and the BDD representation is obtained,

$$\begin{aligned} T &= ite(x_1, 1, 0) + ite(x_2, 1, 0) + ite(x_3, x_4^*, 0) \text{ via Eq. (7.10)} \\ &= ite(x_1, 1, x_2^*) + ite(x_3, x_4^*, 0) \text{ via Eq. (7.12)} \\ &= ite\left(x_1, 1 + \tilde{y}, x_2^* + \tilde{y}\right), \ \tilde{y} = ite(x_3, x_4^*, 0), \text{ via Eq. (7.30)} \\ &= ite(x_1, 1, ite(x_2, 1, \tilde{y})) \text{ via Eqs. (7.31) and (7.12)} \\ &= ite\left(x_1, 1, ite(x_2, 1, ite(x_3, ite(x_4, 1, 0), 0))\right). \end{aligned}$$

The *ite* expression for the TE is converted to a BDD in Fig. 7.9 and properly represents the probability

$$\begin{aligned} P(T) &= P(x_1) + P(\bar{x}_1)\left[P(x_2) + P(\bar{x}_2)P(x_3)P(x_4)\right] \\ &= x_1 + \bar{x}_1\left(x_2 + \bar{x}_2x_3x_4\right). \end{aligned} \quad (7.32)$$

It can be readily verified that the BDD result for the TE probability above is equal to that obtained through the traditional method in Example 6.4. For this simple FT, the conversion between the FT and BDD is rather intuitive and straightforward, and the probability designations for the BDD branches have been left out for compactness, as is done usually in the BDD literature. ◇

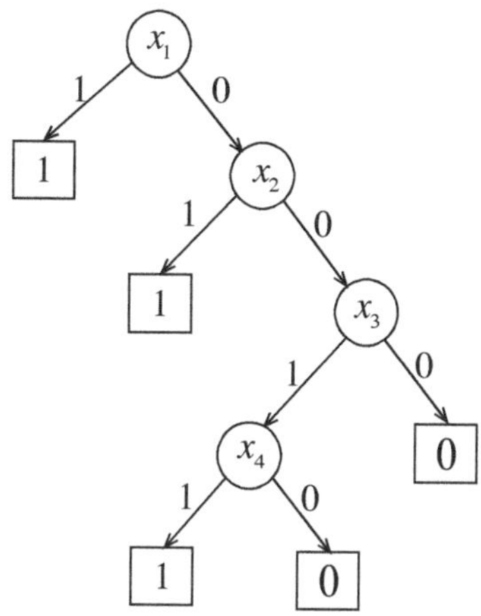

Figure 7.9 BDD illustration of FT $T = x_1 + x_2 + x_3 x_4$.

7.5.3 Zero-Suppressed BDD Algorithm and the FTREX Code

With the examples presented in Sections 7.5.2, two of the key features of the BDD algorithm have been recognized: (a) *ite* expressions and corresponding BDDs can be constructed without the need to identify cut sets and minimal cut sets and (b) overlapping probability corrections for OR gates can be efficiently and precisely represented without the need to truncate the corrections, for complex FTs, according to the order that the user of the PRA codes specifies. With efficient sharing of *ite* nodes and subtrees in a bottom-up structure, the BDD algorithm offers considerable advantages both in accuracy and computational efficiency compared with traditional FT algorithms discussed in Section 7.1.

With increasing emphasis placed on risk-informed licensing and regulations in the nuclear industry, as discussed further in Chapter 12, it has now become routine to perform risk calculations, on a daily basis, for the entire plant represented by several thousand basic events and several million minimal cut sets. For these time-sensitive realistic PRA tasks, active research is under way to develop more efficient methods of developing BDD structures [Rem08] as well as somewhat approximate methods of BDD structures. One such method is the zero-suppressed BDD (ZBDD) algorithm implemented in the Fault Tree Reliability Evaluation eXpert (FTREX) code [Jun09], developed at the Korea Atomic Energy Research Institute and marketed through the Electric Power Research Institute. In this approach, the success path along the 0-branch in the basic *ite* structure of Eq. (7.7) is suppressed, subject to a judicious truncation limit, and Eq. (7.7) is approximated by

$$ite\,(x, e, f) = (x \cap e) \cup (\bar{x} \cap f) \simeq xe + f. \tag{7.33}$$

This appears to be a fruitful area of research that may find useful applications in various PRA applications.

References

[Bar09] A. Barros, A. Grall, and D. Vasseur, "Estimation of Common Cause Failure Parameters with Periodic Tests," *Nucl. Eng. Design* **239**, 761 (2009).

[DoD95] "Reliability Prediction of Electronic Equipment," MIL-HDBK-217, Notice 2, U.S. Department of Defense (1995).

[Eri99] C. A. Ericson II, "Fault Tree Analysis—A History," *Proc. 17th Int. Sys. Safety Conf.* (1999).

[Ful00] R. R. Fullwood, *Probabilistic Safety Assessment in the Chemical and Nuclear Industries*, Butterworth Heinemann (2000).

[Gae89] J. Gaertner, "CAFTA User's Manual Version 2.0," NP-6296, Electric Power Research Institute (1989).

[Han88] S. H. Han, T. W. Kim, and K. J. Yoo, "Development of an Integrated Fault Tree Analysis Computer Code MODULE by Modularization Technique," *Reliab. Eng. Sys. Safety* **21**, 145 (1988).

[IEC04] "Reliability Data Handbook—Universal Model for Reliability Prediction of Electronics Components, PCBs and Equipment," IEC TR 62380, International Electrotechnical Commission (2004).

[Ima84] R. L. Iman and M. J. Shortencarier, "A Fortran-77 Program and User's Guide for the Generation of Latin Hypercube and Random Samples for Use with Computer Models," NUREG-CR-3624, U.S. Nuclear Regulatory Commission (1984).

[Iso09] "Reliability Workbench V10.0 Technical Specification," www.isograph-software.com (2009).

[Jun09] W. S. Jung, "ZBDD Algorithm Features for an Efficient Probabilistic Safety Assessment," *Nucl. Eng. Design* **239**, 2085 (2009).

[Koh89] T. Kohda, E. J. Henley, and K. Inoue, "Finding Modules in Fault Trees," *IEEE Trans. Reliab.* **38**, 165 (1989).

[Kum96] H. Kumamoto and E. J. Henley, *Probabilistic Risk Assessment and Management for Engineers and Scientist*, IEEE Press (1996).

[NRC08] "Systems Analysis Program for Hands-On Integrated Reliability Evaluations (SAPHIRE), Technical Reference," NUREG/CR-6952, vol. 2, U.S. Nuclear Regulatory Commission (2008).

[Par08] E. F. Parsley and L. B. Shanley, "Configuration Risk Management of External Event Risk Using Paragon," *ANS PSA 2008 Topical Meeting—Challenges to PSA During the Nuclear Renaissance*, American Nuclear Society (2008).

[Rau93] A. Rauzy, "New Algorithms for Fault Tree Analysis," *Reliab. Eng. Sys. Safety* **40**, 203 (1993).

[Rel08] "Relex Fault Tree Analysis Software," Relex Software Corporation, www.relex.com (2008).

[Rem08] R. Remenyte-Prescott and J. D. Andrews, "An Enhanced Component Connection Method for Conversion of Fault Trees to Binary Decision Diagrams," *Reliab. Eng. Sys. Safety* **93**, 1543 (2008).

[Wil78] R. R. Willie, "Computer Aided Fault Tree Analysis: FTAP," OC 78-14, University of California Operations Research Center (1978).

[Wor74] R. B. Worrell, "Set Equation Transformation System (SETS)," SLA-73-0028A, Sandia National Laboratories (1974).

Exercises

7.1 For the analysis of the overheating of wire in an electric motor circuitry considered in Example 6.3 and Fig. 6.3, set up an input file for the SAPHIRE code using the reduced fault tree of Fig. 6.5. For the basic events $\{x_1, x_2, x_3, x_4, x_7, x_8, x_9, x_{10}\}$, use the lognormal distribution with mean probabilities $\{0.05,\ 0.08,\ 0.005,\ 0.005,\ 0.08, 0.006, 0.006, 0.006\}$ and error factors $\{3.0, 5.0, 3.0, 3.0, 5.0, 3.0, 3.0, 3.0\}$. (a) Run the SAPHIRE code to obtain the probability of occurrence of the wire overheating incident and (b) verify that the minimal cut sets generated by the code are indeed correct. Identify and discuss two most risk significant cut sets.

7.2 In a simplified PRA study for a LOCA in the Davis-Besse plant, the actuation of a number of engineered safety features (ESFs) is considered to prevent core damage, following the depressurization of the reactor coolant system (RCS): (1) The core flood system (CFS) will inject water into the cold leg of the RCS to ensure that the core remains covered, (2) the low-pressure coolant injection (LPCI) system will inject borated water into the core to maintain a subcritical condition as well as to assist the water from CFS in keeping the core covered, and (3) for long-term cooling, the low-pressure recirculation (LPR) system will recirculate water between the containment sump, the decay heat removal (DHR) heat exchanger, and the RCS cold leg. If any of the above three systems fails to operate, then core damage is assumed to occur. For the PRA study of the LOCA, assume (1) the LOCA frequency is 5×10^{-4}/year, (2) the CFS failure frequency is 2.4×10^{-4}/year, (3) the LPCI system fails if (i) the DHR heat exchanger fails, with a failure frequency of 3.4×10^{-5}/year, or (ii) both of the two motor-driven pumps (MDPs) fail, with a failure frequency of 5.6×10^{-4}/year for each pump, (4) the LPR system fails if (i) the operator fails to initiate the system, with a frequency of 1.0×10^{-3} /year, or (ii) there is a hardware failure in the LPR system, with a failure frequency of 3.5×10^{-4}/year. Use the SAPHIRE code with a mission time of one year to obtain (a) minimum cut sets and the associated frequencies for failure sequences leading to core damage in the event of the LOCA and (b) the total core damage frequency due to the LOCA.

CHAPTER 8

NUCLEAR POWER PLANT SAFETY ANALYSIS

Chapters 2 and 3 discussed basic concepts of probabilities and quantification of reliability data followed by quantitative models in Chapters 4 and 5 to represent the reliability and availability of engineered systems with maintenance duly considered. Chapters 6 and 7 presented the probabilistic risk assessment methodology, with a particular focus on the SAPHIRE code developed for risk assessment of nuclear power plants. We discuss in this chapter key issues and techniques to perform deterministic safety evaluation of nuclear power plants.

8.1 ENGINEERED SAFETY FEATURES OF NUCLEAR POWER PLANTS

Discussion of engineered safety features (ESFs) in this section focuses on the current generation of nuclear power plants utilizing light water reactors (LWRs). The LWRs, both pressurized water reactors (PWRs) and boiling water reactors (BWRs), are classified as Generation II power plants in the Generation IV Roadmap [DOE02]. The ESFs in the current generation of LWR plants comprise primarily active devices and equipment that are designed to mitigate consequences of transients and accidents and protect against the release of radionuclides to the environment.

Generation III and IV nuclear power plants feature a larger number of passive safety systems that rely heavily on natural convection cooling and gravity-driven injection of coolant into the core. Some of the passive safety features for Generation III+ plants, in particular, the Westinghouse AP1000 design and the General Electric Economic Simplified Boiling Water Reactor (ESBWR) design, will be discussed in Chapter 11. The passive safety systems in Generation IV plants will be designed ultimately to provide self-shutdown and short-term cooling capabilities and to maintain long-term coolable geometries.

For Generation II plants in operation in the United States, ESFs are designed with loss of coolant accidents (LOCAs), in particular, large-break (LB) LOCAs, as the main design basis accident (DBA). In LBLOCA analyses, main coolant pipes in the primary system are postulated to undergo an abrupt rupture or guillotine break. Thus, the bulk of the ESFs are designed to handle the overheating or undercooling of the core components subject to various types of LOCAs postulated.

We present in Section 8.1.1 an overview of the PWR system followed by discussions on key ESFs and their functions in both normal operating and accident modes. A brief description of the components of the reactor core and other major equipment then concludes the section. Section 8.1.2 follows essentially the same format for the BWR system but with a somewhat less detailed overview of the entire NPP system.

8.1.1 Pressurized Water Reactor

8.1.1.1 Overview of the PWR System The overall layout [NRC08] of a typical PWR plant is first illustrated in Fig. 8.1, where the primary loop including the core, coolant pump, and steam generator is highlighted. In the secondary loop, feedwater is supplied to the steam generator, which produces steam to turn the turbine generator and generates electricity. The steam discharged from a series of high- and low-pressure turbines is condensed in the condenser, from which the condensed water is returned to the steam generator through the feedwater heaters and pumps. The heat deposited in the condenser is dissipated eventually through a tertiary loop connected to a cooling tower or pond. Figure 8.1 also includes the residual heat removal (RHR) system, which removes the heat produced through the decay of fission products in the core after the reactor is shut down.

A more detailed layout of the overall PWR systems is presented in Figs. 8.2 and 8.3. In Fig. 8.2, developed [Rub79] to explain the 1979 accident at Three Mile Island Unit 2 (TMI-2), an emphasis is placed on presenting key power plant components that played a role in a small-break (SB) LOCA that was initiated by a malfunction in the feedwater system and progressed to opening of the power-operated relief valve (PORV), eventually resulting in a significant meltdown of the core. Thus, for the primary loop, the pressurizer, PORV (labeled electromatic relief valve in the diagram), accumulator (labeled core flood tank), and refueling water storage tank (RWST, labeled emergency water storage tank) are clearly indicated in Fig. 8.2. Likewise, for the secondary loop, the main feedwater pump, demineralizer, auxiliary feedwater pump, and condensate pump are highlighted. Figure 8.2 also shows key components in the containment building, including the drain tank that

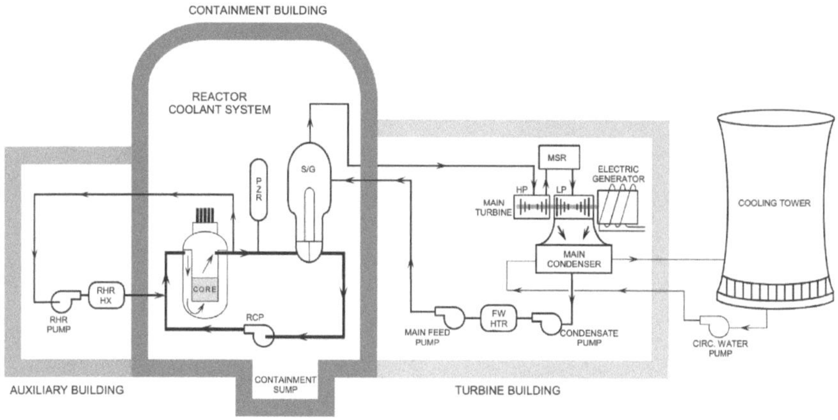

Figure 8.1 Overall layout of a PWR plant. *Source:* [NRC08].

accepts the coolant discharged from the PORV, the containment sump, from which the radioactive primary coolant water was pumped to the waste storage tank in the auxiliary building. Since the auxiliary building was not built leak tight, a limited release of radionuclides occurred in the TMI-2 accident. This diagram will be useful in subsequent discussions of the TMI-2 accident in Chapter 9.

Figure 8.3, borrowed from a PWR training manual [NRC08], presents the overall PWR system with a focus on the ESFs of our primary interest in this section. For the primary loop, the charging and letdown lines connected to the cold and hot legs, respectively, of the reactor coolant system (RCS) and the safety injection (SI) pump, reactor coolant pump (RCP) and accumulator connected to the cold leg are indicated. The diagram also illustrates that the accumulator discharge line has a check valve, with an arrow pointing in the flow direction and a motor-operated valve (MOV) in a normally open position. Note also that the discharge from the accumulator is aided by nitrogen gas pressure. Not shown in Fig. 8.3 is the boron injection tank (BIT), through which the charging pump could be routed for the injection of boric acid to the cold leg. A key primary system shown is the pressurizer, essentially an extension of the hot leg, which regulates and maintains the operating pressure of the RCS. Figure 8.3 also illustrates that the RHR system, delivering coolant to the cold leg, may take suction from the RWST or containment sump as well as from the RCS hot leg. After the reactor is shut down, the primary system is cooled by the RHR heat exchanger, which in turn dissipates heat through the component cooling water (CCW) heat exchanger. Note also that the CCW heat exchanger itself is cooled by the service water system. Thus, similar to the three-level heat transfer loop structure of the plant in the normal operating mode, the CCW and service water systems serve

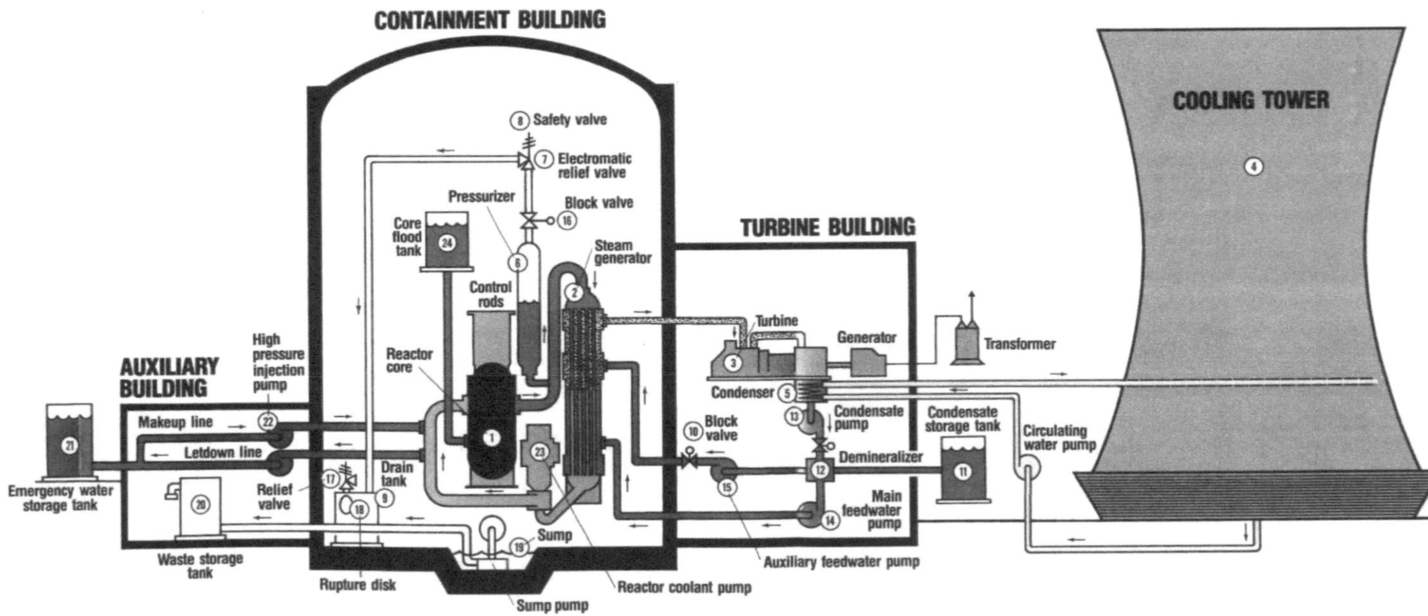

Figure 8.2 Schematic layout of the Three Mile Island Unit 2 plant. *Source:* Reprinted with permission from [Rub79]. Copyright © 1979 The Institute of Electrical and Electronics Engineers.

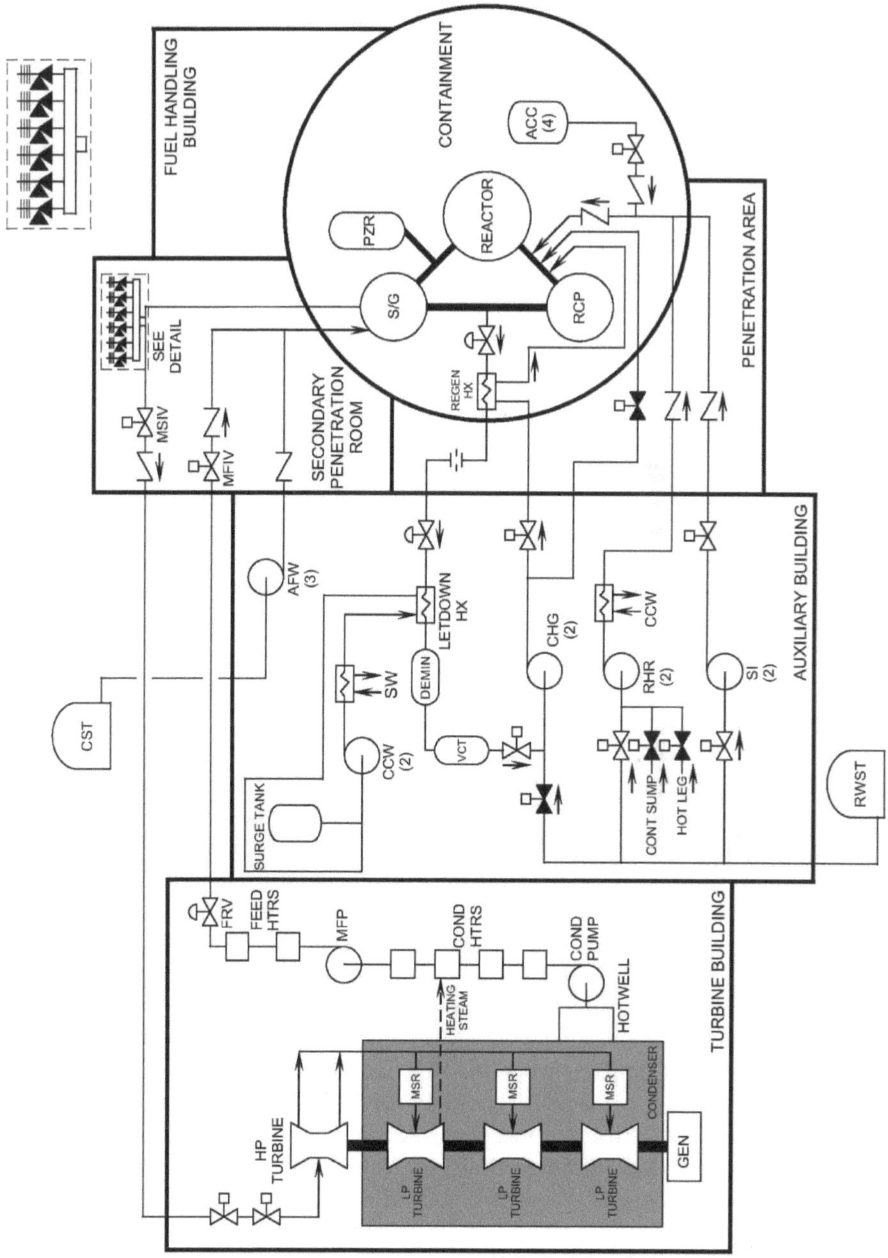

Figure 8.3 Schematic diagram of a PWR plant. *Source:* [NRC08].

as the secondary and tertiary heat transfer loops, respectively, for the RHR system. The two MOVs connecting the RHR charging line to the containment sump and RCS hot leg are blackened, indicating that they are normally in a closed position.

We note also in Fig. 8.3 that, as part of the primary loop, the SI pump takes suction from the RWST via a MOV. The valve is shown in an open position, which is not illustrative of a normal operating mode. The charging pump takes suction normally from the chemical and volume control (CVC) tank (labeled VCT) but may switch to the RWST as necessary. The regenerative and letdown heat exchangers coupled to the demineralizer provides the means to cool down the primary coolant that is discharged from the RCS hot leg and returned to the cold leg via the charging line. The demineralizer and CVC system also serve to filter out unwanted contaminant in the coolant water and maintain the desired soluble boron concentration in the primary loop. If it becomes necessary to increase the soluble boron concentration in an accident situation, the charging flow is switched through the BIT before it is returned to the cold leg.

For the secondary heat transfer loop, Fig. 8.3 shows that the main feedwater line, with a succession of hydraulic or air-operated valves (AOVs) through the turbine and auxiliary buildings, provides feedwater to the shell of a tube-and-shell-type steam generator so that the feedwater picks up heat from a cluster of U-shaped tubes through which the primary coolant circulates. The main steam line delivers hot steam from the secondary or shell side of the steam generator to a series of high-pressure (HP) and low-pressure (LP) turbines in the turbine building. The exhaust steam discharged from the final LP turbine is sent to the hotwell of the steam condenser, from which the condensate and feedwater pumps deliver the condensed water through a series of components in the condensate and feedwater systems to the steam generator. Finally, the auxiliary feedwater (AFW) pump takes suction from the condensate storage tank (CST).

Note also a series of main steam isolation valves (MSIVs) outside the containment but upstream of the pipe tunnel in the auxiliary building. Another ESF, although not indicated explicitly in Fig. 8.3, is the containment spray ring, where the containment spray system (CSS) takes suction from the RWST in accident modes.

Power plant systems and components, including the reactor core, RCS pump, steam generator, and pressurizer, located within the containment building make up the nuclear steam supply system (NSSS), and the rest of the systems making up a nuclear power plant are known as the balance of plant (BOP). Traditionally, the NSSS was supplied by reactor manufacturers or vendors, e.g., Westinghouse Electric Company or General Electric Company, while the BOP was the responsibility of architecture and engineering (AE) companies. Thus, many of the LWR plants currently operating around the world feature a nearly identical NSSS structure, supplied by one particular reactor vendor, but vastly different BOP structures. This has resulted in significant complications in safety and risk assessments of nuclear power plants. It is yet to be seen how much of standardization will be accomplished in the next generation of NPPs currently on the drawing board.

8.1.1.2 PWR Engineered Safety Features We now briefly discuss specific ESFs and their functions for PWR plants, both in normal operation and accident modes, drawing on the descriptions of the NSSS and BOP systems presented via Figs. 8.1 through 8.3. Some of the abbreviations are intentionally reintroduced for clarity.

1. *Residual heat removal (RHR) system*

   - Normal operation
     The RHR system removes decay heat from the reactor core after the reactor is shut down. The system pumps hot reactor coolant system (RCS) water through the RHR heat exchanger and back to the cold leg of the primary system.
   - Accident mode
     In a LOCA, the RHR system pumps cool, borated water from the refueling water storage tank (RWST) into the cold leg as part of the low-pressure coolant injection (LPCI) system. If the RWST inventory is depleted later in the accident, the system can also operate in a recirculating mode to draw suction from the containment sump for sustained supply of coolant for the primary system.

2. *Accumulator*
   As a passive source of coolant water in the LPCI system, the accumulator is maintained in an inert nitrogen environment. Cold water can be supplied from the accumulator by gravity to the RCS through the cold leg.

3. *Safety injection (SI) system*
   The SI system serves essentially the same function as the accident mode of the RHR system, but the SI pumps are used at a high pressure as part of the high-pressure coolant injection (HPCI) system. If the RWST inventory is depleted, it can also operate in a recirculating mode to draw suction from the containment sump.

4. *Charging (makeup) system*

   - Normal operation
     Together with the pressurizer, the makeup system maintains the proper coolant inventory in the primary system and is part of the chemical and volume control (CVC) system. The coolant is taken through the letdown line at the hot leg, cooled through heat exchangers, filtered through a demineralizer, collected in the CVC tank, and eventually pumped by the charging pumps as the makeup flow back to the RCS.
   - Accident mode
     In a LOCA, the charging system can pump borated water from the RWST to the RCS through the boron injection tank.

5. *Auxiliary (emergency) feedwater (AFW) system*

- Normal operation
  The AFW system provides feedwater to the steam generator during startup, up to typically 10 to 15% of rated power, and after shutdown.
- Accident mode
  The system serves as an alternate feedwater delivery system, with water supplied from the condensate storage tank (CST).

6. *Component cooling water (CCW) system*
   The CCW system provides cooling water to the RHR heat exchangers, pumps, and cooling fans for the primary system. The CCW system is in turn cooled in the CCW heat exchanger via the service water system, where the cooling water is finally taken from the cooling pond or cooling tower.
7. *Containment isolation and heat removal system*
   If RCS water spills into the containment, water flashes into steam, thereby increasing the containment pressure. This results in the following remedial actions:
   - The pressure increase automatically closes the containment isolation valves, preventing the release of radioactivity outside the containment.
   - The containment spray system (CSS) is actuated, with the cooling water taken from the RWST. This condenses the steam and minimizes the pressure increase within the containment.
   - The containment fan cooler is turned on, and containment air is filtered and vented outside the containment.
8. *Emergency power*
   - In the case of loss of offsite power (LOOP) leading to a station blackout (SBO) event, emergency diesel generators and batteries provide essential power for the plant.
   - One of the AFW pumps is usually driven by steam turbines, with the steam taken from the exhaust of main turbogenerators. This serves as a passive means of decay heat removal in a LOOP.

8.1.1.3 Brief Description of PWR Components and Equipment To gain clearer understanding of the role of various components in NPP safety and performance, we present a brief description of the structure of the reactor pressure vessel (RPV), reactor core, pressurizer, primary reactor coolant pump, and steam generator via Figs. 8.4 through 8.8. As discussed in the previous two sections, these are the key components making up the NSSS.

Note first in a cutaway RPV view in Fig. 8.4 that the primary coolant water is pumped to the inlet or cold leg nozzle and flows downward in the annulus between the steel reactor vessel wall itself, typically 0.2 m in thickness, and the core barrel. The barrel is essentially a large hollow cylinder that separates the downward flow of the cold coolant from the upward flow of the coolant inside the barrel. Once the coolant water has picked up the nuclear heat generated in the fuel assemblies, it is pumped

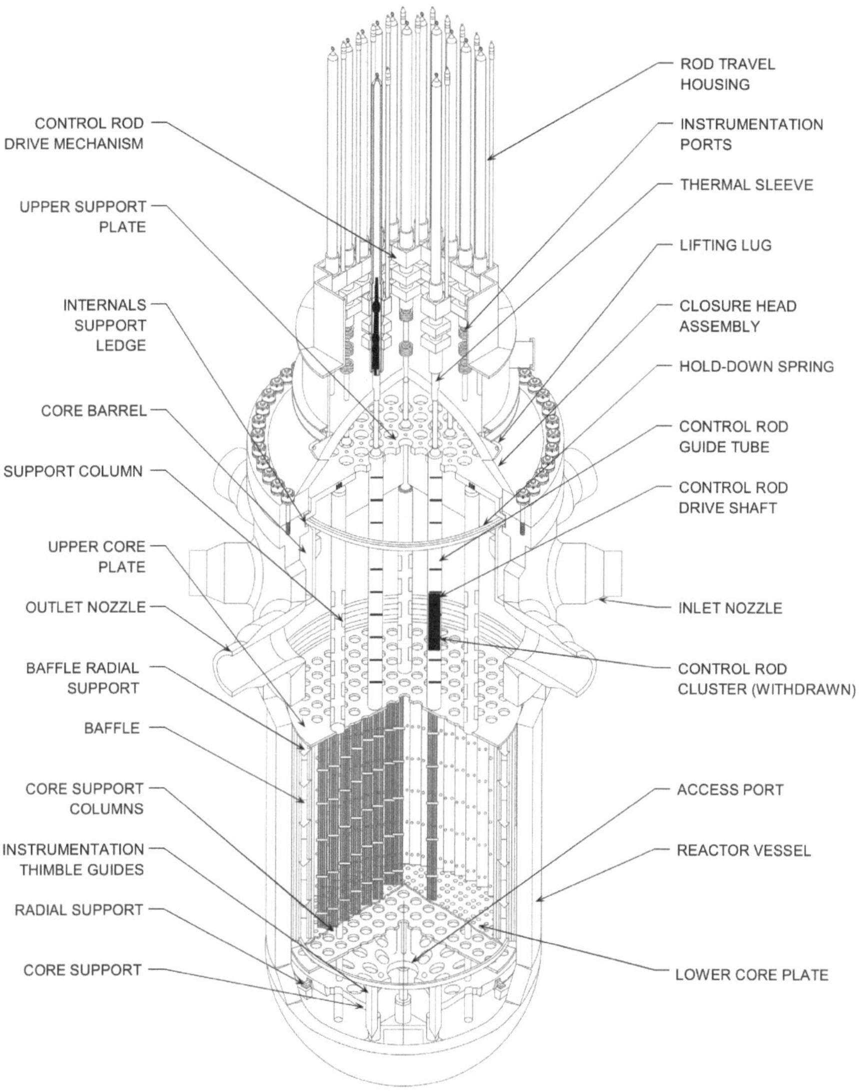

Figure 8.4 PWR pressure vessel. *Source:* [NRC08]

out of the RPV through the outlet or hot-leg nozzle. The incore instrumentation guide tubes penetrate the lower head of the vessel and extend through the lower core support plate with various orifices, while the control rods containing neutron absorbers are inserted through the RPV upper head.

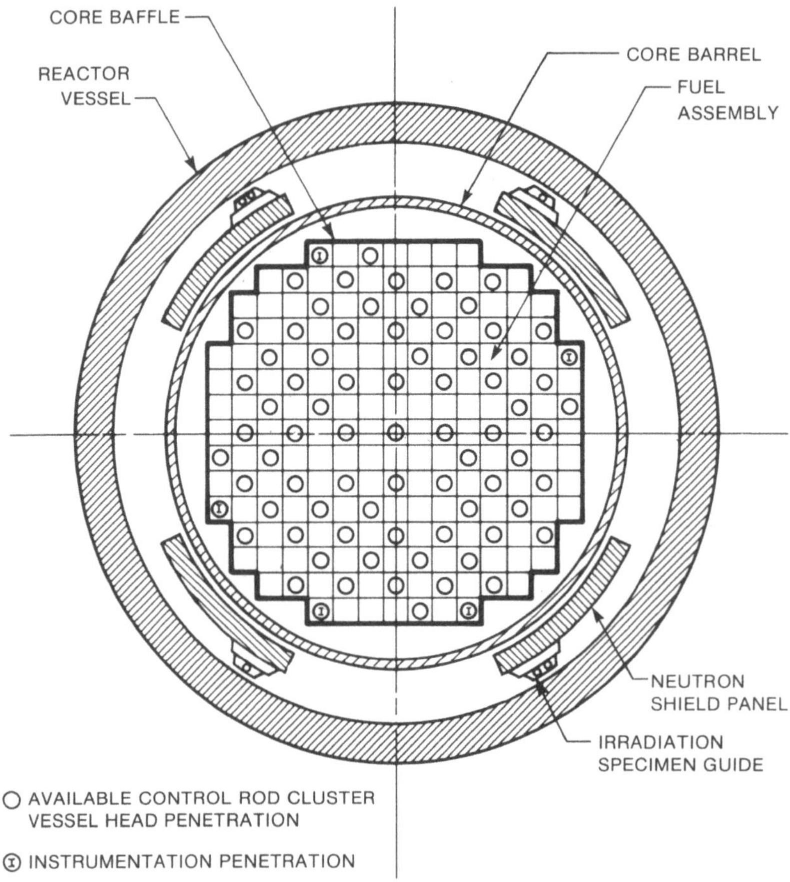

Figure 8.5 Top view inside a PWR pressure vessel. *Source:* Reprinted with permission from [Wes84]. Copyright © 1984 Westinghouse Electric Corporation.

A top view inside the RPV across the midsection of the fuel assemblies shown in Fig. 8.5 clarifies the location of the core barrel within the vessel and indicates the steel baffle, which surrounds the entire cluster of fuel assemblies and directs the upward flow of coolant into the heat producing fuel elements. Indicated also are the neutron shield panels, which protect the RPV wall from both neutron and gamma radiations, and the irradiation specimen guides, where irradiation coupons are stored for periodic evaluation of the cumulative radiation fluence on the RPV wall. Figure 8.5 also illustrates the fuel assemblies where the rod cluster control (RCC) assemblies may be inserted.

Figure 8.6 provides a cutaway view of a typical primary coolant pump, with its suction nozzle and discharge nozzle located near the bottom of the centrifugal pump. Note also a number of seals and a coolant system and a lubricant pump system

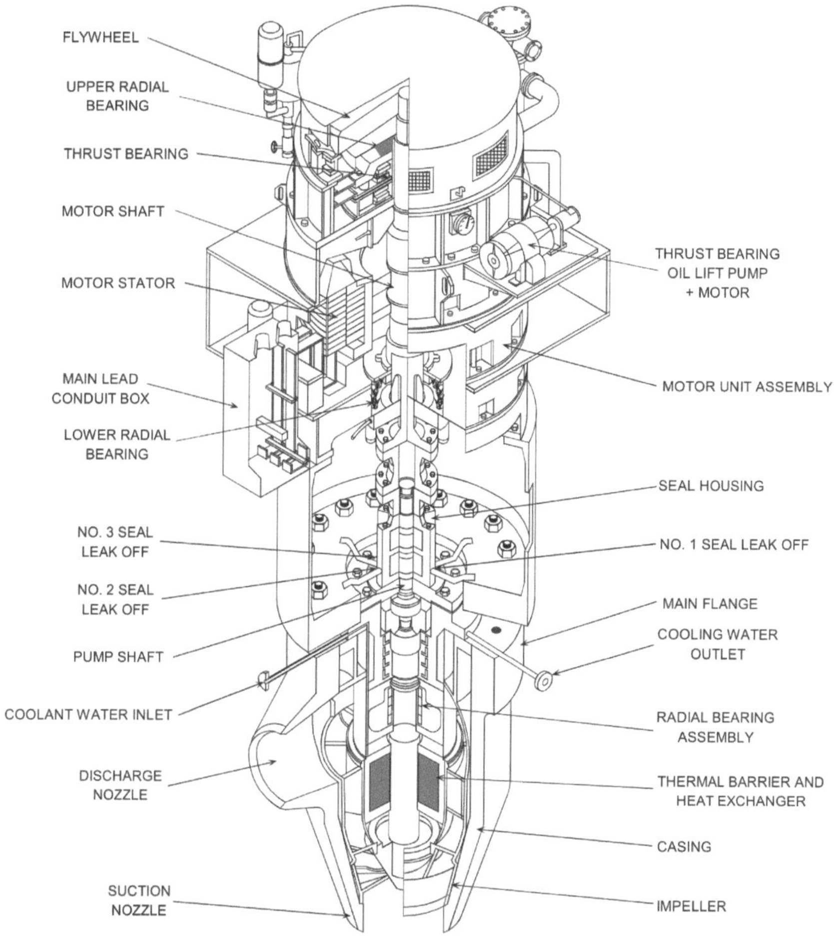

Figure 8.6 Cutaway view of the primary coolant pump for a PWR plant. *Source:* [NRC08].

for the pump motor. A cutaway view of the pressurizer in Fig. 8.7 illustrates key components: (a) safety and relief nozzles, (b) spray nozzle, (c) electrical heater arrays, and (d) surge nozzle. Injection of coolant through the spray nozzle reduces the pressurizer pressure, while the heaters may be turned on to increase the pressure. The surge line delivers coolant to the RCS loop and the safety and relief valves protect the pressurizer. The power-operated relief valve, which was inadvertently left open and misdiagnosed, provided a coolant leakage path in the TMI-2 accident. Finally, we note a detailed structure of a typical U-tube steam generator (UTSG) in Fig. 8.8. In this tube-and-shell-type steam generator, the radioactive primary coolant

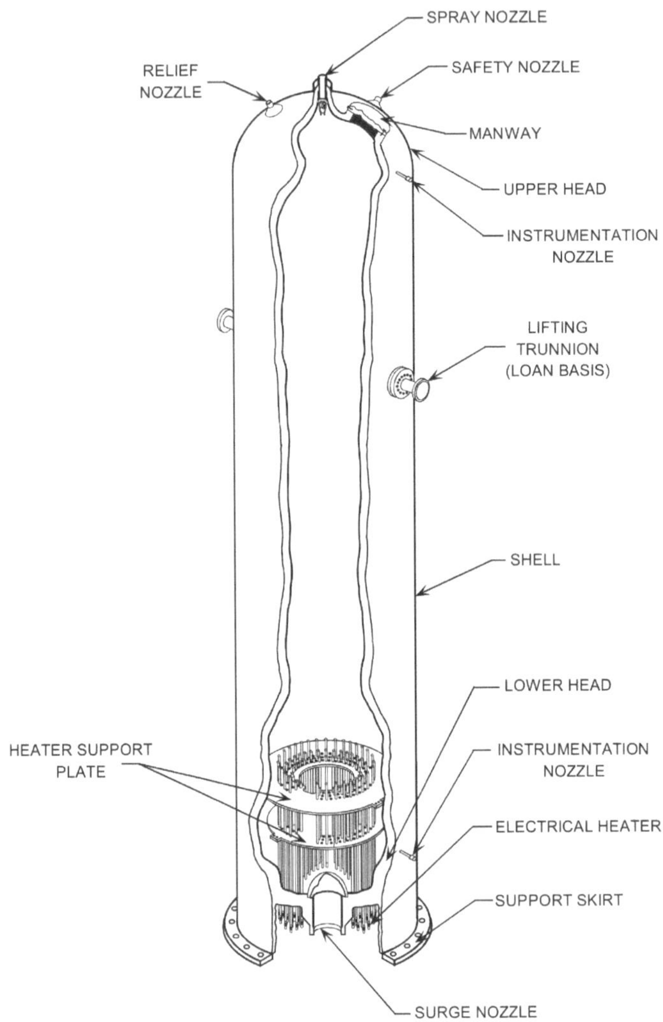

Figure 8.7 Cutaway view of a PWR pressurizer. *Source:* [NRC08].

flows through a bundle of U-shaped tubes, while the feedwater enters above the tube bundle, flows downward in the annulus between the tube wrapper and the steam generator vessel wall, and eventually flows upward through the tube bundle to pick up the heat through the tubes. The feedwater boils along the length of the tube bundle and steam is separated through two stages of steam separation operations: mechanical separation through swirl vane moisture separators followed by steam dryers. The separated steam, containing a small remnant of saturated liquid, is extracted at the steam nozzle at the top and delivered to the turbine generators. The

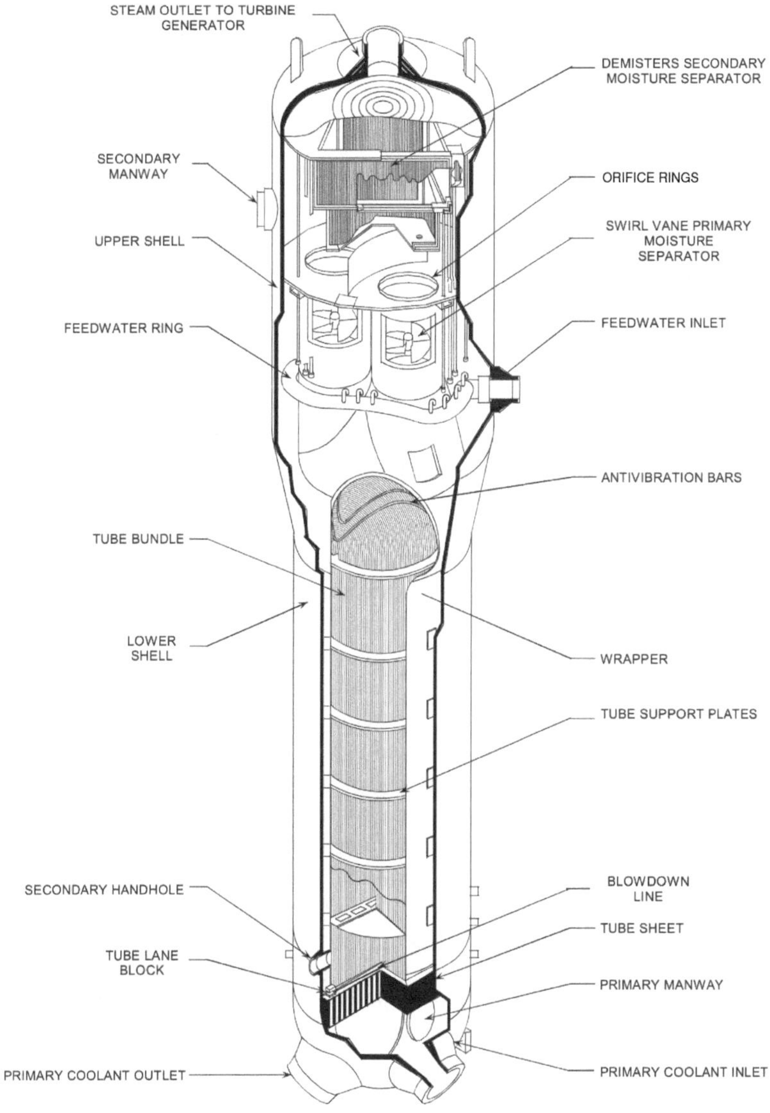

Figure 8.8 Cutaway view of a PWR steam generator. *Source:* [NRC08].

cutaway view also illustrates the tube sheet, through which the tubes are anchored, and tube support plates, which protect the tubes from turbulent flow of feedwater, much like the spacer grids in the fuel elements. Steam generators with straight tubes, known as once-through steam generators (OTSGs), were used in the ill-fated TMI-2 power plant.

8.1.2 Boiling Water Reactor

8.1.2.1 Overview of the BWR System The schematic diagram in Fig. 8.9 presents the overall BWR plant layout starting with the reactor vessel on the far left of the figure. The main difference between the BWR layout and that for the PWR system discussed via Figs. 8.1 through 8.3 is the obvious lack of the steam generator and presence of the steam separation equipment located in the upper region of the reactor vessel. Primary coolant pumps, which are called recirculation pumps in BWR plants, are illustrated, together with the control rod drives located at the bottom of the vessel. The control rods, in the shape of cruciform blades, are inserted through the bottom head of the vessel because of the presence of the steam separation equipment in the upper region of the vessel. An equally important reason for the bottom-entry control blades is to control the axial power distribution, which has to be shaped and controlled allowing for sharp variations in the coolant density due to boiling in the fuel region. The coolant water cleanup system featuring a filtration and demineralization system, cleanup pumps, and heat exchangers are coupled to the recirculation pumps. The BOP structure for BWR plants is fairly similar to that of PWR plants, with one obvious difference due to the use of a direct steam cycle, which does not require steam generators. This implies that the steam is radioactive and hence access to the turbine room has to be limited during operation. The connections between multiple stages of HP and LP turbines are indicated in Fig. 8.9. We also note the MSIVs and safety relief valves in the steam line. The steam discharged from the relief valves is delivered to the pressure suppression pool, or wetwell as it is also called. The reactor vessel itself is located within the primary containment building, known as the drywell. The steel reactor vessel has a wall thickness of 0.15 to 0.18 m.

In the RHR system diagram of Fig. 8.10 for the Mark I type containment, note first the drywell in the shape of an inverted lightbulb connected to the suppression pool located within a torus surrounding the bottom of the drywell. The drywell consists of a 50-mm thick steel shell surrounded by 0.6 to 1.8 m of reinforced concrete. The drywell, wetwell, and other NSSS components are housed in a concrete structure serving as the secondary containment.

In the direct-cycle BWR plant illustrated, feedwater is delivered directly to the feedwater sparger located above the core, mixed with recirculating water through jet pumps, and pumped to the fuel region of the core via the recirculation pumps located within the drywell. Steam is separated from liquid in the upper region of the reactor vessel and delivered to steam turbines. Exhaust steam is condensed in the condenser and returned via the feedwater system to the core, closing the feedwater-steam loop for the BWR plant. A number of AOVs as well as MOVs are noted in various flow paths. The RHR system serves as a normal shutdown cooling system, with 33%

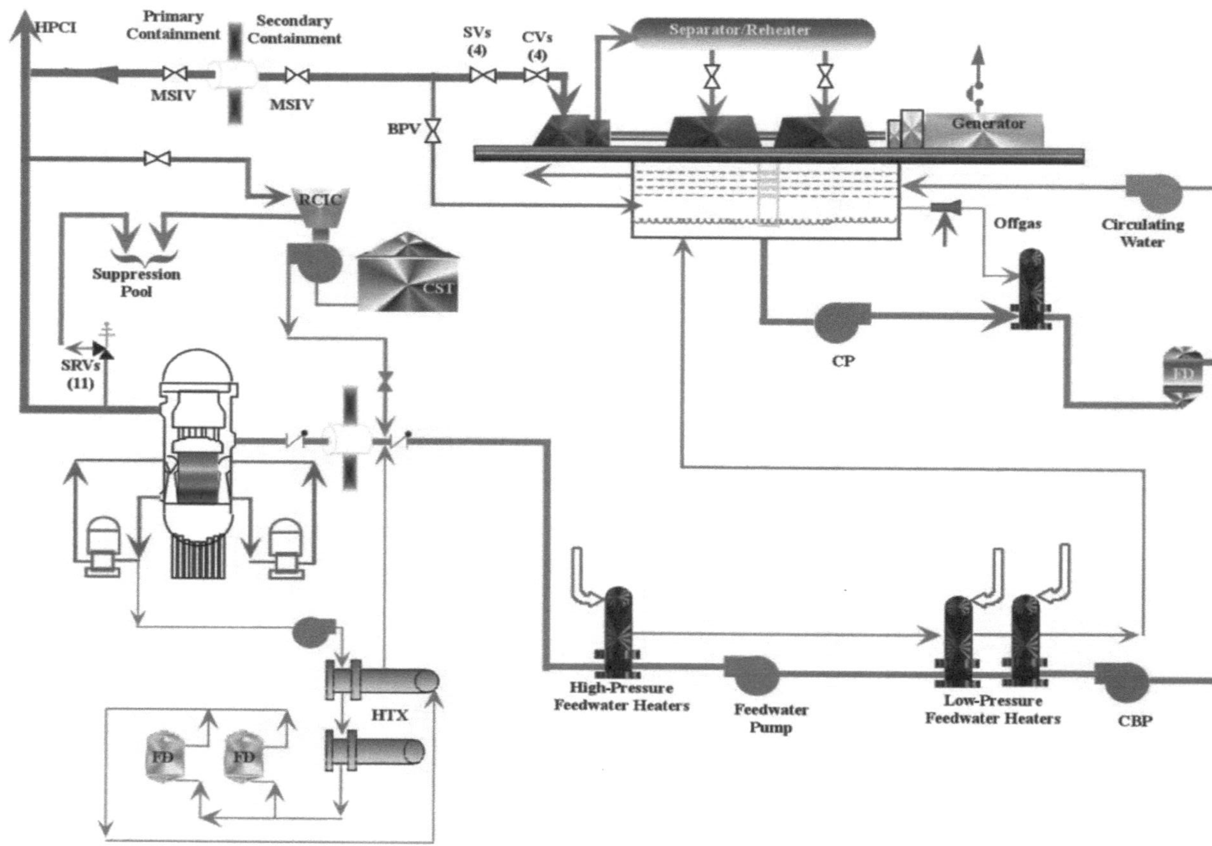

Figure 8.9 Schematic diagram of a BWR plant. Abbreviations: BPV = bypass valve, CV = control valve, CBP = condensate booster pump. CP = condensate pump, FD = filter demineralizer, HTX = heat exchanger, SRV = safety relief valve, SV = stop valve. *Source:* [NRC08].

redundancy and cooled by the service water system in the RHR heat exchanger. As an alternate feedwater delivery system, reactor core isolation cooling (RCIC) pumps take suction from the condensate storage tank and delivers feedwater in case of core isolation transients, explained further in the next section. The RHR system also provides the vessel head spray to the steam dome in the upper region of the reactor vessel above the steam separation equipment.

Figure 8.11 indicates how the RHR system serves as part of the LPCI system in case of a LOCA, where coolant is delivered to the recirculation lines to keep the core cooled. Through the automatic depressurization system (ADS), consisting of safety and relief valves, high-pressure steam is discharged into the suppression pool, where it is condensed, thereby controlling system pressure increases in a LOCA. As part of the emergency core cooling system (ECCS), the suppression pool inventory is used also in the core spray system, while the HPCI pump delivers the coolant inventory in the CST directly to the feedwater line. Note also the containment spray water that the RHR system provides in accident situations.

8.1.2.2 BWR Ennginee red Safety Features Similar to the discussion we had for PWR plants in Section 8.1.1, we now briefly discuss specific ESFs and their functions for BWR plants in both normal operation and accident modes, drawing on the descriptions of the NSSS and BOP systems and the RHR and ECCS layout presented via Figs. 8.9 through 8.11.

1. *Residual heat removal system*

    - Shutdown cooling system

      In normal operation, the RHR pumps take suction from the suppression pool (wetwell torus) and circulate coolant water through the RHR heat exchangers and back to the recirculation line. In a LOCA, the system serves as part of the LPCI system.

    - Vessel head spray

      It delivers some of the RHR system flow to the steam dome to quench steam in the reactor vessel as part of the LPCI system.

2. *Reactor core isolation cooling system*

   For transient events involving loss of feedwater flow coupled with isolation of the reactor core, i.e., closure of the main steam isolation valves (MSIVs), the RCIC system serves as an alternative feedwater delivery system. The RCIC system uses steam-turbine-driven pumps, with suction taken from the CST, and delivers the cooling water to the feedwater sparger located above the core. After passing through the turbine, the steam is discharged to the suppression pool where it is condensed.

3. *High-pressure coolant injection system*

   The HPCI system uses steam-turbine driven pumps, with suction taken from the CST. The emergency coolant water is delivered to the feedwater sparger. The system can take suction also from the suppression pool, if necessary. The HPCI

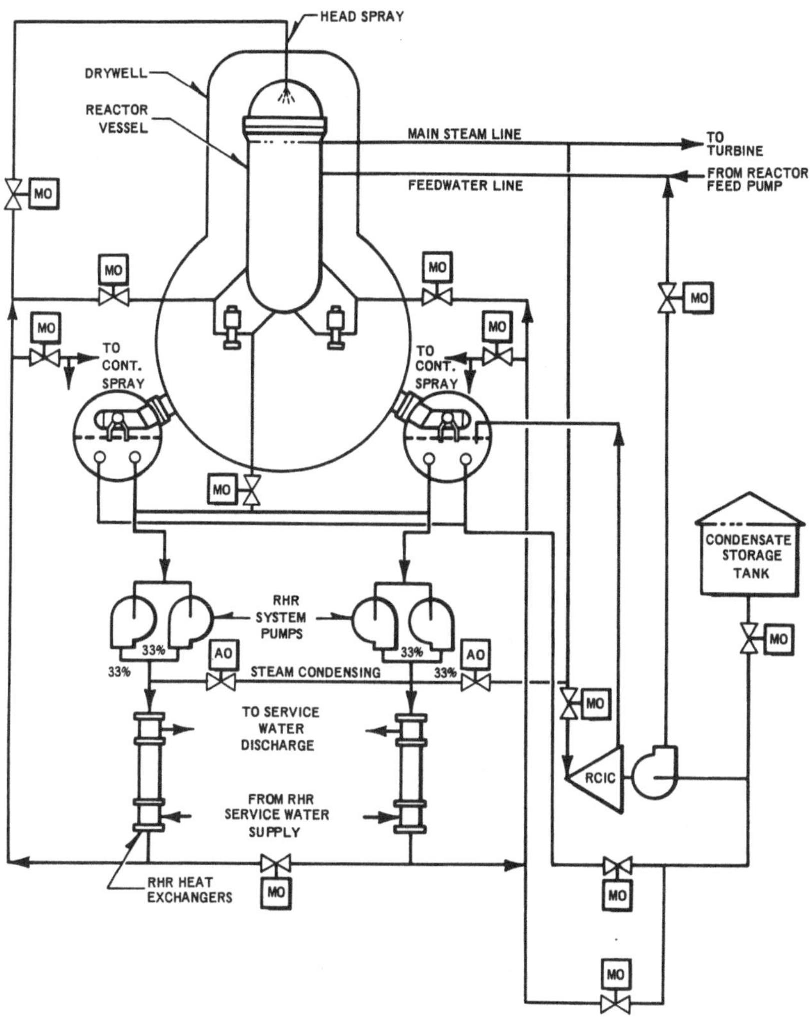

Figure 8.10 Residual heat removal system for a BWR plant. *Source:* Reprinted with permission from [Gen71]. Copyright © 1971 General Electric Company.

system provides cooling water to the RCS in a LOCA similar to the mode of RCIC operation for core isolation events.

4. *Automatic depressurization system*

   The ADS consists of safety and relief valves and the associated piping, which

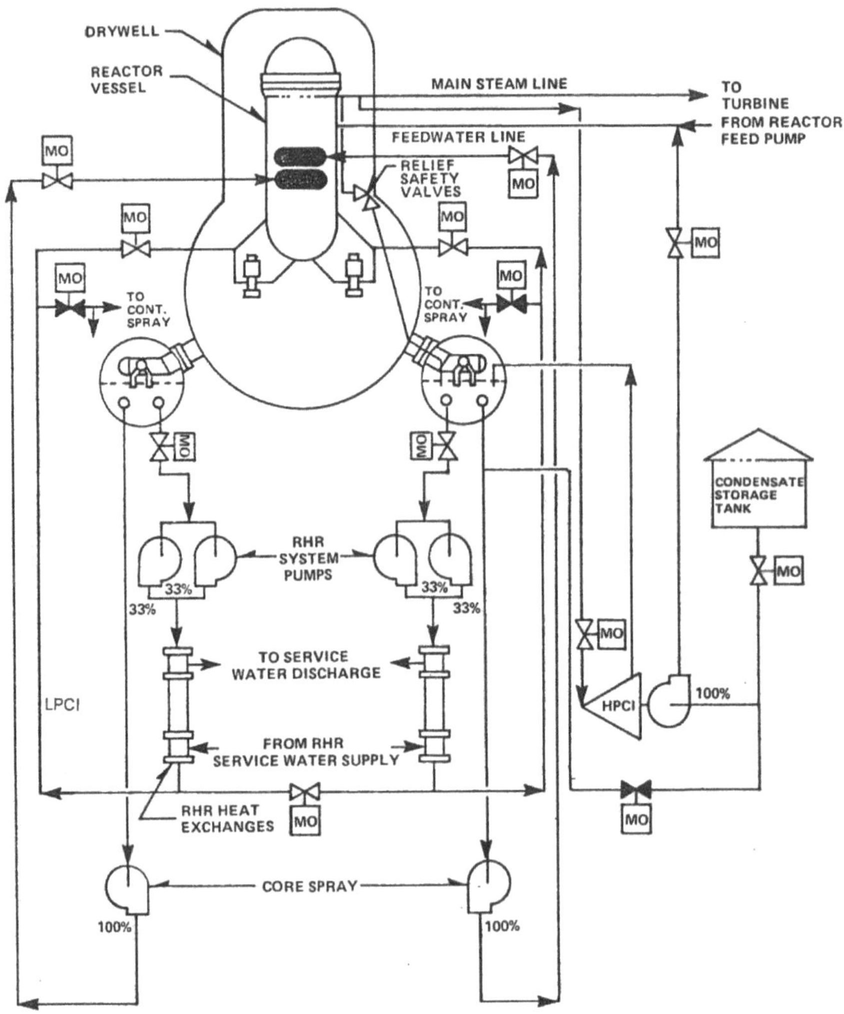

Figure 8.11 Emergency core cooling system for a BWR plant. *Source:* Adapted with permission from [Gen71]. Copyright © 1971 General Electric Company.

discharge high-pressure steam into the suppression pool, minimizing pressure increases in the RCS.

5. *Core spray system*
   The core spray pumps take suction from the suppression pool and deliver cooling water to the core spray nozzles above the top of the active fuel inside the core

shroud, which separates the downcomer annulus from the upward flow of coolant water through the core.

6. *Low-pressure coolant injection system*
   In conjunction with the RHR pumps, the LPCI pumps deliver coolant water from the suppression pool to the recirculation loop, with the heat removed in the RHR heat exchangers.
7. *Containment spray system*
   Coolant water is taken from the suppression pool, passed through the RHR heat exchangers, and delivered to the spray header located in the drywell.
8. *Standby liquid control system (SLCS)*
   High-pressure pumps deliver borated water to the sparger in the inlet plenum below the core. The SLCS is designed primarily for controlling reactivity in the case of scram failure.

8.1.2.3 Brief Overview of BWR Components and Equipment To augment the discussion of the general layout of the NSSS and BOP components presented so far in this section, we now provide a brief description of key components within the BWR reactor vessel. A cutaway view of BWR reactor vessel internals is presented in Fig. 8.12, where two stages of steam separation equipment above the core are clearly illustrated, together with the core spray and sparger lines. The core shroud, which provides the same function as the PWR core barrel, separates the downward flow of coolant flow in the downcomer from the upward coolant water through the core. The funnel-shaped jet pumps pick up the downward flow of liquid, separated from steam in the steam separator and dryer assemblies and mixed with the feedwater delivered through the feedwater sparger and recirculating flow. The recirculation pumps located outside the reactor pressure vessel deliver the mixed flow of coolant through the downcomer and eventually upward through the core. The control rod drive and incore flux monitoring mechanisms are located under the reactor vessel.

8.2 ACCIDENT CLASSIFICATION AND GENERAL DESIGN GOALS

Operational and transient states of a NPP may be classified in a number of ways, depending upon the regulating and licensing agencies of a particular country where the NPP is constructed. In this section, we present two classification systems used often in the United States. The first system [ANS73] is based on the American National Standards Institute (ANSI) standard N18.2 and used as a basic structure for safety analysis reports for LWR plants [Wes03]. The second classification system has been in use for construction permits and operation licenses by the U.S. Nuclear Regulatory Commission (NRC), as stipulated in Title 10 of *Code of Federal Regulations*, Part 50 (10 CFR 50). The two classification systems are discussed in Sections 8.2.1 and 8.2.2, respectively.

Together with the consideration of various operational and accident states for nuclear power plants, any NPP design should follow general guidelines and goals.

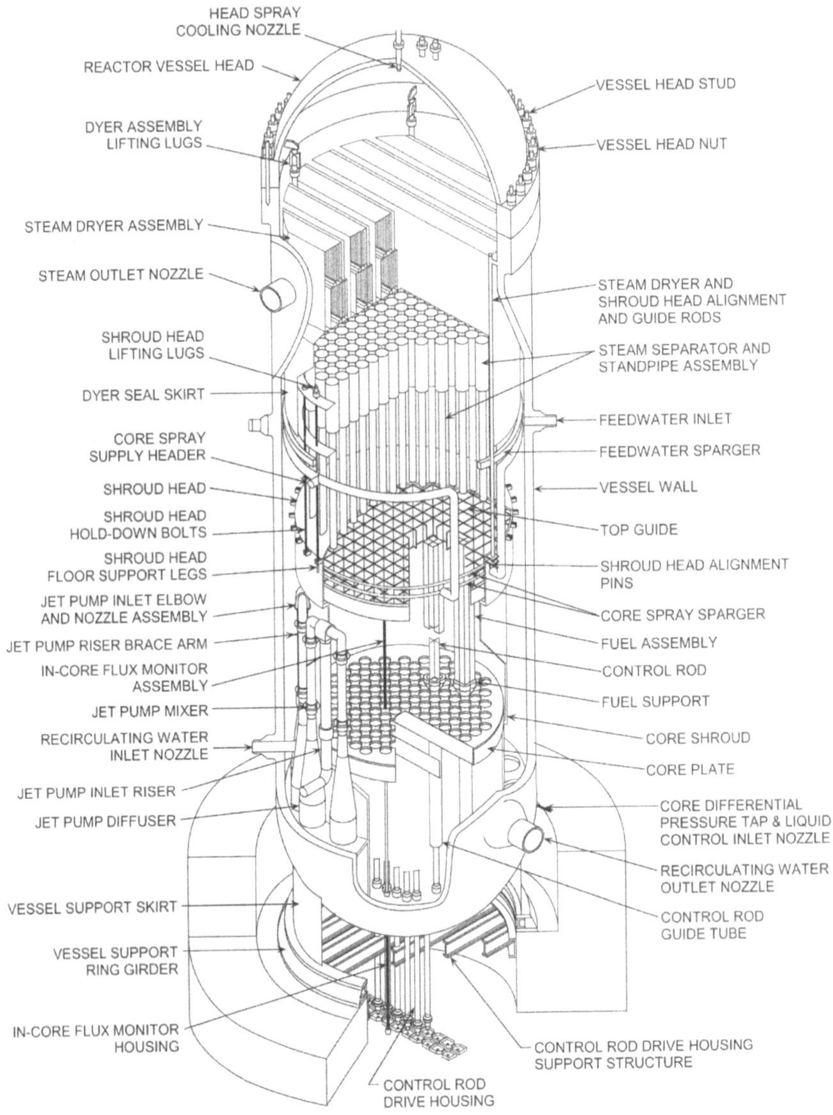

Figure 8.12 Cutaway view of a BWR pressure vessel illustrating detailed coolant flow and core spray arrangement. *Source:* [NRC08].

Among them are (a) the General Design Criteria (GDC) delineated in Appendix A to 10 CFR 50, (b) the Safety Goals documented as a Policy Statement in 10 CFR 50, (c) the Final Acceptance Criteria (FAC) for the design and evaluation of the ECCS spelled out in 10 CFR 50.46, and (d) guidelines for a risk-informed decision making process published as NRC Regulatory Guide 1.174. The FAC will be discussed in connection with the LOCA analysis in Section 8.3. Regulatory Guide 1.174 [NRC02] will be discussed as part of the risk-informed licensing and regulations in Chapter 12. As examples of general design and safety guidelines for NPPs, the GDC and Safety Goals will be discussed in Section 8.2.3.

8.2.1 Plant Operating States

The ANSI N18.2 classification system divides plant operating states into four conditions according to the anticipated frequency of the states and potential radiological consequences to the public. For the structural analysis of NPP systems, including the RPV and RCS pipes, the American Society of Mechanical Engineers (ASME) Boiler and Pressure Vessel (BPV) Code [Rao06] stipulates allowable stress intensity limits according to four service levels, A, B, C, and D. Although the ASME service levels A, B, C, and D generally correspond to ANSI conditions I, II, III, and IV, respectively, the BPC Code recognizes that some components may have to be limited to design conditions more restrictive than those indicated by the ANSI plant operating conditions. The estimated frequencies of the events are from [Hew00].

1. Condition I: normal operation and operational transients
   Events that will occur regularly as part of plant operation, maintenance, and refueling.
   Example: (i) steady-state and shutdown operations, (ii) operation with permissible deviations, (iii) plant heatup and cooldown, and (iv) permissible load rejection.
2. Condition II: faults of moderate frequency or upset conditions
   Events that are expected to occur during the plant lifetime, with frequencies on the order of one occurrence per reactor year. These events are called anticipated transients.
   Example: (i) turbine trip due to lightning, (ii) loss-of-feedwater event leading to steam bypass to condenser, accompanied by reactor trip, and (iii) uncontrolled control rod bank withdrawal.
3. Condition III: infrequent faults or emergency events
   Events that are possible to occur during the plant lifetime, with frequencies on the order of 0.01 occurrences per reactor year.
   Example: (i) SBLOCA, (ii) PORV stuck open, (iii) fires, and (iv) used-fuel cask drop accidents.
4. Condition IV: limiting faults
   Events that are postulated to occur, with frequencies on the order of 10^{-4} occurrences per reactor year, which have to be analyzed as DBAs.
   Example: (i) LBLOCA (200% LOCA), (ii) rod ejection accident, and (iii) steamline break.

8.2.2 Accident Classification in 10 CFR 50

Title 10 of *Code of Federal Regulations*, Part 50, governs the construction and licensing of nuclear power plants in the United States. Appendix I to Regulatory

Guide 4.2 [NRC76], originally proposed as an Annex to Appendix D, 10 CFR 50, presents transient events and postulated accidents grouped into nine classes starting with trivial incidents progressing to core meltdown accidents:

Class 1: trivial incidents, e.g., routine releases of radionuclides inside containment
Class 2: small releases outside containment, e.g., small spills and releases through steamline relief valves
Class 3: radwaste system failures, e.g,, equipment failure and operator error, release of gas or liquid waste
Class 4: events releasing fission products to primary system (BWR), e.g., fuel cladding defects and transients inducing fuel failures
Class 5: events releasing fission products to primary and secondary systems (PWR), e.g., steam generator tube rupture, fuel cladding defects accompanied by steam generator leak
Class 6: refueling accidents inside containment, e.g., fuel element drop, heavy object drop onto fuel in core
Class 7: used-fuel handling accidents, e.g., fuel element drop in fuel storage pool, fuel cask accident
Class 8: accidents considered in design basis evaluation, e.g., break of primary coolant pipe, reactivity transient, steamline break
Class 9: hypothetical accidents more severe than Class 8

Class 8 accidents cover DBAs that each plant is designed for, so that if the ESFs function properly there will be no unacceptable consequences to the public, i.e., no release of radionuclides to the environment. There are two primary groups of DBAs for LWR plants: (a) undercooling of the primary system bounded by LBLOCAs and (b) reactivity-induced accidents caused by rod ejection accidents for PWR plants and rod drop accidents for BWR plants. In PWR cores, control rods containing neutron absorbing material such as an Ag-In-Cd compound are inserted into the core through electromagnets. A rod ejection accident postulates a malfunction in the control rod drive mechanism or a rupture in the control rod housing so that the system pressure of 2250 psia in a PWR core would rapidly eject the control rod out of the core, thereby resulting in a positive reactivity insertion. In BWR cores, the insertion of control blades into the core through the lower RPV head depends on the gravitational head of hydraulic fluid supported in the control rod header. A rod drop accident postulates a malfunction in the control rod header, in which case the hydraulic pressure is lost and the control blades would drop out of the core due to gravity.

Although the reactivity insertion accidents bound by the rod ejection or rod drop accidents could result in overpower transients and overheating of the core, the postulated DBAs would not usually result in unacceptable consequences to the public. On the other hand, LBLOCAs could result in significant overheating of the core, with the potential for substantial release of radionuclides into the environment, if the core cooling is not restored in due time. Thus, LBLOCA events have served as bounding DBAs for LWR plants. A LBLOCA scenario for a typical PWR plant is discussed in Section 8.3 as a key example of the DBA.

Transient calculations representing various classes of postulated events and accidents all the way up to the DBAs are presented in Chapter 15 of the Final Safety Analysis Report (FSAR) submitted to the U.S. Nuclear Regulatory Commission as part of the application for the construction and operation license of nuclear power plants. The format, structure, and contents of the FSAR are stipulated in 10 CFR 52 [NRC09] and Regulatory Guide 1.70 [NRC78a]. Appendix N [NRC07] to 10 CFR 52 provides specific provisions and requirements related to combined construction and operation licenses (COLs) for nuclear power plants of identical design to be located at multiple sites. Design certification rules for four Generation III/III+ designs, System 80+, Advanced Boiling Water Reactor, AP600 and AP1000 plants, are also included as appendices to 10 CFR 52.

8.2.3 General Design Criteria and Safety Goals

As primary examples of general guidelines for the design, safety analysis, and operation of NPPs, we present a brief discussion on the General Design Criteria [NRC71] and Safety Goals [NRC86].

8.2.3.1 General Design Criteria (10 CFR 50, Appendix A, 1971) Consisting of 64 criteria in six categories, the GDC has served since 1971 as a set of guiding principles for the design and evaluation of NPP systems and components and includes the defense-in-depth concept discussed in Section 1.4. We highlight the criteria that are particularly germane to risk and safety analyses.

Category I. Overall requirements, including the criteria for quality assurance and fire protection.

Category II. Protection by multiple fission product barriers, establishing requirements for the defense in depth for NPP designs.

Criterion 10: Acceptable fuel design limits should not be exceeded for anticipated operational occurrences (AOOs).

Criterion 11: Power coefficient of reactivity should be negative in the power operating range for inherent reactor protection.

Criteria 12 and 13: Instrumentation and control systems should be provided so that reactor power oscillations and other AOOs can be detected and controlled.

Criteria 14 and 15: Reactor coolant systems should be designed not to breach the reactor coolant pressure boundary for all accident conditions.

Criterion 16: Containment and engineered safety systems should be designed so that off-site radiation dose will not exceed regulations for all postulated accidents.

Criteria 17 and 18: Electrical power systems should be reliable, with due considerations for independent and redundant components, for all AOOs and postulated accidents.

Criterion 19: Control room should be habitable, with personnel exposure < 5 rem/accident, and functional for all accidents, including loss of coolant accidents.

Category III. Protection and reactivity control systems

Criterion 21: Protection system should be designed for high reliability and testability, satisfying the single failure criterion, discussed in Section 4.1.3.

Criterion 27: Control systems should be capable of shutting down and cooling the reactor with margins for the stuck-rod condition, i.e., with the most reactive control rod stuck out and inoperative.

Criteria 34 and 35: Residual heat removal and ECCS capability should be provided so that fuel design limits and pressure boundary conditions are not exceeded. (See criteria 14 and 15.)

Category IV. Fluid systems
Includes criteria for coolant and containment heat removal systems.
Category V. Reactor containment
Includes design criteria for containment, penetrations, compartment, and containment leakage tests.
Category VI. Fuel and radioactivity control
Includes criteria for fuel and fission product storage and handling and control of radioactivity release.

8.2.3.2 Safety Goals (10 CFR 50, Policy Statement, 1986) The NRC safety goals, established as a policy statement, define an acceptable level of radiological risk to the public associated with the operation of NPPs. A draft policy statement had been released in 1983 for public comments and evaluation by the industry for two years before it was adopted in 1986 formally as part of 10 CFR 50. The policy statement consists of two quantitative safety goals and guidelines for regulatory implementation:

A. Quantitative safety goals for probabilistic risk assessment

- For the vicinity of a plant (1 mile from site boundary), the calculated risk should be $<0.1\%$ of prompt fatality due to all other activities for the people involved.
- Near a plant (10 miles from site boundary), the calculated risk should be $<0.1\%$ of latent cancer due to all other activities for the people involved.

B. Plant performance guideline

- Large radioactive release to the environment should be $<10^{-6}$/reactor-year of plant operation.

8.3 DESIGN BASIS ACCIDENT: LARGE-BREAK LOCA

A typical scenario [Hew00] for a LBLOCA involving a sudden rupture of the cold leg of the primary coolant pipe in a PWR plant is presented in this section, followed by specifications for the emergency core cooling system provided to keep the core cooled and avoid the release of radionuclides to the environment. The coolant pipe rupture is assumed to undergo a double-ended guillotine break, or 200% break, which is to suggest that coolant may leak out of both ends of the broken pipe uninterrupted. The LOCA and associated ECCS analyses are to be performed for a 200% break in the cold leg, because the coolant escaping through the break will not have an opportunity to pick up any heat generated in the core, thereby making the accident consequences more severe than would be the case for a hot-leg break. The sequence of events is fairly similar for a LBLOCA in a BWR plant and will not be considered here explicitly.

8.3.1 Typical Sequence of a Cold-Leg LBLOCA in PWR

The progression of events in a 200% LBLOCA is illustrated in Figs. 8.13 through 8.19. The illustration begins with a diagram indicating the primary and secondary loops, and ECCS components for normal operating condition in Fig. 8.13, and continues through the changes in the system configuration through four phases of the LOCA in Figs. 8.14 through 8.17, culminating in a system configuration for a long-term cooling phase in Fig. 8.18. The corresponding evolutions in the RPV itself are illustrated in Fig. 8.19, starting with full coolant inventory in plot (a).

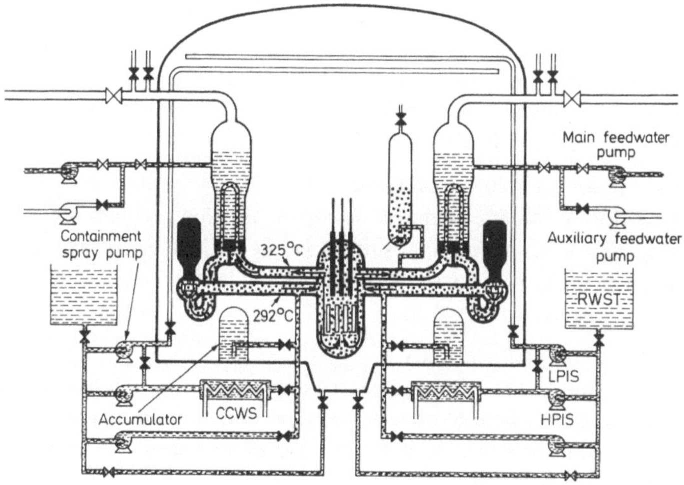

Figure 8.13 Schematic diagram of key PWR engineered safety features in normal operation. *Source:* [Hew00].

1. Blowdown Phase: 0 to 20 seconds
   Coolant is blown down through the break and the system is depressurized. At 10 seconds, the high-pressure coolant injection system actuates around the primary system pressure of 1500 psia and delivers coolant from the refueling water

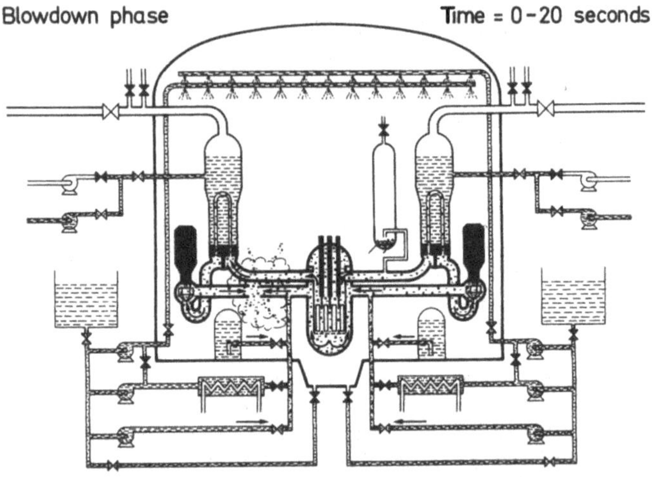

Figure 8.14 Events in a PWR large-break LOCA: blowdown phase (0–20 seconds). *Source:* [Hew00].

storage tank. Later accumulator water discharges into the reactor pressure vessel, following a reduced pressure and water level in the pressurizer shown in Fig. 8.14. The containment spray is also actuated, with water taken from the RWST. Figure 8.19*b* shows how the ECCS water enters through the unbroken cold leg, with the RPV upper plenum filled with steam.

2. Bypass Phase: 20 to 30 seconds
   Upward flow of steam in the downcomer annulus prevents ECCS water from entering the lower plenum, resulting in the ECCS water bypass. The water inventory in the RPV is nearly depleted and coolant is collected in the containment sump, as indicated in Fig. 8.15.

3. Refill Phase: 30 to 40 seconds
   Steam flow out of the RPV decreases accompanied by a further reduction in the system pressure. The low-pressure coolant injection system actuates around the primary system pressure of 450 psia and coolant refills the lower plenum, as illustrated in Figs. 8.16 and 8.19*c*. Heatup of fuel rods is indicated during this phase.

4. Reflood Phase: 40 to 250 seconds
   Fuel elements are reflooded from the bottom up and steam is produced in the upper plenum. Reverse heat flow from the steam generator to the primary loop evaporates liquid droplets in the steam, building up a back pressure in the upper plenum and restricting the reflood rate, illustrated in Figs. 8.17 and 8.19*d*. This phenomenon is called the steam binding. Maximum clad temperature is reached around 120 seconds, and the RWST inventory is nearly depleted.

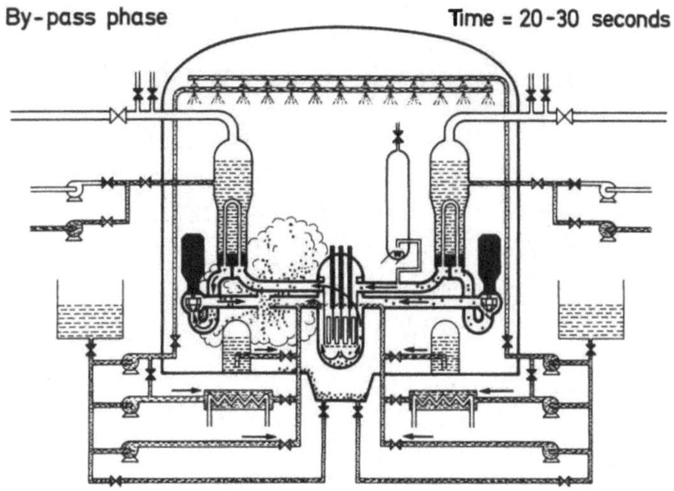

Figure 8.15 Events in a PWR large-break LOCA: bypass phase (20–30 seconds). *Source:* [Hew00].

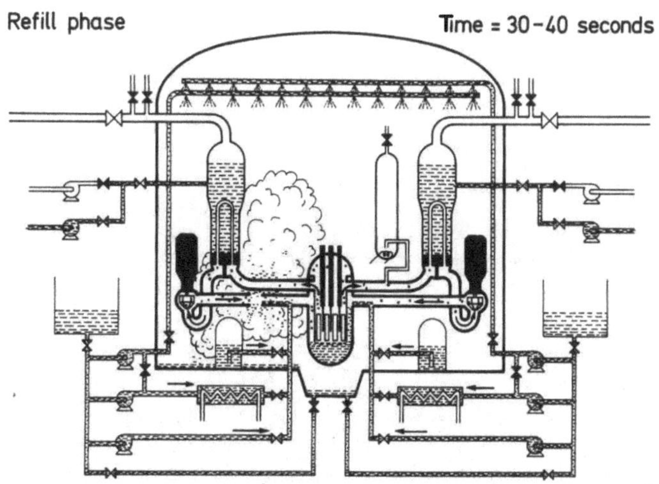

Figure 8.16 Events in a PWR large-break LOCA: refill phase (30–40 seconds). *Source:* [Hew00].

5. Long-Term Cooling
LPCI water enters through the unbroken cold leg and forms a natural circulation path with steam leaking through the break. The break flow accumulates in the

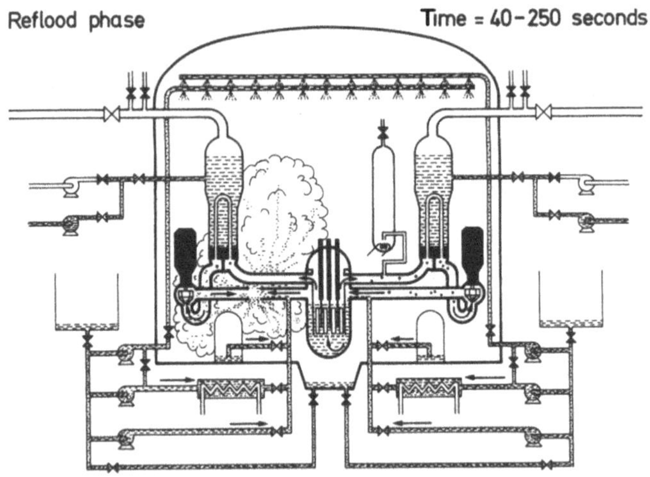

Figure 8.17 Events in a PWR large-break LOCA: reflood phase (40–250 seconds). *Source:* [Hew00].

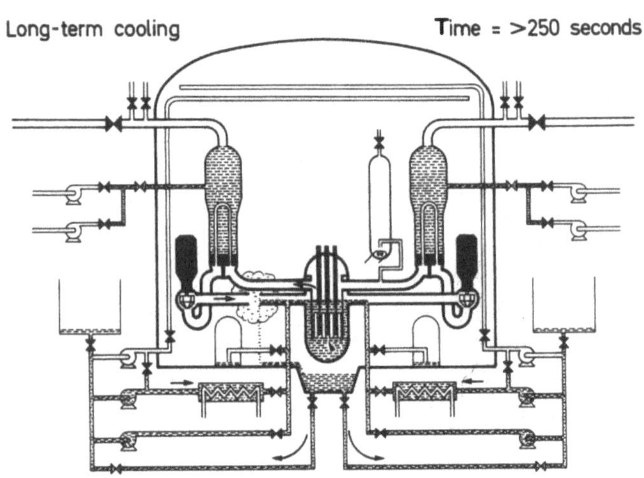

Figure 8.18 Events in a PWR large-break LOCA: long-term cooling phase (>250 seconds). *Source:* [Hew00].

sump and reactor cavity and recirculates via the LPCI pumps. This long-term, recirculating mode of core cooling is illustrated in Fig. 8.18.

We note in summary that the LBLOCA is a rapidly evolving event with a peak clad temperature reached in 2 minutes following the postulated pipe break. Hence,

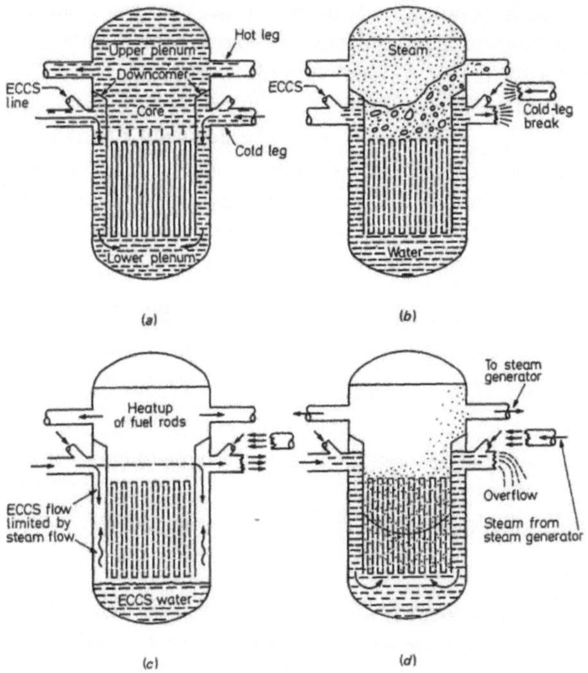

Figure 8.19 Events in the reactor pressure vessel during a PWR large-break LOCA: (*a*) normal operation, (*b*) blowdown phase, (*c*) refill phase, and (*d*) reflood phase. *Source:* [Hew00].

the control and management of the accident should not be left to the manual action of reactor operators, but rather be accomplished through a well-designed and tested ECCS.

8.3.2 ECCS Specifications

A series of long and protracted public hearings were conducted to determine a set of criteria for successful construction and testing of the ECCS in the early 1970s. These hearings were rather unique because the public, in this case many antinuclear activists or environmentalists, was allowed to question the scientists and engineers who participated in the design and analysis of ECCS systems. With official transcripts kept throughout the hearings, it was a painful process for nuclear engineering professionals sometimes to have their credentials and the results of their design calculations and evaluations publicly questioned by nonprofessionals. In many cases, the public would not recognize the difficulty involved in accurate analyses of the complex series of events to follow in postulated LOCA accidents.

At the conclusion of the hearings, the U.S. Nuclear Regulatory Commission adopted and finally promulgated in 1974 a set of acceptance criteria [NRC74a], including detailed specifications for the evaluation model, for the design and analyses of the ECCS for LWR plants. One particular point in the Final Acceptance Criteria (FAC) is the limiting peak clad temperature of 2200°F for Zircaloy clad used in LWR fuel elements. The limiting temperature is substantially below the melting temperature [Hay99] of 2318 K (3713°F) of Zircaloy-4 but was chosen based on the engineering judgment of the nuclear engineering community attained through the lengthy hearing process. The requirements for the evaluation model were somewhat relaxed in a 1988 revision of the FAC.

8.3.2.1 Final Acceptance Criteria (10 CFR 50.46, 1974) The FAC for the design, construction, and operation of the ECCS consists of the following five criteria:

1. Peak clad temperature $< 2200°$F throughout the accident
2. Maximum cladding oxidation $\leq 17\%$ of clad thickness
3. Maximum hydrogen generation $\leq 1\%$ of possible metal-water reaction involving fuel clad
4. Maintain a coolable geometry throughout the accident
5. Provide long-term cooling capability at the end of the accident

8.3.2.2 ECCS Evaluation Models (10 CFR 50 Appendix K, 1974) In a detailed appendix [NRC74b] to Part 50, specifications [NRC74c] for the evaluation and analysis of the performance of the ECCS for LOCA events are presented. In particular, licensees of LWR plants are required to use a set of conservative calculational models, not the best estimate models, including:

1. Assume an instantaneous double-ended pipe break.
2. Assume 102% of rated power, combined with the worst power distribution and maximum power peaking factor, at the initiation of the accident.
3. Use 120% of the heat generation rate due to the decay of fission products (FPs) as calculated by the ANS 5.1 standard [ANS94].

The FP decay heat generation model in the ANS standard begins with the rate $f(t)$ [MeV/fission·s] of energy emitted in the form of β- and γ-rays at t seconds after the fission of one nucleus,

$$f(t) = \sum_{i=1}^{23} \alpha_i \exp(-\lambda_i t), \tag{8.1}$$

where the 23 pairs of coefficients $\{\alpha_i, \lambda_i\}$ are tabulated for each of the fissionable nuclides ^{235}U, ^{238}U, ^{239}Pu, and ^{241}Pu. For use in the ECCS evaluation model, it is necessary to obtain an expression for the FP decay power $P_d(t,T)$ generated at t seconds after the reactor is shut down following the operation at thermal power P for a period of time T seconds. With recoverable energy Q MeV generated per fission, Eq. (8.1) is used to derive

$$P_d(t,T) = \int_{-T}^{0} \mathrm{d}t' \frac{P}{Q} f(t-t') = \frac{P}{Q} F(t,T), \tag{8.2}$$

where P/Q represents the number of fissions per second that take place during the reactor operation and $t - t'$ the elapsed time after the fission events in interval dt' around t'. The integral is performed over the interval $[-T, 0]$ of reactor operation and the function $F(t,T)$ represents the decay heat generated in units of MeV/fission for cooling time t seconds and operating time T seconds. Breaking up the integral over the operating time $[-T, 0]$ into two integrals over the intervals $[-\infty, 0]$ and $[-T, -\infty]$ yields

$$F(t,T) = F(t,\infty) - F(t+T,\infty). \tag{8.3}$$

The ANS 5.1 standard provides, for each of the four fissionable nuclides, a one-dimensional tabulation $F(t,\infty)$, which may be conveniently used twice, once for the cooling time t and the second time for the sum $t + T$ of the cooling and operating times to yield $F(t,T)$. This approach avoids the laborious task of managing two-dimensional tabulations for $F(t,T)$ in terms of t and T separately. Additional guidelines are given in the standard on accounting for the effects of neutron capture in FPs and stepwise variations in the operating power level, together with estimates for uncertainties in the tabulated data. The ECCS evaluation models of 10 CFR 50, Appendix K, suggest the use of 120% of the FP decay power calculated via Eqs. (8.2) and (8.3) in recognition of significant uncertainties in the FP decay heat model. According to the standard, following a long period of reactor operation, ~6% of rated thermal power P is generated through the FP decay immediately after the reactor is shut down, with the decay heat decreasing to ~0.5% of P in a day.

The NRC sponsored the development of a suite of systems analysis computer codes, primarily for the purpose of allowing the NRC staff to perform confirmation calculations for LOCAs. Perhaps best known among them is the RELAP family of codes, with the current version distributed as the RELAP5 Mod 3.3 code [NRC01].

8.3.3 Code Scaling, Applicability, and Uncertainty Evaluation

Following the implementation of the FAC for the evaluation of the ECCS for LWRs discussed in Section 8.3.2, the nuclear engineering community began the task of quantifying uncertainties in the ECCS evaluation models under the auspices of the NRC. This resulted in the release of a revised ECCS rule in 1988, which offered the option to use either the evaluation model (EM) of Appendix K or best estimate (BE) computer codes for ECCS analyses, provided uncertainties in the calculational models are quantified. The NRC issued Regulatory Guide 1.157 [NRC89] detailing the steps and databases that may be used to satisfy the amended ECCS rule and published a report [Tec89] demonstrating the code scaling, applicability, and uncertainty (CSAU) methodology that forms the basis for the revised ECCS rule.

The CSAU methodology represents the combined effort of the Technical Program Group (TPG) and the Peer Review Group comprising a total of 19 experts in reactor safety research and reflects the results of 25 years of thermal-hydraulic research documented [NRC88]. The methodology was discussed in a set of six papers published in a special issue of the journal *Nuclear Engineering and Design* in 1990, with the paper by B. E. Boyack et al. [Boy90] providing an overview of the methodology.

Our discussion on the CSAU methodology borrows heavily from the paper, with an example taken from a more recent paper [Mar05] illustrating the methodology for LBLOCA analyses of PWRs.

8.3.3.1 Objectives of CSAU Evaluation Methodology The objective in developing the CSAU methodology was to provide a technical basis for quantifying uncertainties in the BE computer models for analyzing the performance of the ECCS for postulated LBLOCAs in LWRs. With this basic objective, the CSAU methodology is structured to provide an auditable method for combining quantitative analyses and expert opinions so that meaningful estimates of uncertainties in BE calculations may be obtained. This requires a systematic approach to (a) define the relevant phenomena and scenario represented, (b) evaluate the applicability of the code for the phenomena represented and the scale-up modeling of the relevant test data, and (c) quantify code uncertainties covering both calculational and experimental uncertainties.

8.3.3.2 Three-Element Structure of CSAU Methodology With the requirements discussed above to satisfy the objective to quantify uncertainties in the BE computer models, the CSAU methodology is structured in three elements and 14 steps illustrated in Fig. 8.20.

Element 1: requirements and code capabilities

Specification of the scenario and phenomena to be modeled for a particular NPP is made through a systematic process, together with a complete documentation for a "frozen" version of the computer code including the quality evaluation (QE) for the computational models and empirical correlations. An important process in element 1 is the phenomena identification and ranking table (PIRT) generation in step 3, which provides an efficient way to rank phenomena by evaluating their effects and importance on the primary safety criteria. Through this systematic approach, subsequent effort in the code quantification process can focus on the computational models that are essential to the transients to be modeled. The formal PIRT process is illustrated further in Section 11.3.3.

In the initial CSAU study [Boy90], a total of 41 phenomena for the blowdown phase and 46 phenomena for the reflood phase of the LBLOCA were considered with the TRAC-PF1/MOD1 code [Lil86], from which a formal PIRT process ranked six phenomena as dominant: (a) break flow rate, (b) stored energy and fuel response, (c) reactor coolant pump two-phase flow, (d) steam binding phenomena in the reflood phase, (e) ECCS bypass phenomena in the refill phase, and (f) noncondensable gas representation. In a recent AREVA application [Mar05] of the CSAU methodology, the phenomena ranking process involved "round-table" discussions with nuclear safety and thermal-hydraulic experts. With the peak clad temperature (PCT) limit of 2200°F (1477 K) in the FAC serving as the limiting system parameter for LBLOCA analyses for PWRs, AREVA identified 13 dominant parameters, which included some listed above and some others, e.g., the axial power distribution.

Element 2: assessment and ranging of parameters

In this element, code capabilities to represent processes and phenomena important to the transients of interest are evaluated through an assessment matrix in step 7 representing applicable test data that could be used to verify the scale-up capability

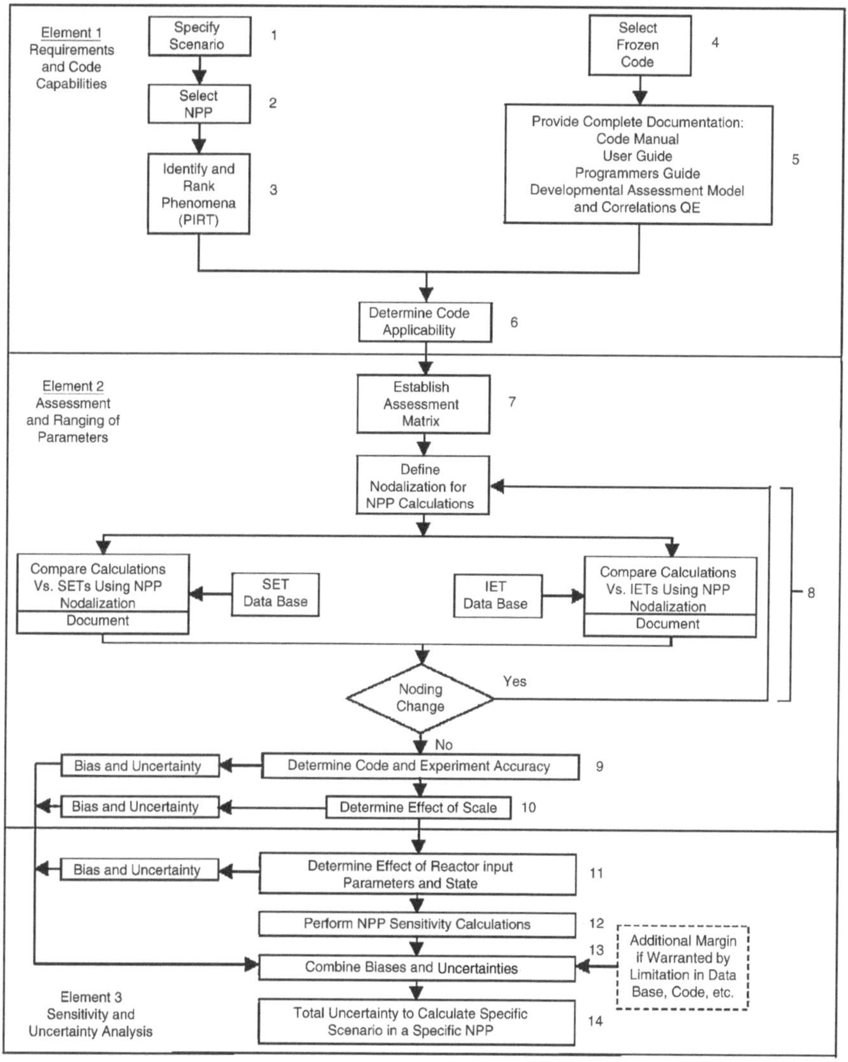

Figure 8.20 Three-element, 14-step CSAU evaluation methodology. *Source:* [Mar05].

of the code. The selection of the code nodalization scheme in step 8 typically requires an iterative process illustrated in Fig. 8.20, involving simulations of the applicable integral effects test (IET) data and separate effects test (SET) data. The IET data should be used to evaluate the overall code accuracy, with the SET data used to determine the code scale-up capability, yielding a clear documentation of the bias and uncertainty in steps 9 and 10.

In the initial CSAU study [Boy90], 18 parameters were identified as key parameters for the TRAC-PF1 code, including 9 parameters associated with the stored energy and fuel response phenomena. Auxiliary calculations and code assessments using applicable IET and SET data, in the form of an assessment matrix, generated a subset of 8 to 10 TRAC-PF1 parameters of significance for PCT calculations together with associated ranges of uncertainties. In the AREVA study [Mar05], the assessment matrix included data from 17 facilities.

Element 3: sensitivity and uncertainty analysis

The bias and uncertainty in the code capability identified in element 2 are combined in step 11 with the uncertainties in code simulations of the NPP operating conditions at the initiation of the postulated transient. The effects on key system parameters, e.g., the PCT, due to the combined uncertainties and biases identified in steps 9, 10, and 11 are evaluated via sensitivity calculations in step 12 and documented in step 13 as a statement of total uncertainty. In this process, the mean and upper 95% values of the PCT may be evaluated through statistical sampling of uncertain parameters and combined with separate biases due to effects not fully represented in the transient simulations. The statement of uncertainty is finally presented in step 14 as an error band or confidence statement about the code calculations.

For the demonstration [Boy90] of the CSAU methodology, 184 limiting PCT values were obtained from TRAC-PF1 sensitivity calculations involving up to quadruple parametric variations. Response surfaces representing the limiting PCT values during the blowdown phase were then generated as a function of seven key parameters including the power peaking factor, break discharge coefficient, and pump characteristics. Monte Carlo sampling of the response surfaces provided statistical estimates of uncertainties in the PCT calculations, in particular, an upper 95% confidence value of 1129 K (1572°F), which shows a significant margin from the FAC limit of 1477 K (2200°F). The rationale behind the uncertainty estimates for the PCT in LBLOCA events was also presented [Cat90] via simplified physical models and engineering correlations.

In the AREVA study [Mar05], a nonparametric method attributed to S. S. Wilks [Wil41,Nut04] was adopted to generate 59 samples, where all 11 major uncertain parameters, including the break size, were randomly and simultaneously sampled in LBLOCA computer simulations so that an upper 95% PCT value could be determined to a 95% confidence level. The statistical basis for this nonparametric method may be derived by following the reliability quantification approach of Eq. (3.49). For a collection of PCT values $y = y_n, n = 1, \ldots, N$, to be calculated, consider a one-sided probability

$$\mu = P\{y < y^*\}, \tag{8.4}$$

where y^* is an unknown upper bound for the collection and the sample size N is yet to be determined. If $\mu = 0.95$, as is assumed often in uncertainty analyses, we desire to determine y^* so that there is a 95% probability that the PCT $y < y^*$ in any simulation. Then, the probability that all N simulations will yield $y < y^*$ is μ^N and the probability that at least one sample will yield $y \geq y^*$ is $1 - \mu^N$. Invoking the binomial distribution of Eq. (2.35) and following Eq. (3.49), we obtain

$$P\{\text{obtaining at least } k \text{ values of } y \geq y^*\} = 1 - \sum_{m=0}^{k-1} \binom{N}{m} \mu^{N-m}(1-\mu)^m = \alpha. \tag{8.5}$$

Given the choice of y^* such that each sample $y_n, n = 1, \ldots, N$, is all bound from above by y^* with the probability μ, Eq. (8.5) provides the probability α of obtaining k or more samples of $y \geq y^*$.

Thus, α may be considered a confidence level associated with at least k samples out of a total of N samples possibly yielding $y \geq y^*$, given the expected fraction μ of all samples $y < y^*$. Equation (8.5) may also be used to determine the sample size N given the upper one-sided fractional range or tolerance interval μ of the samples and confidence level α allowing for the possibility of at least k samples yielding $y \geq y^*$. To establish y^* as an upper 95% sample value with a 95% confidence that at least one sample may be expected to lie above y^*, i.e., $k = 1$, Eq. (8.5) requires a sample size $N = 59$. This sample size corresponding to the 95/95 tolerance/confidence intervals was used in the AREVA LBLOCA study [Mar05] and y^* was approximated by the largest PCT among the 59 cases. To clarify the concept of confidence level α, consider the case where a tighter tolerance interval $\mu = 0.97$ is proposed with $N = 59$. For this case, a confidence level $\alpha = 0.8342$ is obtained, which is substantially reduced from $\alpha = 0.9515$ for $\mu = 0.95$, indicating that the sample size has to be increased to $N = 99$ to restore the confidence level to 95%. To allow for the possibility of two or more samples yielding $y \geq y^*$, a higher value of $k > 1$ should be used with Eq. (8.5). For $\mu = 0.95$ and $\alpha = 0.95$, for example, the required sample size increases from $N = 59$ for $k = 1$ to $N = 93$ for $k = 2$, which would yield a more robust estimate for the upper bound y^*.

Figure 8.21 shows a scatter plot of 59 PCT values as a function of break size, covering both double-ended guillotine and split breaks, in the AREVA study [Mar05]. From the 59 samples, corresponding to $\mu = 0.95$, $\alpha = 0.95$, and $k = 1$, the limiting case yields a PCT of $y^* = 1853$ °F (1285 K) at 87.3 seconds after the postulated break and involves 1.3% of cladding oxidation, indicating a sufficient margin from the FAC limits of 2200°F (1477 K) and 17%, respectively. It should, however, be noted that a simple use of Eq. (8.5) with $k = 1$ may not provide a sufficiently robust statistical estimate of the uncertainties involved and a larger sample size may be necessary.

8.4 SEVERE (CLASS 9) ACCIDENTS

Any potential accidents beyond DBAs, i.e., those grouped into Class 9, were traditionally considered improbable and LWR licenses were not required to analyze and evaluate them, until the TMI-2 accident of 1979. As discussed further in Chapter 9, rather small releases of radionuclides occurred as the result of a misdiagnosed and mismanaged SBLOCA in the TMI-2 accident. Nearly two-thirds of the fuel elements, however, suffered meltdown and the plant was permanently shut down

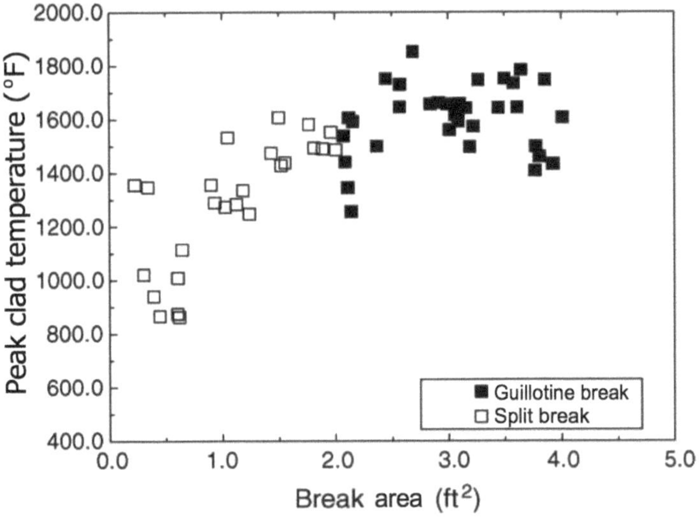

Figure 8.21 Scatter plot for peak clad temperature vs. break area. *Source:* [Mar05].

and decommissioned. Thus, there arose significant interest in understanding the detailed plant behavior for the beyond DBA (BDBA) events, which subsequently have been called severe accidents or core meltdown accidents. In fact, the proper representation of Class 9 accidents in five representative LWR plants was the primary motivation behind a massive probabilistic risk assessment study published as NUREG-1150 [NRC90]. Assessment of risks associated with LWR severe accidents in NUREG-1150 will be discussed further in Chapter 10. Note that in licensing and regulatory applications, *severe accidents* refer not merely to accidents that offer severe consequences but specifically to Class 9 accidents resulting in core damage and meltdown.

Among the events that have been left somewhat in the gray area between Class 8 and Class 9 events, following the NUREG-1150 study, are anticipated transient without scram (ATWS) events and direct containment heating (DCH) events. An ATWS is an otherwise normal anticipated transient event which, however, is accompanied by the failure of the reactor protection system to shut down the reactor. This could result in a significant damage to the core, contributing substantially to the core damage frequency as discussed further in Section 8.5. A DCH event may occur in a PWR when molten corium is ejected from the reactor vessel, which is still at a high pressure, into containment compartments through instrumentation tube penetrations. This could result in significant pressurization of the containment building, potentially contributing substantially to the consequences of the severe accidents, as was discussed as a key item during the NUREG-1150 study.

8.5 ANTICIPATED TRANSIENTS WITHOUT SCRAM

8.5.1 History and Background of the ATWS Issue

A consultant to the Advisory Committee on Reactor Safeguards (ACRS) apparently suggested the possibility of the failure of the reactor protection system (RPS) or the scram system to shut down the reactor following transients that were anticipated to occur during the lifetime of a nuclear power plant. The ACRS is a statutory advisory committee consisting of experts in various areas of nuclear plant safety, established to serve as an independent advisory committee starting from the days of the U.S. Atomic Energy Commission (AEC). The AEC was separated in 1974 into the U.S. Nuclear Regulatory Commission and the Energy Research and Development Administration, which was subsequently restructured into the U.S. Department of Energy.

Following various discussions regarding estimates of RPS failure probability and the need to account for this particular class of accidents, the AEC issued a report, WASH-1270 [AEC73], suggesting requirements for the reliability and testing of scram systems. This was based on a review of various power reactors that had operated until 1973 both in the United States and overseas, which indicated one potential and one actual failure of scram systems during a total accumulated operation time of 1627 reactor years covering 228 reactors. One incident occurred in a U.S.-designed foreign power reactor, where a newly installed scram system, after two weeks of operation, was found inoperative and would have failed if required. The second RPS failure occurred at the N Reactor at the Hanford Reservation, where the normal scram rods failed to actuate but the backup shutdown system was automatically activated to shut down the reactor safely. The failure of the normal scram rods was due to a design deficiency in the scram rod control circuitry that existed since the construction of the plant.

With the assumption that these two RPS incidents were failures, the failure rate of the RPS is estimated as $\bar{x} = 1.23 \times 10^{-3}$/reactor-year. Given this failure rate in 1627 reactor-years, using the χ^2-distribution approach illustrated in Eq. (3.53), with an upper 95% confidence level and with the degree of freedom $\eta = 2\,(n+1) = 6$, the AEC staff obtained [AEC73] an upper bound for the RPS failure rate

$$\lambda < \frac{\chi^2_{6,0.95}}{2N} = \frac{12.59}{2 \times 1627} = 3.9 \times 10^{-3}/\text{reactor-year}. \tag{8.6}$$

Extending the concept of fractional unavailability obtained in Eq. (2.94) to the case of N tests performed during the operating time T yields the fraction ξ of the time the RPS is unable to function,

$$\xi = \frac{\lambda T}{2N}. \tag{8.7}$$

For monthly testing, with $T = 1.0$ year and $N = 12$, Eq. (8.7) yields an estimate for the RPS unreliability $\xi = 1.6 \times 10^{-4}$/demand at the upper 95% confidence level. Although the two RPS failures considered are not necessarily representative of LWR systems at large, WASH-1270 assumed the unreliability estimate as a reasonable starting point for the ATWS discussion.

At the same time, WASH-1270 proposed that the likelihood of all accidents with significant consequences, beyond the DBAs delineated in 10 CFR 100 and discussed in Section 8.2.2, should be

$$P(> \text{DBA}) = P(\text{BDBA}) \leq 10^{-6}/\text{reactor-year}, \tag{8.8}$$

with the rationale that, for a fleet of 1000 LWRs some time in the future, the total BDBA probability would be less than one in 1000 years. It was further proposed that the contributions from ATWS events not be greater than 10% of the total BDBA or severe accident probability: $P(\text{ATWS}) \leq 0.1 \times P(\text{BDBA}) = 10^{-7}/\text{reactor-year}$.

Eventually, after further deliberation and significant input from the industry, the NRC staff released in 1978 a new report, NUREG-0460 [NRC78b], where the requirement was revised to

$$P(\text{ATWS}) \leq 10^{-6}/\text{reactor-year}. \tag{8.9}$$

The ATWS probability may be broken down into three components with

$$P(\text{ATWS}) = P(\text{AT}) \cdot P(\text{WS}|\text{AT}) \cdot P(\text{UC}|\text{WS}), \tag{8.10}$$

where

$P(\text{AT})$ = frequency of anticipated transients,
$P(\text{WS}|\text{AT})$ = conditional probability of scram failure given anticipated transient,
$P(\text{UC}|\text{WS})$ = conditional probability of unacceptable consequences given scram failure.

The anticipated transients occur with a frequency of ~1/reactor-year, i.e., $P(\text{AT})$ may be set to unity, while it is expected that scram failures would result in unacceptable consequences, including core damage, i.e., $P(\text{UC}|\text{WS}) = 1.0$. Thus, we are left with the need to ensure

$$P(\text{WS}|\text{AT}) \leq 10^{-6}. \tag{8.11}$$

Compared with the scram system unreliability $\xi = 1.6 \times 10^{-4}$/demand of Eq. (8.7), subject to monthly testing, the requirement of Eq. (8.11) clearly suggests that the reduction in the scram unreliability could not simply be attained by an increased testing frequency. For nearly 10 years until 1983, when two failures of the automatic control rod trip system at the Salem Unit 1 Plant occurred [Mar83], there was persistent suggestion from the industry that the scram systems in the U.S. nuclear power plants are much more reliable than that suggested by $\xi = 1.6 \times 10^{-4}$/demand. This included PRA studies based primarily on the premise that the mechanical portion of the RPS system is more reliable than the electrical portion and that the scram system has multiple, redundant components. For example, if the scram system consists of three subsystems in series, one may merely have to establish, it could be argued, that each subsystem has unreliability less than one failure in 100 demands. But all of the three subsystems could be subject to common cause failures. The Salem-1 incident, although the reactor was safely shut down through manual scram, involved the failure of a mechanical part of the scram system due to poor maintenance practice, which is a type of common mode failure. Thus, the ATWS events raised significant questions about the reliability of the LWR scram systems, as discussed further in Section 9.5. Soon after the Salem-1 incident, the industry agreed to the NRC Rulemaking process underway, which established several requirements for improved reliability of LWR reactor scram systems in 10 CFR 50.62 [NRC84].

8.5.2 Resolution of the ATWS Issues

Before presenting actual enhancements to the RPS that were adopted in 10 CFR 50.62, it is instructive to discuss representative ATWS scenarios for LWR plants and highlight the limiting system parameters that required remedial actions.

8.5.2.1 Limiting ATWS for PWRs One limiting ATWS for PWR plants may be initiated by the loss of feedwater (LOFW), although with somewhat different consequences depending on the NSSS characteristics that varied among the three reactor vendors, Combustion Engineering, Babcock & Wilcox, and Westinghouse, that provided the systems. The LOFW would trip the turbines, which would normally trigger the scram. If the scram fails to actuate properly and reduce the heat output from the core, steam generators would dry up, resulting in an increase in the primary system pressure and the average moderator temperature. Due to a negative moderator temperature coefficient (MTC) of reactivity, the core reactivity would decrease and there would be no imminent risk of a supercritical transient even with the postulated scram failure.

The primary system pressure increase would drive the pressurizer to be filled with liquid water, rather than with a mixture of steam and liquid water. This state of the pressurizer is known as the *pressurizer becoming solid*. Once the pressurizer becomes solid, the discharge out of the relief and safety valves would be liquid water, rather than steam, which degrades the heat loss through the coolant discharged, thereby further increasing the system pressure. The increase in the system pressure could exceed the ASME Service Level C, corresponding to the emergency events of the operating state classification in Section 8.2.1. This then would require lifting of the upper head of the RPV, which might result in improper reseating of the O-rings in the RPV upper head structure. This could result in significant releases of radioactive nuclides into the containment, which is certainly an event with an unacceptable consequence (UC) and hence should be avoided.

The MTC plays a critical role in the rate of reactivity decrease and the rise of the primary system pressure. In PWR cores, as discussed in Section 8.5.3, the MTC depends on the fuel burnup and the LOFW-initiated ATWS consequences could become unacceptable during the early portion of a typical PWR fuel cycle. Typically, an increase in the MTC of 1 pcm/°F would increase the peak system pressure by as much as 100 psia. Another factor that has a direct impact on the system pressure is the relief capacity of the pressurizer involved; the larger the relief capacity is, the smaller the limiting pressure will be.

8.5.2.2 *ATWS Remedial Actions for PWRs* As a result of the Salem-1 incident, the industry agreed with the NRC to reduce the probabilities P(UC|WS) and P(WS|AT), primarily consisting of:

(i) P(UC|WS)—Improvement of the mitigation circuitries involving PORVs and auxiliary feed water system and installation of a turbine trip actuation system diverse from the RPS
(ii) P(WS|AT)—Installation of a backup to the electrical portion of the existing RPS

Thus, through the adoption of the basic requirements presented in NUREG-0460, all operating PWRs have enhanced the reliability of the system, rather than relying on PRA arguments to support the assumed reliability of the RPS.

8.5.2.3 *Limiting ATWS for BWRs* Limiting ATWS events for BWR plants may be illustrated by the loss of load or loss of turbines, combined with the closure of main steam isolation valves (MSIVs). Such ATWS events could be triggered by a number of events, including a leak of radionuclides to the turbine room. The turbine trip would normally actuate the scram system to shut down the reactor. If the scram were to fail, however, the pressure in the primary system would increase, thereby collapsing the steam voids in the core and increasing the moderator density, which in turn would increase the multiplication factor k_{eff} due to a negative void coefficient of reactivity (VCR). The VCR behavior, together with the MTC for PWRs, will be discussed in Section 8.5.3.

The increase in k_{eff} would drive the reactor power up, thereby further increasing the system pressure. Continuing through the positive feedback cycle results in an increase in the temperature of the wetwell or suppression pool above an acceptable level of 200°F.

8.5.2.4 *ATWS Remedial Actions for BWRs* Similar to the actions taken to ameliorate the ATWS consequences for PWR plants, a number of remedial actions were taken based on the recommendations of NUREG-0460:

(i) Recirculation pump trip upon the indication of scram failure
(ii) Installation of an alternate rod injection (ARI) system diverse from the RPS and from sensor output to the actuation device
(iii) Standby liquid control system with 86 gpm delivery capacity. The SLCS actuation is to be automatic for plants granted construction permit after 1984.

The recirculation pump trip reduces the coolant flow to natural circulation flow, which will increase the void fraction, thereby decreasing k_{eff} and minimizing increases in the RPV pressure and suppression pool temperature. This practice of reducing the recirculation flow, however, may have contributed to an event coupling moderator density variations with power variations, resulting in rapid power oscillations at the LaSalle Unit 2. This event, known as nuclear-coupled density wave oscillations, will be discussed in Section 9.5.

8.5.3 Power Coefficients of Reactivity in LWRs

8.5.3.1 Two-Group Representation of Reactivity feedback To gain a clear physical understanding of the reactivity feedback effects in LWR cores, primarily associated with moderator density changes, we present a two-group model for the effective multiplication factor $k_{eff} = k \simeq k_\infty$, ignoring small leakage probabilities in large LWR cores:

$$k \simeq k_\infty = \frac{\nu\Sigma_{f1}}{\Sigma_{a1}+\Sigma_r} + \frac{\nu\Sigma_{f2}}{\Sigma_{a2}}\frac{\Sigma_r}{\Sigma_{a1}+\Sigma_r} = k_1 + k_2 = k_1 + pf\eta, \qquad (8.12)$$

where k_1 and k_2, representing the contributions to k from fast and thermal fissions, respectively, are defined in terms of two-group cross sections:

Σ_{a1} and Σ_{a2} = fast and thermal absorption cross sections,
$\nu\Sigma_{f1}$ and $\nu\Sigma_{f2}$ = fast and thermal fission cross sections times number ν of neutrons released per fission, and
Σ_r = slowing down cross section.

The thermal fission contribution k_2 is further broken down into the *resonance escape probability* p, *thermal utilization* f, and *number* η *of neutrons released per thermal neutron absorption in fuel.* The resonance escape probability is rewritten in terms of the *effective resonance integral I,*

$$p = \exp\left[-\frac{N_F}{\xi\Sigma_s}I\right] = \exp\left[-\frac{N_F}{\xi\Sigma_s}\int_0^u du'\sigma_a(u')\,\phi(u')\right], \qquad (8.13)$$

where I physically represents the *flux-weighted effective absorption cross section.* Here, the absorption of neutrons in the fast group is approximately represented by the fuel number density N_F multiplied by I, together with the average lethargy gain per collision ξ and scattering cross section Σ_s [Dud76].

The effects of moderator temperature changes in a PWR may first be represented in terms of the thermal utilization written explicitly for a fuel-moderator mixture,

$$f = \frac{\Sigma_{a2}^F}{\Sigma_{a2}^F + \Sigma_{a2}^M}, \qquad (8.14)$$

so that f represents the fraction of thermal neutron absorptions that take place in the fuel. Suppose we experience a moderator temperature increase during a power maneuver or due to an accident. Due to this temperature increase, we expect a decrease in the moderator density and hence a decrease in the number density of water and hydrogen. This decrease in the water number density results in a decrease in the thermal absorption cross section of the moderator, Σ_{a2}^M, without much change in the thermal absorption cross section of fuel, Σ_{a2}^F. Thus, a decrease in the neutron moderation, due to a decrease in the H/U atomic ratio, results in an increase in thermal utilization f.

The other parameter that is affected by an increase in moderator temperature T_M is the resonance escape probability p. Since the scattering cross section Σ_s in Eq. (8.13) is mostly associated with moderator scattering, an increase in moderator temperature

T_M results in a decrease in Σ_s. Due to spectral hardening, the resonance integral I may decrease slightly, but this effect is smaller than the decrease in Σ_s, with the result that p itself decreases as T_M increases.

Returning to Eq. (8.12), we note that a change in T_M hardly affects the parameters k_1 or η and hence that k_{eff} or k_∞ will decrease or increase as a result of competing changes in p and f. These competing trends in p and f are sketched in the left-hand plot of Fig. 8.22 as a function of the H/^{235}U and H/U atomic ratios, the moderator-to-fuel number density ratio N_M/N_F, and equivalently in terms of moderator density ρ_M and the inverse of moderator temperature T_M. Due to the competition between p and f, for some value of the H/U atomic ratio or moderator density ρ_M, the effective multiplication factor $k_{eff} = k$ will reach a maximum, as illustrated by a bell-shaped curve on the right-hand plot of Fig. 8.22. The MTC can be obtained as the slope of the k_{eff} curve with respect to T_M,

$$\alpha_M = \frac{\Delta k/k}{\Delta T_M} = \frac{\partial \ln k}{\partial T_M} = \frac{\partial \ln k}{\partial \rho_M}\frac{\partial \rho_M}{\partial T_M}. \tag{8.15}$$

The left-hand half of the bell-shaped curve corresponds to an *undermoderated regime* so that any increase in T_M or a decrease in ρ_M will result in sliding down the k_{eff} curve, yielding a negative value of the MTC. This can be understood by noting that the slope of the curve yields the fractional change in reactivity with respect to the density change, yielding a positive value for the first derivative in the product expression of Eq. (8.15). The second derivative in the product expression simply represents the derivative of the moderator density with respect to the moderator temperature, which is simply negative. Thus, as long as the operating point is in the undermoderated regime, we are guaranteed to have a negative MTC. Furthermore, α_M itself becomes more negative as T_M increases, since this corresponds to evaluating the slope further down the k_{eff} curve.

Likewise, the VCR is defined as

$$\alpha_V = \frac{\partial \ln k}{\partial \ln V_M} = -\frac{\partial \ln k}{\partial \ln \rho_M} \text{ with } \rho_M \propto \frac{1}{V_M}, \tag{8.16}$$

where we note that the moderator density is inversely proportional to the fraction V_M of steam or void in the coolant/moderator. Thus, as long as the BWR design locates the operating point in the undermoderated region in Fig. 8.22, the VCR will always be negative too.

Thus, LWR designs in the United States have always been chosen in the undermoderated regime marked by a plus sign to guarantee a negative α_M or α_V. This key inherent safety feature was apparently violated in the ill-fated Chernobyl design, where a positive value of α_V was possible at low power with a small number of control rods inserted, and that is where the 1986 accident was initiated. The bell-shaped curve in Fig. 8.22 is a succinct way of visualizing the moderator temperature feedback effects in LWRs. The negative VCR, however, contributes to the severity of the ATWS events in BWR plants, as discussed in Section 8.5.2.3. We note in passing that as the fuel temperature increases the resonance escape probability p of

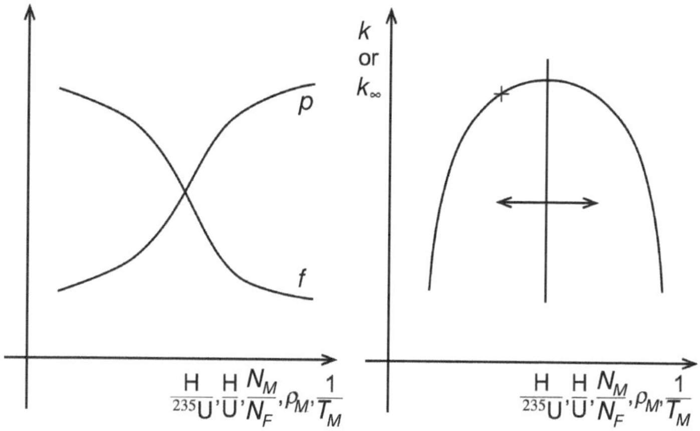

Figure 8.22 Moderator temperature feedback effects on reactivity.

Eq. (8.13) decreases due to the decreased self-shielding of absorption resonances and the resulting increase in the effective resonance integral I. This is known as the negative Doppler or fuel temperature feedback, which is yet another inherent safety mechanism built into LWR fuel elements. The sum of the reactivity coefficients associated with the fuel temperature and moderator density feedback effects is called the *power coefficient of reactivity*.

8.5.3.2 Parametric Dependences of the Moderator Temperature Feedback Now that we have discussed how the power level and associated fuel temperature and moderator density variations affect the reactivity, we are ready to examine how reactivity coefficients are influenced by key reactor physics parameters, e.g., fissile enrichment, soluble boron concentration, lumped neutron poison, and fuel burnup.

(a) Fissile enrichment of the fuel has a direct effect on neutron moderation and flux spectrum. In terms of the moderator temperature feedback effects illustrated in Fig. 8.22, an increase in the fissile enrichment is equivalent to decreasing the H/^{235}U atomic ratio and hence making the system more undermoderated and the flux spectrum harder. This means that the slope of the k_{eff} curve becomes more negative, yielding a larger magnitude of the negative MTC. In passing, it should be mentioned that 0.1 wt% of ^{235}U corresponds to approximately 1.0 $\%\Delta k/k$ of reactivity and is worth about a month or so of full power operation in current LWR designs.

(b) The concentration of ^{10}B dissolved in coolant water as a chemical shim in PWRs influences the MTC through its effect primarily on thermal utilization f. Suppose an increase in the moderator temperature takes place during a power maneuver. This results in a decrease in the moderator absorption cross section Σ_{a2}^M due to a decrease in water number density N_M, as discussed earlier in this section. When ^{10}B atoms, with a large thermal absorption cross section, are homogeneously dissolved in water, Σ_{a2}^M decreases further as the ^{10}B number density decreases together with the water number density. Hence, the increase in thermal utilization f, associated with an increase in T_M, will be larger with ^{10}B dissolved in water and the MTC will become less negative. With a large ^{10}B concentration, it is even possible to have a positive MTC, which is equivalent to operating the reactor in an overmoderated regime.

(c) The presence of lumped neutron absorbers, e.g., lumped burnable poison rods or control rods, in LWRs affects the behavior of MTC in a manner distinct from that of soluble neutron absorbers. To study the effects of lumped absorbers, we extend the definition of thermal utilization in Eq. (8.14) to include the contribution Σ_{a2}^P from the absorbers to the thermal absorption cross section,

$$f = \frac{\Sigma_{a2}^F}{\Sigma_{a2}^F + \Sigma_{a2}^M + \Sigma_{a2}^P}. \tag{8.17}$$

For an increase in moderator temperature T_M, as usual Σ_{a2}^M will decrease. But due to the presence of Σ_{a2}^P, the soluble boron concentration is decreased and Σ_{a2}^M itself is reduced, thereby lessening the effect of any T_M increase on MTC. At the same time, the thermal diffusion length L_2 increases for the core, due to a reduction in the absorption cross section, $\Sigma_{a2}^F + \Sigma_{a2}^M$, for the core material. Since L_2 is proportional to the distance thermal neutrons travel between collisions on average, an increase in L_2 has the effect of increasing the likelihood that thermal neutrons encounter lumped absorbers during the migration. Thus, an increase in T_M will increase *effectively* the parasitic absorption term Σ_{a2}^P in Eq. (8.17), countering the decrease in Σ_{a2}^M. This is another reason why the MTC of a PWR becomes more negative as lumped neutron absorbers are added. Because Eq. (8.17) does not fully reflect a heterogeneous lattice consisting of fuel, moderator, and parasitic absorbers, it cannot be used directly to explain why lumped absorbers act to cancel the increase in f due to an increase in T_M; instead we have explained this effect in terms of an effective increase in Σ_{a2}^P.

(d) The void coefficient of reactivity α_V for BWRs represents the coolant density feedback effect, as discussed in connection with Eq. (8.16). Hence, we may use the k_{eff} curve of Fig. 8.22 again to illustrate the dependence of α_V on void fraction itself. As the void fraction increases, moderator density ρ_M decreases, thus yielding a negative slope of the k_{eff} curve and a negative value of α_V in an undermoderated regime. Furthermore, around a higher void fraction, i.e., further down the curve, the slope will become steeper, and hence an increase in the magnitude of the negative α_V as a function of void fraction itself.

(e) Fuel depletion in a reactor core also influences reactivity coefficients in a complex manner. In general, the evolution in fuel isotopics, especially the production of plutonium isotopes with low-lying resonances, could have a significant impact on reactivity coefficients. In LWRs, however, the primary fuel depletion effects on reactivity coefficients are those associated with control poisons. We illustrate the burnup dependence of α_M for PWRs and α_V for BWRs in Fig. 8.23. Since PWRs operate with control rods essentially fully withdrawn, we observe

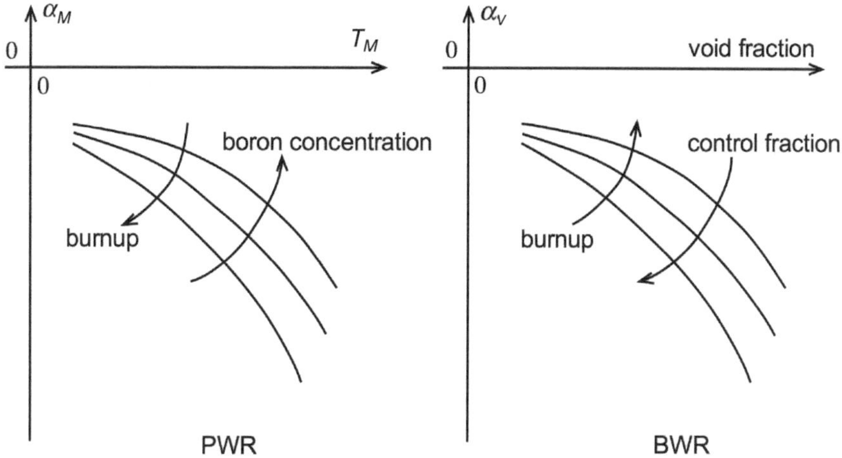

Figure 8.23 Burnup dependence of reactivity coefficients in LWRs.

that the MTC is influenced primarily by the soluble boron concentration, which decreases as the fuel burnup increases and excess reactivity decreases, and hence that the MTC itself becomes more negative as a function of fuel burnup. The situation is reversed for BWRs, because a BWR core typically has about 25% full-length equivalent of control blades inserted into the core at the beginning of cycle (BOC) and hence has the largest magnitude of the negative void coefficient at the BOC. As the fuel depletes and the excess reactivity decreases, the control blades are gradually withdrawn, making the void coefficient less negative.

We now conclude our discussion on the impact of MTC on ATWS events in PWRs by remembering that the magnitude of the negative MTC becomes larger as the fuel burnup increases in the core. Thus, in the LOFW-initiated ATWS for PWRs considered in Section 8.5.2.1, the consequences of the postulated accident become less severe as the fuel burnup increases.

8.6 RADIOLOGICAL SOURCE AND ATMOSPHERIC DISPERSION

A key objective of safety and risk analyses of nuclear power plants is to determine the source of radionuclides that could be released as a result of various accidents, especially core damage accidents coupled with other system failures or leakage in the containment building. It is desirable to determine the radiological source term, which represents the amounts and species of radionuclides released, and the probability of these releases in various accidents. Once the radiological source term is determined, the next step in the NPP safety and risk analyses entails the calculation of the atmospheric dispersion of radionuclides to determine the offsite concentration of the radioactivity and eventually determine the radiation dose and health effects of the radionuclide releases.

Basic approaches for determining the radiological source term are discussed in Section 8.6.1, with a brief presentation on the siting criteria and containment leakage or failure analysis. This is followed by an analytical model describing the atmospheric dispersion of radioactive plumes in Section 8.6.2. Finally, Section 8.6.3 concludes with a simple method for a dose rate calculation given the offsite radioactivity concentration.

8.6.1 Radiological Source Term

The amount of radioactivity, radionuclide species, and probability of releases in an NPP accident should in general be determined through probabilistic methods accounting for various initiating events and progression of the accidents leading to radionuclide releases out of the containment building. The PRA-based approach for detailed source term calculations will be discussed in Chapter 10.

The licenses for the construction and operation of the current generation of LWRs, however, have been granted via a deterministic approach presented in 10 CFR 100 as part of the reactor siting criteria [NRC04], originally released as TID-14844 [DiN62] by the U.S. Atomic Energy Commission in 1962. In this license base, a set of conservative criteria is stipulated for the analysis of DBAs, which assumes that 100% of the inventory of noble gases and 50% of the ^{131}I inventory in the fuel elements and core is released to the containment. The ^{131}I inventory is further assumed to comprise 91% elemental, 5% particulate, and 4% organic iodide (methyl iodide) forms of iodine.

Together with the conservative source term, an exclusion zone is established at a radius of 0.8 to 1.0 km and a low population zone (LPZ) at 5.0 km from the plant. Dose limits of 25 rem for the whole body and 300 rem for the thyroid within 2 hours of a postulated accident at the boundary of the exclusion zone should not be exceeded. The preceding dose limits apply to the entire accident duration for the LPZ. The NRC published a number of regulatory guides (RGs) to implement the siting and radiological dose criteria, including RG 1.3, 1.4, and 4.7.

For the analysis of containment failures and leakage, the prevailing practice calls for minimizing leakage through design and surveillance and monitoring through periodic testing. As part of the containment analysis, it is necessary to account for radioactive decay of radionuclides released into the containment building and the decontamination factor associated with the removal of the radionuclides through containment fan sprays and filters.

Several different guidelines for calculating radiological source terms have been published in recent years. Among them are NUREG-1465 [Sof95] and RG 1.183 [NRC00]. Detailed specifications, including the composition and magnitude of the radioactive material, the chemical properties of the material, and the timing of the release to the containment, are specified separately for PWR and BWR plants for future licensing applications. In contrast to the instantaneous releases of radionuclides postulated in RG 1.3 and 1.4, the source term guidelines in NUREG-1465 allow the releases distributed in time to reflect the degree of fuel melting and relocation, the integrity of the reactor pressure vessel, and the interaction of molten core materials and the concrete basemat. Regulatory Guide 1.183 presents practical guides for alternative radiological source terms, including more realistic specifications for the release fractions for risk-significant radionuclides. Specific considerations are allowed for the chemical form of iodine in the containment building so that $\leq 3\%$ of the airborne iodine will be in organic form. When pH ≥ 7 is maintained in the containment, $\leq 5\%$ of the total I is assumed to be elemental, with $\leq 0.15\%$ in organic form. These new and alternate source terms have been developed for regulatory applications to future LWR plants but have been adopted to assess regulatory issues for existing plants.

8.6.2 Atmospheric Dispersion of Radioactive Plume

An analytical model is first developed that can be used to model the dispersion of contaminants, i.e., radionuclides, in an infinite homogeneous atmosphere. The simple model is then modified to account for the contaminant releases at or near ground, followed by the applications of empirical factors for estimating offsite radionuclide concentrations.

8.6.2.1 Dispersion of Contaminants in Infinite Homogeneous Atmosphere For concentration χ of radionuclides expressed, for convenience, in units of Ci/m^3, we begin with a time-dependent diffusion equation [Lew77]

$$\frac{\partial \chi(\mathbf{r}, t)}{\partial t} = -\nabla \cdot \mathbf{J}(\mathbf{r}, t) + S(\mathbf{r}, t), \tag{8.18}$$

which is in the form of the time-dependent heat conduction equation with source $S(\mathbf{r}, t)$ or time-dependent neutron diffusion equation without the absorption term. With Fick's law of diffusion introduced for current $\mathbf{J} = -k\nabla\chi$, Eq. (8.18) is rewritten first as the one-dimensonal (1-D) diffusion equation in plane geometry,

$$\frac{\partial \chi(x, t)}{\partial t} = k\frac{\partial^2 \chi(x, t)}{\partial x^2} + S(x, t), \tag{8.19}$$

where k is the diffusion coefficient in units of m^2/s. For the 1-D source of the contaminants, we represent a *puff* source of radionuclides, released at position $x = 0$ and time $t = 0$, using Dirac delta functions:

$$S(x, t) = q\delta(x)\delta(t). \tag{8.20}$$

Here the source intensity q is given in units of Ci/m^2 so that $S(x, t)$ is expressed in units of Ci/m^3·s, i.e., a volumetric source released per unit time. We note here that the delta functions $\delta(x)$ and $\delta(t)$ have the units of the inverse of the distance x from the source and inverse of time, respectively.

The solution of the 1-D diffusion equation (8.19) with a puff source may be readily obtained [Lew77] using a combination of the Fourier and Laplace transforms as

$$\chi(x, t) = \frac{q}{\sqrt{4\pi k t}} \exp\left(-\frac{x^2}{4kt}\right). \tag{8.21}$$

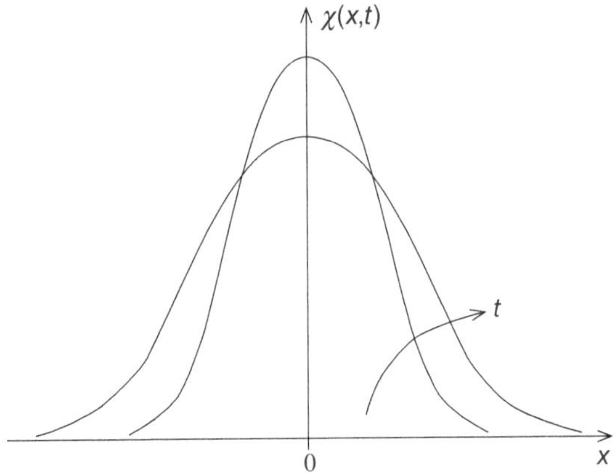

Figure 8.24 Gaussian plume distribution evolving as a function of time.

This can be rewritten in the form of the Gaussian distribution by introducing $\sigma^2 = 2kt$,

$$\frac{\chi(x,t)}{q} = \frac{1}{\sqrt{2\pi}\sigma}\exp\left(-\frac{x^2}{2\sigma^2}\right). \tag{8.22}$$

The Gaussian distribution becomes flatter as σ increases with increasing time, as illustrated in Fig. 8.24. The 1-D solution may be readily extended to 3-D geometry to yield an expression for the radioactivity concentration at distance r and at time t, following an instantaneous radioactivity release of magnitude q (Ci) at the origin,

$$\frac{\chi(\mathbf{r},t)}{q} = \frac{1}{(4\pi kt)^{3/2}}\exp\left(-\frac{r^2}{4kt}\right). \tag{8.23}$$

The Gaussian plume solution of Eq. (8.23) may now be converted into the Cartesian coordinates with the introduction of wind speed u in the x-direction and the specification that the source is located at $z = h$ and $u \gg$ diffusion speed $\sqrt{k_x/t}$ in the x-direction:

$$\frac{\chi(x,y,z,t)}{q} = \lim_{k_x\to 0}\frac{\exp[-(x-ut)^2/4k_xt]}{(4\pi k_xt)^{1/2}}\frac{\exp[-y^2/4k_yt]}{(4\pi k_yt)^{1/2}}\frac{\exp[-(z-h)^2/4k_zt]}{(4\pi k_zt)^{1/2}}, \tag{8.24}$$

with

$$\lim_{k_x\to 0}\frac{\exp[-(x-ut)^2/4k_xt]}{(4\pi k_xt)^{1/2}} = \delta(x-ut) = \frac{1}{u}\delta\left(\frac{x}{u}-t\right). \tag{8.25}$$

The distance in the x-direction diffusion term is replaced by $x - ut$ in Eq. (8.24) to account for the wind that carries the plume. To convert the x-direction diffusion term in Eq. (8.25) a definition for the Dirac delta function as a limiting form of the Gaussian distribution [Mar56] was used,

$$\delta(x-\xi) = \lim_{a\to 0}\frac{1}{a\sqrt{\pi}}\exp\left[-\frac{(x-\xi)^2}{a^2}\right]. \tag{8.26}$$

Furthermore, for the final form of the delta function, we made use of the property [Fri56]

$$\delta(ax) = \frac{1}{a}\delta(x),\ a \neq 0, \tag{8.27}$$

which can be understood by noting that

$$\int_{-\infty}^{\infty} \delta(ax)\,\mathrm{d}x = \frac{1}{a}\int_{-\infty}^{\infty} \delta(ax)\,\mathrm{d}(ax) = \frac{1}{a} = \frac{1}{a}\int_{-\infty}^{\infty} \delta(x)\,\mathrm{d}x. \tag{8.28}$$

We then rewrite Eq. (8.24) with Eq. (8.25) to obtain

$$\frac{\chi(x,y,z,t)}{q} = \frac{1}{2\pi\sigma_y\sigma_z u}\exp\left[-\frac{y^2}{2\sigma_y^2} - \frac{(z-h)^2}{2\sigma_z^2}\right]\delta\left(\frac{x}{u} - t\right) \tag{8.29}$$

by introducing the dispersion coefficients, with $t = x/u$,

$$\sigma_y^2 = 2k_y t = 2k_y x/u,\ \sigma_z^2 = 2k_z t = 2k_z x/u. \tag{8.30}$$

Equation (8.30) represents the radioactivity concentration $\chi(x,y,z,t)$ following a release of radioactivity of magnitude q (Ci) at the origin at $t = 0$, where the plume travels with the wind of speed u, thereby arriving at position x at time $t = x/u$.

We may finally extend the Gaussian plume solution to the case of constant release rate Q (Ci/s) by summing up or integrating, over the interval $[-\infty, t]$, the contributions represented by Eq. (8.29) from a puff source $Q\,\mathrm{d}t'$ (Ci) released in differential interval $\mathrm{d}t'$ around time t',

$$\chi(y,z) = \int_{-\infty}^{t} \frac{\chi(x,y,z,t-t')}{q} Q\,\mathrm{d}t', \tag{8.31}$$

where the elapsed time $t - t'$ for a puff source released at t' replaces the elapsed time t for a puff source released at $t = 0$ in Eq. (8.29). Carrying on the integral indicated in Eq. (8.31) removes the x-directional dependence, represented by the Dirac delta function, to yield

$$\frac{\chi(y,z)}{Q} = \frac{1}{2\pi\sigma_y\sigma_z u}\exp\left[-\frac{y^2}{2\sigma_y^2} - \frac{(z-h)^2}{2\sigma_z^2}\right]. \tag{8.32}$$

Equation (8.32) is the *Gaussian plume model* providing the radioactivity concentration in a plume at position (y,z) carried by a wind of speed u in the x-direction, which is significantly larger than the diffusion speed in that direction. Note here that the ratio χ/Q is in units of inverse volumetric flow rate in s/m^3.

8.6.2.2 Contaminant Release At or Near Ground For radionuclides released at a low altitude, we treat the ground as a perfect reflector of the contaminants and set up an image source for the source of radionuclides released at elevation $z = h$,

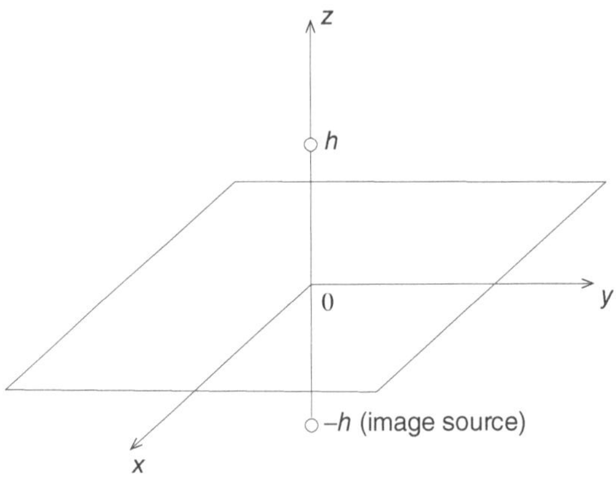

Figure 8.25 Image source for radionuclides released at elevation h.

as illustrated in Fig. 8.25. The Gaussian plume model of Eq. (8.32) yields the total radioactivity concentration as

$$\frac{\chi(y,z)}{Q} = \frac{1}{2\pi\sigma_y\sigma_z u}\left[\exp\left\{-\frac{y^2}{2\sigma_y^2} - \frac{(z-h)^2}{2\sigma_z^2}\right\} + \exp\left\{-\frac{y^2}{2\sigma_y^2} - \frac{(z+h)^2}{2\sigma_z^2}\right\}\right]. \tag{8.33}$$

Now introduce the assumptions:

(a) The detector or observer is located at ground, $z = 0$.
(b) The source is also located at ground, i.e., $h = 0$.
(c) The detector is located on a line from the source along the wind direction, i.e., $y = 0$.

The assumptions simplify Eq. (8.33) to

$$\frac{\chi}{Q} = \frac{1}{\pi\sigma_y\sigma_z u}. \tag{8.34}$$

Equation (8.34) is the simple, final form of the *Gaussian plume model*, expressed in terms of the wind speed u and the dispersion coefficients

$$\sigma_y^2 = 2k_y x/u, \; \sigma_z^2 = 2k_z x/u. \tag{8.35}$$

Note that the radioactivity concentration χ is in units of Ci/m^3 and the constant source Q in units of Ci/s so that the ratio χ/Q is obtained in units of inverse volumetric flow rate $\mathrm{s/m^3}$. The ratio is known as the *atmospheric diffusion* or *dispersion factor*, representing the offsite concentration of radioactivity normalized by the release rate. Hence, χ/Q decreases as the actual dilution of the radioactivity increases as the plume travels.

The dispersion factor χ/Q should be distinguished from another parameter D often called the *dilution factor* that represents the actual dilution of the radioactivity concentration C within the containment in units of Ci/m^3 to the offsite concentration χ (Ci/m^3). Thus, introduce the volumetric flow rate W of radionuclides in gaseous, volatile form in units of m^3/s so that the radionuclide release rate Q (Ci/s) may be simply obtained as

$$Q = WC, \tag{8.36}$$

which allows for the determination of the offsite concentration in terms of Eq. (8.34),

$$\chi = \left(\frac{\chi}{Q}\right) WC \equiv \frac{C}{D}. \tag{8.37}$$

The definition of D introduced in Eq. (8.37) may be rewritten explicitly for the *dilution factor*:

$$D = \frac{1}{W}\frac{1}{(\chi/Q)}. \tag{8.38}$$

Unlike the dispersion factor, the dilution factor D is dimensionless and is always greater than unity, increasing as the dilution increases.

8.6.2.3 Empirical Factors for Real Atmosphere To render the simple analytical expression, Eq. (8.34), applicable to represent the radionuclide dispersion in a meaningful way, we determine the dispersion coefficients σ_y and σ_z of Eq. (8.35) through Pasquill's semiempirical curves [Sla68] reproduced in Figs. 8.26 and 8.27. Since σ_y and σ_z are obtained as a function of the distance x from the source and wind speed u, as in Eq. (8.35), the dispersion factor χ/Q of Eq. (8.34) is evaluated as a function of x and u for different turbulence categories. For high-turbulence conditions, vertical diffusion is important for $x > 1$ km, while for low-turbulence conditions and large distances, horizontal diffusion plays a more important role in determining χ/Q.

For a more realistic determination of the dispersion factor, RG 1.111 [NRC77] and RG 1.4 [NRC74c] provide guidelines for measuring the wind speed and other meteorological conditions. Two summary figures from RG 1.4 are presented in Figs. 8.28 and 8.29 for direct determination of the dispersion factor χ/Q as a function of the distance from the source and of the time following the accident. The basis for selecting Pasquill's turbulence categories is also provided in the regulatory guide, together with a correction factor for building wake effects. For the NUREG-1150 offsite consequence analysis, discussed in Chapter 10, detailed models were developed to account for site-specific meteorological data and buildings and structures in the surrounding areas.

8.6.3 Simple Models for Dose Rate Calculation

The Gaussian plume model presented in Section 8.6.2 provides a means to determine the offsite radionuclide concentration χ, given the release rate Q. Once χ is obtained,

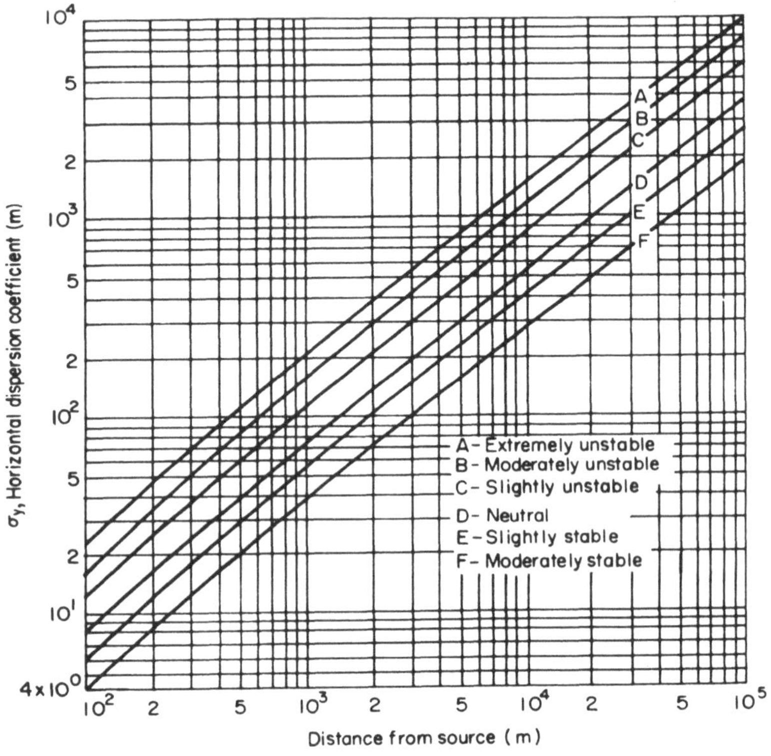

Figure 8.26 Horizontal dispersion coefficient versus downwind distance from source for different turbulent categories. *Source:* [Sla68].

we may use a simple model, known as the *infinite cloud model*, presented in RG 1.4 to determine the dose rate due to exposure to the radionuclides. In the infinite cloud model, we visualize an infinite cloud comprising uniformly distributed radionuclides, which allows us to equate the rate of energy absorption in air, i.e., the dose rate, with the rate of energy release due to the radioactive decay. Thus, for the dose rate in air from β-particles we obtain

$$\begin{aligned}\frac{\mathrm{d}D_\beta}{\mathrm{d}t} &= [\text{production rate of radiation, } \chi \ (\mathrm{Ci/m^3})] \\ &\times \frac{[\text{average energy of } \beta\text{-particles, } \bar{E}_\beta \ (\mathrm{MeV/dis})]}{[\text{air density, } 1293 \ \mathrm{g/m^3}]} \times \frac{3.7\times 10^{10}\ \mathrm{dis/s}}{\mathrm{Ci}} \\ &\times \frac{1.6\times 10^{-6}\ \mathrm{erg}}{\mathrm{MeV}} \times \frac{\mathrm{rad}}{100\ \mathrm{erg/g}} = 0.458\chi\bar{E}_\beta\ (\mathrm{rad/s}),\end{aligned} \tag{8.39}$$

with the recognition that the average energy of β-particles (MeV) is given by

$$\bar{E}_\beta = E_{\max}/3, \text{ where } E_{\max} = \text{ endpoint energy of the beta spectrum.} \tag{8.40}$$

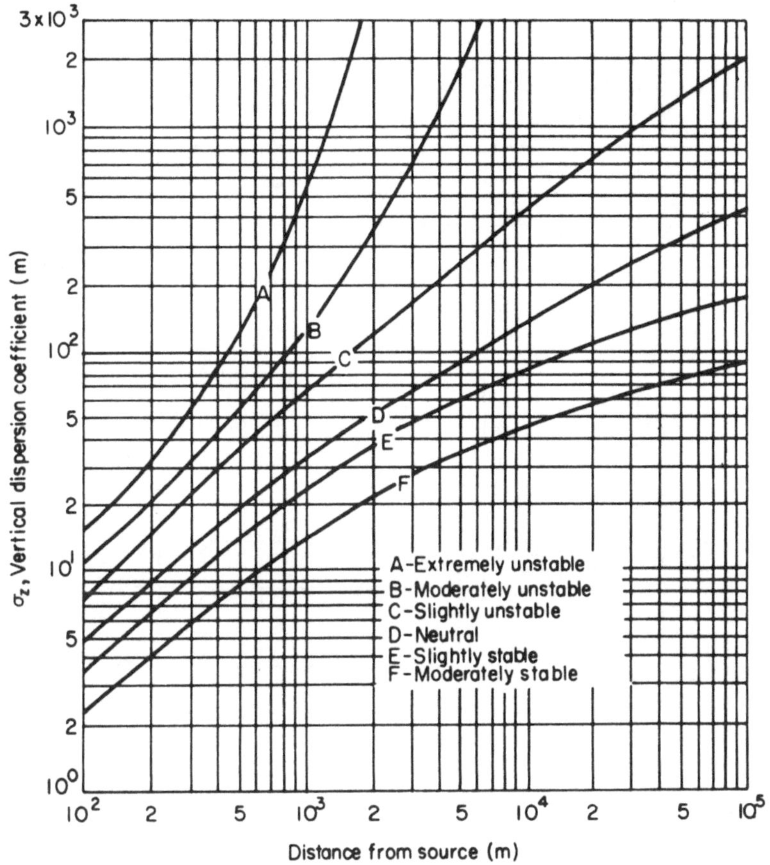

Figure 8.27 Vertical dispersion coefficient versus downwind distance from source for different turbulent categories. *Source:* [Sla68].

For the skin dose rate due to β-particles, with a typical range $\simeq$ 4 mm for tissue for 1.0-MeV β-particles, the exposure due to emitters below the skin is negligible. This suggests the use of a semi-infinite cloud model:

$$\frac{\mathrm{d}D_\beta}{\mathrm{d}t} = 0.23\chi\bar{E}_\beta(\mathrm{rad/s}). \tag{8.41}$$

For exposure to γ-rays, we may use Eq. (8.39) for the infinite cloud model with $\bar{E}_\beta$ replaced by $\bar{E}_\gamma$ and the conversion factor of tissue dose $\simeq 1.11 \times$ air dose to obtain

$$\frac{\mathrm{d}D_\gamma}{\mathrm{d}t} = 0.507\chi\bar{E}_\gamma(\mathrm{rem/s}). \tag{8.42}$$

Finally, for ground-level releases, a semi-infinite cloud model is suggested:

$$\frac{\mathrm{d}D_\gamma}{\mathrm{d}t} = 0.25\chi\bar{E}_\gamma(\mathrm{rem/s}). \tag{8.43}$$

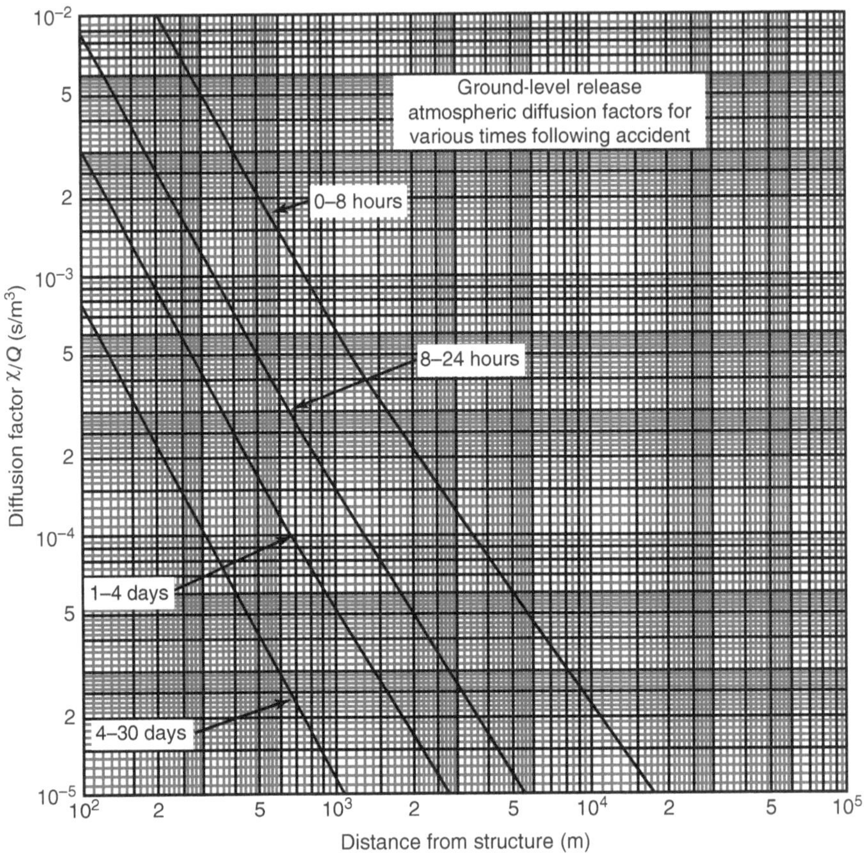

Figure 8.28 Atmospheric dispersion factor for ground-level release at various times following an accident. *Source:* [NRC74c].

Realistic dose rate calculations require detailed accounting of the direct and indirect pathways for radionuclides resulting in external radiation exposure, deposit of radionuclides on the skin, and radionuclide ingestion. Dose conversion factors for various body organs may then be used to determine the dose rate and the eventual dose for the postulated accident. Further discussion on realistic dose calculations is given in terms of the NUREG-1150 models in Chapter 10.

8.7 BIOLOGICAL EFFECTS OF RADIATION EXPOSURE

The final step in the evaluation of risk associated with the operation of a nuclear power plant is to determine the biological effects due to exposure to the radiation released from the plant. Since the radioactivity released in routine plant operation is negligible, most of the radiological risk due to the operation of a NPP comes from

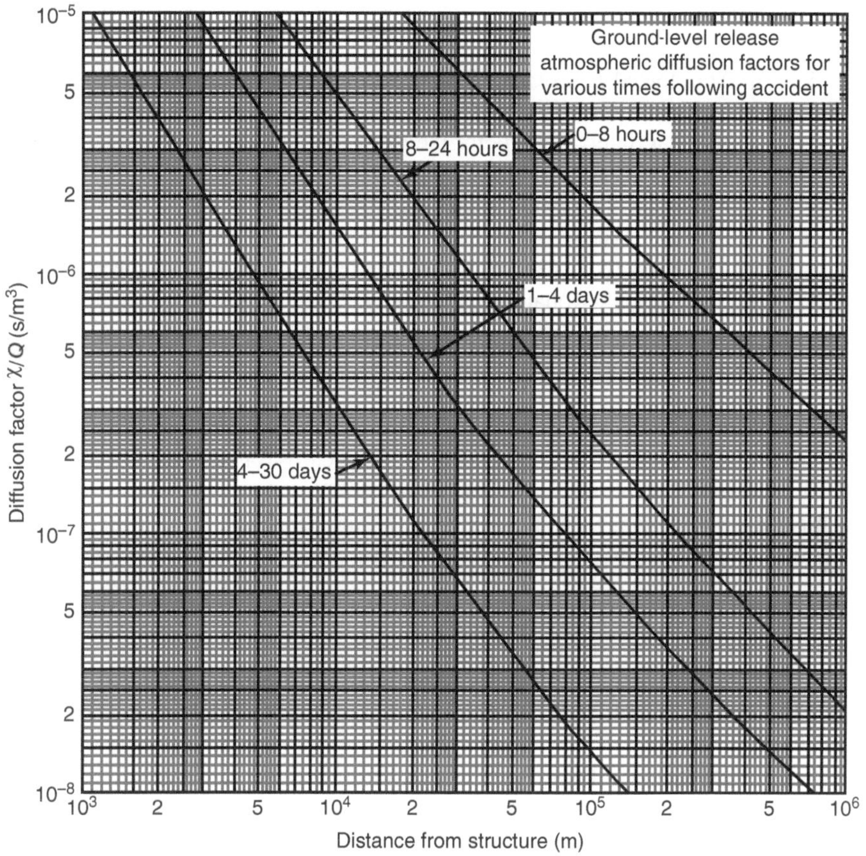

Figure 8.29 Atmospheric dispersion factor for ground-level release at various times following an accident. *Source:* [NRC74c].

the release of radionuclides in postulated accidents. Thus, radionuclide release rate Q is determined from accident and transient analyses, which then is used to calculate the offsite radionuclide concentration χ through the Gaussian plume model of Eq. (8.34) and finally to determine the dose rate via Eqs. (8.40) through (8.43) associated with the postulated accidents. The process involved in obtaining Q in PRA studies of NPPs is discussed further in Chapter 10.

The final conversion of the radiation dose to biological effects has to be performed accounting specifically for individual nuclides and for the possibility of ingestion as well as external exposure. In addition, for the case of a specific accident, biological effects due to acute radiation exposure as well as latent health effects including increased cancer risk should be considered.

References

[AEC73] "Anticipated Transients Without Scram for Water-Cooled Power Reactors," WASH-1270, U.S. Atomic Energy Commission (1973).

[ANS73] "Nuclear Safety Criteria for the Design of Stationary PWR Plants," ANSI-N18.2, American National Standards Institute (1973).

[ANS94] "American National Standard for Decay Heat Power in Light Water Reactors," ANSI/ANS-5.1-1994, American Nuclear Society (1994).

[Boy90] B. E. Boyack, I. Catton, R. B. Duffey, P. Griffith, K. R. Katsma, G. S. Lellouche, S. Levy, U. S. Rohatgi, G. E. Wilson, W. Wulff, and N. Zuber, "Quantifying Reactor Safety Margins, Part 1: An Overview of the Code Scaling, Applicability, and Uncertainty Evaluation Methodology," *Nucl. Eng. Des.* **119**, 1 (1990).

[Cat90] I. Catton, R. B. Duffey, R. A. Shaw, B. E. Boyack, P. Griffith, K. R. Katsma, G. S. Lellouche, S. Levy, U. S. Rohatgi, G. E. Wilson, W. Wulff, and N. Zuber, "Quantifying Reactor Safety Margins, Part 6: A Physically-Based Method of Estimating LBLOCA PCT," *Nucl. Eng. Des.* **119**, 109 (1990).

[DiN62] J. Di Nunno, R. E. D. Baker, F. D. Anderson, and R. L. Waterfield, "Calculation of Distance Factors for Power and Test Reactors," TID-14844, U.S. Atomic Energy Commission (1962).

[DOE02] "A Technology Roadmap for Generation IV Nuclear Energy Systems," GIF-002-00, U.S. Department of Energy (2002).

[Dud76] J. J. Duderstadt and L. J. Hamilton, *Nuclear Reactor Analysis*, Wiley (1976).

[Fri56] B. Friedmann, *Principles and Techniques of Applied Mathematics,* Wiley (1956).

[Gen71] "BWR Power Plant Training—BWR Technology," NEDO-10260, General Electric Company (1971).

[Hay99] P. J. Hayward and I. M. George, "Determination of the Solidus Temperatures of Zircaloy-4/Oxygen Alloys," *J. Nucl. Mater.* **273**, 294 (1999).

[Hew00] G. F. Hewitt and J. G. Collier, *Introduction to Nuclear Power*, 2nd ed., Taylor and Francis (2000).

[Lew77] E. E. Lewis, *Nuclear Power Reactor Safety*, Wiley (1977).

[Lil86] D. R. Liles, et al., "TRAC-PF1/MOD1: An Advanced Best Estimate Computer Program for PWR Thermal-Hydraulic Analysis," NUREG/CR-3858, U.S. Nuclear Regulatory Commission (1986).

[Mar56] H. Margenau and G. M. Murphy, *The Mathematics of Physics and Chemistry*, 2nd ed., Van Nostrand (1956).

[Mar83] E. Marshall, "The Salem Case: A Failure of Nuclear Logic," *Science* **220**, 280 (1983).

[Mar05] R. P. Martin and L. D. O'Dell, "AREVA's Realistic Large Break LOCA Analysis Methodology," *Nucl. Eng. Des.* **235**, 1713 (2005).

[NRC71] "General Design Criteria for Nuclear Power Plants," Title 10, *Code of Federal Regulations*, Part 50, Appendix A, U.S. Nuclear Regulatory Commission (1971).

[NRC74a] "Acceptance Criteria for Emergency Core Cooling Systems for Light-Water Nuclear Power Reactors," Title 10, *Code of Federal Regulations*, Part 50.46, U.S. Nuclear Regulatory Commission (1974).

[NRC74b] "ECCS Evaluation Models," Title 10, *Code of Federal Regulations*, Part 50, Appendix K, U.S. Nuclear Regulatory Commission (1974).

[NRC74c] "Assumptions Used for Evaluating the Potential Radiological Consequences of a Loss of Coolant Accident for Pressurized Water Reactors," Regulatory Guide 1.4, U.S. Nuclear Regulatory Commission (1974).

[NRC76] "Preparation of Environmental Reports for Nuclear Power Stations," Regulatory Guide 4.2, rev. 2, U.S. Nuclear Regulatory Commission (1976).

[NRC77] "Methods for Estimating Atmospheric Transport and Dispersion of Gaseous Effluents in Routine Releases for Light-Water Cooled Reactors," Regulatory Guide 1.111, U.S. Nuclear Regulatory Commission (1977).

[NRC78a] "Standard Format and Content of Safety Analysis Reports for Nuclear Power Plants," Regulatory Guide 1.70, rev. 3, U.S. Nuclear Regulatory Commission (1978).

[NRC78b] "Anticipated Transients Without Scram for Light Water Reactors," NUREG-0460, vols. 1–3, U.S. Nuclear Regulatory Commission (1978).

[NRC84] "Requirements for Reduction of Risk from Anticipated Transients Without Scram (ATWS) Events for Light-Water-Cooled Nuclear Power Plants," Title 10, *Code of Federal Regulations*, Part 50.62, U.S. Nuclear Regulatory Commission (1984).

[NRC86] "Safety Goals for the Operation of Nuclear Power Plants," Title 10, *Code of Federal Regulations*, Part 50, Policy Statement, U.S. Nuclear Regulatory Commission (1986).

[NRC88] "Compendium of ECCS Research for Realistic LOCA Analysis," NUREG/CR-1230, U.S. Nuclear Regulatory Commission (1988).

[NRC89] "Best Estimate Calculations of Emergency Core Cooling System Performance," Regulatory Guide 1.157, U.S. Nuclear Regulatory Commission (1989).

[NRC90] "Severe Accident Risks: An Assessment for Five U.S. Nuclear Power Plants," NUREG-1150, vol. 1, U.S. Nuclear Regulatory Commission (1990).

[NRC00] "Alternative Radiological Source Terms for Evaluating Design Basis Accidents at Nuclear Power Reactors," Regulatory Guide 1.183, U.S. Nuclear Regulatory Commission (2000).

[NRC01] "RELAP5/MOD3.3 Code Manual, Volume 1: Code Structure, Systems Models, and Solution Methods," NUREG/CR-5535, rev. 1, U.S. Nuclear Regulatory Commission (2001).

[NRC02] "An Approach for Using Probabilistic Risk Assessment in Risk-Informed Decisions on Plant-Specific Changes to the Licensing Basis," Regulatory Guide 1.174, rev. 2, U.S. Nuclear Regulatory Commission (2002).

[NRC04] "Reactor Site Criteria," Title 10, *Code of Federal Regulations*, Part 100, U.S. Nuclear Regulatory Commission (2004).

[NRC07] "Standardization of Nuclear Power Plant Designs: Combined Licenses to Construct and Operate Nuclear Power Reactors of Identical Design at Multiple Sites," Title 10, *Code of Federal Regulations*, Part 52, Appendix N, U.S. Nuclear Regulatory Commission (2007).

[NRC08] "NRC Reactor Systems Training Manual," U.S. Nuclear Regulatory Commission (2008).
[NRC09] "Licenses, Certifications, and Approvals for Nuclear Power Plants," Title 10, *Code of Federal Regulations*, Part 52, U.S. Nuclear Regulatory Commission (2009).
[Nut04] W. T. Nutt and G. B. Wallis, "Evaluation of Nuclear Safety from the Outputs of Computer Codes in the Presence of Uncertainties," *Reliab. Eng. Sys. Safety* **83**, 57 (2004).
[Rao06] K. R. Rao, ed., *Companion Guide to the ASME Boiler & Pressure Vessel Code*, vol. 1, ASME Press (2006).
[Rub79] E. Rubinstein, "Three Mile Island and the Future of Nuclear Power," *IEEE Spectrum* **16**, 30 (1979).
[Sla68] D. L. Slade, ed., "Meteorology and Atomic Energy," TID-24190, U.S. Atomic Energy Commission (1968).
[Sof95] L. Soffer et al., "Accident Source Terms for Light-Water Nuclear Power Plants," NUREG-1465, U.S. Nuclear Regulatory Commission (1995).
[Tec89] Technical Program Group (TPG), "Quantifying Reactor Safety Margins," NUREG/CR-5249, U.S. Nuclear Regulatory Commission (1989).
[Wes84] "The Westinghouse Pressurized Water Reactor Nuclear Power Plant," Westinghouse Electric Corporation (1984).
[Wes03] "AP1000 Design Control Document," APP-GW-GL-700, rev. 3, Westinghouse Electric Company (2003).
[Wil41] S. S. Wilks, "Determination of Sample Sizes for Setting Tolerance Limits," *Ann. Math. Stat.* **12**, 91 (1941).

Exercises

8.1 The average energy $f(t)$ emitted per second in the form of β- and γ-rays, at t seconds after the fission of one ^{235}U nucleus, can be estimated by the Way-Wigner formula

$$f(t) = 2.66t^{-1.2}(\text{MeV/fission} \cdot \text{s}),$$

which is approximately valid for the range $t = [1, 10^6]$ seconds for thermal neutron fission of ^{235}U. (a) Determine the time, after a single fission of ^{235}U, over which we can expect the release of one-half of the total amount of energy that can be ultimately released by the decay of fission products, (b) obtain an expression for thermal power released due to the decay of FPs in a PWR at T seconds of operation at a constant power level of 1000 MWe, and (c) determine the equilibrium FP decay power for the reactor in units of MWt. What fraction of the rated reactor power does this amount to?

8.2 A nuclear fission reactor has been operating with ^{235}U fuel at a steady-state thermal power of P watts for a period of T seconds. It is then shut down essentially instantaneously. (a) Use the Way-Wigner formula of Exercise 8.1 to determine the heat generation rate $P_d(t, T)$ due to the FP decay at t seconds after shutdown. Compare the result with Eq. (8.2). (b) Show that Eq. (8.3) is valid for any arbitrary functional relationship $f(t)$, not merely for the Way-Wigner formula. (c) Based on the data of the ANS-5.1 standard, make a plot of the fraction $g(t, \infty)$ of the steady-state operating power that is released as the FP decay heat versus cooling time $t = [1, 10^6]$ seconds following a long period of reactor operation at constant power. Compare with the corresponding results from the Way-Wigner formula of Exercise 8.1 for a few representative values of cooling time t.

8.3 The FP decay heat data of the ANS 5.1 standard may be approximated as the heat generation rate $f(t)$ at t seconds following the fission of one ^{235}U nucleus: $f(t) = \alpha\lambda \exp(-\lambda t)$ (MeV/fission·s), with $\alpha = 13.0$ MeV/fission and $\lambda = 3 \times 10^{-7}$ s^{-1}. Calculate (a) the equilibrium FP decay power for a 1.0-GWe boiling water reactor and (b) the fraction ξ of the equilibrium FP decay power reached after 10 days of full-power operation.

8.4 A critical PWR configuration may be described by the following two-group model: (1) thermal utilization f = 0.76, resonance escape probability p = 0.80, number of neutrons released per thermal absorption in fuel η = 1.25, (2) fast-fission contribution to the infinite multiplication factor k_1 = 0.27, (3) nonleakage probability P_{NL} = 1/1.03, (4) thermal neutron absorption in nonfuel material is evenly divided between moderator and lumped burnable absorbers (BAs), i.e., $\Sigma_{a2}^P = \Sigma_{a2}^M$ in Eq. (8.17), and the BA absorption cross section is independent of moderator density, and (5) scattering cross section Σ_s for resonance neutrons can be assumed entirely due to that of water. If the water density is reduced by 3% due to a moderator temperature increase of 7 K, calculate the MTC for the reactor. Assume that changes in the microscopic cross sections may be ignored and list any other assumptions introduced in your analysis.

8.5 In the two-group PWR model considered in Exercise 8.4, obtain the maximum concentration of soluble boron that can be added to the coolant water without making the moderator temperature coefficient of reactivity positive. Each addition of 100 ppm by weight of natural boron in water increases thermal absorption cross section Σ_{a2}^M of water by 10%, which is to be compensated for by a 10% reduction in thermal absorption cross section Σ_{a2}^P of lumped BAs to retain criticality at rated condition. Assume that boron is a pure thermal absorber and consider a 3% decrease in the density of water due to a temperature increase of 7 K as in Exercise 8.4.

8.6 To estimate the radiation dose rate for a large release of radioactivity into the atmosphere, assume that the radionuclides can be approximated by ^{60}Co and that the radioactive nuclei are distributed uniformly in an infinite volume of air at a concentration of 1.0 Ci/m^3 of air at STP. Estimate the dose rate in air in units of rad/s and Gy/s due to the radioactivity release. Compare the result with Eq. (8.42).

8.7 Using the procedures presented in Regulatory Guide 1.4, calculate the dispersion factor χ/Q for atmospheric diffusion for 0 to 8 hours following a ground-level release of radionuclides at 5 km from the release. Compare the result with Fig. 8.28.

8.8 A 3300-MWt PWR is operating with leaks in 0.01% of the fuel rods. The fractional fission yield of ^{131}I is 0.028 and ^{131}I has a half-life of 8.02 days. Assume that each hour 0.1% of the ^{131}I nuclei inside the cladding leaks out of the rods and is dissolved in the coolant water. Assume also that, due to leakage and water purification, there is complete replacement of coolant water in the reactor vessel every 30 days, with the replacement taking place at a constant rate. Calculate the equilibrium ^{131}I activity of the coolant water.

8.9 For the ^{131}I leakage issue studied in Exercise 8.8, it turns out that 10% of the coolant water replacement rate is associated with the leakage of coolant water into the containment. To obtain an upper bound estimate of the radiological consequences of the coolant leakage, assume that the contaminated coolant water evaporates completely and uniformly mixes with the containment air. The contaminated air eventually leaks out of a total containment air volume of 5.0×10^4 m^3 at a rate of 500 m^3/hr into the atmosphere. (a) Determine the equilibrium ^{131}I radioactivity concentration in the containment air, (b) use the dispersion factor calculated in Exercise 8.7 to determine the equilibrium ^{131}I radioactivity concentration χ at a monitoring station located 5 km from the release, and (c) determine the dilution factor D of Eq. (8.38) for the ^{131}I released from the containment.

8.10 A PWR has been operating for nine months at rated power of 3560 MWt when 0.05% of the fuel rods start leaking. The leakage rate of ^{131}I from the leaking fuel rods is estimated to be 0.002%/min. The coolant water leaks out of the primary system at a rate of 0.1%/day and the chemical and volume control system replaces the coolant inventory every 10 days at a constant replacement rate. ^{131}I decays with a half-life of 8.02 days. (a) Using Appendix A, calculate the production rate (MCi) and concentration N (atoms) of ^{131}I nuclei in the core and (b) using the result of part (a), set up a balance equation for the concentration C (atoms) of ^{131}I in the coolant water and obtain the equilibrium concentration C_{eq} and equilibrium radioactivity of ^{131}I in the coolant water.

8.11 Assuming that the ^{131}I nuclei leaking out of the primary system in Exercise 8.10 are directly released to the atmosphere at the ground level, calculate the ^{131}I concentration χ (μCi/m^3) at the ground level and the boundary of the low population zone (LPZ). Assume that the wind speed is 1.0 m/s and unidirectional, subject to a neutral meteorological condition.

8.12 Solve the balance equation obtained in part (b) of Exercise 8.10 for the ^{131}I concentration as a function of time and calculate the ^{131}I dose received by an individual located at the LPZ boundary during two weeks following the initiation of the fuel rod leakage. For the average γ- and β-energy released from the decay of ^{131}I, consult the website: nucleardata.nuclear.lu.se.

8.13 The saturation activity of ^{133}I in a PWR core is 37.5 MCi. Assume that, due to a LOCA, 25% of the saturated ^{133}I inventory is released instantaneously into the containment with a volume of 5.0×10^4 m^3. The contaminated air leaks out of the containment at the rate of 500 m^3/hr at the ground level and ^{133}I decays with a half-life of 20.8 hours. (a) Determine the atmospheric diffusion factor χ/Q at a monitoring station 5 km from the containment. The wind speed is 2.0 m/s and unidirectional, subject to a *slightly unstable* meteorological condition. (b) Set up a balance equation for the ^{133}I activity $q(t)$ in the containment at time t following the LOCA. Solve the differential equation for $q(t)$ and, using χ/Q of part (a), determine the ^{133}I concentration $\chi(t)$ at the monitoring station. (c) Calculate the ^{133}I dose received by an individual located at the monitoring station during the first 4 hours following the LOCA.

8.14 The PWR containment considered in Exercise 8.13 is refurbished so that the leakage rate of contaminated air is reduced to 0.2 volumetric percent per day. (a) Determine the ^{133}I concentration $\chi(t)$ at $t = 0.5$ hour into the LOCA at the monitoring station 5 km from the containment and compare with the permissible effluent concentration of 0.001 μCi/m^3 of ^{133}I in air, stipulated in 10 CFR 20, Appendix B. (b) Calculate the minimum removal efficiency required for the containment spray and air filter system in order to satisfy the 10 CFR 20 requirements at $t = 0.5$ hour following the LOCA. (c) With the containment spray and filter activated, determine the overall dilution factor of Eq. (8.38) at $t = 0.5$ hour.

8.15 In a design study for a monitored retrievable storage (MRS) facility for spent fuel assemblies of a PWR plant, it is proposed that, after cooling in the spent fuel storage pool for five years, the fuel assemblies be stored in dry concrete casks, vented but without forced cooling mechanism provided. If the fuel assemblies, with an irradiation time of three years, are to be stored in the concrete casks, calculate the heat generation rate of each fuel assembly using the ANS-5.1 standard. Would you approve the proposed MRS design without forced cooling? Assume that the PWR core, operating at rated power of 1000 MWe, consists of 200 fuel assemblies.

8.16 A postulated rupture of the heavy water tank with a subsequent release of tritium into the pool floor room is to be considered, in connection with a proposed license amendment for a swimming-pool type research reactor. In this analysis, assume that an equilibrium tritium activity of 8 mCi is uniformly mixed with 1600 m^3 of air in the pool floor room. The contaminated air would be released at a rate of 2 m^3/s through an exhaust stack to the unrestricted outside atmosphere surrounding the reactor building. (a) Determine the atmospheric diffusion factor χ/Q for a ground release using Regulatory Guide 1.4, with additional dilution by a factor of 5 included to reflect the tritium release through the stack, and calculate the tritium concentration χ in air at 0.25 km from the stack during the first hour following the postulated accident. Compare it with the maximum permissible tritium concentration of 0.1 μCi/m^3, stipulated in 10 CFR 50, Appendix B. (b) Using the results of part (a), determine the overall dilution factor of Eq. (8.38).

CHAPTER 9

MAJOR NUCLEAR POWER PLANT ACCIDENTS AND INCIDENTS

This chapter presents a discussion on a few major accidents or incidents that have occurred at commercial nuclear power plants, including the Three Mile Island Unit 2 and Chernobyl accidents and an updated summary of the station blackout accident of March 2011 at the Fukushima Daiichi nuclear complex in Japan. The Salem anticipated transient without scram, the LaSalle transient event, and the Davis-Besse potential LOCA event also are considered. For each case the focus of the discussion is on the events that are of interest primarily from the reactor core or NSSS side of the plant. Thus, we will not dwell, e.g., on the 1975 cable fire [Dav75] at the Browns Ferry Unit 1 power plant, which had serious implications on nuclear power plant safety but had to do primarily with the wiring and layout of the plant. For each event covered, we will discuss the sequence and chronology, together with its causes, implications, and follow-up actions taken. In addition, in-vessel melt progression phenomena are discussed to gain a better understanding of the Three Mile Island Unit 2 (TMI-2) accident of 1979, which resulted in a large-scale meltdown of the core and eventual decommissioning of the plant.

9.1 THREE MILE ISLAND UNIT 2 ACCIDENT

9.1.1 Sequence of the Accident—March 1979

The TMI-2 accident was initiated [Col80] as a loss of feedwater transient, due to a malfunction in a condensate cleaning (polishing) operation, in a PWR that had operated for a relatively short period of time. In a normal LOFW transient, the auxiliary feedwater system would have kicked in, but it was unavailable due to a block valve left shut during the previous maintenance outage. This resulted in overheating the primary coolant system, with the increased primary system pressure opening up the power-operated relief valve in the pressurizer, as it should. The PORV failed to close and remained in a stuck-open state, however, when the system pressure eventually decreased to the closure setpoint. This resulted in a small-break LOCA. The coolant discharged through the PORV eventually collected in the drain tank. The overflowing in the drain tank opened the rupture disk, sending the coolant to the containment sump. The sump pump eventually pumped the radioactive coolant to the waste storage tank in the nonsafety class auxiliary building, from which some radioactive coolant evaporated and eventually radionuclides leaked to the environment. Figure 8.2 illustrates the overall schematics of the TMI-2 plant during the accident.

The reactor operators misdiagnosed the stuck-open PORV, partly due to previous malfunctions of the PORV and incorrect readings of the valve position, and throttled off the ECCS pumps that had been properly activated. The primary concern among the operators was that the pressurizer might be going solid, which was to be avoided for fear that the reactor pressure vessel might develop leaks through overpressurization. Due to the lack of coolant water delivery, the core was uncovered and melting of several fuel assemblies occurred, primarily through the U-Zr-O eutectic formation, which has a liquefaction temperature $\sim$1000 K below the UO_2 melting point (MP) of 3123 K. The melting of fuel assemblies further blocked the flow of coolant through the rest of the core, eventually resulting in the damage and melting of nearly two-thirds of the core. The extent of the core damage is illustrated in Fig. 9.1 [Bro89].

Hydrogen was generated from the zirconium-water reaction and collected in a bubble in the upper plenum of the RPV, which raised significant concern and anxiety among the public that the hydrogen bubble might ignite. This turned out to be ill-founded. The Zr-water reaction results in a reducing environment, which would not allow a sufficient amount of oxygen available for the ignition of the hydrogen bubble.

9.1.2 Implications and Follow-Up of the Accident

Despite the extent of the damage in the TMI-2 core, the maximum exposure anyone in the public received is estimated to be around 90 mrem, which is equivalent to a typical exposure expected from three well-administered chest X-rays. This worst exposure of 90 mrem among the public should also be compared with the average background exposure of 240 mrem/year worldwide, which includes 120 mrem due to radon exposure, as discussed in Appendix A and the BEIR-VII report [NAP05]. The

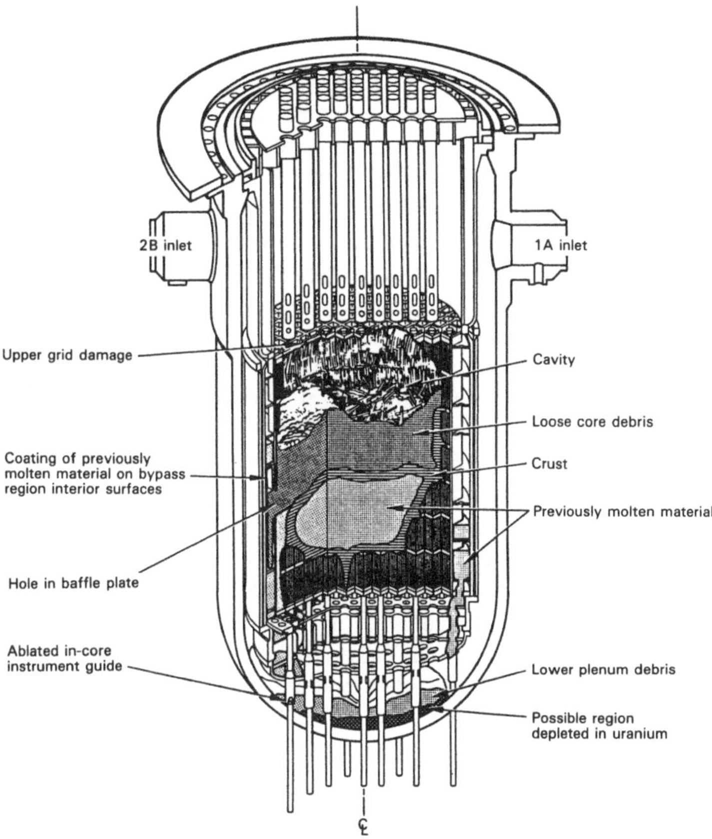

Figure 9.1 Final TMI-2 debris configuration showing significant damage to the core. *Source:* [Bro89].

average annual background exposure in the United States is 300 mrem, primarily due to a relatively high radon burden in the country.

The TMI-2 accident, however, imposed significant anxiety and psychological trauma among the general public. As a result, the U.S. nuclear industry lost the confidence of the public for continued development of nuclear energy in the country. This, together with the Chernobyl accident seven years later, contributed to canceling a number of plants on order and slowed down the development of nuclear energy in other countries as well.

The accident revealed the lack of emphasis on safety culture and of communication about significant events among the nuclear utility companies. This prompted the nuclear industry to form the Institute of Nuclear Power Operation (INPO) for improved NPP operator training. Before the TMI-2 accident, with a heavy emphasis on providing the ECCS for large-break LOCAs, there was a sense among the community that it was highly unlikely to have core meltdown accidents, let alone from a small-break LOCA. Indeed, there was a mindset among the nuclear community that serious accidents were unlikely and it was often suggested that the first major PRA study on nuclear power plants, WASH-1400 [NRC75], overestimated the likelihood of accidents and risks associated with NPP operation. As discussed further in Chapter 10, WASH-1400, or the Reactor Safety Study as it is often called, indicated that the most risk-significant NPP accidents were SBLOCAs, exemplified by the TMI-2 accident. Furthermore, the eutectic formation of liquefied U-Zr-O at temperatures significantly below the MP of ceramic UO_2 is one of the important lessons that the nuclear engineering community learned from the ill-fated TMI-2 accident, since it had been previously assumed that fuel melting would not occur below the UO_2 MP.

With improved operator training and maintenance practices, 104 NPPs operating in the United States in 2010 had improved capacity factors from somewhere around 70% in the 1970s to a recent fleet average of >90%, which is equivalent to adding to the national grid approximately 20 new 1.0-GWe power plants. As a result of the TMI-2 accident, a number of measures [Pet06] were developed to manage severe accidents in NPPs. They include provisions for portable electric generators, e.g, jet engines that can be started without any independent electric source at the Fermi Unit 2 plant, together with procedures to supply electricity to vital equipment in case of station blackout events. Some plants also installed a filtered containment venting system that could have alleviated concerns about possible ignition of the hydrogen bubble during the TMI-2 accident. The accident also suggested the need to provide a means to reduce the primary system pressure early in a potential LOCA so that the plant may be able to passively tap into the reservoir of coolant in the PWR accumulator. This eventually led to a core depressurization system in the AP600 and subsequently in the AP1000 design, coupled with an increased capacity for the in-containment refueling water storage tank (IRWST) and for the pressurizer, as discussed further in Chapter 11.

Despite the extent of the core meltdown in the TMI-2 accident, which resulted in the decommissioning of the plant, the radioactivity release was relatively small, perhaps on the order of 15 to 20 Ci [Bal96]. This raised a number of questions about the prevailing source term discussed in Section 8.6.1, which assumes in the case of DBAs that 100% of the inventory of noble gases and 50% of the ^{131}I inventory are released to the containment. The recovery effort for the TMI-2 resulted in the following observations:

- Volatile fission products are mostly in the form of CsI and CsOH, instead of elemental I_2 and Cs, and hence are less likely to leak out to the containment.

- Fission products form aerosols, which plate out on the cold surface of the containment and are mostly retained in the containment.

Furthermore, the molten corium, i.e., the mixture of fuel and structural materials in the core, was retained in the lower head of the RPV. Thus, the possibility of penetration of the RPV and the basemat or bottom of the containment building, nicknamed the China Syndrome, appears rather remote. Furthermore, early containment failures are rather unlikely, as the RPV turned out stronger than expected. The failure threshold for the RPV is now expected to be >1.5 GJ of kinetic energy, which is equivalent to the explosive potential of 0.4 ton TNT. It is estimated that a steam explosion involving 25 Mg of corium, approximately comprising 17% of molten core, and a sufficient amount of coolant water could result in such a magnitude of kinetic energy release. Because such a large fraction of the molten core is not likely to form during the early phase of a postulated LOCA, when the overall radioactivity level is high and an early containment failure could likely be avoided, the consequences of postulated accidents will be significantly less than that evaluated with the assumptions of the conventional source term. This is discussed further in Section 9.2.

All these issues raised the need to revisit the radiological source term in 10 CFR 100 [NRC04] and the U.S. Nuclear Regulatory Commission initiated a major study on the consequences of core meltdown or severe accidents, which resulted in the publication of the NUREG-1150 report [NRC90]. The NUREG-1150 report will be discussed in detail in Chapter 10. Alternate guidelines, NUREG-1465 [Sof95] and RG 1.183 [NRC00], for calculating radiological source terms have also been published. These new source terms have been already adopted by a number of operating NPPs [Bla06] and will be applied in the combined construction and operation license (COL) applications for new NPPs on the drawing board. One question that agonized the nuclear engineering community during and shortly after the TMI-2 accident was whether the plant operators should be allowed to dump cold coolant into a degraded, molten core, because of the potential for major steam explosions that could rupture the RPV. The general consensus now seems to have settled down to the suggestion that we should try to use any available source of coolant during an undercooling accident, with the understanding that the potential for steam explosions is relatively small and could be dealt with in due course.

9.2 PWR IN-VESSEL ACCIDENT PROGRESSION

To develop an increased understanding of the progression of the events in the TMI-2 accident that resulted in a significant meltdown of the core, we present a simplified step-by-step analysis of a core damage accident based on a detailed discussion in [Has02]. The discussion begins with an accident that has reached a point of core uncovery, which leads to (a) overheating of fuel, (b) oxidation of the cladding, (c) liquefaction of fuel, (d) interaction between molten fuel and water in the reactor vessel, and (e) vessel breach. Table 9.1 summarizes starting conditions for the five stages of the core meltdown accident and approximate estimates of the associated time intervals for a PWR plant with a total failure of coolant injection postulated.

The accident progression phases in Table 9.1 may be compared with a summary of the TMI-2 accident scenario in [Bro89], where the accident is divided into seven major periods following the initiation:

1. 0–100 minutes: Cooling of core maintained with primary coolant pumps on, despite loss of some coolant.

Table 9.1 In-Vessel Accident Progression Stages

Stage	Starting Condition	Description	PWR Durations, No Injection[a]
1	Accident Initiator	Initiation	0–90 min
2	Core uncovering begins	Core uncovering and heatup	5–35 min
3	Hottest cladding reaches 1832°F (1273 K, 1000°C)	Cladding oxidation, melting of structural and control materials	5–10 min
4	Hottest cladding reaches its melt temperature, 3200°F (2033 K, 1760°C)	Clad melting, fuel liquefaction, holdup in core region	10–30 min
5	Core materials first enter lower plenum	Core slumping, quenching, reheating	0–80 min
6	Vessel breach	Vessel breach and materials discharge to containment	—

[a] Approximate ranges for duration with total failure of coolant injection.
Source: Adapted from [Has02]

2. 100–174 minutes: Shutdown of reactor coolant pumps resulting in coolant boiloff and core uncovery, cladding oxidation, partially molten region of core material (corium).
3. 174–180 minutes: RCP 2B restarted at 174 min pumping water into the reactor vessel and cooling peripheral fuel assemblies and forming upper core debris bed.
4. 180–200 minutes: Heatup of the consolidated molten corium region continues.
5. 200–224 minutes: High-pressure coolant injection initiates quenching and cooling of the upper debris bed, but the growth of molten corium region continues.
6. 224–226 minutes: Molten corium flows to core support assembly, eventually forming a lower plenum debris bed.
7. 226 minutes–15.5 hours: Forced coolant flow resumed, quenching and cool-down of the lower plenum debris bed.

The reactor coolant system (RCS) pressure history [Bro89] corresponding to the first six periods of the accident is plotted in Fig. 9.2. Stage 1 of Table 9.1 corresponds to the TMI-2 accident period 1, while stages 2 through 4 cover TMI-2 periods 2 through 5, lasting up to 224 minutes. Stage 5 of Table 9.1 corresponds to TMI-2 periods 6 and 7. The final vessel breach stage in Table 9.1 is a sequence that fortunately did not materialize at TMI-2. A more detailed discussion of the TMI-2 accident scenario is presented in [EPR80].

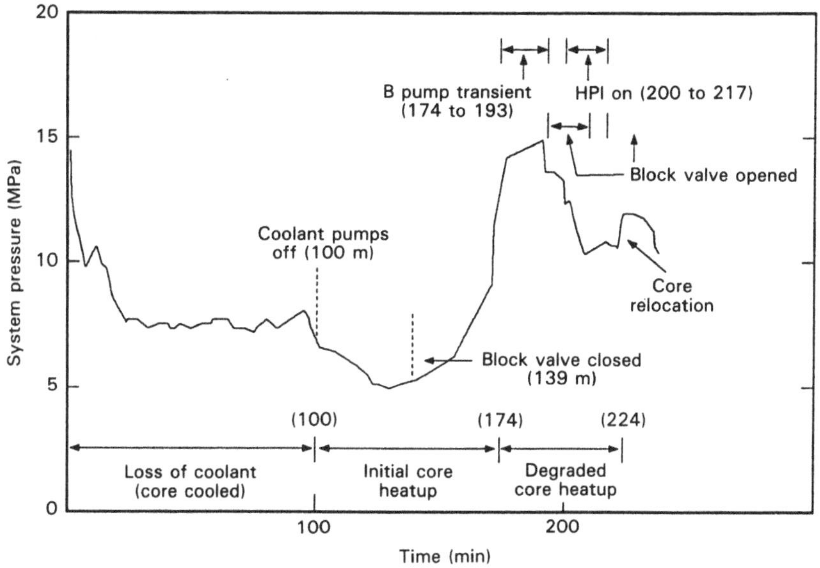

Figure 9.2 RCS pressure history during the TMI-2 accident. *Source:* [Bro89].

9.2.1 Core Uncovery and Heatup

Following the reactor shutdown, the failure of coolant delivery to the core results in the entire core essentially attaining the system saturation temperature, with the fission product decay power continuing to add heat to the inventory of coolant water. We assume that all FP decay power generated in the water-covered fuel region contributes to the evaporation of water, resulting in a continual decrease in length z of the covered region. The decrease Δz in the covered length in time interval Δt is determined by equating the FP decay energy deposited in the remaining inventory of water to the energy required to evaporate the water of volume $A\,\Delta z$,

$$P_d \frac{z}{H} \Delta t = \rho h_{fg} A\,\Delta z, \tag{9.1}$$

where P_d = FP decay power for the core assumed constant during the fuel uncovery, H = active core height, ρ = liquid water density, h_{fg} = latent heat of vaporization of water, and A = cross sectional area of water in the active core region. Rearranging Eq. (9.1) yields

$$\frac{\mathrm{d}z(t)}{\mathrm{d}t} = \frac{P_d}{\rho h_{fg} AH} z \equiv \frac{z}{\tau}, \tag{9.2}$$

where $\tau = \rho h_{fg} AH/P_d$ is a time constant for the boiloff process. Integrating Eq. (9.2), with $z(0) = H$, provides an expression for the covered fuel length at time t during stage 2 of Table 9.1:

$$z(t) = H \exp(-t/\tau). \tag{9.3}$$

The boiloff time constant τ is a function of the system pressure and may be estimated for a typical PWR configuration corresponding to the relief valve setpoint of 2500 psig:

$$\begin{aligned}\tau &= \frac{(\rho = 558\ \mathrm{kg/m^3}) \cdot (h_{fg} = 830\,\mathrm{kJ/kg}) \cdot (A = 4.96\,\mathrm{m}^2) \cdot (H = 3.66\,\mathrm{m})}{P_d = 3.25 \times 10^4\,\mathrm{kW}} \\ &= 259\,\mathrm{s}. \end{aligned} \tag{9.4}$$

The governing equation (9.3) for the boiloff process is valid for approximately $2\tau \simeq$ 8.6 minutes, in reasonable agreement with an estimate of 5 to 35 minutes in Table 9.1 obtained via more realistic calculations with the MARCH code [Riv81].

As the fuel region is uncovered through the evaporation of water, the uncovered region heats up because the water vapor surrounding the region does not provide much of cooling. To the first approximation, the heatup of the uncovered region may be represented as an adiabatic addition of the FP decay heat to determine the fuel temperature $T(z,t)$ at height z and at time t:

$$T(z,t) = T(z,0) + \frac{p(z)P_d}{C_p}\left[t - t^*(z)\right], \tag{9.5}$$

where $t = 0$ chosen at the initiation of the core uncovery and $t^*(z)$ = time at which the water level decreases to height z, C_p = heat capacity of the entire core, and $p(z)$ = normalized or relative axial distribution for decay power. Thus, the term $p(z)P_d$ represents the decay heat power per unit length at elevation z and the time interval, $t - t^*(z)$, accounts for the fuel heatup after the fuel is uncovered at z.

The fuel temperature $T(z,t)$ calculated with Eq. (9.5) for the heatup stage following the fuel uncovery is compared with MARCH calculations [Has02] in Fig. 9.3 at three times expressed in terms of the uncovery time constant τ of Eq. (9.4). The temperature distributions plotted as a function of the normalized axial elevation z/H display significant axial variations, apparently representing both the underlying axial distribution of the decay power and the increases in $t^*(z)$ as z decreases. The good agreement between the approximate results of Eq. (9.5) and MARCH calculations indicates that the adiabatic heating assumption is reasonable for this phase of the accident progression. The heatup model is valid up to the point when the peak fuel temperature reaches about 1273 K, when it becomes necessary to represent the cladding oxidation in the energy balance. Indeed, from Fig. 9.3 the peak temperature at $t = 1.58\tau = 409$ s = 6.8 minutes reaches 1730°F = 1216 K, with the saturation temperature T_{sat} = 650°F. We note again that our estimate of t = 6.8 minutes for the end of the heatup process is in general agreement with the 5~35 minutes indicated for stage 2 in Table 9.1.

9.2.2 Cladding Oxidation

When the peak fuel temperature reaches about 1273 K, following the uncovery, cladding oxidation initiates with the exothermic reaction

$$\mathrm{Zr} + 2\mathrm{H_2O} \rightarrow \mathrm{ZrO_2} + 2\mathrm{H_2}, \tag{9.6}$$

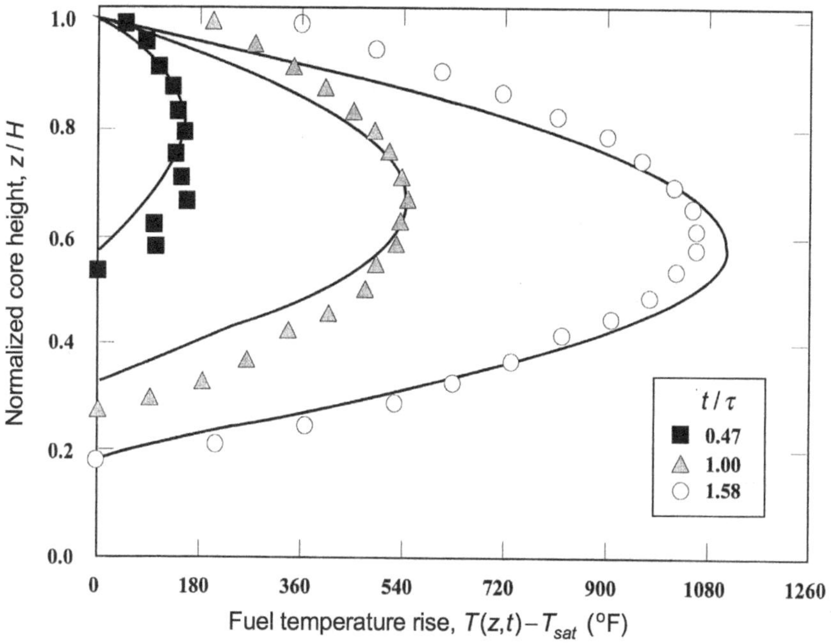

Figure 9.3 Fuel temperature distributions at three different times during the fuel uncovery. The MARCH results are compared with solid curves representing Eq. (9.5). *Source:* [Has02].

with the release of 6.5 MJ/kg of Zr reacted. If adequate steam is available, the mass W of Zr oxidized per unit area exposed to steam at temperature T in time interval t may be determined by

$$W^2 = Ae^{-B/RT}t, \tag{9.7}$$

where R = universal gas constant, 8.314 kJ/kg·mol·K, A = 294 kg^2/m^4·s, and B = 167 MJ/kg·mol. For the case when all of the fuel cladding with a combined surface area of 5400 m^2, corresponding to the Zion PWR plant, is exposed to steam at 1473 K for 5 minutes, Eq. (9.7) yields W = 0.322 kg/m^2 and a total of 1740 kg of Zr oxidized, which is 14.2% of Zr in the core. Since 2 mol of hydrogen is produced per mole of Zr reacted, we obtain the corresponding mass of hydrogen produced,

$$m_H = 1740\,\text{kg Zr} \cdot \frac{2\ \text{kg·mol H}_2}{91.22\ \text{kg Zr}} \cdot \frac{2.016\ \text{kg H}_2}{\text{kg·mol H}_2} = 76.9\ \text{kg H}_2, \tag{9.8}$$

and a total reaction energy of 11.3 GJ released through the Zr-water reaction during the 5-minute interval. This corresponds to a reaction energy release rate of 37.7 MWt, which is somewhat larger than the total FP decay power P_d = 32.5 MWt assumed in Eq. (9.4).

With a significant reaction energy released, the cladding temperature increase results in a further increase in the oxidation process. For example, the reaction rate calculated above at T = 1473 K could double with the oxidation temperature increased by 200 K. Furthermore, a significant amount of energy transfer from the uncovered region, via thermal radiation, heat conduction, or the movement of debris, will enhance the production of steam beyond the rate calculated with Eq. (9.7), which in turn will increase the oxidation rate. Thus, during stage 3, virtually all of the vapor produced could participate in the Zr oxidation process and the oxidation energy release rate can substantially exceed the FP decay power.

Together with the cladding oxidation and increased boiloff of water, deformation of the clad as well as embrittlement and spallation of ZrO_2 from the surface of the clad can be expected at this stage of accident progression. If coolant water is reintroduced into the core during the oxidation stage, the core damage process may initially increase due to additional Zr-steam reactions. In addition, significant fracturing of cladding may occur during the reflooding of the core, leading to the formation of coarse rubble comprising fractured cladding, fuel, and control absorber materials. Indeed, such rubble formation was observed in the upper region of the damaged TMI-2 core.

9.2.3 Clad Melting and Fuel Liquefaction

Due to sustained oxidation of the cladding combined with the FP decay heat, the Zircaloy-4 cladding may approach the MP of 2033 K. The liquefaction of the cladding may, however, occur below the MP due to the formation of a eutectic with structural and control absorber materials. During the TMI-2 accident, the Ni-Zr eutectic formation at 1473 K due to the interaction between the Inconel spacer grids and fuel cladding near the core center perhaps resulted in the onset of melt formation. In addition, the Ag-In-Cd control rod material with an MP of 1073 K, and stainless steel cladding for control rods with an MP of 1723 K, contributed to eutectic formation with Zr in the initial molten mixture at TMI-2. Figure 9.4 illustrates a postulated TMI-2 core configuration [Bro89] shortly after the initiation of clad melting at 150 to 160 minutes into the accident, where the molten metallic mixture of control, cladding, and structural materials froze at the steam/liquid interface and formed a crust that blocked coolant channels between fuel rods.

Together with the eutectic formation of the Zircaloy cladding with structural and control materials, fuel could undergo eutectic formation with Zircaloy at its MP of 2033 K, which is over 1000 K below the UO_2 MP of 3123 K. This eutectic process produces a downward flow of liquefied U-Zr-O, destroying the UO_2 matrix and accelerating the release of FPs from the fuel. Figure 9.5 shows how the TMI-2 accident could have progressed further at 173 minutes into the accident, as the molten flow of U-Zr-O mixture is contained by the crust of control, cladding, and structural materials, essentially blocking the coolant flow to the central region of the core. The configuration corresponds to the state just prior to a brief restart of the RCP 2B and the crust remains cooled by the water covering the bottom of the core.

Activation of RCP 2B at 174 minutes injected approximately 28 m^3 of coolant into the RPV, which generated, upon contact with hot surfaces in the core, a significant amount of steam and oxidation of metallic Zircaloy in the upper core region. This

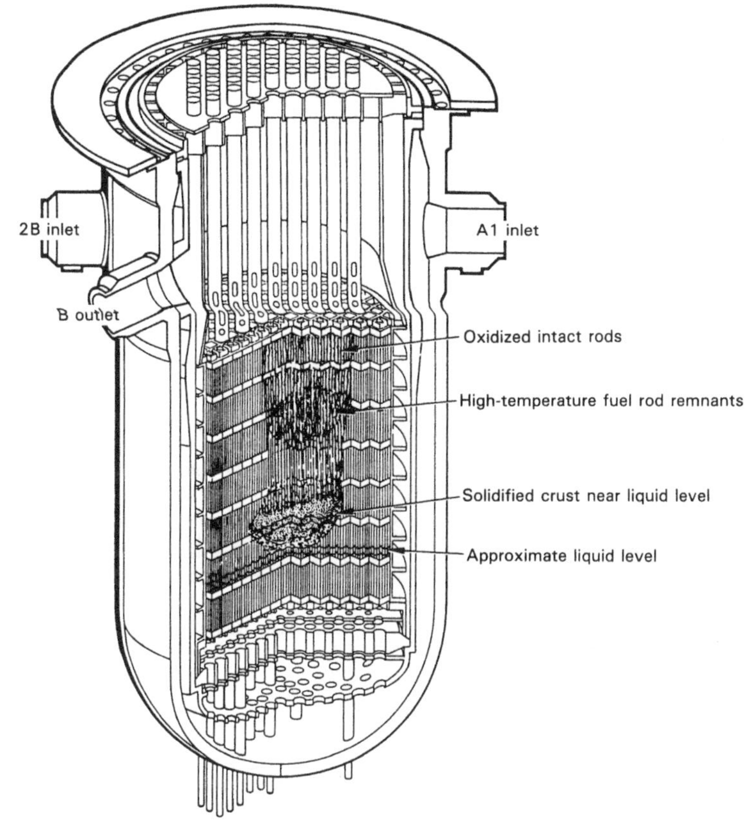

Figure 9.4 Hypothesized TMI-2 core configuration during 150 to 160 minutes. *Source:* [Bro89].

caused a rapid increase in the system pressure, indicated in the RCS pressure history of Fig. 9.2. It is postulated that the resulting thermal-mechanical forces fragmented fuel pellets and oxidized cladding and damaged the upper core support grid, as illustrated in Fig. 9.6.

Reactor coolant pump 2B operated only 19 minutes and the water level in the core continued to decrease with the FP decay heat evaporating the water during 180 to 200 minutes into the accident. The HPCI system was manually actuated during 200 to 217 minutes and emergency cooling water refilled the RPV by 207 minutes into the accident. Analyses indicate that by 230 minutes the upper debris bed was fully quenched. Figure 9.7 indicates a hypothesized configuration at 224 minutes, where water covered the upper debris bed but was unable to cool the consolidated molten region between the upper and lower crusts. Despite the meltdown of a large fraction of the core, the TMI-2 accident indicates that the injection of sufficient coolant water even after the liquefaction of cladding and fuel can successfully terminate the

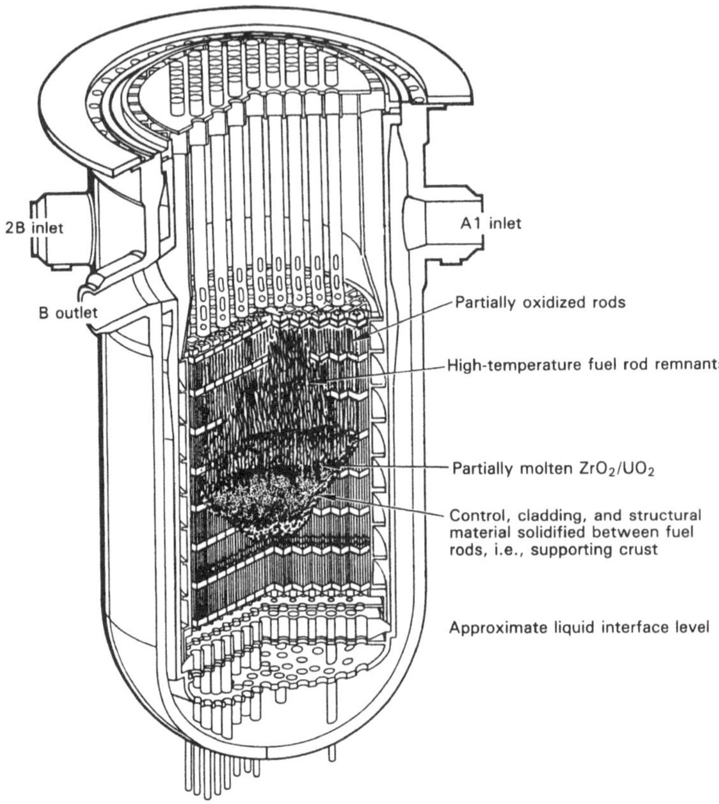

Figure 9.5 Hypothesized TMI-2 core configuration at 173 minutes. *Source:* [Bro89].

meltdown progression within the RPV. Thus, periods 4 and 5 of the TMI-2 accident correspond to stage 4 in Table 9.1.

9.2.4 Molten Core Slumping and Relocation

In period 6 of the TMI-2 accident, following the RPV reflooding, 19.2 Mg of molten core material was relocated into the RPV lower head during 224 to 226 minutes into the accident, with another increase in the system pressure indicated in Fig. 9.2. This corresponds to stage 5 of Table 9.1. Neutron instrumentation and thermocouple data also confirmed this event. The slumping and relocation of the molten corium could have resulted from continued heating of the molten pool, combined with a decrease in the system pressure due to the opening of the pressurizer block valve at 220 minutes. The final RPV configuration in Fig. 9.1 shows the failure of the crust near the core periphery and a probable relocation path for the molten corium. A large void region above the damaged upper core support is also indicated.

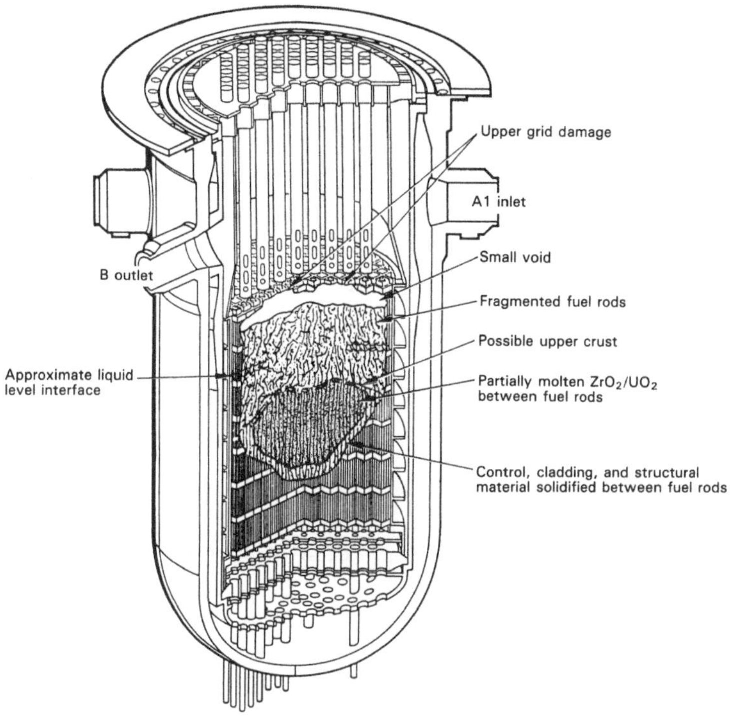

Figure 9.6 Hypothesized TMI-2 core configuration during 174 to 180 minutes. *Source:* [Bro89].

9.2.5 Vessel Breach

Although the TMI-2 accident terminated without the penetration of the RPV lower head, the possibility existed for fuel coolant interactions (FCIs) or steam explosions that could have breached the RPV. An upper bound estimate of 170 GJ for the thermal energy release from the FCI process is obtained [Has02] by calculating the energy required to quench an entire core inventory of 100 Mg of UO_2 plus 30 Mg of Zr and Fe, at 2873 K to the atmospheric boiling temperature of 373 K for water. A thermal-to-work energy conversion efficiency of 5% is considered probable [The81,Cor83], although an isentropic thermodynamic efficiency could be as high as 30%. The 5% conversion efficiency would yield 8.5 GJ of kinetic energy available, compared with 1.5 GJ of kinetic energy usually estimated as the minimum required to breach the PRV lower head. This implies that unless $>$17% of the entire corium inventory is postulated to undergo rapid interactions with a large inventory of water, the breach of the RPV lower head is not likely to occur. Furthermore, additional studies [The81,Cor83,The89] suggest that $<$5 Mg of molten corium would likely participate actively in FCIs in a degraded core.

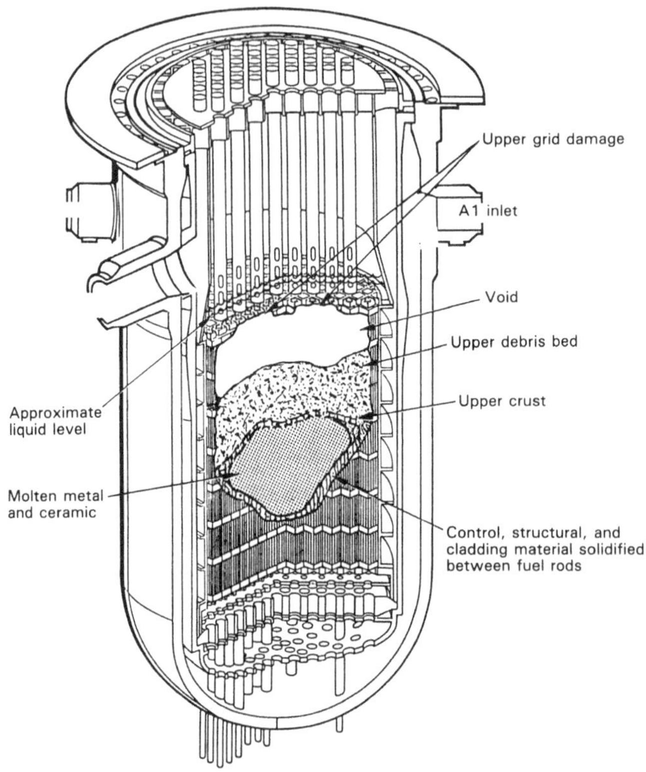

Figure 9.7 Hypothesized TMI-2 core configuration at 224 minutes, just prior to the molten corium relocation. *Source:* [Bro89].

In-vessel steam explosions leading to containment failure were introduced as the α-mode in the WASH-1400 Reactor Safety Study [NRC75]. An α-mode probability [Has02] in the range of $[0.001, 0.01]$ was used in the NUREG-1150 PRA study [NRC90].

9.3 CHERNOBYL ACCIDENT

9.3.1 Cause and Nature of the Accident—April 1986

The Chernobyl plant was a graphite-moderated, pressure tube or channel-type BWR of the RBMK-1000 design comprising a large core, with a $(12 \text{ m} \times 12 \text{ m})$ cross-sectional area and height of 7 m, as schematically illustrated in Fig. 9.8. During a test of turbogenerator coastdown, which was to provide power to feedwater pumps and the ECCS, a sequence of transient events resulted in a prompt critical accident. The accident occurred due to a number of peculiar features including:

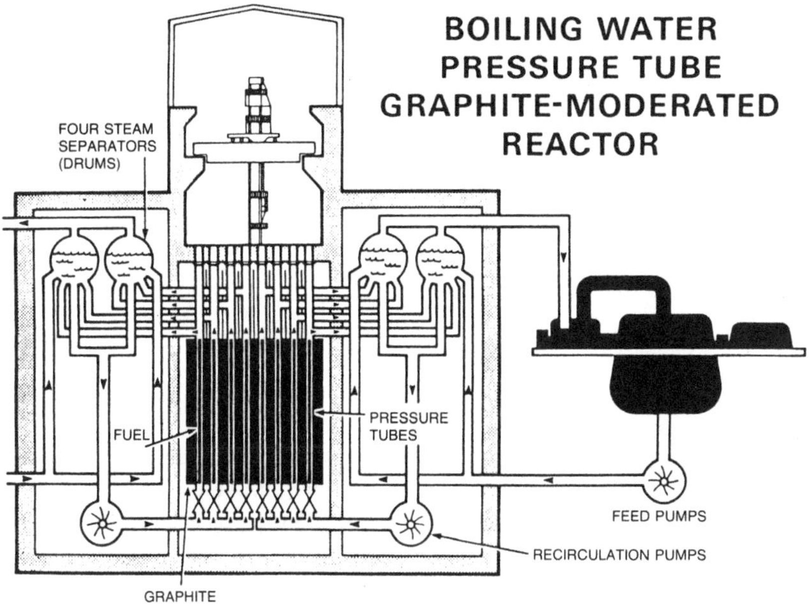

Figure 9.8 Schematic diagram of the RBMK-1000 Chernobyl plant. *Source:* [Nuc86].

(a) The Chernobyl core operated in an overmoderated regime of Fig. 8.22 and had a positive void coefficient of reactivity so that the overheating of the coolant resulted in an increase in the fraction of steam void and an increase in reactivity. This positive void feedback drove the core beyond the prompt criticality, which was not possible for the plant operators to control and contain.

(b) The initiation of the ill-fated turbogenerator coastdown test was in flagrant violation of operating procedures and willful disregard for safe operating practice. One particular operating procedure is related to the operating reactivity margin, which required the insertion of a sufficient number of control rods into the core. Recall from the reactivity coefficient discussion of Section 8.5.3 that the insertion of a sufficient amount of lumped neutron absorbers, e.g., control rods, could have made the VCR less positive or could even have placed the reactor in the undermoderated regime of Fig. 8.22.

(c) The large core size implies neutronically that different parts of the core are largely decoupled from one another. This makes controlling the power distribution difficult, especially at low power, when the fraction of control rods inserted is relatively small.

Table 9.2 Estimated Release of Radionuclides and Fuel in the Chernobyl Accident

Nuclide/Fuel	Release Fraction (%)	Radioactivity (MCi)
Noble gases	100	190±20
^{131}I	55±5	45±5
^{137}Cs	33±10	2.3±0.7
^{90}Sr, ^{90}Y	4.0	2.8±0.8
Fuel	3.5±0.5	

Source: [Has02].

9.3.2 Sequence of the Accident

The reactor power was reduced from 3200 to 200 MWt over 24 hours, resulting in a large reactivity poisoning due to the buildup of fission product ^{135}Xe and control rod withdrawal beyond the operating reactivity margin. At the same time, the reactor was brought to a point where the core inlet temperature was near the saturation point, where flashing of coolant water to steam was very likely. The operators initiated the turbine isolation, which triggered the reactor coolant pump rundown. This caused a reduction in the core coolant flow, which increased the production of steam voids and the reactivity, again due to the positive VCR of the core.

The reactor operator then initiated a reactor scram, in an effort to control the reactivity increase. In an overmoderated Chernobyl core, the replacement of water by control rods increased the reactivity, again due to the positive VCR, into a superprompt critical configuration. Furthermore, the leading edge of the control rods contained graphite followers, containing no neutron poison, which, upon insertion into the core, could have physically increased the reactivity. This triggered two prompt neutron pulses [Ahe87]. The first pulse raised the power level from 200 to 3800 MWt over 2.5 seconds, followed by a second pulse over another 1.5 seconds, resulting in an estimated peak power of 120 times the rated power of 3200 MWt, which is a nearly 2000-fold increase from the starting power level of 200 MWt.

As a result of the superprompt critical transient with a large release of thermal energy, molten fuel and fission gases were ejected into the coolant, leading to an explosive formation of steam through energetic fuel-coolant interactions. With a positive VCR, this resulted in a further increase in the core reactivity in a positive-feedback loop. The ensuing massive thermal explosion lifted the 1000-Mg reactor cover assembly releasing a large inventory of fission products into the open atmosphere. A fire soon started on the reactor structural components and graphite moderator, eventually burning 10%, or 250 Mg, of the graphite [Ahe87]. The accident released >250 MCi of radionuclides, out of a total inventory of 1000 MCi [IAE86], and 3.5% of the initial fuel inventory of 190 Mg. A breakdown of major radionuclide releases is given [Has02] in Table 9.2. The large quantities of radionuclides released in the accident are partly due to the particular containment structure that did not fully cover the fuel channel heads and the reactor itself.

9.3.3 Estimate of Energy Release in the Accident

Given the estimate for the two-pulse prompt critical power burst of Section 9.3.2 that the Soviets provided, we may use a simple point kinetics equation to estimate the energy released in the accident, ignoring delayed neutrons and treating the power burst as a single pulse. For core power level $n(t)$, the prompt kinetics equation with a step insertion of reactivity K_0 yields

$$\frac{\mathrm{d}n(t)}{\mathrm{d}t} = \frac{K_0 - \beta}{\Lambda} n(t), \tag{9.9}$$

where the neutron generation time $\Lambda = 0.64$ ms and effective delayed neutron fraction $\beta = 0.0057$ are given in a recent study [Moc07]. This study as well as other simulations [Fle88] of the Chernobyl accident suggest that the reactivity increased rapidly to $\sim 1.5\$$ at the peak of the power excursion, due to the positive void coefficient of 20 to 30 pcm/%void and the insertion of control rods, mitigated partly by the negative Doppler reactivity.

In our simple analysis, we infer an effective step reactivity K_0 from the peak power level estimated by the Soviets. Integrating Eq. (9.9) with the power level $n(0) = 200$ MWt at the beginning of the pulse yields

$$n(t) = n(0) \exp\left(\frac{K_0 - \beta}{\Lambda} t\right). \tag{9.10}$$

With the peak power $n(T) = 384$ GWt at $T = 4$ seconds, Eq. (9.10) suggests an effective step reactivity $K_0 = 1.2\$$, which is reasonable compared with a peak reactivity of $\sim\$1.5$ estimated [Moc07,Fle88]. Integrating Eq. (9.10) over time for $T = 4$ seconds yields the total energy $Q(T)$ released in the power excursion:

$$Q(T) = n(0) \frac{\Lambda}{K_0 - \beta} \exp\left(\frac{K_0 - \beta}{\Lambda} T\right) = 203 \text{ GJ}, \tag{9.11}$$

which is somewhat smaller than the Soviet estimate of 239 to 279 GJ, corresponding to the core average energy density [Ahe87] of 1.26 to 1.47 MJ/kg. RELAP5 calculations [Fle88] indicate $n(T) = 391$ GWt and $Q(T) = 169$ GJ. The NRC estimated [Ahe87] that UO_2 melts at an energy density of 1.09 MJ/kg and vaporizes at 2.93 MJ/kg. This suggests that parts of the Chernobyl core likely reached well above the MP. Since the bulk of the molten fuel dropped to the bottom of the reactor building, we may assume that only 5% of the fuel was ejected upward with a thermal-to-mechanical energy conversion efficiency of 5% considered in Section 9.2.5. The resulting mechanical energy >0.5 GJ could have lifted the 1000-Mg reactor shield block by >50 m! The actual disruption and lifting of the shield block, albeit quite visible, was not as spectacular as our simple energetics analysis would indicate.

9.3.4 Accident Consequences

As a result of the massive fire and huge releases of radioactive nuclides in the atmosphere, the immediate consequences of the accident include:

- Casualties—31 deaths, 500 hospitalized (with 203 persons receiving >100 rem).
- 135,000 persons were evacuated from within a 30-km radius.
- 24,000 persons evacuated from within a radius of 15 km from the plant received radiation exposures of 35 to 50 rem each.

The Soviet authority's initial estimate for a collective dose of 1.6×10^6 person-rems suggests 912 excess cancer deaths, according to the BEIR-VII recommendation [NAP05] of 5.7×10^{-4} additional cancer deaths per person-rem of radiation exposure above background, as discussed in Appendix A. This is to be compared with natural cancer deaths of 27,000 expected for the population of 135,000 persons evacuated. Thus, the 912 excess cancer deaths estimated may be on the order of 3% above the natural cancer death and may well be within statistical fluctuations in the estimates for natural cancer deaths.

One significant health effect of the radiation exposure of the population in the Chernobyl region of Ukraine is a sharp increase in the childhood thyroid cancer rates [Bal96,Bav02]. An increase in the childhood cancer incidences by roughly a factor of 15 is generally attributed to two factors. The increase by a factor of 4 is first attributed to an iodine-deficient diet among the children in the region, which resulted in a rapid uptake of radioactive iodine in the thyroid of the children. The second increase by a factor of 4 is generally attributed to the increased medical screening of the thyroid among the children, leading to the diagnosis of natural incidences which otherwise would have been left undetected.

Recognizing the importance of operating any water-cooled reactor in an undermoderated regime, soon after the Chernobyl accident the Soviets redesigned the fuel elements for the remaining plants of the same RBMK design to increase the fuel enrichment from 2.0 to 2.4 wt% ^{235}U and added additional control rods in the critical region of the reactor [Wil87,Afa93]. The bell-shaped curve of Fig. 8.22 indicates that these changes would have moved the operating point from the overmoderated toward the undermoderated regime, thereby eliminating one of the key reasons for the runaway reactivity accident at Chernobyl.

9.3.5 Comparison of the TMI and Chernobyl Accidents

The TMI-2 and Chernobyl accidents occurred and resulted in serious consequences due largely to operator errors and the prevailing mindset that no serious accidents were possible. In particular, the TMI-2 accident was due largely to poor operator training, poor maintenance practice, and the lack of communication that could have alerted TMI-2 operators about a similar incident at the Davis-Besse plant.

In comparison, the Chernobyl accident was the result of willful violations of safe operating procedures and lack of understanding of the effects of positive void coefficient of reactivity and minimum control rod requirements. The Chernobyl accident resulted in a massive release of radioactivity to the environment with serious health consequences, while the TMI-2 accident had minimal actual radiological consequences, causing nonetheless serious psychological trauma to the public at large.

9.4 FUKUSHIMA STATION ACCIDENT

9.4.1 Overview of the Accident–March 2011

On March 11, 2011, a massive earthquake of Richter scale 9.0 followed within an hour by a tsunami with waves of 10 to 14 m struck the Fukushima Daiichi (FD) nuclear complex operated by Tokyo Electric Power Company (TEPCO). The FD complex had Units 1, 2, and 3 in operation and Units 4, 5, and 6 in refueling outage. All six units are BWRs of General Electric design and started operation between 1971 and 1979 with power ratings of 439 to 1067 MWe. Units 1 through 5 feature the Mark I containment discussed in Section 8.1.2 and Unit 6 has an alternate design known as the Mark II containment. Within seconds of the earthquake, the reactor was shut down in all three operating units. The turbogenerators also tripped and main steam isolation valves closed. The earthquake, however, disrupted the electrical supply from the grid that resulted in a loss of offsite power for all six units. As designed, the emergency diesel generators (EDGs) started providing essential power for all safety systems including the residual heat removal (RHR) system discussed in Section 8.1.2. When the tsunami waves hit the FD complex, however, the EDGs were disabled, with the exception of one air-cooled EDG at Unit 6, which supplied AC power to Units 5 and 6. The EDGs were disabled due to the loss of cooling water, flooding of the switchgear, or flooding of the EDG rooms [Ins11]. This resulted in a station blackout (SBO) event for Units 1 through 4. Station batteries were also lost at Units 1, 2, and 4 due to the tsunami but some batteries survived initially at Unit 3. The tsunami also damaged the coolant intake structures and turbine and reactor buildings, resulting in a loss of ultimate heat sink (LUHS) for all six units. It took nine days to restore offsite power to the site.

Following the loss of the EDGs, core cooling for Units 2, 3, 5, and 6 was provided by the reactor core isolation cooling (RCIC) system, as discussed in Section 8.1.2 and illustrated in Fig. 8.10. All fuel elements in Unit 4 had been off-loaded into the used fuel pool (UFP) and the upper head had been removed from the reactor pressure vessel (RPV) during the outage before the earthquake and tsunami. For Unit 1, the emergency core cooling was provided by isolation condensers (ICs) of the type described for the ESBWR containment in Section 11.2.3.1. In the primary side of the IC, steam from the main steam line is condensed and the water is returned to the RPV via the recirculation line, while the secondary side may be cooled by the plant demineralizer or fire main water. The reactor operators manually turned on and off the IC for Unit 1 and the RCIC and HPCI systems for Units 2 and 3 to keep the system parameters within technical specifications. The emergency cooling systems degraded gradually due to increases in the temperature of the coolant water that circulated through the reactor vessel and suppression pool [Ins11,Lev11], as a consequence of the LUHS event. Following the loss of the IC, RCIC, and HPCI systems, Unit 1 suffered fuel uncovery and core damage as early as 3 hours into the accident, while Units 2 and 3 avoided core damage for 72 and 36 hours, respectively.

Despite the effort to inject seawater into the reactor core via fire hoses, fuel uncovery resulted in the exothermic reaction involving steam and the zirconium fuel cladding and generated hydrogen, as discussed in Section 9.2.2. The hydrogen generated appears [Jap12] to have leaked into the reactor building since the sealing materials for the flange gaskets, manholes, or cable penetrations of the containment are vulnerable to high temperature and pressure reached in severe accidents. The hydrogen and other volatile radionuclides leaked or vented through the relief valves were collected in the secondary containment structure and eventually reacted with oxygen in the containment air. This resulted in explosions that destroyed the roof of the secondary containment structure for Units 1 and 3 during the first week following the SBO event and released a significant amount of radionuclides to the surrounding atmosphere. Hydrogen gases generated in Unit 3, which shared facilities with Unit 4, appear to have migrated to the Unit 4 reactor building and resulted in an explosion.

The BWR containments were inerted with N_2 gas and equipped with hydrogen recombiners, which, however, could not be actuated due to the loss of station power. There were also provisions made for hardened venting of the containments and the effort to manually initiate venting through the suppression pool was successful for Units 1 and 3. It may be noted that the venting through the suppression pool is almost equivalent to filtered venting due to the high decontamination factors of pool scrubbing.

All UFPs were exposed to atmosphere, as the result of various accidents, and active cooling was lost. The water level, however, was never lowered below the top of fuel in any of the pools and none of the pools suffered any significant structural damage [Ins11], contrary to the concerns raised during the weeks following the accident. The common UFP storing 1500 fuel assemblies was designed such that the fuel elements would retain integrity for 30 days without active cooling.

The TEPCO employees made valiant efforts to control and contain the unanticipated severe accidents, which were well beyond the design basis for each plant, so that the radiation releases to the environment could be minimized. Until the offsite power was restored, they carried out most of the recovery activities in complete darkness and improvised to accomplish containment venting and other difficult tasks. One example is the heroic effort by 40 workers to manually connect a 0.1-m diameter, 200-m long power cable weighing one ton to a mobile generator across a debris-covered field, only to have the cable destroyed unfortunately by a hydrogen explosion soon after the connection was completed. Many workers volunteered to get access to high radiation areas with the full knowledge of the radiation exposure expected.

9.4.2 Radiological Consequences of the Accident

High dose rates during the accident include 10$\sim$30 rem/hr on site near Units 1 through 3 and 0.8 mrem/hr at 60 km from the site [Ins11]. The first air samples from the site boundary on March 22 and 23 showed ^{131}I concentrations that were equivalent to

80 mrem/hr inhaled [Ins11]. The evacuation zone for the residents initially was set at 2 km from the site boundary but eventually was increased to 20 km and involved 150,000 residents [Die12].

Initial radiation exposure estimates [Jap11] indicate that 7800 individuals received significant doses, with an average dose of 0.77 rem, while a subsequent report [Die12] suggests that 167 workers received doses >10 rem, with two control room operators [Ins11] exposed to combined external and internal doses of 68 and 64 rem, respectively. Workers wore breathing apparatus and cycled in and out of the high radiation areas to limit the radiation exposure to 10 rem [Ins11]. At one point, only 70 workers were allowed on site, with 650 individuals evacuated off site [Ins11]. It is estimated that approximately 900 PBq or 25 MCi of iodine equivalent radionuclides, representing a sum of the ^{131}I activity and 40 times the ^{137}Cs activity, were released from the Fukushima accident [Die12]. This may be compared with a total release of 250$\sim$380 MCi from the Chernobyl accident. The Fukushima accident was rated as a level 7 accident on the International Nuclear and Radiological Event Scale (INES) as was the case for the Chernobyl accident. Approximately 1800 km^2 of land surrounding the FD complex was contaminated to an exposure level of 500 mrem/yr [Die12]. A June 2011 study [Bue12] of radionuclides released into an area of 150,000 km^2 in the ocean and biota surrounding Japan indicates that the Cs activity increased by a factor of 10$\sim$1000 from the pre-existing level. The total gamma activity of ^{134}Cs, ^{137}Cs, and ^{110m}Ag in biota was, however, 1$\sim$54% of the total natural activity, mostly due to ^{40}K, and is considered to be well below the level harmful to marine animals and human consumers.

9.4.3 Implications and Follow-Up of the Fukushima Accident

Despite the valiant effort made by TEPCO to keep the reactor cores and UFPs replenished with water, the reactor cores were severely damaged with a significant meltdown of the fuel rods and contamination of the entire FD site, which will require decommissioning all six units. Although it is unclear if any of the RPVs in the FD units suffered breaches, the suppression chamber for Unit 2 and several valves and pipes for Units 1 through 4 appear to have been damaged. A significant amount of radionuclides leaked into the secondary containments, which eventually were damaged through hydrogen explosions at Units 1, 3, and 4, releasing radionuclides into the air and the sea. It took $1.0 billion and ten years to decommission the Three Mile Island plant that suffered partial fuel meltdown in 1979. Despite significant recovery effort under way, it may take 30$\sim$40 years to clean up and decommission the FD plant [IAE15]. All of the 50 NPPs in Japan were shut down following the 2011 accident and only two PWRs resumed operation in 2015.

The Fukushima accident was clearly the result of the largest earthquake ever experienced in Japan followed by equally historic tsunami waves that caused long-term SBO and LUHS events for all six units. The design basis tsunami of 5.7 m for the Fukushima site was, however, woefully inadequate, especially in light of a massive earthquake in AD 869 during Japan's Jogan era [Oni11], which produced a tsunami that reached nearly a mile inland just north of the FD site.

No acute health effects due to the radiation exposure from the FD accident have been reported and indeed the long-term health effects of radiation exposure due to the FD accident may be almost negligible. Thus, the FD accident could perhaps be evaluated in the bigger context of the natural disaster that caused possibly as many as 17,500 deaths [IAE15] and countless residents to become homeless. As of December 2015, however, many of the evacuees have yet to return to their ancestral homes and have suffered death due to causes indirectly linked to the Fukushima accident. The psychological trauma for the society may indeed be the most significant and lasting impact of the FD accident.

The National Diet Fukushima Accident Investigation Commission report [Die12] summarizes the first ever independent investigation conducted by the Japanese Parliament and is highly critical of TEPCO, regulatory agencies, and the central government in the licensing and regulations, preparations for accidents, and management of the March 2011 accident. The report states unambiguously that the FD accident was a man-made disaster that should have been avoided with proper preparations and organization. In this regard, it cites the availability of a 66-kV backup transmission line for Unit 1 after the tsunami attack, which, however, could not be connected due to a mismatched socket. The second INPO report [Ins12] also suggests that the IC system for Unit 1 apparently had never been tested for at least 25 years, since a shift supervisor indicated that he had never seen the system in operation during his 25-year career. These two observations succinctly illustrate the lack of safety culture in general and the lack of preparations for low-probability high-consequence nuclear emergencies at TEPCO. In addition, the operators tended to follow operation guidelines without questioning attitude and lack the ability to improvise in safety critical maneuvers [Ins12].

The amount of the radioactivity released in this long-term SBO event for the FD complex is a million times larger than that experienced in the 1979 TMI-2 accident discussed in Section 9.1. This is partly because of the hydrogen explosions that damaged the secondary containment structures. The higher radiation level experienced is also due to the direct-cycle steam generation structure inherent in BWR plants, where all coolant water and steam generated are radioactive in normal operation. The NUREG-1150 PRA study for five LWRs, discussed in Chapter 10, indicates that BWR plants are much more vulnerable to SBO events than PWR plants although the overall early fatality or latent cancer risk is lower for the two BWR plants than the three PWR plants studied.

As a diverse and flexible (FLEX) industry initiative following up on the Fukushima accident, sheltered facilities have been built at all U.S. nuclear plants where portable equipment including power sources, hardened vents, pumps, hoses, and communication devices are stored and shared among the plants. Actions taken by the Japanese government include the consolidation of the Nuclear and Industrial Safety Agency and the Nuclear Safety Commission into the Nuclear Regulation Authority under the Ministry of the Environment, which is expected to provide more independent oversight and regulations for NPPs. A comprehensive review on the FD accident is available in an an International Atomic Energy Agency report comprising five technical volumes and a summary by the director general [IAE15].

9.5 SALEM ANTICIPATED TRANSIENT WITHOUT SCRAM

9.5.1 Chronology and Cause of the Salem Incident

During February 1983, the failure of the automatic scram system occurred twice over a period of three days at the Salem Unit 1 PWR plant of Westinghouse design [Mar83]. In the first scram failure, the operator did not notice the failure, because the reactor was manually scrammed and the operator assumed a sensor problem. The operator noted the second failure and manually initiated a scram, which safely shut down the reactor. Investigation of the second scram failure revealed that the automatic scram system also failed to function three days earlier.

A schematic diagram [Boe83] of the reactor trip system in Fig. 9.9 indicates a double 2/4 logic for the actuation circuitry, which could have provided a high reliability for the automatic scram system. The instrument channels provide multiple sensors with signals generated by diverse events, including high neutron flux. The trip system has two DB-50 circuit breakers in series, either of which may be opened by one of the 2/4 actuation logics in an automatic scram. The DB-50 breaker is actually a rather complex system [Boe83], as illustrated in Fig. 9.10. It includes an undervoltage (UV) electromagnet that is normally energized during reactor operation so that the breaker is closed, working against a spring, and current flows to hold the control rods in place. An automatic scram signal would deenergize the UV magnets, thereby releasing the spring, opening the breaker, and dropping the control rods. In contrast, the manual scram switches would send signals both to deenergize the UV magnets and to energize the shunt magnets, which are normally deenergized. The shunt magnet when energized would open the scram breaker. The function of the manual scram switches is illustrated in Fig. 9.9, which also indicates the manual breaker controls that would energize the shunt magnets and the test circuitry for the scram breakers. It turned out that, in the DB-50 circuit breaker, the shunt system has a stronger magnet than the UV system. Thus, in both of the automatic scram failure events at the Salem Unit 1 plant, the UV systems failed to deenergize but the manual scram switches were able to energize the shunt magnets, which opened the scram breakers and shut down the reactor.

When the automatic trip signal reached the DB-50 scram breakers, the UV systems at both breakers failed to open because the breakers were stuck together due to poor maintenance over a long period of plant operation. The breakers were worn and bound by friction, possibly due to the use of inappropriate lubricant. Prior to the two scram failure events in February 1983, there had been several UV system failures at the Salem plant and failed UV coils had been often swapped between the two units at the plant. Apparently the plant personnel had not put sufficient emphasis on good maintenance practice and record keeping. One simple example of poor maintenance practice perhaps is that the knob for the manual scram switch came off when the reactor operator attempted to initiate a manual scram in the first scram failure event. Following a trip breaker malfunction at the H. B. Robinson plant in 1973, Westinghouse Electric Corporation issued a service letter that specified the proper inspection and service procedures for the scram system. The maintenance

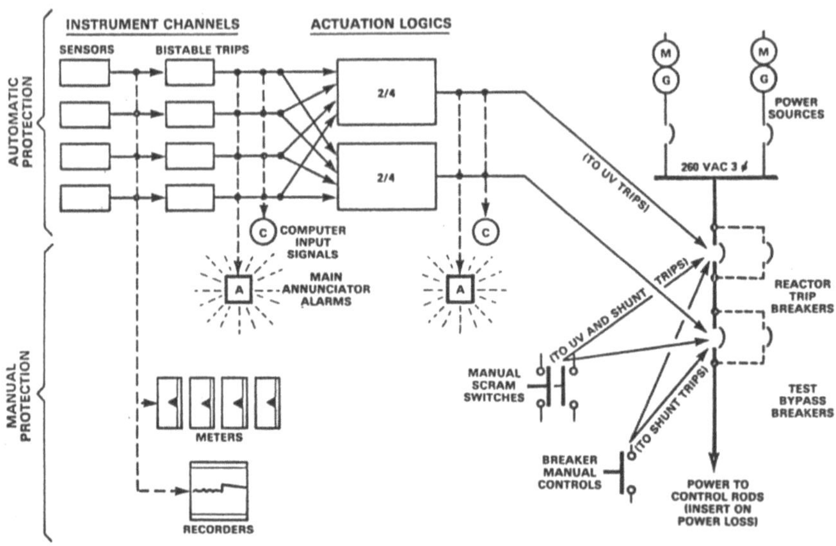

Figure 9.9 Typical PWR trip system similar to the Salem Unit 1 system. *Source:* [Boe83].

procedures had not been followed, because the Salem plant had not apparently received the Westinghouse service letter. The brief review of the DB-50 scram breaker design presented above also clearly indicates the deficiencies in the design, especially in regard to a small margin of error allowed in the UV system that the automatic scram relied on.

9.5.2 Implications and Follow-Up of the Salem ATWS Event

The reactor was shut down in both incidents within 30 seconds of the receipt of the automatic scram signal. According to NRC calculations [Mar83], a delay of 100 seconds could have led to a serious accident. As discussed in connection with the anticipated transient without scram (ATWS) rulemaking in Section 8.5.1, the scram failure probability was supposed to be as small as one in a million reactor-years of operation. The ability to shut down a nuclear reactor whenever required is perhaps the first and most important requirement in the defense-in-depth approach for nuclear reactor safety discussed in Sections 1.4 and 8.2.3. This point was indeed stressed in the 1983 *Science* magazine article [Mar83] that reported on the Salem incident, with the heading: *A failure of nuclear logic—the "impossible" happened twice in three days when a fail-safe device failed at a New Jersey plant.*

Of course, one could argue, as indeed some in the Salem management and the nuclear industry initially tried, that the February 1983 Salem event was not a truly ATWS event, because the manual scram was available as a backup to safely shut down the reactor. As we recall from our discussion of ATWS events in Section 8.5, the industry had resisted for a full decade the NRC's suggestion that the scram

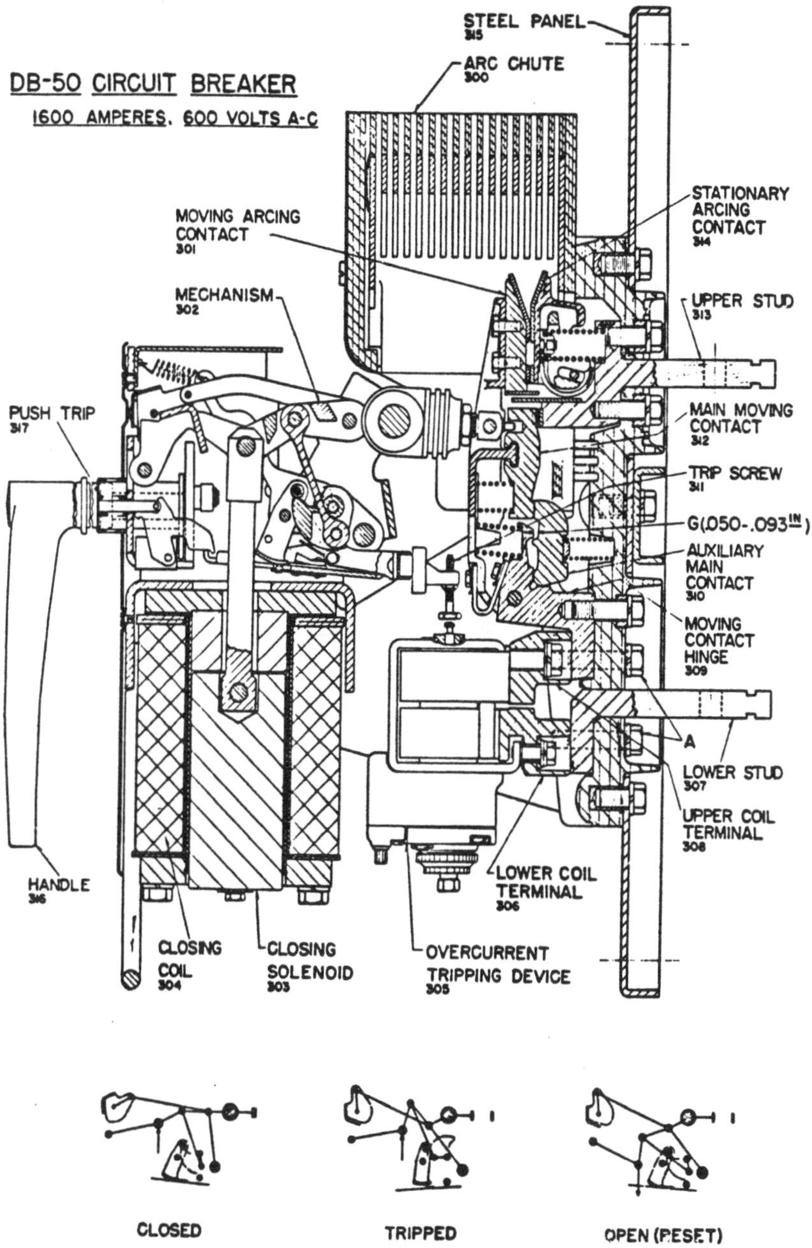

Figure 9.10 Cross-sectional view of type DB-50 circuit breaker. *Source:* [Boe83].

system reliability is not as high as the NPP owners and NSSS manufacturers would suggest. Indeed, barely four years after the tragic TMI-2 accident of 1979, there still persisted the mindset among the industry that another NPP accident was not likely to happen. Soon after the Salem incident, however, the nuclear industry agreed with the NRC that the distinction between failures of automatic and manual scram systems should not be made and steps be taken without delay to reduce the probabilities and consequences of ATWS events. A number of remedial requirements, primarily focusing on hardware improvements, were adopted in 10 CFR 50.62 [NRC84], as discussed in Section 8.5.2. One regulatory decision-making case where PRA applications were questioned is the ATWS issue. A recent review [Rau03] emphasizes that the uncertainty in the calculated values of the reactor scram system reliability requires maintaining defense-in-depth regarding ATWS, with reliable engineered systems, rather than relying heavily on PRA results.

The limitation of PRA applications in safety-significant decision processes may be illustrated in a nonnuclear field. The aerospace industry has to deal with possible catastrophic accidents, similar to the nuclear industry, and adopted PRA techniques [Sta02] for the evaluation of risk associated especially with the space shuttle program. In the aftermath of the tragic Columbia disaster of 2003, it was revealed [Cha03] that a simple model had been used to assess the integrity of protective tiles for the space shuttle vehicle, which played an important role [NAS03] in the disintegration of the vehicle with its seven crew members during the reentry into the earth atmosphere. The crater equation given in Fig. 9.11 was apparently used for predicting the depth of the gouge expected in the tiles due to the impact of pieces of insulating foam. Serious questions were raised about relying on the simple equation for assessing the integrity of a key component of the vehicle, no matter how sophisticated the rest of the risk calculation model could have been. This is a simple reminder that the nuclear industry must maintain due vigilance in its applications of PRA techniques in all future risk evaluations.

9.6 LASALLE TRANSIENT EVENT

9.6.1 LaSalle Nuclear-Coupled Density-Wave Oscillations

The ATWS rules adopted in 10 CFR 50.62 for BWR plants require that recirculation pumps should be tripped upon the indication of a scram failure. This particular requirement was to reduce the coolant flow in the core to natural circulation flow, which would increase the void fraction in the core and reduce the reactivity, thereby minimizing increases in the reactor vessel pressure and suppression pool temperature. It was recognized, however, during the deliberation on ATWS events that a BWR core operating with a low flow rate but at a substantially high power level may result in high-frequency oscillations in the reactor power coupled to oscillations in the core average coolant density. The coolant density oscillations take the form of oscillating waves [Mar86] represented by movements of the boiling boundary, i.e., the interface between the liquid and vapor phases of the two-phase mixture making up the

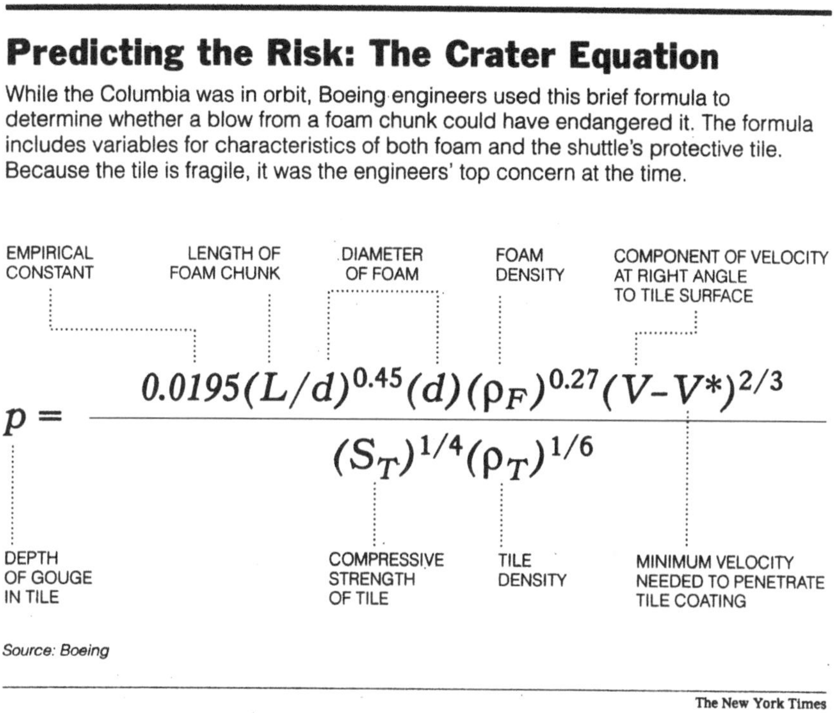

Figure 9.11 The crater equation for estimating the integrity of protective tiles for the Columbia space shuttle. *Source:* [Cha03].

coolant. Hence the phenomenon is called nuclear-coupled density-wave oscillations (NCDWOs) [War87b,Lee89]. The physics of the coupling and interplay between the core power and coolant flow and density feedback effects will be discussed in Section 9.6.2.

It has always been, indeed, an important operating guideline of BWR plants that the power escalations during plant startup maneuvers follow the power flow map illustrated in Fig. 9.12. General Electric Company, the NSSS manufacturer for all BWR plants, performed extensive analyses for the NCDWO phenomena as part of the NUREG-0460 report [NRC78]. The coupled nuclear-thermal-hydraulic phenomena are, however, quite nonlinear in nature and did not allow for simple definitive analyses. General Electric, however, fully recognized the potential for the nonlinear oscillations that could lead to unstable oscillations terminating in high-frequency limit cycle oscillations in reactor power.

A transient event [Rin88] involving large-amplitude oscillations in neutron flux and power occurred following a recirculation pump trip at the LaSalle Unit 2 plant in March 1988 and dramatically indicated the safety implications of the nuclear-coupled thermal-hydraulic instabilities. The reactor was operating at a steady-state condition

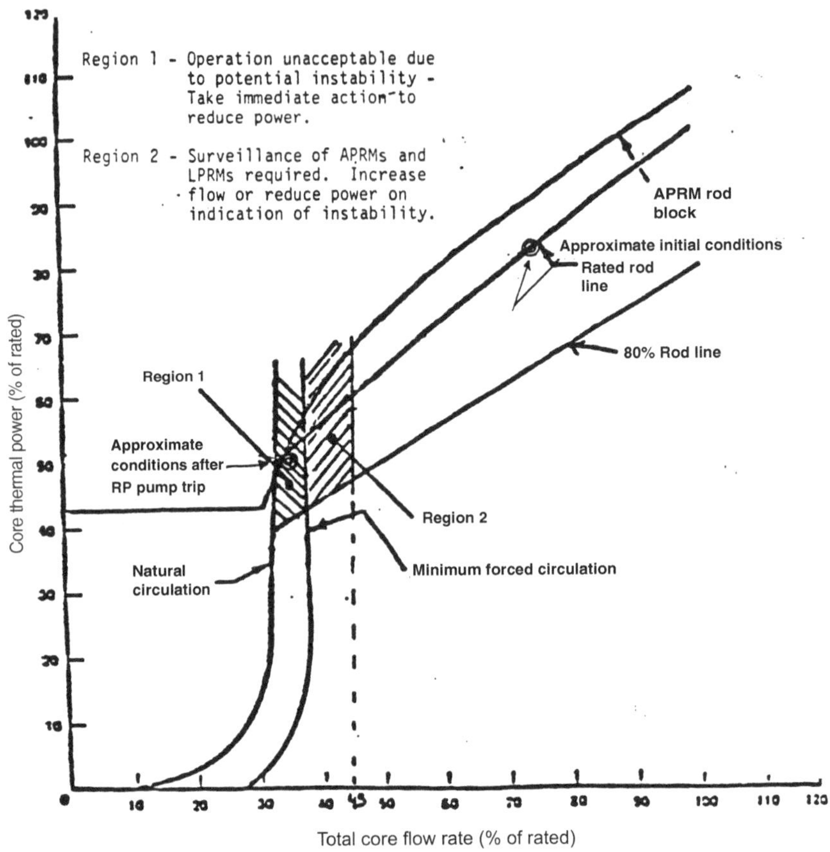

Figure 9.12 Power flow map for the LaSalle Unit 2.

Figure 9.12 Power flow map for the LaSalle Unit 2 indicating the conditions that initiated the March 1988 NCDWO event. *Source:* [Rin88].

with 84% of the rated power and 76% of the rated flow, when instrumentation personnel made a valving error. This resulted in a pressure pulse that tripped both recirculation pumps and rapidly reduced the power to 45% at natural circulation flow. The feedwater controller was unable to handle the large magnitude of the resulting steam flow and load reduction, causing the feedwater temperature to decrease by 45°F in 4 minutes and thereby inserting a positive reactivity. The plant went through hundreds of unstable oscillatory cycles before the reactor scrammed automatically on a high neutron flux level of 118%. The NCDWO event was initiated in region 1 of Fig. 9.12, which should have been avoided due to the potential for unstable oscillations.

Based on average power range monitor (APRM) indications representing the average of incore neutron detector signals, the operators assumed the core power was oscillating between 25 and 50% of rated power every 2 to 3 seconds. Subsequent analysis of the Startup Transient Recorder (Startrec) traces indicated APRM peak-to-peak oscillations ranged from 20 to 95% of rated power. The Startrec is a high-speed, multichannel recording system that is used in startup testing and other maneuvers when selected parameters exceed predetermined limits. The augmented inspection team (AIT) concluded [Rin88], based on an extrapolation of the traces to the time of the scram, that the oscillations actually were at least 100% peak to peak when the scram occurred, with a frequency of 0.45 Hz. The associated temperature oscillations were considerably smaller in magnitude, because a thermal time constant of 6 to 7 seconds connecting the neutron flux to clad temperature filtered out the neutron flux spikes. Hence the heat flux oscillations for the event were estimated to be $<10\%$ of neutron flux oscillations and no damage to the plant equipment occurred. Nonetheless, this event essentially simulated an ATWS event and raised concerns regarding the potential limit cycle oscillations in BWRs. The Startrec traces for 5.8 to 6.8 minutes after the pump trip, shown in the raw data with hand-written notes in Fig. 9.13, indicate APRM oscillations with an amplitude of up to 40% and period of 2.2 seconds.

The amplitude of the unstable oscillations rapidly increased to an asymptotic or limiting value, characteristic of nonlinear limit cycles. The reactor operators were unprepared for the high-frequency power oscillations with a frequency of $\sim$0.5 Hz and watched helplessly for several minutes. Finally, realizing the unusual nature of the oscillations, the operators initiated steps to shut down the reactor, when the reactor scrammed automatically.

The AIT report [Rin88] for the LaSalle incident revealed that the plant personnel received notification from General Electric about the possibilities of the NCDWOs but the operators had not been trained on the issue. This is because General Electric apparently advised the LaSalle and other BWR plants that the NCDWO possibility is not large enough for the operations personnel to be concerned about. This is perhaps a good example of the lack of proper vigilance in the industry that prompted the occurrence of this unfortunate incident. Although there was no damage to the plant, high-frequency uncontrollable power oscillations raised significant concerns about the stability and safety of BWR plants, especially among the residents in the vicinity of the LaSalle plant in Illinois.

9.6.2 Simple Model for Nuclear-Coupled Density-Wave Oscillations

Among various modes of potential instabilities in BWR plants, the density-wave oscillation (DWO) is most significant for the system stability. The DWOs occur as a result of regenerative interactions among the overall channel pressure drop, flow rate, and vapor generation rate. If the pressure drop across the boiling channel remains approximately constant during the oscillation, as is the case in BWR coolant channels with a large flow rate, a perturbation in inlet flow rate will cause a change in outlet flow rate in the opposite direction. In addition to this momentum feedback

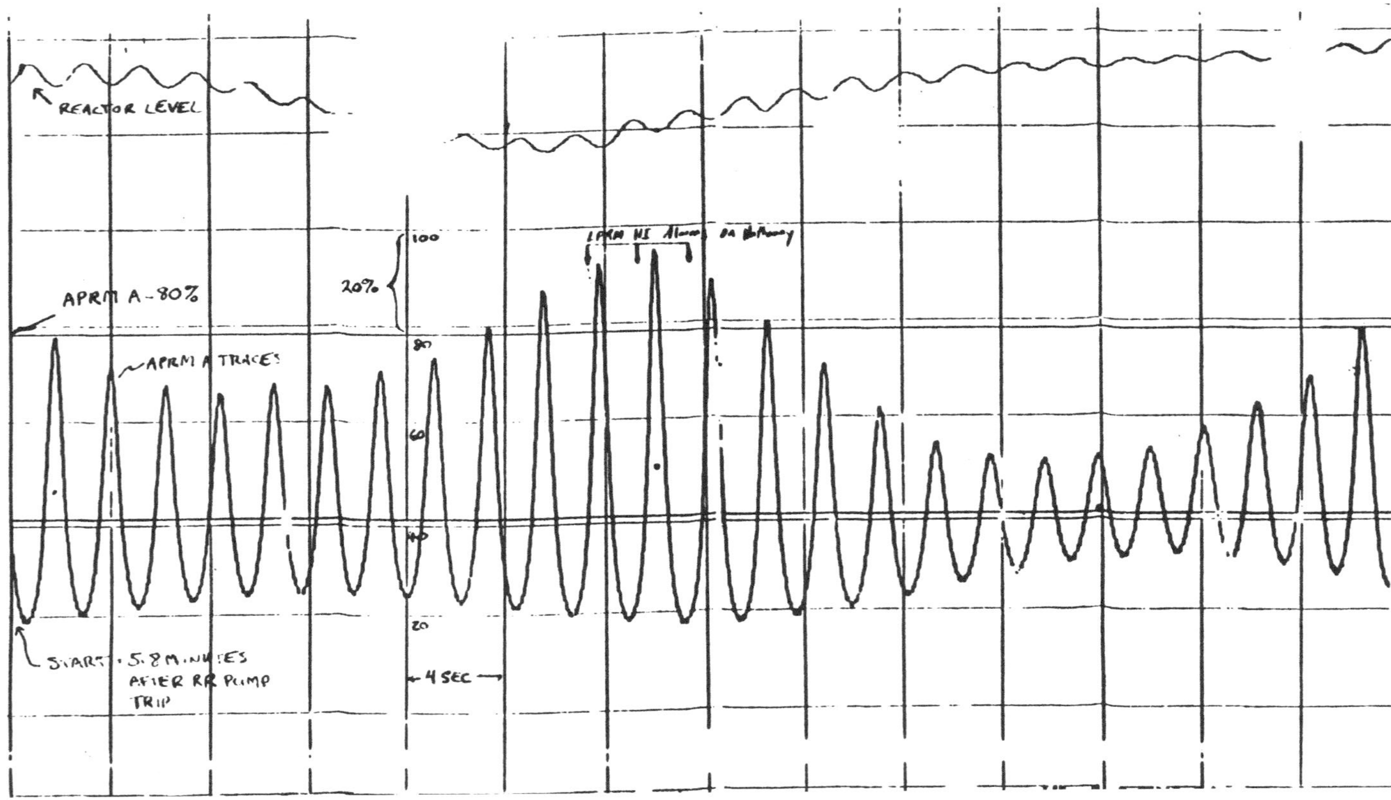

Figure 9.13 Startrec traces indicating power oscillations in the March 1988 LaSalle event. The abscissa indicates time with tick marks at 4.0-second intervals beginning at 5.8 minutes, while the ordinate indicates core power level with tick marks at 20% intervals. *Source:* [Rin88].

effect, the inlet flow rate perturbation will also result in perturbations in the vapor generation rate, boiling boundary, and void fraction in the two-phase region, with a time delay associated with the fluid motion. These delayed effects will eventually be propagated to the outlet, which will, in turn, cause a reversal of the initial inlet perturbation due to the momentum feedback effect and result in oscillations in the boiling boundary and density waves. In BWR channels, the DWO behavior will be reflected in neutronic power and flux oscillations due to the void reactivity feedback. Thus, the characteristics of NCDWOs, including the stability and oscillation period, are determined essentially by the DWO phenomena. In unstable NCDWOs, the oscillation amplitude grows as the heat generation exceeds the dissipation and the fuel temperature increases. With the resulting increase in the heat transferred to the coolant channel, the coolant density decreases, thereby reinforcing the ongoing DWOs. The limit cycle is reached when the heat generation equals the dissipation over each cycle.

A simple model [War87b] may be constructed by considering a single coolant channel representing the entire core coupled to a point kinetics model of the core, illustrated in Fig. 9.14. Since the DWO period is on the order of 2 seconds in BWR plants, we may use an infinite delayed approximation so that the production rate of delayed neutrons is fixed at the steady-state value. The core is represented by the change $T(t)$ in the average fuel temperature and total core heat flux $q(t)$. The coolant void feedback is represented in terms of the change $\rho(t)$ in the channel average coolant density, which is a function of the boundary $z(t)$ between the liquid water and vapor regions.

In terms of normalized reactor power $n(t)$, neutron generation time Λ, and delayed neutron fraction β, the core power is calculated from

$$\frac{\mathrm{d}n(t)}{\mathrm{d}t} = \frac{K(t) - \beta}{\Lambda} n(t) + \frac{\beta}{\Lambda}, \tag{9.12}$$

where reactivity $K(t)$ represents linear fuel and coolant density feedback effects via VCR α_V and fuel temperature coefficient of reactivity α_F,

$$K(t) = \alpha_V \rho(t) + \alpha_F T(t). \tag{9.13}$$

A lumped representation of the reactor fuel with heat capacity C_p and heat flux $q(t)$, given as a linear function of fuel temperature change $T(t)$ from the steady-state value, provides an energy balance for the core,

$$C_p \frac{\mathrm{d}T(t)}{\mathrm{d}t} = q(0)n(t) - q(t) = q(0)[n(t) - 1 - T(t)]. \tag{9.14}$$

The boiling channel is represented by a simple mass balance in terms of the inlet flow rate W_{in} and outlet flow rate W_{out}. An energy balance for the subcooled region with heat flux $q(t)$ provides a solution for the boiling boundary $z(t)$, while an average void fraction $\alpha(t)$ for the two-phase region is obtained from a separate energy balance. The channel average density change $\rho(t)$ is then determined from $z(t)$ and $\alpha(t)$, together with the appropriate phase densities, for the void feedback in Eq. (9.13). An

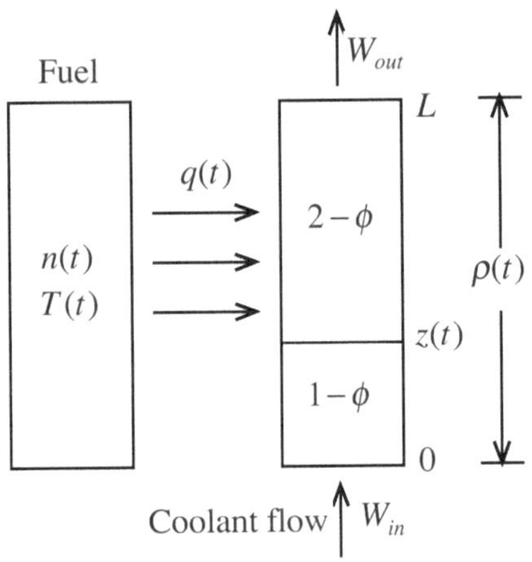

Figure 9.14 Simplified boiling channel of length L for NCDWO modeling.

approximate but physically meaningful momentum balance is established by setting the total pressure drop across the channel constant during the transient. This duly reflects the interaction between single- and two-phase pressure drops, which drives the DWOs.

The evolution of the DWOs that excite the oscillations in reactor power $n(t)$ and core average fuel temperature $T(t)$ is illustrated in Fig. 9.15. Small-amplitude sinusoidal oscillations in inlet and outlet flow rates and boiling boundary $z(t)$ grow and, coupled to the core through a strong void feedback, drive the coupled nuclear-thermal-hydraulic oscillations eventually to large-amplitude power pulses. Note that the flow oscillations are centered around the initial steady-state mass velocity G_0 and normalized value of the boiling boundary $z(t)$ in Fig. 9.15. In this numerical simulation of the NCDWO behavior based on Vermont Yankee tests [San83], the limit cycle is attained after ~200 cycles to yield 400% power oscillations and fuel temperature oscillations of ~17 K. The simple two-region coolant channel model underpredicts the fluid transit time and hence the oscillation period somewhat but accurately represents the physics of NCDWOs. The actual limit-cycle NCDWOs experienced at the LaSalle plant were, of course, a lot less severe than the simulation summarized in Fig. 9.15.

Note that the void feedback contributes much more significantly than the fuel temperature to the severity of NCDWOs. In fact, if the fuel temperature feedback is suppressed, Eq. (9.12) shows that the maximum power level may be determined by the largest value of the coolant density decrease $\rho_{\min} < 0$:

$$n_{\max} = \frac{\beta}{\beta - \alpha_V \rho_{\min}}. \tag{9.15}$$

Hence, the larger the magnitude of the negative VCR α_V is, the larger is the maximum power amplitude. This is then one of the few cases where the large negative values of VCR are detrimental to the safe operation of BWR plants.

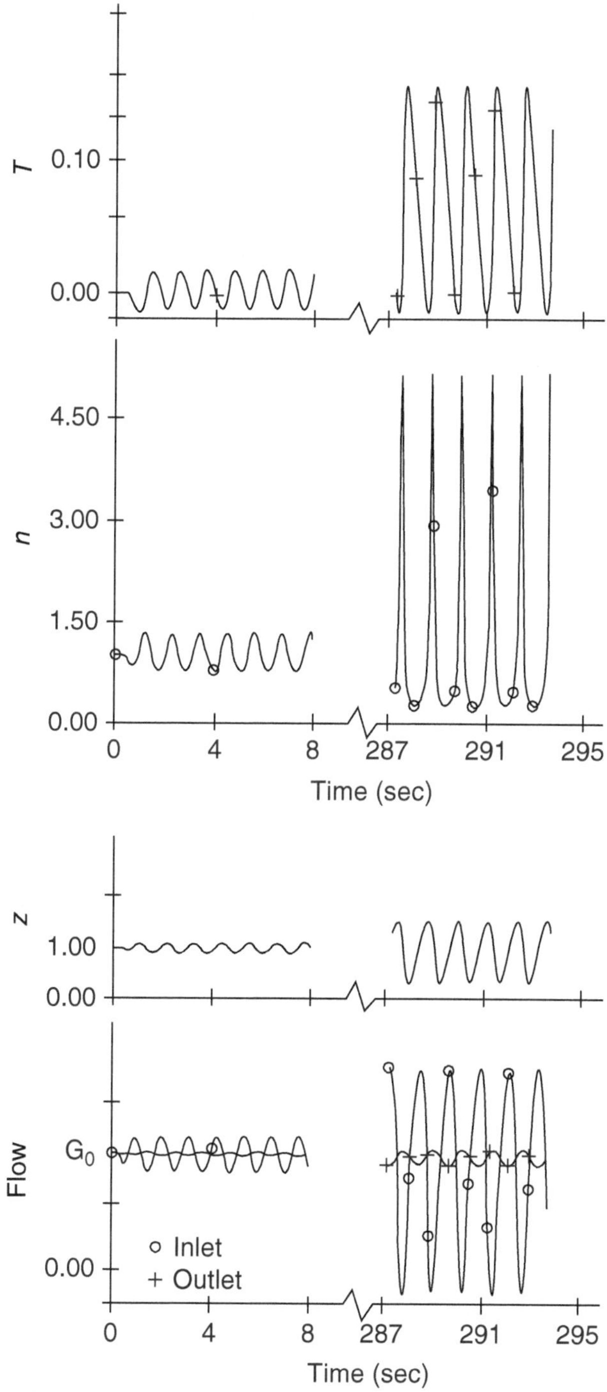

Figure 9.15 Evolution of NCDWOs to limit cycle oscillations of fuel temperature, core power, boiling boundary, and core flow (from top to bottom). *Source:* [War87b].

9.6.3 Implications and Follow-Up of the LaSalle Incident

The need to pay close attention to the relationship between the core power level and flow rate had been well recognized from the early days of BWR development, as exemplified by the power flow map of Fig. 9.12. The core power response to core flow rate changes is determined by the operating conditions so that the fractional decrease in power associated with a flow reduction is a decreasing function of the starting power level indicated by three different load lines, including the 100 and 80% lines, in the power flow map. The possibility of high-amplitude, high-frequency power and flow oscillations was not, however, fully communicated to BWR plant personnel. In fact, some time in the middle 1980s, before the LaSalle incident, the BWR Owners Group invited a licensed senior reactor operator (SRO) to attend a subcommittee meeting of the NRC's Advisory Committee on Reactor Safeguards (ACRS). After displaying his expert knowledge on various BWR operational issues, however, the SRO simply stated that he was unaware of the potential for any oscillatory events in BWRs.

Following the 1988 LaSalle NCDWO incident, the BWR Owners Group developed plans to monitor the onset of NCDWO events and avoid an entry into region 1, bound by the natural circulation line and minimum forced circulation line, in Fig. 9.12. The monitoring involves essentially determining, via a combination of time- and frequency-domain methods, the eigenvalue ξ characteristic of the oscillatory mode so that the core average neutron flux variations are represented by

$$\phi(t) = \phi_0 \exp(\xi t) = \phi_0 \exp(at) \cos \omega t. \tag{9.16}$$

Thus, determining the ratio of the flux amplitudes of two successive cycles, which is known as the *decay ratio*, the real part a of the eigenvalue ξ may be monitored on a continuous basis. At the same time, strict guidelines have been established to avoid the region of instabilities in the power flow map of Fig. 9.12. Monitoring the onset of unstable NCDWOs is performed through the APRM system.

The NCDWO mechanism discussed in Section 9.6.2 involves the oscillations in the total core power level and core average flow rate. There exists another mode of NCDWOs, however, that involves parallel-channel oscillations [Ony92,Zho05], where coolant channels in different parts of the core would oscillate out of phase from each other. The monitoring of the out-of-phase NCDWOs requires judicious uses of local power range monitors (LPRMs) comprising groups of incore neutron detectors. Several NCDWO events have happened in BWR plants in the United States and overseas since the 1988 LaSalle event, and General Electric Company maintains continuing support activities in this area for BWR plants. It is anticipated that the NCDWO issues will receive due attention in full development and deployment of the ESBWR design, which received the NRC design certification in 2014.

9.7 DAVIS-BESSE POTENTIAL LOCA EVENT

9.7.1 Background and Chronology of the Incident

Corrosion and cracking of steel structures used in NPPs have been a concern throughout the history of nuclear energy development in the world. In particular, cracking of control rod drive mechanism (CRDM) nozzles, made of alloy 600 carbon steel, in PWR pressure vessel upper heads was observed over the years. This prompted the replacement of vessel upper heads in a number of PWR plants in France. The NRC issued Generic Letters in 1988 and again in 1997 alerting NPP owners of the corrosion and cracking of vessel head penetrations.

In the spring of 2001, large circumferential cracking in several CRDM nozzles were found at the Oconee plant. In August 2001, the NRC issued a bulletin requesting that licensees of 12 PWRs, deemed highly susceptible to stress-corrosion cracking of CRDM nozzles, provide plans to conduct nozzle inspections before December 31, 2001. In September 2001, the Davis-Besse (DB) Nuclear Power Station requested that the vessel head inspection be delayed until after its planned March 31, 2002, outage. Through various negotiations with and deliberation among the NRC regulatory staff, a compromise was made to delay the vessel head inspection until February 16, 2002. On March 7, 2002, during the outage for maintenance and refueling, FirstEnergy Nuclear Operating Company, the owner of Davis-Besse, discovered a pineapple-sized cavity in the vessel head, leaving only a 5-mm-thick corrosion-resistant steel liner [GAO04]. The reactor vessel head is an 80-Mg cap with a diameter of 18 feet and thickness of 6 inches. The vessel head is an integral part of the reactor coolant pressure boundary that serves as a vital barrier to contain radionuclides in all PWRs. Arrangement of the CRDM nozzles in the vessel upper head is shown in Fig. 9.16, together with a diagram of the DB cavity in Fig. 9.17. A photograph of the cavity is included as Fig. 9.18.

Following the mid-February 2002 shutdown, FirstEnergy removed about 900 pounds of boric acid crystals and powder from the reactor vessel head and subsequently discovered that three central nozzles developed through-wall axial cracks and one nozzle had a circumferential crack. The inspection also revealed that the boric acid corrosion had penetrated the 6-inch-thick steel head, exposing the thin steel liner to withstand the primary system pressure of 2250 psia. The probability for circumferential cracking of 65 CRDM nozzles, out of a total of 69 nozzles, had been estimated based on visual inspections during the three previous refueling outages. The central four nozzles were, however, judged not to be susceptible to circumferential cracks and had not been included in the inspections, which turned out to be an erroneous decision.

Furthermore, the DB personnel had to periodically enter the containment building and remove large quantities of boric acid deposits from containment cooling fans and other equipment before the February 2002 outage. The FirstEnergy management, however, apparently gave little consideration to the possibility that wet boric acid leaking from the CRDM nozzles could induce corrosion of the vessel upper head. This clearly indicates a gross lack of the proper attention to safe operation of the plant.

9.7.2 NRC Decision to Grant DB Shutdown Delay

The NRC staff relied heavily on a Standardized Plant Analysis Risk (SPAR) study [Sat00] for Davis-Besse that Idaho National Engineering and Environmental Labora-

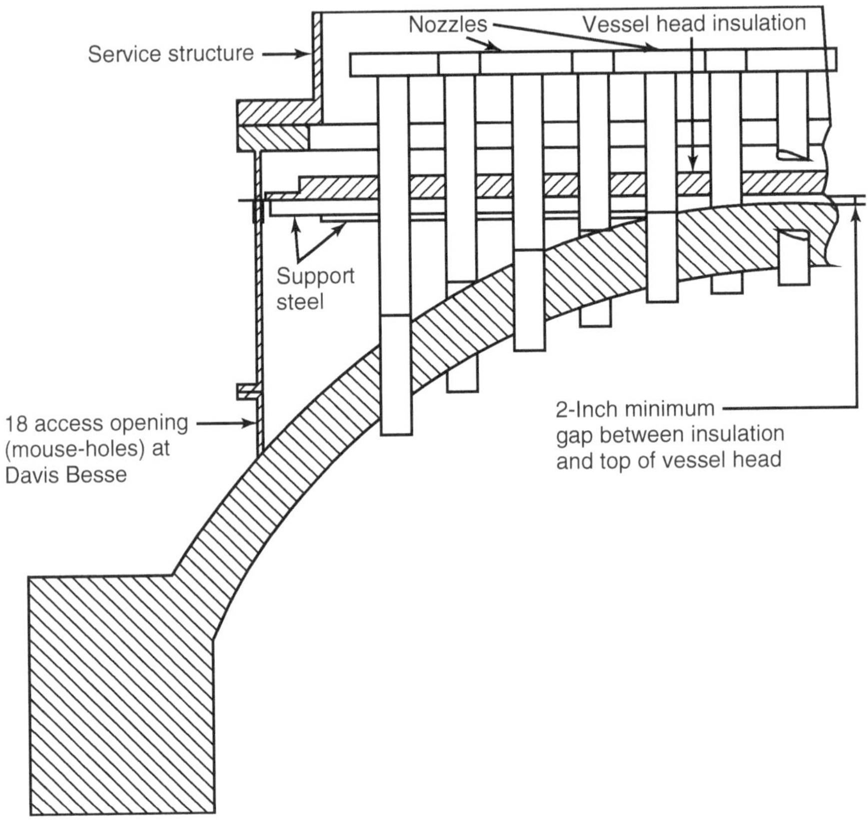

Figure 9.16 Arrangement of CRDM nozzles in the PWR vessel upper head. *Source:* [GAO04].

tory performed. The SAPHIRE code [NRC08], discussed in Chapter 7, provided the PRA tools and database for key system failure rates and human error probabilities in the SPAR study. The PRA study provided the core damage frequency (CDF) and large early release frequency (LERF) of radioactivity associated with the DB operation.

A medium-break (MB) LOCA, assumed to occur following the failure and ejection of CRDM nozzles at Davis-Besse, was analyzed in the SPAR report [Sat00] as one of 12 major internal events postulated to lead to core damage and radioactivity release. A baseline CDF of 1.0×10^{-7}/year for MBLOCA results from a generic value [Pol99] of the initiating event frequency of 4.0×10^{-5}/year for the MBLOCA combined with the failure probabilities of a number of engineered safety features, including the HPCI and LPCI systems. This results in an estimate of 2.5×10^{-3} for the conditional core damage probability (CCDP) for MBLOCA. The CCDP of 2.5×10^{-3} is almost entirely due to the failure of low-pressure recirculation pumps, which in turn depends

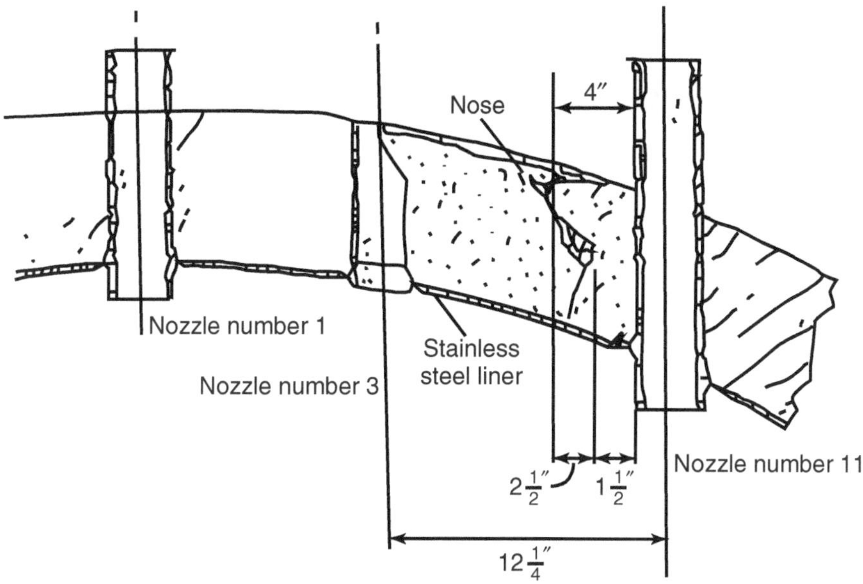

Figure 9.17 Diagram of the cavity in the Davis-Besse reactor vessel head. *Source:* [GAO04].

heavily on the ability of the operator to properly align and start the pumps. Based on human factor analysis, an estimate of 1.0×10^{-3} for the operator error is included in determining the CCDP of 2.5×10^{-3}. The baseline or point estimate CDF of 1.0×10^{-7}/year for MBLOCA contributes 0.5% toward the total baseline CDF of 2.5×10^{-5}/year, with uncertainties represented as CDF = {5th percentile, median, mean, 95th percentile | 6.3×10^{-6}, 1.6×10^{-5}, 5.1×10^{-5}, 9.6×10^{-5}} per year. The SPAR report for Davis-Besse provides only baseline CDF estimates for individual core damage events; hence no uncertainty estimates are available for the MBLOCA event. The mean overall CDF = 5.1×10^{-5}/year for Davis-Besse compares well with the those for internal initiating events for three PWR plants analyzed extensively as part of the NRC's severe accident evaluation project in NUREG-1150 [NRC90], discussed further in Chapter 10: Surry Unit 1, 4×10^{-5}/year; Sequoyah Unit 1, 6×10^{-5}/year; and Zion Unit 1, 6×10^{-5}/year. The CDF estimates for the four PWRs are, however, an order of magnitude larger than those for two BWRs analyzed in NUREG-1150: Peach Bottom Unit 2, 5×10^{-6}/year, and Grand Gulf Unit 1, 4×10^{-6}/year.

FirstEnergy also performed an event tree analysis, beginning with the CRDM leak frequency, accounting for crack growths and failures during subsequent operation and CRDM nozzle inspection failures, and culminating with a total CDF. The event tree analysis included CCDP $= 2.7\times10^{-3}$ for all 65 CRDM nozzles, again excluding four central nozzles that had been erroneously judged to be not susceptible to corrosion and cracking. The resulting total CDF summed over 65 nozzles was 6.97×10^{-6}/year.

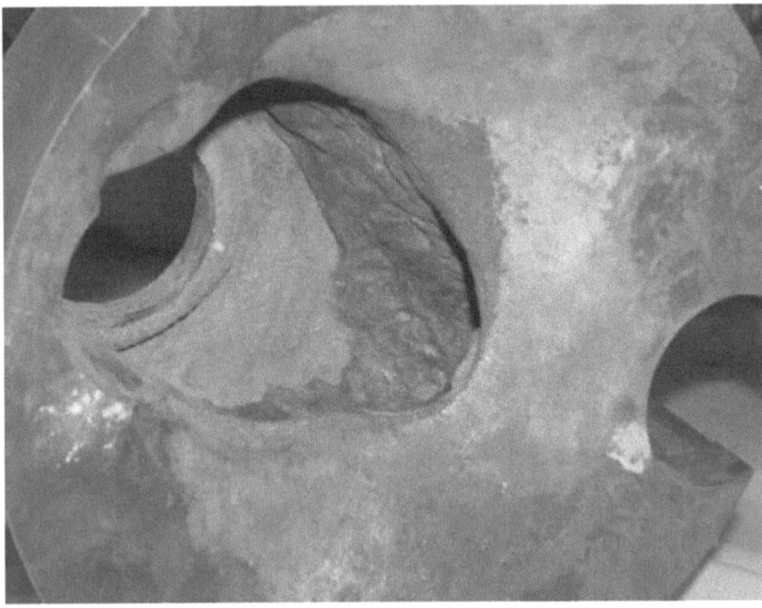

Figure 9.18 The cavity in the Davis-Besse vessel head after the March 7, 2002, discovery. *Source:* [GAO04].

Dividing by the CCDP yielded a value of the initiating event (IE) frequency of 2.58×10^{-3}/year representing an MBLOCA due to CRDM nozzle ejection. Using the IE frequency, one would then calculate an IE probability of 3.4×10^{-4} for continued DB operation for another 0.13 year, representing the period of shutdown delay between December 31, 2001, and February 16, 2002. Note here also that the DB estimation of CCDP = 2.7×10^{-3} agrees closely with the SPAR estimate of 2.5×10^{-3} discussed earlier.

In their final decision-making process, however, the NRC staff decided to use the IE frequency of 2.0×10^{-2}/year for MBLOCA, apparently citing engineering judgment and not allowing full credit to discover the nozzle cracking during inspections [GAO04]. Thus, combining the MBLOCA frequency and CCDP upon MBLOCA, the NRC estimated an incremental CDF due to CRDM nozzle failure at Davis-Besse:

$$\begin{aligned}\Delta\text{CDF} &= (\text{MBLOCA frequency } = 0.02/\text{year}) \times (\text{CCDP } = 0.0027/\text{year}) \\ &= 5.4 \times 10^{-5}/\text{year}. \qquad (9.17)\end{aligned}$$

Among various perspectives they considered, the NRC staff brought into discussion RG 1.174 [NRC02], which was introduced as a key guide for risk-informed regulations and licensing. In particular, they considered a chart copied in Fig. 9.19 that illustrates the criteria for accepting proposed licensing changes in terms of incremental CDF and incremental LERF. According to the RG 1.174 guidelines, any

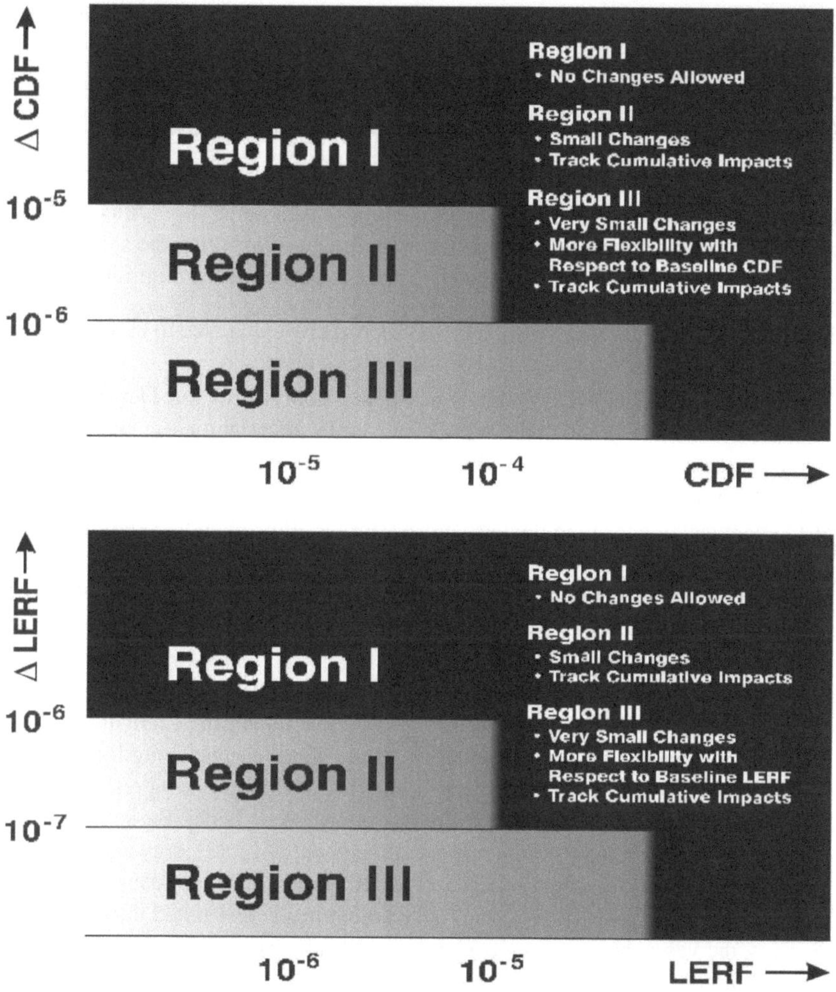

Figure 9.19 Numerical PRA guidelines for accepting proposed licensing changes. *Source:* [NRC02].

licensing changes resulting in either incremental change $\Delta\text{CDF} \geq 10^{-5}$/year or $\Delta\text{LERF} \geq 10^{-6}$/year would land in region I of the respective chart and should not be allowed. Thus, the NRC estimate of $\Delta\text{CDF} = 5.4 \times 10^{-5}$/year given in Eq. (9.17) would have rendered the decision to delay the shutdown unacceptable according to RG 1.174. Furthermore, a simple comparison of $\Delta\text{CDF} = 5.4 \times 10^{-5}$/year with the total mean $\text{CDF} = 5.1 \times 10^{-5}$/year for the baseline case, excluding the MBLOCA associated with the CRDM nozzle failure, would indicate that the additional MBLOCA would double the baseline CDF. Either consideration should have served as a warning that the shutdown delay request should not be granted. This is perhaps an example where PRA approaches have not served well in an important NPP safety and risk decision.

9.7.3 Causes for the Davis-Besse Incident and Follow-Up

A committee [Lee04] that reviewed the NRC oversight for Davis-Besse for the General Accounting Office, now the Government Accountability Office, provided a detailed analysis of the NRC decision-making process regarding the shutdown delay and other related issues. Among the key findings of the review committee are:

1. Risk due to CRDM nozzle failures was incorrectly calculated to be small, because the possibility for vessel corrosion was never considered.
2. The NRC did not perform any uncertainly analysis in the Davis-Besse PRA application and should have recognized large uncertainties in the incremental CDF estimated. The NRC relied too heavily on very uncertain PRA results to grant Davis-Besse a shutdown delay.
3. Coolant leakage through flanges and valves was allowed under Davis-Besse Technical Specifications, leading the plant personnel and NRC resident inspectors to treat boric acid deposits as routine events, and hence not risk significant. Note that in one outage alone, 15 five-gallon buckets of boric acid deposits were removed from the containment building.
4. Communication was sorely lacking between the NRC inspectors, region III, and headquarters. On the part of FirstEnergy Nuclear Operating Company and Davis-Besse Nuclear Power Station, a safety culture was clearly lacking for a number of years prior to the vessel head corrosion event of 2002. A congressional hearing was held in May 2004 to review the issues involved. Davis-Besse restarted only after a complete change in the upper management of FirstEnergy and replacement of the vessel upper head. In March 2010, however, significant indications of CRDM nozzle cracking with the replacement head were noted during an outage. Thus, the vessel head and CRDM nozzle corrosion still remains a concern and the handling of coolant leakage as part of technical specifications should be resolved in a more transparent manner.

References

[Afa93] A. A. Afanasieva, E. V. Burlakov, A. V. Krayushkin, and A. V. Kubarev, "The Characteristics of the RBMK Core," *Nucl. Technol.* **103**, 1 (1993).

[Ahe87] J. F. Ahearne, "Nuclear Power After Chernobyl," *Science* **236**, 673 (1987).

[Bal96] M. Balter, "Chernobyl: 10 Years After; Thyroid Cancer—Children Become the First Victims of Fallout," *Science* **272**, 357 (1996).

[Bav02] K. Baverstock and D. Williams, "Chernobyl: An Overlooked Aspect?" *Science* **299**, 44 (2002).

[Bla06] E. M. Blake, "Alternative Source Term Amendments: Limited Interest, Slow Adoption," *Nucl. News,* 20 (April 2006).

[Boe83] P. Boehnert, "Commission Meeting—Scram Failure Incident at Salem Unit 1—March 2, 1983," Memorandum to ACRS members, Advisory Committee on Reactor Safeguards (1983).
[Bro89] J. M. Broughton, P. Kuan, D. A. Petti, and E. L. Tolman, "A Scenario of the Three Mile Island Unit 2 Accident," *Nucl. Technol.* **87**, 34 (1989).
[Bue12] K.O. Buesseler, S. R. Jayne, N. S. Fisher, I. I. Rypina, H. Baumann, Z. Baumann, C. F. Breler, E. M.Douglass, J. George, A. M. Macdonald, H. Miyamoto, J. Nishikawa, S. M. Pike, and S. Yoshida, "Fukushima-derived radionuclides in the ocean and biota off Japan," *Proc. Nat. Acad. Sci.*, early ed., www.pnas.org/cgi/doi/10.1073/pnas.1120794109 (2012).
[Cha03] K. Chang,"Questions Raised on Equation NASA Used on Shuttle Peril," *The New York Times* (June 8, 2003).
[Col80] J. G. Collier and L. M. Davies, "The Accident at Three Mile Island," *Heat Transfer Eng.* **1**, 56 (1980).
[Cor83] M. L. Corradini and G. A. Moses, "A Dynamic Model for Fuel-Coolant Mixing," in *Proc. Int. Meeting LWR Severe Accident Evaluation*, Cambridge, MA (1983).
[Dav75] J. G. Davis, "Cable Fire at Browns Ferry Nuclear Plant," IE Bulletin No. 75-04A, U.S. Nuclear Regulatory Commission (1975).
[Die12] "The Official Report of the Fukushima Nuclear Accident Independent Investigation Commission," The National Diet of Japan (July 2012).
[EPR80] "Analysis of Three Mile Island Unit 2 Accident," NSAC-80 (NSAC-1 rev.), Electric Power Research Institute (1980).
[Fle88] C. D. Fletcher, R. Chambers, M. S. Bolander, and R. J. Dallman, "Simulation of the Chernobyl Accident," *Nucl. Eng. Design* **105**, 157 (1988).
[GAO04] "NUCLEAR REGULATION: NRC Needs to More Aggressively and Comprehensively Resolve Issues Related to the Davis-Besse Nuclear Power Plant's Shutdown," GAO-04-415, U.S. General Accounting Office (2004).
[Has02] F. E. Haskin, A. L. Camp, S. A. Hodge, and D. A. Powers, "Perspectives on Reactor Safety," NUREG/CR-6042, rev. 2, U.S. Nuclear Regulatory Commission (2002).
[IAE86] "INSAG [International Safety Advisory Group] Summary Report on the Post-Accident Review Meeting on the Chernobyl Accident," International Atomic Energy Agency (1986).
[IAE15] "The Fukushima Daiichi Accident," STI/PUB/1710, International Atomic Energy Agency (2015).
[Ins11] "Special Report on the Nuclear Accident at the Fukushima Daiichi Nuclear Power Station," INPO 11-005, Institute of Nuclear Power Operations (November 2011).
[Ins12] "Lessons Learned from the Nuclear Accident at the Fukushima Daiichi Nuclear Power Station," INPO 11-005 Addendum, Institute of Nuclear Power Operations (August 2012).
[Jap11] "Report of the Japanese Government to the IAEA Ministerial Conference on Nuclear Safety: The Accident at TEPCO's Fukushima Nuclear Power Stations," Nuclear Emergency Response Headquarters, Government of Japan (June 2011).

[Jap12] "Final Report of the Investigation Committee on the Accident at Fukushima Nuclear Power Stations of Tokyo Electric Power Company," Government of Japan (July 2012).

[Lee89] J. C. Lee and A. Onyemaechi, "Phase Plane Analysis of Nuclear-Coupled Density-Wave Oscillations," in *Noise and Nonlinear Phenomena in Nuclear Systems*, J. L. Munoz-Cobo and F. C. Difilippo, eds., 399, Plenum Press (1989).

[Lee04] J. C. Lee, T. H. Pigford, and G. S. Was, "Report of the Committee to Review the NRC's Oversight of the Davis-Besse Nuclear Power Station," Appendix II, GAO-04-415, U.S. General Accounting Office (2004).

[Lev11] S. Levy, "Fukushima Daiichi crisis. How would US units fare?," *Nucl. Eng. International*, 14 (Nov. 2011).

[Mar83] E. Marshall, "The Salem Case: A Failure of Nuclear Logic," *Science* **220**, 280 (1983).

[Mar86] J. March-Leuba, D. G. Cacuci, and R. B. Perez, "Nonlinear Dynamics and Stability of Boiling Water Reactors: Part 1—Qualitative Analysis," *Nucl. Sci. Eng.* **93**, 111 (1986).

[Moc07] H. Mochizuki, "Analysis of the Chernobyl Accident from 1:19:00 to the First Power Excursion," *Nucl. Eng. Design* **237**, 300 (2007).

[NAP05] *Health Risks from Exposure to Low Levels of Ionizing Radiation*, BEIR VII—Phase 2, Biological Effects of Ionizing Radiation Committee, National Academies Press (2005).

[NAS03] "The Columbia Accident Investigation Board Report," National Aeronautics and Space Administration (2003).

[NRC75] "Reactor Safety Study—An Assessment of Accident Risks in U.S. Commercial Nuclear Power Plants," WASH-1400, U.S. Nuclear Regulatory Commission (1975).

[NRC78] "Anticipated Transients Without Scram for Light Water Reactors," NUREG-0460, vols. 1–3, U.S. Nuclear Regulatory Commission (1978).

[NRC84] "Requirements for Reduction of Risk from Anticipated Transients Without Scram (ATWS) Events for Light-Water-Cooled Nuclear Power Plants," Title 10, *Code of Federal Regulations*, Part 50.62, U.S. Nuclear Regulatory Commission (1984).

[NRC90] "Severe Accident Risks: An Assessment for Five U.S. Nuclear Power Plants," NUREG-1150, U.S. Nuclear Regulatory Commission (1990).

[NRC00] "Alternative Radiological Source Terms for Evaluating Design Basis Accidents at Nuclear Power Reactors," RG 1.183, U.S. Nuclear Regulatory Commission (2000).

[NRC02] "An Approach for Using Probabilistic Risk Assessment in Risk-Informed Decisions on Plant-Specific Changes to the Licensing Basis," Regulatory Guide 1.174, U.S. Nuclear Regulatory Commission (2002).

[NRC04] "Reactor Site Criteria," Title 10, *Code of Federal Regulations*, Part 100, U.S. Nuclear Regulatory Commission (2004).

[NRC08] "Systems Analysis Program for Hands-On Integrated Reliability Evaluations (SAPHIRE), Technical Reference," NUREG/CR-6952, vol. 2, U. S. Nuclear Regulatory Commission (2008).

[Nuc86] "Chernobyl: The Soviet Report," *Nucl. News*, 59 (October 1986).

[Oni11] N. Onishi and J. Glanz, "Japanese Rules for Nuclear Plants Relied on Old Science," *The New York Times* (March 27, 2011).
[Ony92] A. C. Onyemaechi and J. C. Lee, "Parallel Channel Instability of Boiling Water Reactors," *Trans. Am. Nucl. Soc.* **66**, 606 (1992).
[Pet06] G. Petrangeli, *Nuclear Safety*, Elsevier (2006).
[Pol99] J. P. Poloski, et al., "Rates of Initiating Events at U.S. Nuclear Power Plants: 1987–1995," NUREG/CR-5750, U.S. Nuclear Regulatory Commission (1999).
[Rau03] W. S. Raughley and G. F. Lanik, "Regulatory Effectiveness of the Anticipated Transient Without Scram Rule," NUREG-1780, U.S. Nuclear Regulatory Commission (2003).
[Rin88] M. A. Ring, "Dual Recirculation Pump Trip Event of March 9, 1988, at the LaSalle County Station Unit 2," Augmented Inspection Team Report, U.S. Nuclear Regulatory Commission (1988).
[Riv81] J. B. Rivard et al., "Interim Technical Assessment of the MARCH Code," NUREG/CR-2285, U.S. Nuclear Regulatory Commission (1981).
[San83] S. A. Sandoz and S. F. Chen, "Vermont Yankee Stability Tests During Cycle 8," *Trans. Am. Nucl. Soc.* **45**, 754 (1983).
[Sat00] M. B. Sattison, J. K. Knudsen, L. M. Wolfram, and S. T. Beck, "Standardized Plant Analysis Risk Model for Davis-Besse," ASP PWR D, rev. 3i, Idaho National Engineering and Environmental Laboratory (2000).
[Sof95] L. Soffer et al., "Accident Source Terms for Light-Water Nuclear Power Plants," NUREG-1465, U.S. Nuclear Regulatory Commission (1995).
[Sta02] M. Stamatelatos, "Probabilistic Risk Assessment Procedures Guide for NASA Managers and Practitioners," Version 1.1, Office of Safety of Mission Assurance, National Aeronautics and Space Administration (2002).
[The81] T. G. Theofanous and M. Saito, "An Assessment of Class 9 (Core-Melt) Accidents for PWR Dry Containment Systems," *Nucl. Eng. Design* **66**, 301 (1981).
[The89] T. G. Theofanous, W. H. Amarasooriya, B. Najafi, M. A. Abolfadl, G. E. Lucas, and E. Rumble, "An Assessment of Steam-Explosion-Induced Containment Failure," NUREG/CR-5030, U.S. Nuclear Regulatory Commission (1989).
[War87a] M. E. Ward and J. C. Lee, "Singular Perturbation Analysis of Relaxation Oscillations in Reactor Systems," *Nucl. Sci. Eng.* **95**, 47 (1987).
[War87b] M. E. Ward and J. C. Lee, "Singular Perturbation Analysis of Limit Cycle Behavior in Nuclear-Coupled Density-Wave Oscillations," *Nucl. Sci. Eng.* **97**, 190 (1987).
[Wil87] R. Wilson, "A Visit to Chernobyl," *Science* **236**, 1636 (1987).
[Zho05] Q. Zhou and Rizwan-uddin, "In-Phase and Out-of-Phase Oscillations in BWRs: Impact of Azimuthal Asymmetry and Second Pair of Eigenvalues," *Nucl. Sci. Eng.* **151**, 95 (2005).

Exercises

9.1 As discussed in connection with the Salem-1 incident of Section 9.4, the reactor protection system for a typical PWR plant consists of two circuit breakers in series.

Each of two 2-out-4 actuation logic circuits delivers a trip signal to one of the circuit breakers. Consider a reactor protection system, where the first circuit breaker is connected to an undervoltage (UV) coil and shunt device, while the second breaker is to open up only through a shunt device. The unreliability of each of the actuation logic circuits and shunts is estimated to be 0.01 per demand, while the failure rate for the UV coil is 0.03 per demand. (a) Draw a fault tree for the top event, failure to scram, and determine the minimal cut sets for the tree and calculate the probability for the top event to occur. List any assumptions you make in your analysis. (b) If the fail-safe rates for the logic circuits and the UV coils are assumed the same as their respective fail-danger rates but the fail-safe rate for the shunts is negligibly small, obtain the spurious scram probability. Why is this assumption regarding the shunt fail-safe rate reasonable?

9.2 A component with a failure rate λ is monitored N times at regular intervals during the operating time T and repaired online if faults are detected. The time required for the test and repair is negligibly small compared with T. Starting from the fractional unavailability of Eq. (2.94), derive an expression for the fraction ξ of the operating time during which the component is in a failed state.

9.3 U.S. nuclear power plants are expected to undergo, on average, one anticipated transient a year that requires reactor scram. In addition, the scram system is tested another five times a year on average. No complete scram failure has occurred over 2500 reactor-years of nuclear power plant operation in the United States. Using the result of Exercise 9.2, determine, to a 90% confidence level, the probability of ATWS events per reactor-year. Justify any assumptions you make.

9.4 Obtain an alternate estimate for the total energy release using the Ergen-Weinberg model [War87a] for power excursion and compare it with Eq. (9.11).

9.5 The reactor protection system for a typical PWR, studied in connection with the Salem-1 incident in Section 9.4, includes two 2-out-of-4 bistable trip logic circuits, each of which sends the trip signal to a scram breaker. The unreliability of each of the bistable units making up the two 2/4 logic circuits is estimated to be 0.05 per demand. Determine the probability of each of the 2/4 trip circuits failing to provide the necessary trip signals to the scram breaker.

9.6 For the analysis of CRDM nozzle failures that resulted in severe corrosion of the pressure vessel head at the DB plant discussed in Section 9.7, it is suggested that the probability of crack initiation in the nozzles at t years of operation may be represented by a Weibull distribution of Eq. (2.124), with the probability density function given by

$$f(t) = \frac{\alpha}{\beta}\left(\frac{t}{\beta}\right)^{\alpha-1} \exp\left[-\left(\frac{t}{\beta}\right)^{\alpha}\right] \text{ with } \alpha = 1.5, \beta = 211 \text{ years.}$$

(a) Determine the probability for any of the nozzles to develop cracks during 20 years of operation and (b) given that the DB pressure vessel head has 69 nozzles, calculate the expected number of nozzles that have cracked during 20 years of operation.

9.7 Determine the cladding oxidation rate W of Eq. (9.7) with the cladding surface temperature increased from 1473 to 1673 K. Discuss the result.

CHAPTER 10

PRA STUDIES OF NUCLEAR POWER PLANTS

Building upon the discussions on general PRA methodology and representative PRA programs in Chapters 6 and 7, we present here specific approaches taken for two key PRA studies, WASH-1400 and NUREG-1150, performed for nuclear power plants over the past three decades. Our discussion on the PRA studies will also benefit from our review of key issues related to safety analyses of NPPs in Chapter 8. The review and analysis in Chapter 9 of several accidents and incidents that occurred at NPPs, including the TMI-2 accident, also provide the appropriate background for the study of WASH-1400 and NUREG-1150.

Section 10.1 begins with a discussion of key features of the general PRA methodology introduced in WASH-1400, highlighting the specific terminology and conventions adopted and summary results. An in-depth study of the more recent PRA project published as NUREG-1150 is then presented in Section 10.2, followed by a simplified analysis in the structure of the NUREG-1150 methodology in Section 10.3.

It should be mentioned, perhaps for the sake of completeness, that there was an earlier study on nuclear reactor safety published as WASH-740 [AEC57]. In this report prepared by Brookhaven National Laboratory, theoretical consequences of a core meltdown accident in a 500-MWt reactor with no containment building were considered, with pessimistic assumptions including the release of 50% of the entire fission product inventory to the atmosphere. With these assumptions, WASH-740 suggested the possibility of 3400 early fatalities, 43,000 acute injuries, and $7 billion (1957) damages resulting from a meltdown accident. No attempt was made, however, to determine credible estimates of the probability of such an accident.

10.1 WASH-1400 REACTOR SAFETY STUDY

The Reactor Safety Study (RSS) [NRC75], published as WASH-1400, was the first major study in the early 1970s that developed a PRA methodology for complex engineered systems in two nuclear power plants representing a fleet of light water reactors to be deployed in the United States. The study established the basic PRA framework combining fault and event tree structures and assembled the relevant reliability data for all major components of the NPPs. The basic methodology consists of quantifying the risk in terms of a probability per year,

$$\begin{aligned}\text{Risk} &= (\text{probability of accidents per year}) \\ &\quad \times(\text{consequences of accidents}),\end{aligned} \tag{10.1}$$

where the probability and consequence measures may further be broken down into two blocks of calculations each:

$$\begin{aligned}\text{Risk} &= \{(\text{probability of initiating events per year}) \\ &\quad \times (\text{probability of containment failure})\} \\ &\quad \times\{(\text{radionuclide release rate [Ci/accident]}) \\ &\quad \times (\text{probability of damage [deaths/Ci]})\}.\end{aligned} \tag{10.2}$$

Here the last three probabilities on the right-hand-side are conditional on the preceding probabilities in each initiating event. Each of the four blocks in Eq. (10.2) would involve detailed event tree analyses, supported by fault tree calculations for each event analyzed, as discussed further in Section 10.2. Note also that in Eq. (10.2) the number of early fatalities is suggested as an example of various consequence measures for illustrative purposes although other consequences such as latent-cancer fatalities, thyroid nodule incidence, and property damage also could be considered.

System dynamics and accident simulation models, together with atmospheric dispersion and dose calculation models for radionuclides released to the environment in postulated accidents, were developed in WASH-1400 to perform ET calculations indicated for each of the four blocks in Eq. (10.2). The study also introduced a set of terminology and conventions that facilitated a succinct and compact representation of various accident scenarios involving a large number of systems and components. The WASH-1400 terminology and conventions have been used, with minor modifications, in numerous PRA studies for individual NPPs and in the more recent PRA study covering five NPPs published as NUREG-1150 [NRC90]. The basic PRA methodology has been extensively used in other disciplines [Mor93], especially in

Table 10.1 Key to PWR Accident Sequence Symbols

Symbol	Meaning
A	Intermediate to large LOCA
B	Failure of electric power to ESFs
B′	Failure to recover either on-site or off-site electric power within about 1 to 3 hr following an initiating transient which is a loss of off-site ac power
C	Failure of the containment spray injection system
D	Failure of the emergency core cooling injection system
F	Failure of the containment spray recirculation system
G	Failure of the containment heat removal system
H	Failure of the emergency core cooling recirculation system
K	Failure of the reactor protection system
L	Failure of the secondary system steam relief valves and the auxiliary feedwater system
M	Failure of the secondary system steam relief valves and the power conversion system
Q	Failure of the primary system safety relief valves to reclose after opening
R	Massive rupture of the reactor vessel
S_1	A small LOCA with an equivalent diameter of about 5 to 15 cm
S_2	A small LOCA with an equivalent diameter of about 1 to 5 cm
T	Transient event
V	Failure of low-pressure injection system check valve
α	Containment rupture due to a reactor vessel steam explosion
β	Containment failure resulting from inadequate isolation of containment openings and penetrations
γ	Containment failure due to hydrogen burning
δ	Containment failure due to overpressure
ϵ	Containment vessel melt-through

Source: [NRC75].

risk analyses of space flight missions [Sta02] by the National Aeronautics and Space Administration.

Because a detailed review of a more recent application of the PRA methodology is presented in Section 10.2, this section focuses on a set of WASH-1400 terminology and conventions, reproduced for convenience as Tables 10.1 and 10.2. The symbols given in the English alphabet in Tables 10.1 and 10.2 are chosen to represent key initiating events and subsequent failures of engineered safety features that could result in the release of radionuclides to the environment in postulated NPP accident sequences. The events represented by Greek letters summarize different categories of containment failures.

The RSS analyzed the Surry Power Station Unit 1, a 778-MWe PWR of Westinghouse design, and the Peach Bottom Atomic Power Station Unit 2, a 1065-MWe BWR of General Electric design. The two plants were chosen to represent a fleet of 100 LWR plants in operation, under construction, and on order in the United States in the early 1970s. The PRA study of these two representative LWR plants resulted in grouping various accidents into 38 and 24 key accident sequences covering 9 and

Table 10.2 Key to BWR Accident Sequence Symbols

Symbol	Meaning
A	Rupture of reactor coolant boundary with an equivalent diameter of greater than 15 cm
B	Failure of electric power to ESFs
C	Failure of the reactor protection system
D	Failure of vapor suppression
E	Failure of emergency core cooling injection
F	Failure of emergency core cooling functionability
G	Failure of containment isolation to limit leakage to less than 100 vol % per day
H	Failure of core spray recirculation system
I	Failure of low-pressure recirculation system
J	Failure of high-pressure service water system
M	Failure of safety/relief valves to open
P	Failure of safety/relief valves to reclose after opening
Q	Failure of normal feedwater system to provide core make-up water
S_1	Small pipe break with an equivalent diameter of about 5 to 15 cm
S_2	Small pipe break with an equivalent diameter of about 1 to 5 cm
T	Transient event
U	Failure of high-pressure coolant injection system or reactor core isolation cooling system to provide core make-up water
V	Failure of low-pressure ECCS to provide core make-up water
W	Failure to remove residual core heat
α	Containment failure due to steam explosion in vessel
β	Containment failure due to steam explosion in containment
γ	Containment failure due to overpressure–release through reactor building
γ'	Containment failure due to overpressure–release direct to atmosphere
δ	Containment isolation failure in drywell
ϵ	Containment isolation failure in wetwell
ζ	Containment leakage greater than 2400 vol % per day
η	Reactor building isolation failure
θ	Standby gas treatment system failure

Source: [NRC75].

5 radioactive release categories of decreasing severity for the Surry PWR and Peach Bottom BWR plants, respectively.

For the PWR plant, release category 1 represents a steam explosion (α sequence) resulting from a large volume of molten corium dropped into a pool of coolant in the bottom of the reactor vessel. Such an explosion could release sufficient thermal and kinetic energy to rupture the reactor vessel and the containment building, thereby releasing a large amount of radionuclides into the environment. Category 2 releases also are associated with core melt, followed by the rupture of the containment due to hydrogen burning (γ sequence) and steam overpressurization (δ sequence). Categories 3, 4, and 5 represent core melt but with varying success levels of the radioactivity removal systems. Categories 6 and 7 involve accidents where the molten core melts through the bottom of the containment or basemat (ε sequence), without and with radioactivity removal systems operating, respectively. The above-ground structure of the containment remains intact in categories 6 and 7. Finally, in categories 8 and 9, the core does not undergo melting but releases a limited amount of radionuclides into the containment, without and with the containment isolation, respectively. Likewise, for the BWR plant, release categories 1, 2, and 3 involve containment rupture with varying levels of radionuclide releases. Categories 4 and 5 for the BWR plant do not involve containment rupture, with different amounts of release.

To illustrate the main results of WASH-1400, we reproduce in Table 10.3 a summary of risk-significant accident sequences for the Surry PWR plant [NRC75]. We note that the accident sequences in Table 10.3 represent five sequences of accidents from Table 10.1:

1. A: large-break LOCAs involving double-ended guillotine break of the primary coolant pipe
2. S_1 and S_2: small-break LOCAs of different break sizes
3. R: reactor pressure vessel rupture
4. V: interfacing system LOCAs
5. T: transient events

Sequence V originally was introduced to primarily represent the failure of the LPCI check valve but later generalized to represent various system failures that could release radionuclides without any containment failure or leakage. The interfacing system LOCA events also are known as the containment bypass events, as discussed further in Section 10.2.

Table 10.3 indicates that some of the sequences resulting in release categories 1 through 5 and involving core melt and containment failures make only small contributions to the overall release frequency. On the other hand, categories 8 and 9 incur large frequencies but would entail minimal radionuclide releases. Thus, we concentrate on category 7, which involves core melt and containment basemat melt-through but with the radioactivity removal system functioning properly. This release category results in a median frequency of 4×10^{-5}/year. A review of the events making up the category reveals that two SBLOCAs and two transient events account for 3×10^{-5}/year or 75% of the total frequency for this release category:

1. Sequences S_2D-ε and S_2H-ε, SBLOCA followed by the failure of the ECCS and the ECCS recirculation, respectively, both resulting eventually in the containment basemat melt-through, account for a frequency of nearly 2×10^{-5}/year.
2. Sequence TML-ε, anticipated transient accompanied by a turbine trip and the failure of the auxiliary feedwater system and steam generator relief valves, and sequence TKQ-ε, anticipated transient accompanied by a scram failure and the failure of the PORV to reclose, both resulting in the containment basemat melt-through, account for a frequency of nearly 1×10^{-5}/year.

It is noteworthy that the TMI-2 accident discussed in Section 9.1 involved a succession of events resembling sequence S_2D-ε, which according to WASH-1400

Table 10.3 PWR Dominant Accident Sequences with Frequencies in yr^{-1}

Accident	Release Categories								
	Core Melt							No Core Melt	
	1	2	3	4	5	6	7	8	9
Large LOCA dominant accident sequences A	AB-α 1×10^{-11}	AB-γ 1×10^{-10}	AD-α 2×10^{-8}	ACD-β 1×10^{-11}	AD-β 4×10^{-9}	AB-ϵ 1×10^{-9}	AD-ϵ 2×10^{-6}	A-β 2×10^{-7}	A 1×10^{-4}
	AF-α 1×10^{-10}	AB-δ 4×10^{-11}	AH-α 1×10^{-8}		AH-β 3×10^{-9}	AHF-ϵ 1×10^{-10}	AH-ϵ 1×10^{-6}		
	ACD-α 5×10^{-11}	AHF-γ 2×10^{-11}	AF-δ 1×10^{-8}			ADF-ϵ 2×10^{-10}			
	AG-α 9×10^{-11}		AG-δ 9×10^{-9}						
A frequencies	2×10^{-9}	1×10^{-8}	1×10^{-7}	1×10^{-8}	4×10^{-8}	3×10^{-7}	3×10^{-6}	1×10^{-5}	1×10^{-4}
Small LOCA dominant accident sequences S_1	S_1B-α 3×10^{-11}	S_1B-γ 4×10^{-10}	S_1D-α 3×10^{-8}	S_1CD-β 1×10^{-11}	S_1H-β 5×10^{-9}	S_1DF-ϵ 3×10^{-10}	S_1D-ϵ 3×10^{-6}	S_1-β 6×10^{-7}	S_1 3×10^{-4}
	S_1CD-α 7×10^{-11}	S_1B-δ 1×10^{-10}	S_1H-α 3×10^{-8}		S_1D-β 6×10^{-9}	S_1B-ϵ 2×10^{-9}	S_1H-ϵ 3×10^{-6}		
	S_1F-α 3×10^{-10}	S_1HF-γ 6×10^{-11}	S_1F-δ 3×10^{-8}			S_1HF-ϵ 4×10^{-10}			
	S_1G-α 3×10^{-10}		S_1G-δ 3×10^{-8}						
S_1 frequencies	3×10^{-9}	2×10^{-8}	2×10^{-7}	3×10^{-8}	8×10^{-8}	6×10^{-7}	6×10^{-6}	3×10^{-5}	3×10^{-4}
Small LOCA dominant accident sequences S_2	S_2B-α 1×10^{-10}	S_2B-γ 1×10^{-9}	S_2D-α 9×10^{-8}	S_2DG-β 1×10^{-12}	S_2D-β 2×10^{-8}	S_2B-ϵ 8×10^{-9}	S_2D-ϵ 9×10^{-6}		
	S_2F-α 1×10^{-9}	S_2HF-γ 2×10^{-10}	S_2H-α 6×10^{-8}		S_2H-β 1×10^{-8}	S_2CD-ϵ 2×10^{-8}	S_2H-ϵ 6×10^{-6}		
	S_2CD-α 2×10^{-10}	S_2B-δ 4×10^{-10}	S_2F-δ 1×10^{-7}			S_2HF-ϵ 1×10^{-9}			

	S_2G-α 9×10^{-10} S_2C-α 2×10^{-8}		S_2C-δ 2×10^{-6} S_2G-δ 9×10^{-8}						
S_2 frequencies	1×10^{-7}	3×10^{-7}	3×10^{-6}	3×10^{-7}	3×10^{-7}	2×10^{-6}	2×10^{-5}		
Reactor vessel rupture dominant accident sequences R	RC-α 2×10^{-12}	RC-γ 3×10^{-11} RF-δ 1×10^{-11} RC-δ 1×10^{-12}	R-α 1×10^{-9}				R-ϵ 1×10^{-7}		
R frequencies	2×10^{-11}	1×10^{-10}	1×10^{-9}	2×10^{-10}	1×10^{-9}	1×10^{-8}	1×10^{-7}		
Interfacing systems LOCA (check valve) V		V 4×10^{-6}							
V frequencies	4×10^{-7}	4×10^{-6}	4×10^{-7}	4×10^{-8}					
Transient event dominant accident sequences T	TMLB′-α 3×10^{-8}	TMLB′-γ 7×10^{-7} TMLB′-δ 2×10^{-6}	TML-α 6×10^{-8} TKQ-α 3×10^{-8} TKMQ-α 1×10^{-8}		TML-β 3×10^{-10} TKQ-β 3×10^{-10}	TMLB′-ϵ 6×10^{-7}	TML-ϵ 6×10^{-6} TKQ-ϵ 3×10^{-6} TKMQ-ϵ 1×10^{-6}		
T frequencies	3×10^{-7}	3×10^{-6}	4×10^{-7}	7×10^{-8}	2×10^{-7}	2×10^{-6}	1×10^{-5}		
Summation of all accident sequences per release category									
Median (50% value)	9×10^{-7}	8×10^{-6}	4×10^{-6}	5×10^{-7}	7×10^{-7}	6×10^{-6}	4×10^{-5}	4×10^{-5}	4×10^{-4}
Lower bound (5% value)	9×10^{-8}	8×10^{-7}	6×10^{-7}	9×10^{-8}	2×10^{-7}	2×10^{-6}	1×10^{-5}	4×10^{-6}	4×10^{-5}
Upper bound (95% value)	9×10^{-6}	8×10^{-5}	4×10^{-5}	5×10^{-6}	4×10^{-6}	2×10^{-5}	2×10^{-4}	4×10^{-4}	4×10^{-3}

Source: [NRC75].

could occur with a frequency of 9×10^{-6}/year, or nearly 25% of the entire category 7 frequency. Of course, the TMI-2 accident did not progress to the melt-through of the containment basemat but resulted nonetheless in the release of radionuclides through a different leakage path involving containment bypass.

A summary of the WASH-1400 calculations of the NPP risks is presented in Fig. 10.1. The plots for the relevant failure probability density functions (PDFs) $f(x)$ are converted into complementary cumulative distribution functions (CCDFs):

$$\begin{aligned} \text{Probability}(X > x) &= \int_x^{\infty} f(x')\mathrm{d}x' = 1 - F(x) \\ &= \text{probability of event } X \text{ greater than } x, \quad (10.3) \end{aligned}$$

where the random event X represents the number of early fatalities in Fig. 10.1. Note that $F(x)$ is the usual cumulative distribution function introduced in Section 2.7. Thus, the plots in Fig. 10.1 should be interpreted as representing the frequency per reactor year of the number of early fatalities greater than the number indicated on the abscissa. For example, for the Surry plant, the frequency of accidents resulting in more than 100 fatalities is 2×10^{-7}/year. The average of the frequencies for a PWR and a BWR core, scaled to 100 NPPs, is comparable to that associated with meteorites resulting in fatalities and much lower than those associated with other natural events, e.g., earthquakes and tornadoes, as shown in Fig. 10.2.

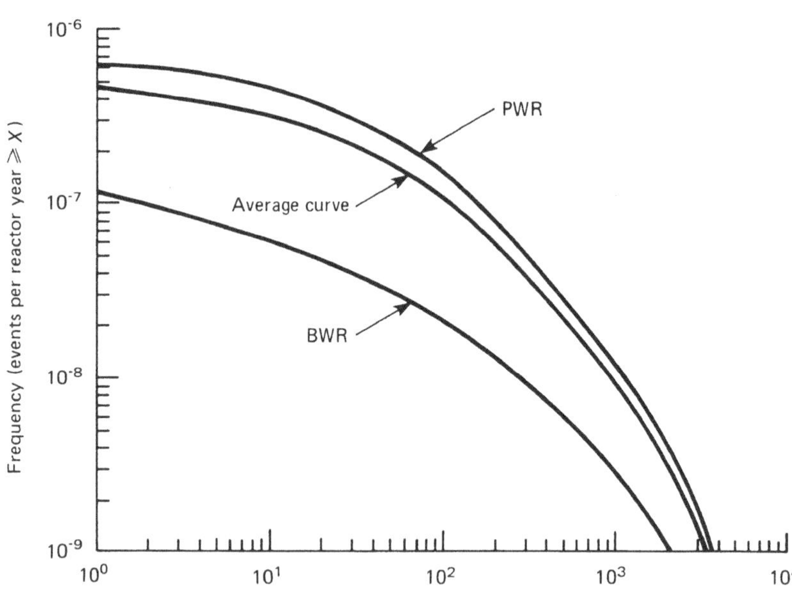

Figure 10.1 WASH-1400 estimates of risk of early fatalities for LWRs. *Source:* [NRC75].

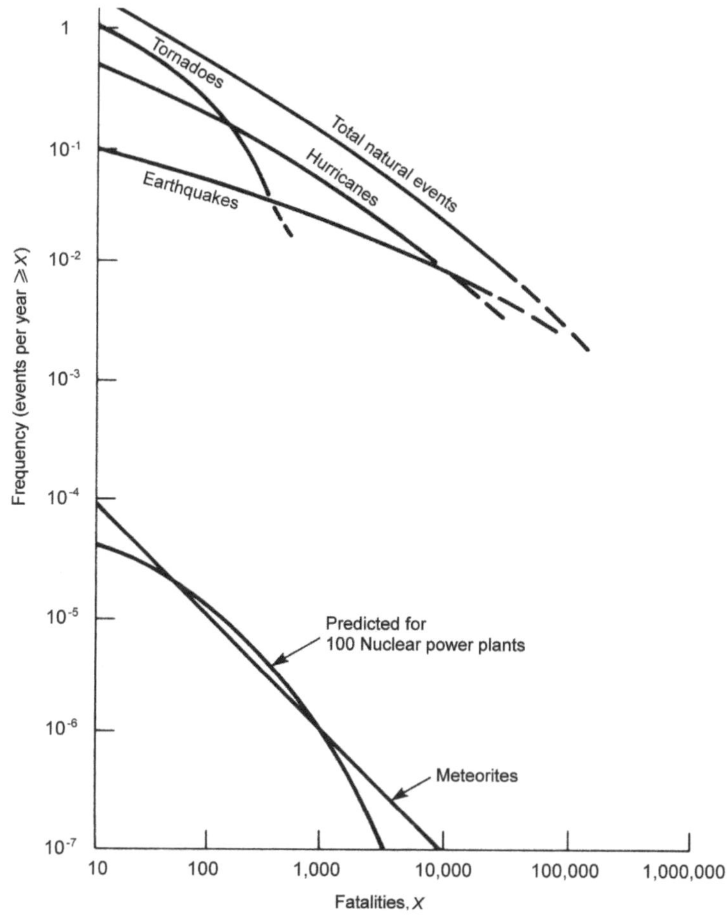

Figure 10.2 Comparison of WASH-1400 estimate of NPP risk with those for natural events involving fatalities. *Source:* [NRC75].

10.2 ASSESSMENT OF SEVERE ACCIDENT RISKS: NUREG-1150

10.2.1 Background and Scope of the NUREG-1150 Study

As discussed in Section 10.1, WASH-1400 is the first full-scale application of PRA techniques to the risk assessment of nuclear power plants. In addition to providing the first realistic estimates of risks associated with the operation of NPPs, this landmark study developed the basic approaches that should be taken to combine the fault trees and event trees in complex systems while accounting for uncertainties associated with the likelihood of various failures and events in terms of PDFs. It also introduced the basic terminology and nomenclature that could succinctly represent key system components, with various operational status and events leading to malfunctions or failures cascading through multiple layers of ESFs built into NPPs. The ground-breaking nature of the PRA study in WASH-1400 is well recognized by PRA practitioners in other fields, including aerospace engineering [Sta02].

In the 1970s, when the study was conducted, there were significant concerns about the possibility and impact of large-break LOCAs. Thus, one important conclusion of WASH-1400 indicating that a large contribution to the NPP risk might be due to small-break LOCAs, rather than large-break LOCAs, was viewed with a degree of surprise, if not skepticism, among the nuclear engineering community. The TMI-2 accident, however, unfortunately occurred within a few years following the release of WASH-1400 and essentially established the validity of this particular conclusion of the PRA study.

As discussed in Section 9.1, the severity of the TMI-2 accident and the extent of the meltdown of the core, due to the misguided, deliberate attempts by the reactor operators to secure the functioning ECCS pumps—thereby starving the core of the much needed coolant—were certainly unanticipated. In fact, there was general belief among the people involved closely with the post-accident recovery operation and analysis of the accident that the extent of the core damage was relatively small, certainly much less than the actual meltdown that covered nearly two-thirds of the core. Despite the severity of the accident and the extent of the core meltdown, however, the amount of radionuclides released to the environment and the resulting radiation exposure to the public was relatively small, perhaps largely because the partially molten, heavily damaged core did not drop to the bottom of the pressure vessel.

In parallel with the effort to remove the damaged core and the eventual decommissioning of the plant, endeavor was made in the 1980s to understand what contributed to the nearly complete containment of the radionuclides released from the heavily damaged core. Recall that the design basis accidents (DBAs) evaluated as Class 8 accidents for LWRs did not, and still do not, include core meltdown accidents. As discussed at the beginning of this section, it was largely the intent of the WASH-1400 study to understand and reflect the consequences of large-break LOCAs as the limiting DBA in LWRs. This resulted in extensive studies in the efficacy of the ECCS and significant investments to enhance the ECCS in all LWRs in the United States and other countries as well. Thus, the TMI-2 accident presented a new challenge to the nuclear engineering community to explicitly consider beyond-DBAs in assessing and managing the risk of NPP operations. This triggered the U.S. NRC to embark on a study to evaluate and analyze the consequences of core meltdown accidents, which became known as either Class 9 accidents or more commonly as severe accidents, as discussed in Section 8.4. Another motivation behind the study was to explore the possibility of relaxing the prevailing radionuclide source term requirements for LWR safety analyses, as discussed in Section 8.6.1.

The NRC-sponsored study eventually resulted in the release of a report known as NUREG-1150 [NRC90] and the term *severe accident* has taken on a specific reference to beyond-DBAs resulting in core damage and meltdown. The report, titled *Severe Accident Risks: An Assessment for Five U.S. Nuclear Power Plants*, entailed the reevaluation of the risk of operating two LWRs, Surry Unit 1 and Peach

Bottom Unit 2, studied in WASH-1400 and three other plants representing a fleet of approximately 100 NPPs operating in the United States in the 1980s. Key attributes of the five LWRs included in the NUREG-1150 study are:

1. Surry Unit 1, 788-MWe Westinghouse three-loop PWR with a subatmospheric containment
2. Zion Unit 1, 1100-MWe Westinghouse four-loop PWR with a large dry containment
3. Sequoyah Unit 1, 1148-MWe Westinghouse four-loop PWR, with an ice-condenser containment
4. Peach Bottom Unit 2, 1065-MWe General Electric BWR-4 with a Mark I containment
5. Grand Gulf Unit 1, 1250-MWe General Electric BWR-6 with a Mark III containment

The five LWRs were selected to represent different core sizes as well as diverse containment structures that would have significantly different consequences in the case of core meltdown accidents. The Surry Unit 1 and Peach Bottom Unit 2 NPPs would also provide fiducial points so that the results of the new PRA study could be checked for consistency to the extent possible. The study turned out to be more complex and involved than initially envisioned, resulting in three different sets of massive volumes of reports, the first draft published in 1987 and the second draft in 1989, followed by the release of the final report in 1990 [NRC90]. The study involved several national laboratories, with lead efforts by the Sandia National Laboratories, and is estimated to have cost the NRC approximately $100 million over a period of nearly a decade. The first two drafts of the report received significant reviews by a number of organizations, including the NRC's Advisory Committee on Reactor Safeguards. These reviews prompted revisions and enhancements to the study and the report in various ways, including new accident sequences and phenomena that required detailed studies, the use of expert opinions, and calculations and presentations of uncertainties associated with various stages of risk estimates.

10.2.2 Overview of NUREG-1150 Methodology

The NUREG-1150 study followed the basic PRA structure of multiplying the frequency of accidents and consequences resulting from the accidents for the determination of the risk of operating each of the five NPPs. As discussed in a qualitative manner in Eqs. (10.1) and (10.2), the accident frequency is calculated by enumerating core damage events leading to containment failures, while the release of radionuclides to the environment in containment failure events and the eventual fatality and property damage estimates are reflected in determining the overall public risk associated with operating each plant. Our discussion of the methodology as well as sample results is borrowed heavily from volume 1 of the main NUREG-1150 report.

Thus, we begin with a flow chart summarizing the overall methodology in Fig. 10.3. Similar to the basic building blocks of a PRA model illustrated in Eq. (10.2),

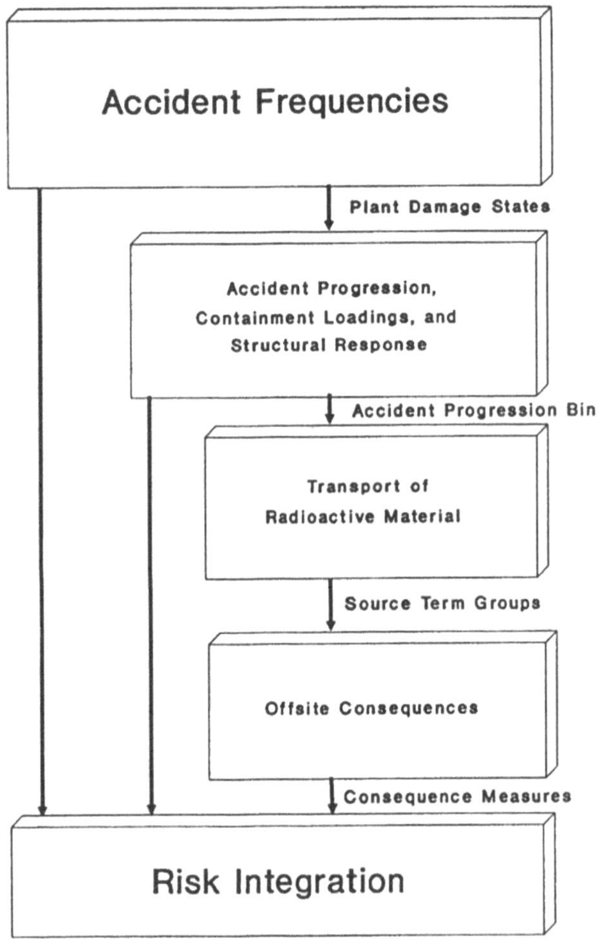

Figure 10.3 Elements of NUREG-1150 risk analysis process. *Source:* [NRC90, Vol. 1].

the NUREG-1150 risk analysis starts with the enumeration of initiating accidents or events under the *accident frequency analysis* task, which follows through *initiating events* (IEs) that could result in *plant damage states* (PDSs). Each PDS is represented in terms of the mode of core damage and the associated probability. This is followed by *accident progression* (AP) analyses under the second task of Fig. 10.3, which generate *AP bins* according to containment failure mode, together with its probability. The AP analyses essentially complete the first building block of Eq. (10.1) for the PRA methodology yielding the *overall accident frequency* resulting in the release of radionuclides.

The second PRA building block representing the *consequence analysis* begins with *radionuclide transport* analyses performed under the third task summarized in Fig. 10.3. The results from this task are represented as radionuclide *source term groups*, according to the release category and fraction. Offsite consequence analyses are then performed under the fourth task of Fig. 10.3 to determine *consequence measures* entailing the number of fatalities and other damage estimates, which completes the second PRA building block. With the evaluation of consequence measures completed, the fifth and final task of the NUREG-1150 methodology involves the *risk integration*, which determines the overall risk to the public of operating each of the five NPPs according to diverse measures. The risk integration task does not merely present the quantitative risk measures but also reflects on the IEs and AP bins that contribute to significant risk measures. This last step is indicated by the lines connecting the first and second boxes to the last box in Fig. 10.3.

We illustrate more explicitly in Fig. 10.4 the five tasks and the intermediate results of the NUREG-1150 methodology summarized in Fig. 10.3. In Fig. 10.4, each square box indicates the calculation of a conditional probability or consequence, while rounded boxes indicate the event probabilities or consequences calculated in the PRA methodology. The analysis begins with estimates of the probabilities of IEs lumped into $P(I)$ representing as many as 30 to 50 events that could potentially result in PDSs of up to 20 different characteristics. Accident and transient calculations are performed beginning with the IEs and following through every possible mode of core damage events to determine the probabilities $P(D)$ of PDSs. The set of calculations evaluating possible core damage events is summarized in the first square box, *accident frequency analysis*, and is represented in terms of conditional probability $P(D|I)$. Given the PDS probabilities $P(D)$, the *accident progression analysis,* represented in the second box as $P(A|D)$, is performed to yield AP bins with the associated probabilities $P(A)$. Note that the grouping of containment failure modes depends heavily on the NPP analyzed and results in as many as 2000 distinct AP bins, indicated by O(1000) bins in Fig. 10.4. Given the delineation of AP bins, the ET analysis continues with the *radionuclide transport analysis* $P(S|A)$ in box 3, which yields radionuclide sources $P(S)$ binned into 30 to 60 groups. The fourth block of ET calculations performs the *offsite consequence analysis* $P(C|S)$ providing eight different consequence measures $P(C)$. The *risk integration* task makes a combined use of $P(C)$ and other probabilities and measures $P(I)$, $P(D)$, $P(A)$, and $P(S)$ in the final risk estimates. The matrix manipulations indicating the progression of ET calculations are shown at the bottom of Fig. 10.4 and will be explained further as we discuss the detailed ET steps involved in the next few sections.

10.2.3 Accident Frequency Analysis

The task of enumerating initiating transients that could result in core damage events begins with the familiarization of the plant systems and components. This involves the review of the final safety analysis report (FSAR), piping and instrumentation diagrams (P&IDs), technical specifications, operating procedures, and maintenance records for the plant, followed by a site visit by the PRA study team. Involvement of

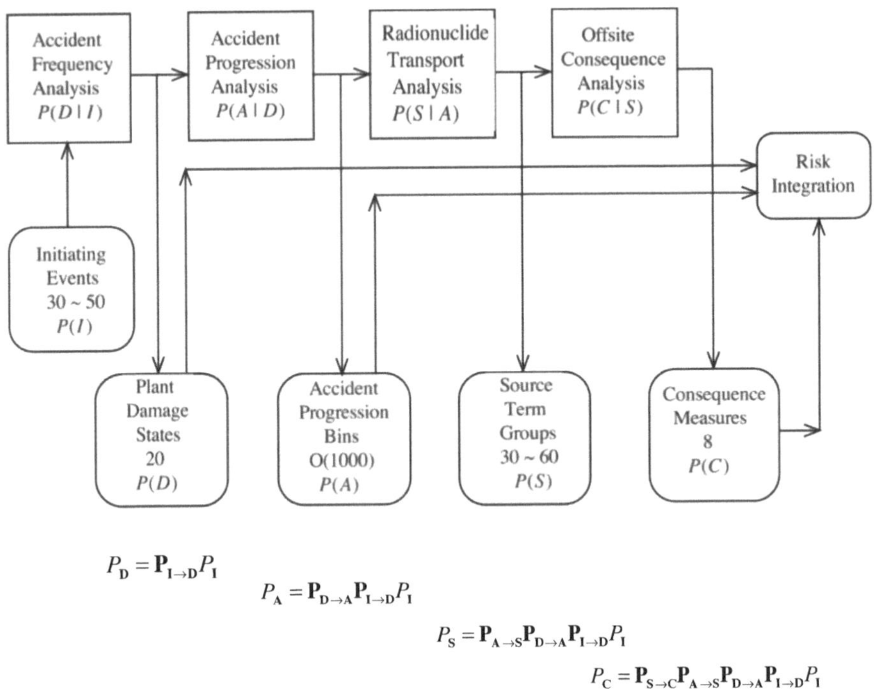

Figure 10.4 Event tree structure of the NUREG-1150 PRA study.

the plant personnel is maintained throughout the study and is an important step, as is the case for any PRA study. Based on the plant data gathered, ETs are constructed and the IEs of similar characteristics leading to core damage are grouped together to obtain

$$P(I_i) = \text{ frequency of initiating event of group } i, \; i = 1, \ldots, n_I, \qquad (10.4)$$

with the total number of IE groups n_I typically chosen between 30 and 60. In this *accident sequence or frequency ET analysis,* sometimes also known as *front-end ET analysis*, important contributors to the failure of key systems and components, e.g., pumps and valves, are evaluated using standard FT methods. A generic database of equipment and human failure rates and IE frequencies, reflecting commercial NPP operating experience, was used for all five NPPs studied, but due consideration was given for plant-specific data whenever necessary.

The accident frequency ET analysis, represented as $P(D|I)$ in Fig. 10.4, follows through IEs resulting in core damage and groups the accident sequences into PDSs, according to the operability of systems, e.g., the availability of containment spray systems and key system parameters, e.g., reactor coolant system pressure. Thus, in terms of

$$P(D_i|I_j) = \text{conditional probability for PDS occurring in the } i\text{th group, given IE in the } j\text{th group, } i = 1, \ldots, n_D, \tag{10.5}$$

the probability that the accidents progress to PDS group *i* is represented in terms of the *PDS frequency*

$$P(D_i) = \sum_{j=1}^{n_I} P(D_i|I_j)P(I_j), \; i = 1, \ldots, n_D, \tag{10.6}$$

with the number of PDS groups typically chosen as $n_D = 20$. Equation (10.6) may be written compactly in matrix form as

$$P_{\mathbf{D}} = \mathbf{P}_{\mathbf{I} \to \mathbf{D}} P_{\mathbf{I}} \tag{10.7}$$

as illustrated at the bottom of Fig. 10.4.

The PDS frequencies $P(D_i)$ for $n_D = 20$ are condensed into a few groups for the purpose of a summary report for each plant analyzed. The mean core damage frequencies (CDFs) for five summary PDS groups are illustrated in Fig. 10.5 for the Surry plant for a total mean CDF of 4×10^{-5}/reactor-year due to internal IEs. In addition to the five PDS groups that contribute to the CDF, additional effort was made in the NUREG-1150 study to account for the plant risk due to external events, including earthquakes, floods, and fires, as summarized as column headings in Fig. 10.6. Although fires that could result in core damage may properly be considered internal events, they were classified as external events, partly because these events were analyzed only for select plants in the NUREG-1150 study.

The summary PDS groups illustrated in Figs. 10.5 and 10.6 represent sets of internal events that have been considered routinely in NPP safety analyses:

1. Loss of all AC electric power to the plant or *loss of station power* (LOSP), more usually known as the *station blackout* (SBO) event
2. Anticipated transient without scram events representing transients with failure of the reactor protection system, i.e., failure of the reactor shutdown system
3. Other transient events that are not accompanied by scram failures
4. LOCAs that occur within the containment building, due to failures in the reactor coolant system, including pipe ruptures and failures of RCS seals and relief valves
5. LOCAs that bypass the containment building or the *interfacing system LOCAs*

It is worth remembering that the *containment bypass* event represented in PDS group 5 is designated as an extended definition of accident sequence V in the WASH-1400 nomenclature presented in Table 10.1. Although sequence V was originally introduced to represent the failure of check valves of the ECCS low-pressure injection system in the WASH-1400 study, this class of events would have effects similar to those associated with interfacing system LOCAs, e.g., failure of the main steam isolation valves (MSIVs) to close in BWRs and steam generator tube rupture (SGTR)

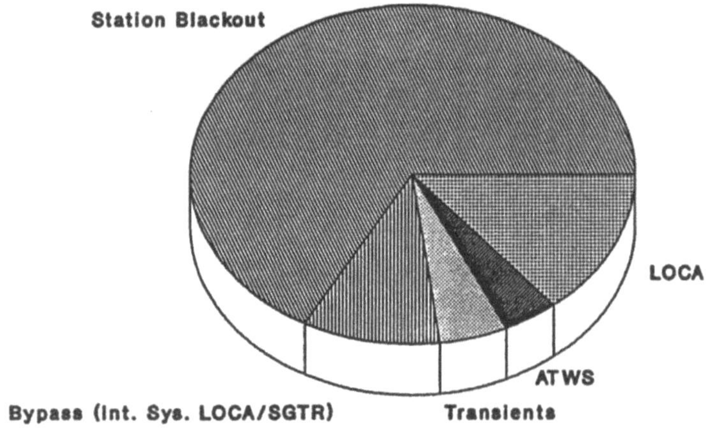

Figure 10.5 Contributions of summary PDS groups to core damage frequencies for the Surry plant due to internal IEs, with a total mean CDF of 4×10^{-5}/reactor-year. *Source:* [NRC90, Vol. 1].

events in PWRs. Although the containment bypass events, sequence V, do not necessarily involve either failures of the containment building or core damage, they could result in leakage of radioactive water outside the containment building and hence could contribute significantly to the overall risk that is equivalent to core damage events. This is one of the significant findings of the NUREG-1150 study and hence sequence V, labeled *bypass* events, is treated separately as a PDS in Figs. 10.5 and 10.6.

The bulk of the accident frequency analysis entails ET evaluations via the SETS code [Sta84] with the help of the top event probability of the FT representing every component or subsystem in the ESFs of the plant that is triggered as a result of a postulated IE. Significant effort was made to develop detailed FT models for key ESF systems and key support systems. Common mode failure and human reliability analyses also were included in the accident frequency ET analysis, with nominal human error probabilities evaluated via modified THERP techniques [Swa87]. A formal structure also was developed to elicit expert opinions to estimate certain system behaviors and component failure probabilities. Uncertainties in the estimates of IE frequency $P(I)$ and conditional probability $P(D|I)$ are generally represented through PDFs for the top events of FTs and the overall uncertainties for the PDS frequency $P(D)$ are evaluated through Monte Carlo convolutions of PDFs at every stage of the ET analysis.

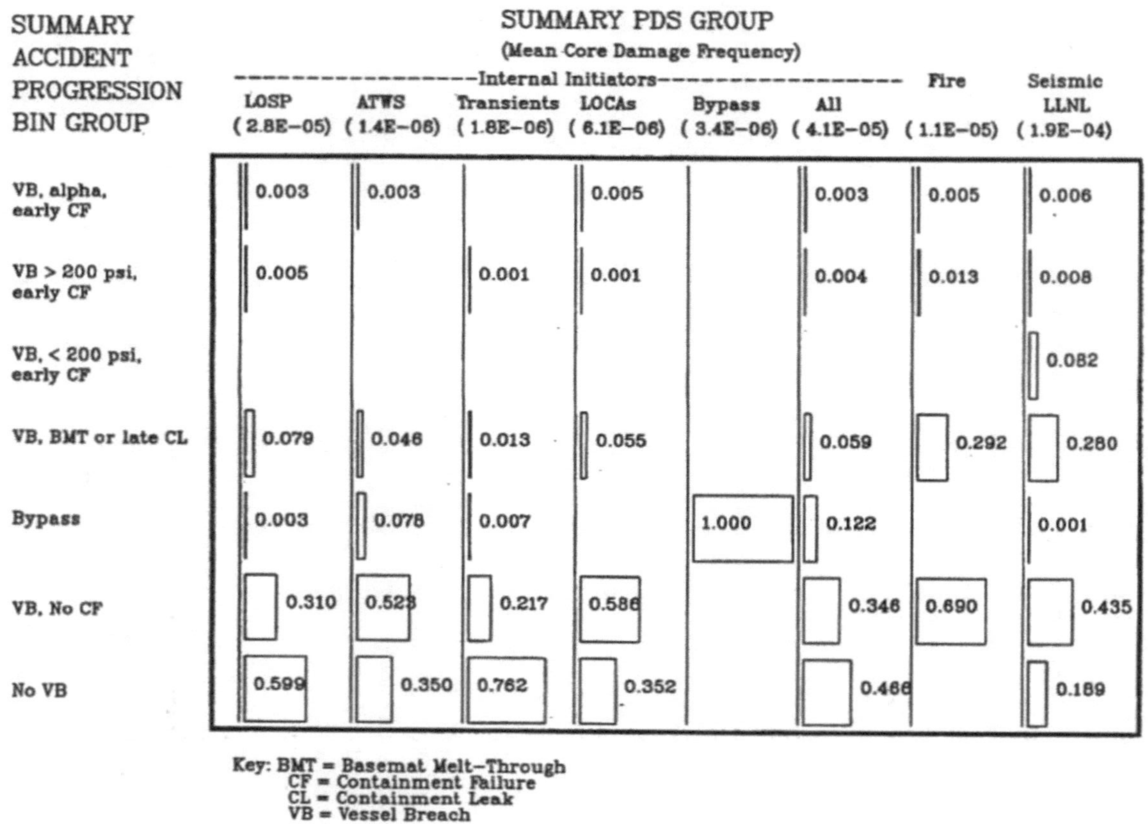

Figure 10.6 Mean PDS frequencies and conditional probabilities $P(A|P)$ for AP bins for the Surry plant. *Source:* [NRC90, Vol. 1].

10.2.4 Accident Progression Analysis

The accident progression event tree (APET) analysis began with the PDSs and associated frequencies to represent the progression of severe accidents that could result in containment failures and eventual release of radionuclides into the environment. The APET calculations made use of relevant accident and experimental data, accident simulation codes, and analyses of containment building structures. Among the accident simulation codes used are MELCOR [Sum95], MELPROG [Dos89], CONTAIN [Was91], and a suite of codes known as the Source Term Code Package (STCP) [Gie86], which was developed specifically for the study.

The MELCOR and MELPROG codes are two examples of comprehensive codes that were developed to follow the progression of core meltdown accidents using phenomenological models, empirical correlations, and experimental data to the maximum extent possible. The MELPROG code uses two-dimensional geometries of the core in its AP calculations and has not seen much developmental effort in recent years. The MELCOR code uses somewhat approximate geometrical representations but accounts for numerous phenomena and events relevant for AP analyses. The code is still under development for expanded applications in severe accident analyses. The CONTAIN code was developed to simulate performance of containment structures, simultaneously treating thermal hydraulics and mixing of water, aerosols, and fission products in severe accidents. The code does not represent complex in-vessel phenomena but rather relies on other simulation codes, e.g., the RELAP5 code [NRC01], to provide time-dependent mass and energy flow rates as boundary conditions. The STCP was developed to represent primary system behavior in core meltdown accidents and includes improved versions of simulation codes, in particular, the MELT code [Gie79].

For APET analyses that involve a large number of complex paths and branches, MELPROG, MELCOR, and STCP—in a descending order of complexity and detail—were used selectively for computational efficiency. This was especially necessary to quantify uncertainties in system parameters via PDFs. In addition, panels of experts were assembled and consulted on accident progression and containment structural issues for:

1. In-vessel accident progression, including temperature-induced failures of the RCS hot leg, steam generator tubes, and reactor vessel bottom head, and in-vessel hydrogen generation
2. Containment loading, including the containment pressure increase due to vessel breach and hydrogen combustion in the reactor building
3. Molten core-containment interactions for BWRs, including the pedestal erosion due to core-concrete interaction and melt-through of the drywell shell
4. Containment structural performance, including the containment failure pressure and modes and effects of hydrogen denotation on the containment

Substantial effort was made to formalize the process of eliciting expert opinions and quantifying them in the whole PRA process, as exemplified by the four AP expert panels. This elaborate process was one of the reasons for the three different versions of the NUREG-1150 documentation.

The complex accident scenarios simulated in the APET analyses result in a large set of alternate branches in the ET analyses. Even with efforts to group diverse outcomes into AP bins of similar characteristics, as many as 2000 AP bins were represented, especially for BWR plants. Each AP bin consists of a group of postulated accidents with similar consequences for risk analysis, characterized primarily by the containment failure time and mode. The binning process generates a two-dimensional conditional probability matrix with elements

$$P(A_i|D_j) = \text{conditional probability for the } i\text{th group, resulting from PDSs in the } j\text{th group, } i = 1, \ldots, n_A, \tag{10.8}$$

with $n_A = \mathrm{O}(10^3)$, as discussed above. Similar to the evaluation of the PDS frequency $P(D)$ of Eq. (10.6), the frequency of the ith AP bin is determined by

$$P(A_i) = \sum_{j=1}^{n_D} P(A_i|D_j)P(D_j),\ i = 1, \ldots, n_A, \tag{10.9}$$

which may be written in matrix notation as

$$P_{\mathbf{A}} = \mathbf{P}_{\mathbf{D}\rightarrow\mathbf{A}} P_{\mathbf{D}} = \mathbf{P}_{\mathbf{D}\rightarrow\mathbf{A}} \mathbf{P}_{\mathbf{I}\rightarrow\mathbf{D}} P_{\mathbf{I}}. \tag{10.10}$$

Equation (10.10) is displayed at the bottom of Fig. 10.4. Note here that each matrix element $P(A_i|D_j)$ of Eq. (10.8) is a PDF, with $n_A = \mathrm{O}(10^3)$ and $n_D = 20$. This quickly illustrates the computational burden associated with Monte Carlo evaluations of the matrix manipulations represented by Eq. (10.9) or (10.10), which required heavy use of supercomputers during the NUREG-1150 study in the late 1980s.

To convey the essence of the APET analysis, without the full machinery indicated in Eqs. (10.9) and (10.10), the NUREG-1150 summary report provides *summary AP bins*, listed here in the WASH-1400 nomenclature of Table 10.1:

1. R-α: Reactor vessel breach (VB) followed by containment rupture due to an in-vessel steam explosion, resulting in an early containment failure (CF)
2. R with pressure $>$ 200 psia: VB when the RCS pressure is greater than 200 psia, resulting in an early CF
3. R with pressure $<$ 200 psia: VB when the RCS pressure is less than 200 psia, resulting also in an early CF
4. R-ε: VB followed by containment basemat melt-through, resulting in a late containment leak (CL)
5. V: containment bypass event, including SGTRs for PWRs and MSIV failures for BWRs
6. R with no CF: VB not resulting in CF and hence no radionuclide leakage to the environment
7. No R: no VB, hence no radionuclide leakage to the environment

The summary AP bins are listed in the left column of Fig. 10.6, together with the *mean* values of conditional probabilities $P(A_i|D_j)$ for the five internal PDS groups, together with the sum of the internal PDSs and two external events. This means, of course, that summing the probability values down each PDS column should yield unity,

$$\sum_{i=1}^{n_A} P(A_i|D_j) = 1.0, \ j = 1, \ldots, n_D, \tag{10.11}$$

which can be readily verified. Note here that the structure of summary AP and PDS bins, for internal PDS events only, is represented by $n_A = 7$ and $n_D = 5$. In Fig. 10.6 note also that for all PDSs, except for the seismic events, the probability $P(A_3|D_j)$ of the accident landing in summary AP bin 3, i.e., R with pressure < 200 psia, is either negligibly small or zero. For the PDS representing sequence V or containment bypass events, $j = 5$, the conditional probability is simply unity for $i = 5$, i.e., $P(A_5|D_5) = 1.0$, and is identically zero for all other AP bins.

A sample display of matrix elements for APs resulting in early CF is presented in Fig. 10.7 to clarify the point that every element of the conditional probability matrix $P(A|D)$ is a PDF. We first note that each PDF displays a rather distinct distribution, with a different mean value and other characteristics. After considering a number of other display modes, the NUREG-1150 study group arrived at the particular display pattern of Fig. 10.7 for all PDFs used in the report, partly to highlight the long low-probability tails apparent for most of the PDFs involved. We clarify briefly the characteristics of the PDFs displayed in Fig. 10.7 by considering a general PDF $f(x)$ with the definitions:

$$\begin{aligned}
\text{Mean of } f(x) &= M = \int_0^\infty x f(x) \mathrm{d}x, \\
\text{Median of } f(x) &= m = \left\{ \nu \,\middle|\, \int_0^\nu f(x) \mathrm{d}x = \int_\nu^\infty f(x) \mathrm{d}x = \frac{1}{2} \right\}, \\
n\text{th percentile of } f(x) &= \left\{ \nu \,\middle|\, \int_0^\nu f(x) \mathrm{d}x = \frac{n}{100} \right\}.
\end{aligned} \tag{10.12}$$

A comparison of the last two of Eqs. (10.12) shows that median *m* is simply equal to the 50th percentile value of the PDF. The mean *M* of each PDF illustrated in Fig. 10.7, when compared with those summarized in Fig. 10.6, indicates that each PDF of the former corresponds to the sum of PDFs for three early CF AP bins, $i = 1, 2, 3$, of the latter.

Throughout the NUREG-1150 reports, 5th and 95th percentile values of each PDF are displayed together with its mean and median to characterize the distribution. Besides the significantly different shapes of the distributions noted earlier, the 5th and 95th percentile values span up to five orders of magnitude in the PDFs illustrated in Fig. 10.7. This succinctly illustrates the large degree of uncertainties that must be dealt with in PRA studies of NPPs. The different shapes of the PDFs displayed also indicate the need to sample each PDF directly through Monte Carlo techniques, rather

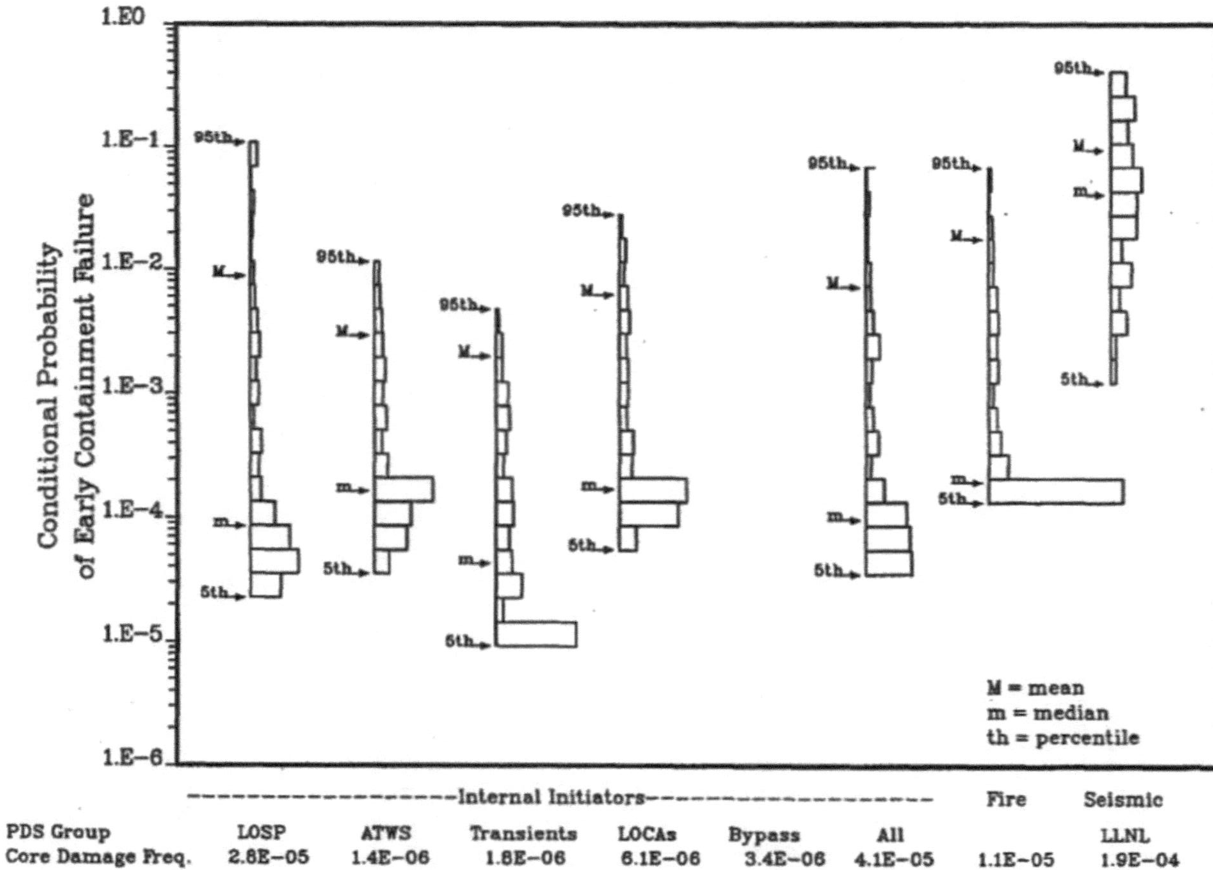

Figure 10.7 Conditional probability $P(A|D)$ of early containment failures for internal and external summary PDS groups. *Source:* [NRC90, Vol. 1].

than rely on some analytical convolutions, which explains the large computational requirements experienced in the NUREG-1150 study.

10.2.5 Radionuclide Transport Analysis

The determination of the frequency $P(A)$ of AP bins via Eq. (10.9) completes the first building block of Eq. (10.1) for the PRA methodology by providing the *overall accident frequency*. The AP bins and associated frequencies now will be used in the second building block, the *consequence analysis*, to calculate the release of radionuclides to the atmosphere and resulting health effects. This requires tracking the transport of radioactive materials from the fuel to the RCS and all the way through the containment building while accounting for all the potential leak paths. The fractions of core radionuclide inventories released and the times at which the releases occur comprise the *source term* for the third key step, the *radionuclide transport analysis*, illustrated in Fig. 10.4. The source term data are finally used in the fourth task, the *offsite consequence analysis*, to determine the impacts of radionuclides released to the environment.

The source terms were determined through a combination of detailed mechanistic computer models, including the CONTAIN and MELCOR codes and the STCP, and simplified algorithms developed as a XSOR family of codes [Jow93]. Based on a limited number of CONTAIN and MELCOR runs, key parameters were determined to represent the release fractions and transmission factors, for nine groups of radionuclides, in simplified XSOR functional fits at successive AP stages for various release paths. Figure 10.8 illustrates representative leakage pathways modeled in XSOR algorithms. For each pathway, a radioactive material balance is set up with the constituent parameters, together with associated PDFs, obtained from CONTAIN and MELCOR mechanistic calculations, relevant experimental data, and expert judgments. Similar to the elicitation of expert opinions in the APET analysis, a source term expert panel was consulted on a number of issues, including (a) in-vessel retention and release of radioactive material, (b) revolatization of radionuclides from the reactor vessel and RCS, (c) radioactive releases during high-pressure melt ejection resulting in direct containment heating, and (d) radioactivity releases during core-concrete interaction. These are the issues that relate directly to the leakage paths represented in Fig. 10.8. Because source term calculations are significantly different from plant to plant, XSOR models were developed to address the radionuclide transport analysis for each plant, e.g., SURSOR for the Surry plant.

The XSOR parametric models merely calculate the source term as the product of release fractions and transmission factors, without representing any detailed physical or chemical mechanisms. For example, the fraction of radionuclide (RN) releases from the fuel that occur within the reactor pressure vessel (RPV), before the RPV breach, is calculated through a simple ET structure illustrated in Fig. 10.9. In the simple ET structure of Fig. 10.9, for nuclide group i, we begin with the fraction F_{RPV} of the RNs released in the RPV and consider separate paths for the RNs escaping through the steam generators (SGs) and through the rest of the RCS. This branching in the RN transport process accounts for the containment bypass events due to SGTRs

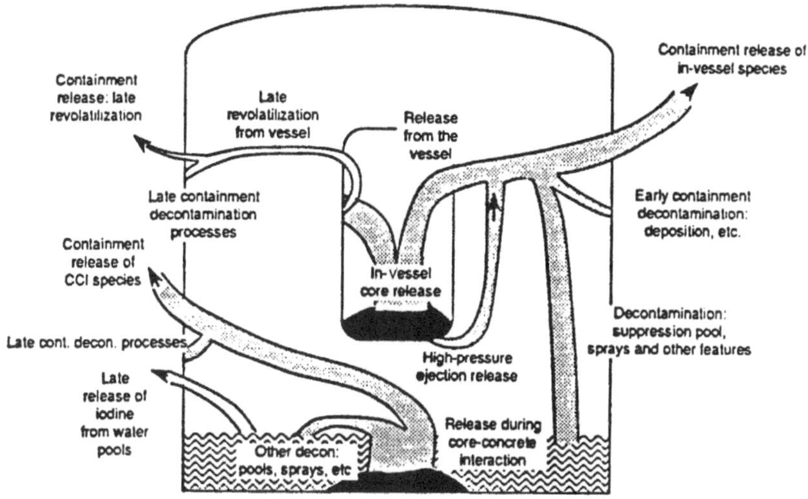

Figure 10.8 Simplified leakage pathways represented in XSOR algorithms. *Source:* [NRC90, Vol. 2].

considered in Figs. 10.5 and 10.6. For the SG leakage path, the sequence evolves with the probability F_{ISG} that the RNs enter the SGs and eventually, with probability F_{OSG}, are released from the SGs into the environment. For the leakage path not involving the SGs, the branch entails the probability $1 - F_{ISG}$ that the RNs enter the RCS, excluding the SGs, followed by the probability F_{RCS} that they are released from the RCS and the probability F_{CMT} that the RNs eventually leak out of the containment (CMT) building into the environment. The latter probability is reduced by the decontamination or dilution factor D_{CMT} that the containment sprays and filters provide. Recall that the dilution factor was defined in Eqs. (8.37) and (8.38). Thus, summing up the RN release fractions for the two branches yields, for nuclide group i, the total fraction of the RNs released into the environment in this particular AP scenario:

$$f = F_{RPV}\left[F_{ISG}\,F_{OSG} + (1 - F_{ISG})\,F_{RCS}\,F_{CMT}/D_{CMT}\right]. \tag{10.13}$$

The leakage and transmission probabilities in Eq. (10.13), together with uncertainty estimates represented through suitable PDFs, are obtained via a combination of MELCOR and CONTAIN calculations and expert judgments, as discussed earlier. The *source term event tree* (STET) analysis thus makes heavy use of the simple parametric models of the XSOR family of codes to determine source terms partitioned according to the potential for causing early and latent cancer fatalities and the warning time associated with the events. Summation of the RN release fractions illustrated in

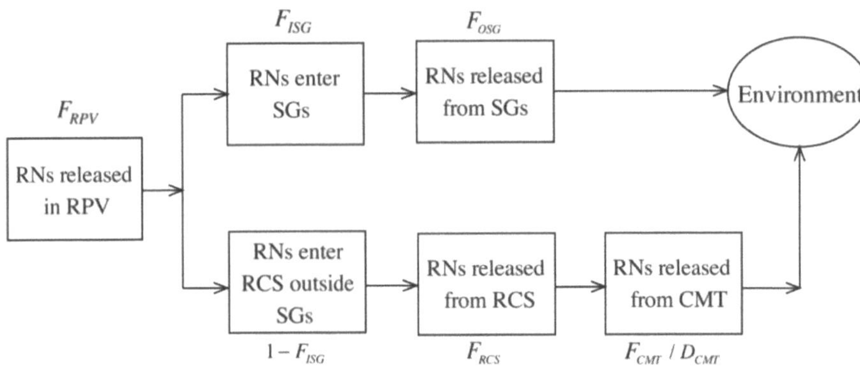

Figure 10.9 Event tree structure for determination of the fraction of radionuclides released in reactor pressure vessel that escape to the environment.

Eq. (10.13) yields

$$f(S_i|A_j) = \text{fraction of radionuclides released into source term group } i, \text{ resulting from AP bin } j, \qquad (10.14)$$

for each of the nine RN groups shown in Fig. 10.10. The groups

{noble gas iodine cesium tellurium strontium ruthenium lanthanum cerium barium}

are structured to reflect the chemical activity and volatility of 60 major fission products that would be released in postulated core melt accidents.

The release fraction $f(S_i|A_j)$, combined with the RN inventory Q_i, yields the inventory of RNs released for source term group i resulting from AP bin j:

$$P(S_i|A_j) = Q_i f(S_i|A_j), \quad i = 1, \ldots, n_S, \qquad (10.15)$$

where $n_S = 30$ to 60. It should be emphasized that the matrix elements $P(S_i|A_j)$ represent inventories, not probabilities, although for notational convenience they are written in the same mathematical form as the conditional probabilities of Eqs. (10.5) and (10.8).

For one source term group resulting in early RN releases due to containment bypass events, Fig. 10.10 shows the release fraction $f(S_i|A_j)$ in a PDF form for all of the nine RN groups, where fifth percentile values are not indicated when they fall below 1×10^{-5}. For the iodine group, the release fractions $f(S_i|A_j)$ are plotted as CCDFs in Fig. 10.11, thus graphically providing the frequency per year of the RN release fraction $f(S_i|A_j)$ exceeding the value indicated on the abscissa. In the NUREG-1150 study, the CCDF is sometimes referred to as the *exceedance frequency*.

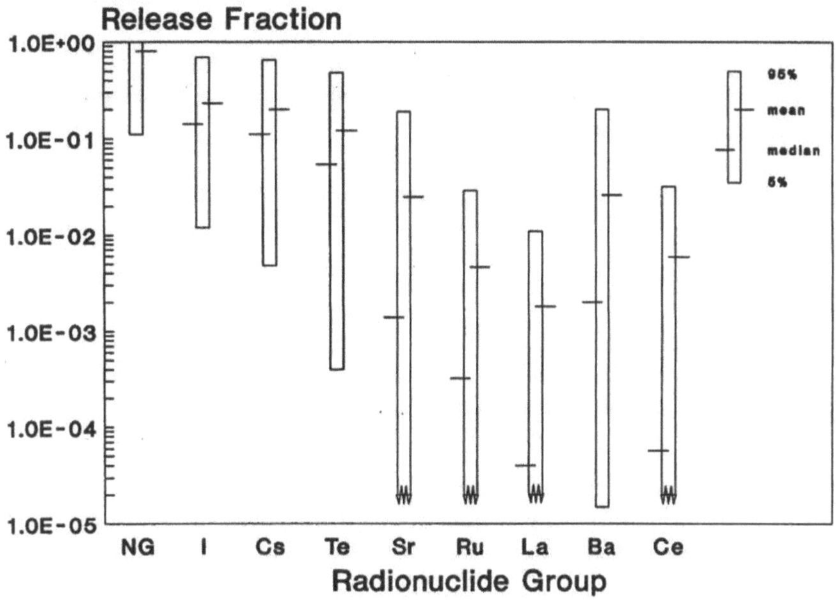

Figure 10.10 Probability density functions for release fraction $f(S_i|A_j)$ in nine radionuclide groups for early leakage due to Surry containment bypass events. *Source:* [NRC90, Vol. 1].

The RN inventory released for source term group *i*, as given in Eq. (10.15), may now be formally combined with the AP frequency of Eqs. (10.9) and (10.10) to yield the RN source vector:

$$P(S_i) = \sum_{j=1}^{n_A} P(S_i|A_j)P(A_j),\ i = 1, \ldots, n_S;\ n_S = 30 \text{ to } 60, \tag{10.16}$$

or in matrix notation, as included at the bottom of Fig. 10.4,

$$P_{\mathbf{S}} = \mathbf{P}_{\mathbf{A}\to\mathbf{S}} P_{\mathbf{A}} = \mathbf{P}_{\mathbf{A}\to\mathbf{S}} \mathbf{P}_{\mathbf{D}\to\mathbf{A}} \mathbf{P}_{\mathbf{I}\to\mathbf{D}} P_{\mathbf{I}}. \tag{10.17}$$

10.2.6 Offsite Consequence Analysis

The fourth step in the NUREG-1150 risk calculation is to determine the consequences of the RN releases to the atmosphere. The consequences or impacts of the RN releases on the surrounding environment and population are classified in 8 different *consequence measures*:

1. Number of early fatalities expected within one year of incident
2. Number of early injuries expected within one year of incident

3. Number of latent cancer fatalities expected to occur over the lifetime of the exposed individuals
4. Total radiation dose imparted to the population within 50 miles
5. Total radiation dose imparted to the population in the entire site region
6. Economic cost of the accidents
7. Average individual early fatality probability within 1 mile of the site boundary
8. Average individual latent cancer fatality probability within 10 miles of the site boundary

Consequence measures 1 and 3 are chosen to compare with the corresponding risk estimates from the WASH-1400 study, while measures 7 and 8 are evaluated to compare with the NRC safety goals [NRC86] discussed in Section 8.2.3.

To determine the above eight consequence measures, the *offsite consequence analysis*, the fourth key task illustrated in Fig. 10.4, consists of a sequence of calculations for each of the source term group with the RN inventory $S_i, i = 1, \ldots, n_S$:

1. Transport and dispersion of the RNs are calculated using the Gaussian plume model [Cha90], together with wake effects [Bri75] due to buildings and structures, and site-specific meteorological data for approximately 160 representative weather conditions.
2. Deposition of the RNs from the plume on the ground is determined via experimental deposition rates.
3. Radiation doses are calculated using dose conversion factors [Koc81,Int77,Int78] for various body organs and for direct and indirect pathways, with site-specific population data.
4. Health effects of radiation exposures are determined via the BEIR-III model [Eva85,NRC80].

Recall that an introduction to the Gaussian plume model was presented in Section 8.6.2. The sequence of consequence analyses was performed through the MELCOR Accident Consequence Code System (MACCS) [Cha90] to yield the eight consequence measures $C_i, i = 1, \ldots, 8$, with the MACCS analyses providing the conditional probability:

$$f(C_i|S_j) = \begin{array}{l}\text{probability of consequence measure } C_i \\ \text{resulting from source term group } j.\end{array} \tag{10.18}$$

In the MACCS calculations of consequence measures, several scenarios were considered to represent the effects of dose mitigation by emergency response actions, with the base case assumption that 99.5% of the population within the 10-mile emergency planning zone (EPZ) participate in an evacuation. In addition, the variability in weather, including wind directions and weather sequences, was represented to quantify the uncertainties in the risk estimated.

The actual consequence measures resulting from each source term group are obtained by weighting the measure C_i itself with the conditional probability $f(C_i|S_j)$:

$$P(C_i|S_j) = C_i f(C_i|S_j), \; i = 1, \ldots, n_C, \tag{10.19}$$

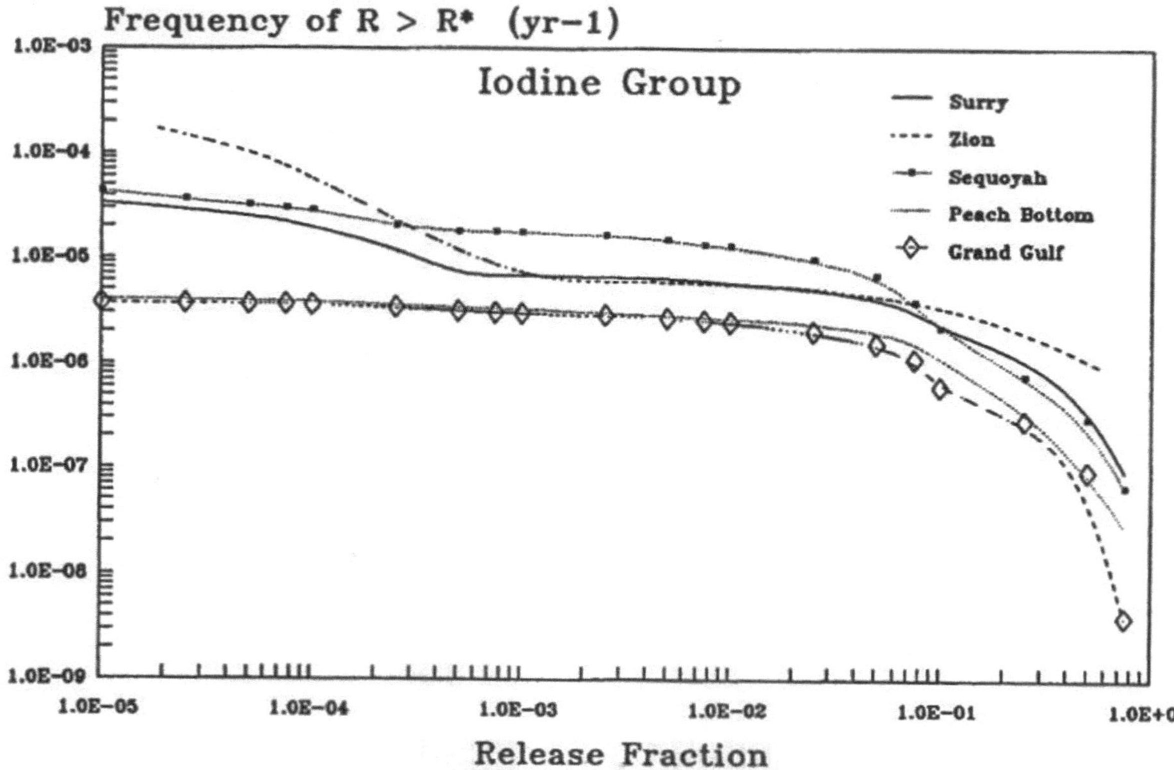

Figure 10.11 Sample CCDF plots of radionuclide release fraction $f(S_i|A_j)$ for five NPPs analyzed in NUREG-1150. *Source:* [NRC90, Vol. 1].

with $n_C = 8$. Finally, the eight consequence measures representing the risk due to the entire set of accidents analyzed are obtained by duly accounting for the source term inventories $P(S_j)$:

$$P(C_i) = \sum_{j=1}^{n_S} P(C_i|S_j)P(S_j),\ i = 1, \ldots, n_C, \tag{10.20}$$

or in matrix notation, as included at the bottom of Fig. 10.4,

$$P_{\mathbf{C}} = \mathbf{P}_{\mathbf{S}\to\mathbf{C}} P_{\mathbf{S}} = \mathbf{P}_{\mathbf{S}\to\mathbf{C}} \mathbf{P}_{\mathbf{A}\to\mathbf{S}} \mathbf{P}_{\mathbf{D}\to\mathbf{A}} \mathbf{P}_{\mathbf{I}\to\mathbf{D}} P_{\mathbf{I}}. \tag{10.21}$$

Note that, similar to $P(S_i|A_j)$ of Eq. (10.15), matrix elements $P(C_i|S_j)$ and vector elements $P(C_i)$ in Eq. (10.20) do not represent probabilities but rather consequence measures, e.g., number of early fatalities for $i = 1$ and economic cost of accidents for $i = 6$. The matrix equation (10.21) clearly indicates that the evaluation of eight consequence measures requires quadruple summations involving IE bins, PDS bins, AP bins, and source term (ST) bins. As discussed in Section 6.4.2, the sequence of PRA steps is often classified in three levels given by:

Level 1 PRA: Usually known as the *system analysis*: the accident frequency ET analysis represented by $P(D|I)$ based on system and human factor evaluations and core damage frequency.

Level 2 PRA: Usually known as *the containment analysis*: the performance of the damaged core and the radionuclide release to the environment. This step in the risk determination is accomplished with the APET calculations represented by $P(A|D)$ and the radionuclide transport analysis represented by $P(S|A)$.

Level 3 PRA: The *consequence analysis* to represent the offsite dispersion and transport of radionuclides released to the environment and the health effects and other consequences of the postulated accidents. This stage of the risk calculation is represented by $P(C|S)$, finally yielding the eight consequence measures $P(C)$.

10.2.7 Uncertainty Analysis

One of the important tasks in the NUREG-1150 study of the risk of operating five representative LWRs was to evaluate the uncertainties associated with the risk estimated. To represent uncertainties in the overall risk calculations involving the four key tasks illustrated in Fig. 10.4 and by the four matrices in Eq. (10.21), Monte Carlo calculations were performed to sample various PDFs. For computational efficiency, a stratified Monte Carlo method, known as the Latin hypercube sampling (LHS) technique [Ima84], was used to perform the series of matrix manipulations represented by Eq. (10.21). Even with this approximate Monte Carlo technique, a significant use of supercomputers was required to statistically sample a large number of key variables in the entire risk estimation process. Furthermore, full-blown uncertainty analyses for the offsite consequence part of the study were not performed, although the variability in meteorological conditions was accounted for.

Both modeling and data uncertainties were represented throughout the risk analysis via 150 to 250 LHS samples, each of which models approximately 2500 variations of key parameters. The PDF for each of the eight consequence measures resulting from each of the LHS samples is summarized as a CCDF in the form of Eq. (10.3). The collection of CCDFs for each of the eight consequence measures is sorted for each value of the consequence measure and plotted as a set of four CCDF curves, as illustrated in Fig. 10.12. Thus, the four CCDF plots for each of consequence measures 1, 3, 4, and 5 do not represent actual CCDFs corresponding to any particular LHS samples but represent the overall distribution of the 105 to 250 CCDFs generated in the uncertainty analysis process.

10.2.8 Risk Integration

The fifth and final step in the NUREG-1150 PRA study is the *risk integration* indicated in the last box of Fig. 10.3. In this step, effort is made to integrate the eight consequence measures calculated through Eq. (10.21) with other risk estimates, in particular, the PDS frequency $P(D)$ and AP frequency $P(A)$. The integration step provides valuable insights to risk-dominant accident sequences and could highlight vulnerabilities in the particular hardware and safety features of a power plant.

We begin in Fig. 10.13 with PDFs representing frequencies of consequence measures 1 and 3, i.e., the number of early and latent cancer fatalities per reactor-year (ry), respectively, for the five LWRs. We note that the early fatality frequencies are generally lower for BWRs, the Peach Bottom and Grand Gulf plants, than those for PWRs and that the latent cancer fatality frequencies are somewhat more even among the PWR and BWR plants. For both fatality estimates, the NUREG-1150 risks are at least an order of magnitude lower than the corresponding WASH-1400 or RSS estimates, but the uncertainties range over several orders of magnitude. We may note that system improvements had been made over the period of 15 years or so between the two PRA studies, but a number of new accident sequences uncovered during the NUREG-1150 study contributed to some increased risk estimates. The points marked by plus (+) signs for the Zion plant present mean frequency estimates recalculated with system modifications during the NUREG-1150 study as discussed further in the section. Figure 10.14 presents similar comparisons for consequence measures 7 and 8, i.e., the average number of individual early and latent cancer fatalities per reactor year, respectively, where we note again the risks for the BWR plants are lower than the PWR risks. Note also that all of the five NUREG-1150 LWRs easily meet the NRC safety goals for individuals, calculated as 5×10^{-7} early fatalities per reactor-year and 2×10^{-6} latent cancer fatalities per reactor-year.

Figures 10.15 and 10.16 present one particular example for integrating consequence measures 1 and 3 with the dominant PDSs. Similar integrations between consequence measures and accident progression bins are presented in NUREG-1150, Volume 1. Relative contributions of plant damage states to early and latent cancer fatalities for all five LWR plants are presented in terms of the mean fatality estimates. For each pie chart, the actual mean fatality frequency is indicated. Note first that the mean fatality estimates match the corresponding values in Fig. 10.13. For the Zion plant, the fatality frequencies given in Fig. 10.15 correspond to the values before the system modifications discussed for Figs. 10.13 and 10.14.

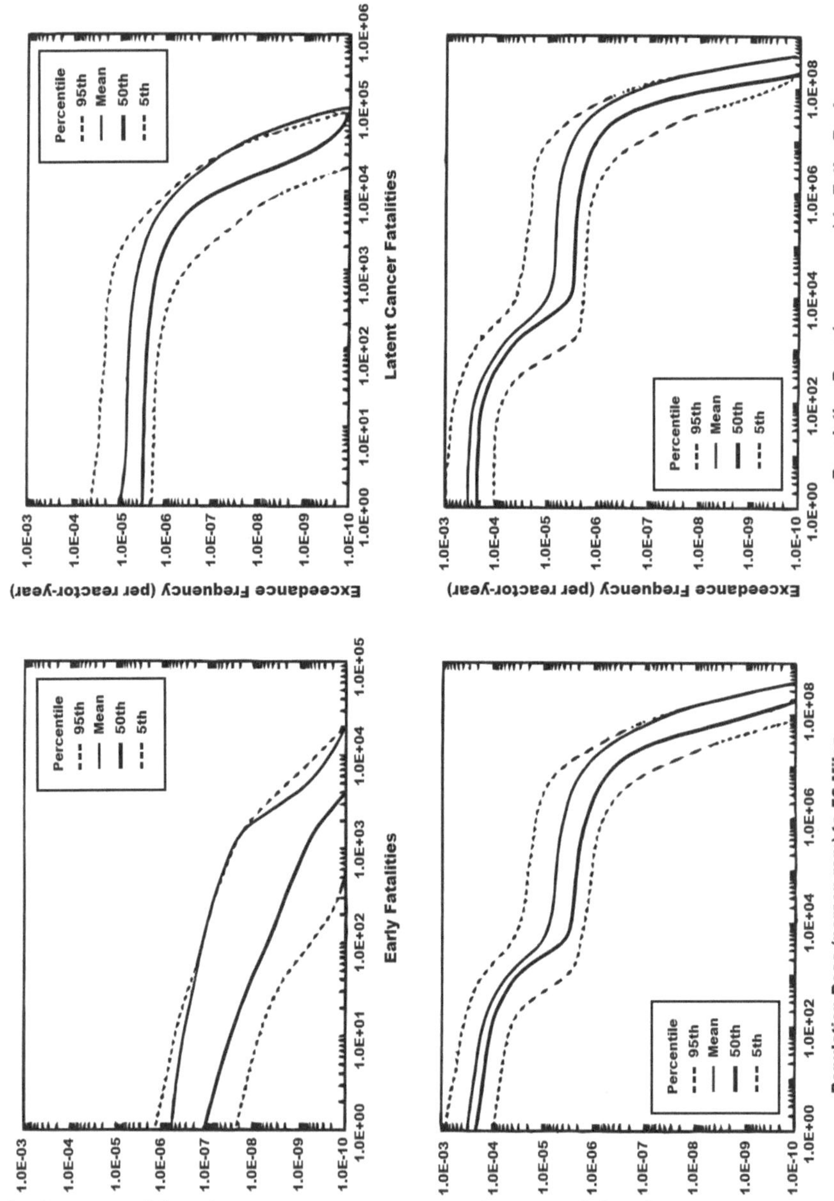

Figure 10.12 Sample CCDF plots of four representative offsite consequence measures. *Source:* [NRC90, Vol. 1].

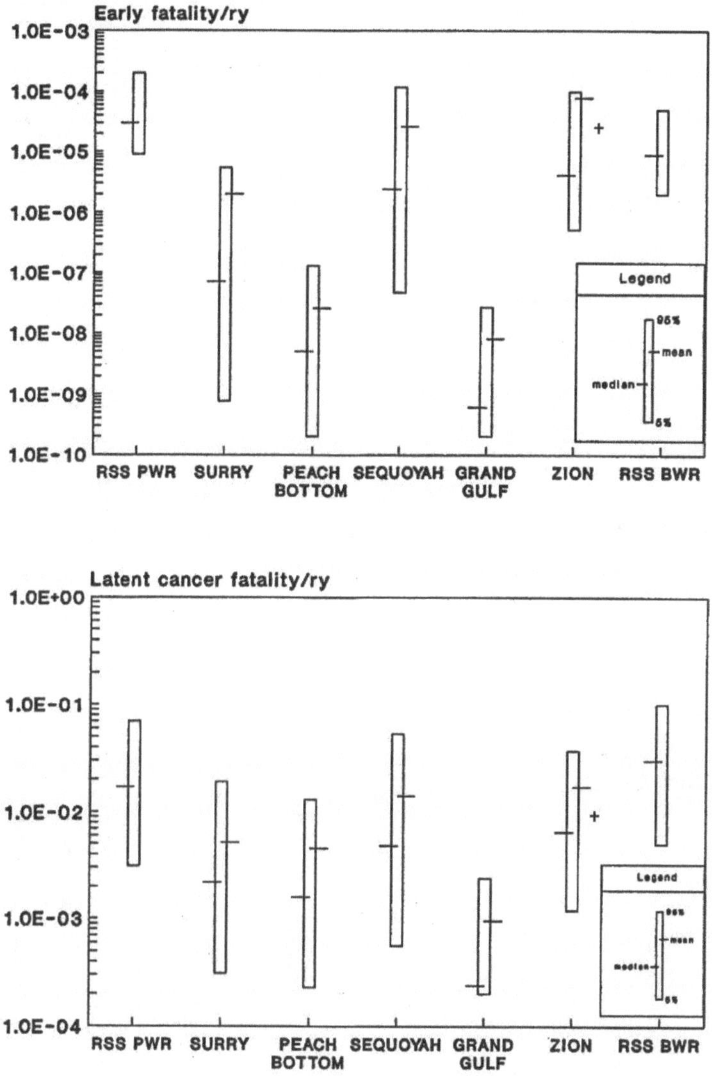

Figure 10.13 Comparison of early and latent cancer fatality risks for five LWR plants. *Source:* [NRC90, Vol. 1].

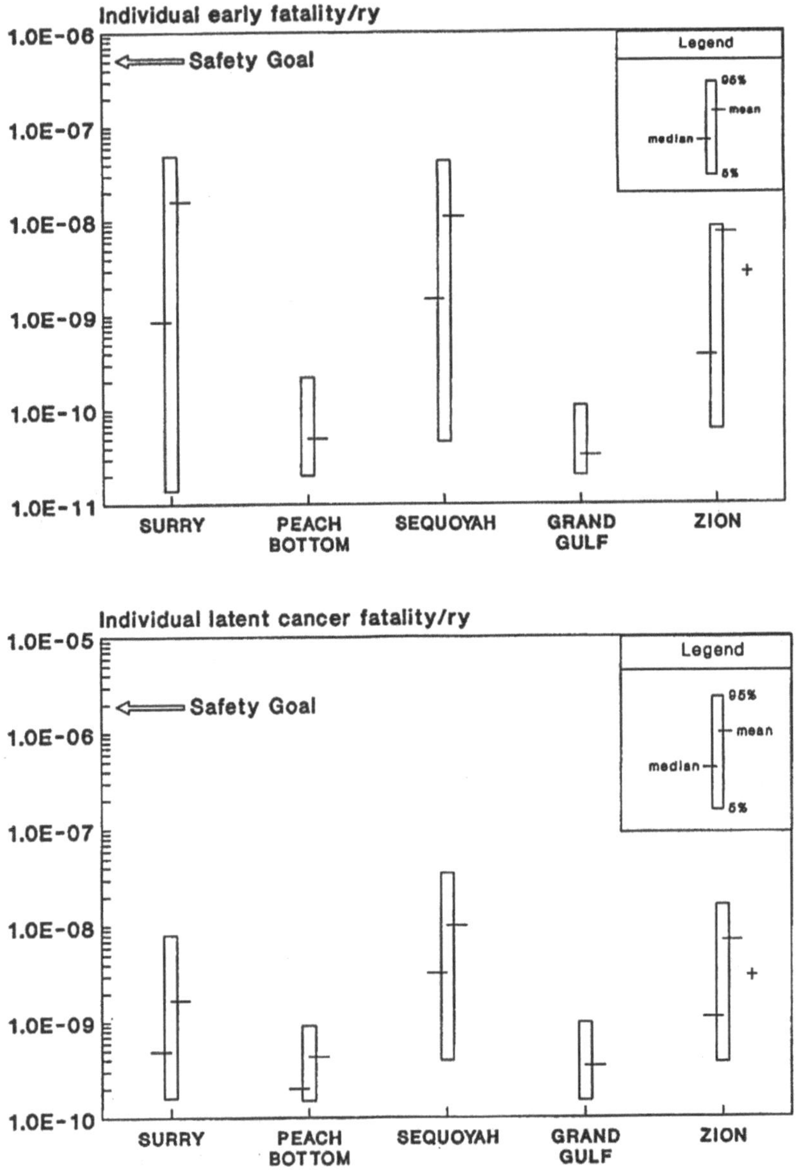

Figure 10.14 Comparison of individual early and latent cancer fatality risks for five LWR plants. *Source:* [NRC90, Vol. 1].

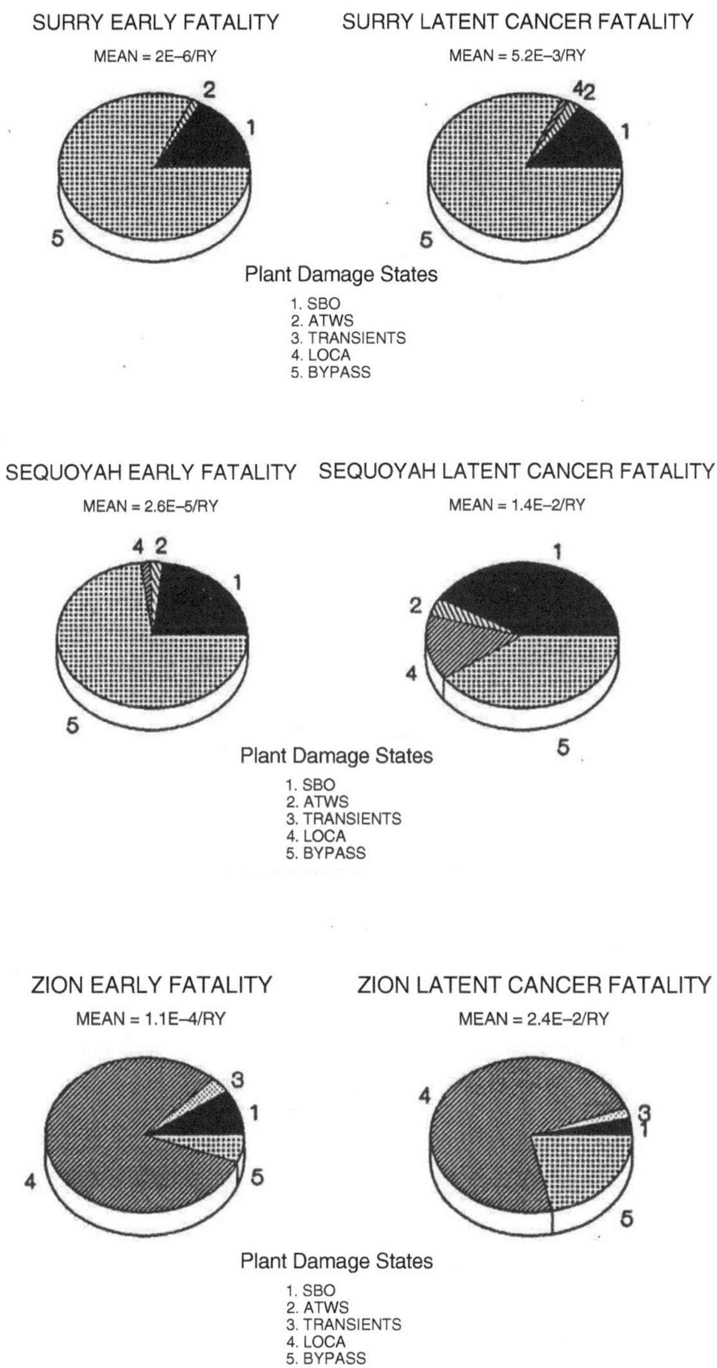

Figure 10.15 Contributions of plant damage states to mean early and latent cancer fatality risks for PWR plants. *Source:* [NRC90, Vol. 1].

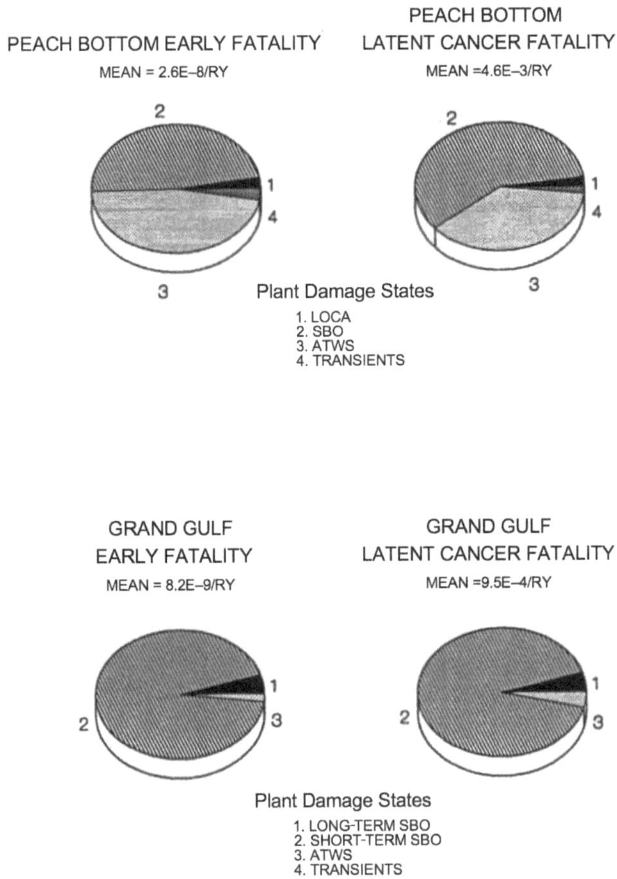

Figure 10.16 Contributions of plant damage states to mean early and latent cancer fatality risks for BWR plants. *Source:* [NRC90, Vol. 1].

Comparison of the relative PDS contributions to both early and latent cancer fatality risks for the Surry and Sequoyah PWR plants in Fig. 10.15 indicates that the major contributors to the overall plant risk are the containment bypass events, followed by the SBO scenarios. Other accidents including LOCA and ATWS events make relatively small contributions. This is one of the major differences we note in comparison with the WASH-1400 summary of Table 10.3, where SBLOCA and transient events are the major contributors to the PWR plant risk. The Zion pie charts, however, indicate that the main contributors to the plant risk are LOCAs, followed by the bypass and SBO events. This prompted a quick evaluation of the reasons for this surprising difference. A review of the system interactions at the Zion plant revealed that the high risk due to LOCAs is attributed to a single component cooling water system (CCWS) providing coolant water to both reactor coolant pumps and HPCI pumps. Before the final version of NUREG-1150 report was issued, system modifications and procedure changes were made at the Zion plant, which reduced the contributions from LOCAs and the overall risk. In particular, the early fatality frequency was reduced from 1.1×10^{-4}/reactor-year to 2.0×10^{-5}/reactor-year with the modifications. The revised risk values are shown with plus (+) signs in Figs. 10.13 and 10.14.

A similar comparison of the major PDS contributors to health risks for BWR plants in Fig. 10.16 indicates that the SBO events are most risk significant for both plants. The ATWS events, however, make much larger contributions for the Peach Bottom plant than for the Grand Gulf plant. No simple explanation was readily available for this difference in the NUREG-1150 report, although there apparently are a number of differences in the safety systems between the two BWR plants.

10.2.9 Additional Perspectives and Comments on NUREG-1150

The NUREG-1150 study on severe accident risks for three PWR and two BWR plants produced a massive volume of documents and detailed PRA results and provided numerous valuable insights to nuclear plant safety. The report has been used effectively for general risk assessment and for developing strategies for the management of severe accidents. We summarize some of the perspectives discussed in Section 10.2.8 and provide additional comments in this section.

1. The overall risk estimates obtained in NUREG-1150 are smaller than the corresponding estimates in WASH-1400 for the Surry PWR plant and the Peach Bottom BWR plant. This may be in part due to various improvements and backfits made to the plant systems during the span of 15 years between the two PRA studies but also to different assumptions made in the risk assessment. One particular difference that was pointed out by a review committee [Kou90] for NUREG-1150 is the reduction in the fraction of the core radionuclide inventory eventually released to the environment. To illustrate the point, the release fractions calculated for two key elements, I and Cs, are compared in Figs. 10.17 and 10.18. The CCDF plots clearly show that the median NUREG-1150 fractions both for I and Cs are significantly lower than the WASH-1400 fractions calculated for the Surry plant. The reductions in the release fractions are attributed to three factors in NUREG-1150: (a) core damage probability is lower, (b) higher containment failure pressure, and (c) greater retention of fission products, in particular, I and Cs, within the containment. The last factor represents an observation from the TMI-2 accident that iodine would combine with cesium as cesium iodide, which is soluble in water. In contrast, WASH-1400 assumed that iodine would remain as insoluble elemental iodine vapor.

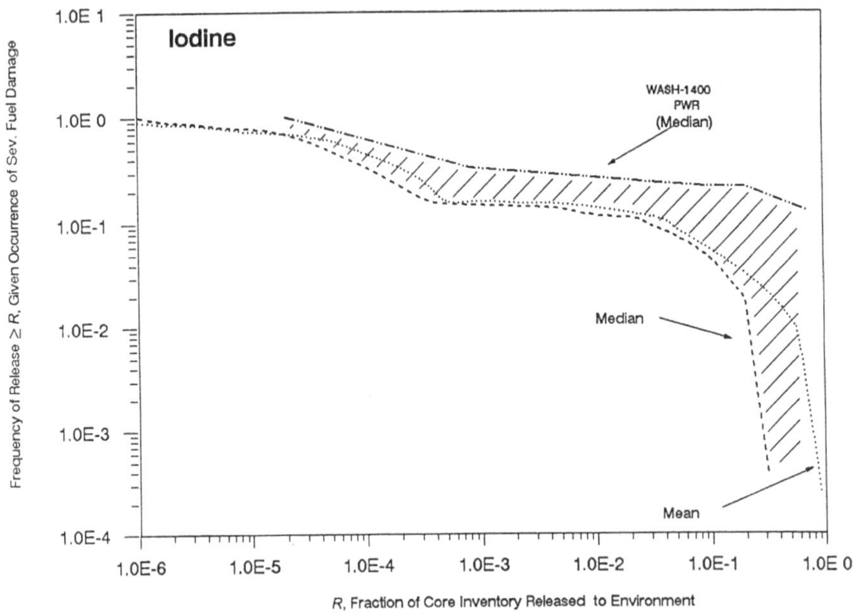

Figure 10.17 Comparison of iodine release fractions calculated for the Surry plant in NUREG-1150 with WASH-1400 release fractions. *Source:* [Kou90].

2. Despite the reduction in radionuclide release fractions achieved, the overall risk has not significantly decreased, partly because new risk-significant accident scenarios were discovered, e.g., DCH events and interfacing system LOCAs or containment bypass events. The risk significance of station blackout events also was recognized and emphasized as a result of the NUREG-1150 study. The DCH events discussed in Section 8.4 as part of Class 9 accidents appear to remain as somewhat of an unresolved safety issue. This also raises a general question regarding PRA studies if calculated risk estimates would keep increasing as increasing details and outliers are included in the risk calculations.

3. Despite significant effort made to quantify and reduce uncertainties in the risk calculations, the uncertainties persist over several orders of magnitude. With long low-probability tails inherent in all PDFs plotted in Figs. 10.13 and 10.14, there are significant differences noted between the median and mean values. This suggests that mean values should be used in general rather than the median values extensively used in WASH-1400.

4. The calculated consequences depend heavily on the details of the balance of plant (BOP) as well as on the nuclear steam supply steam (NSSS). This was evident in the CCWS issue for the Zion plant that was promptly corrected before the completion of NUREG-1150. The significant differences noted between the PDS contributions

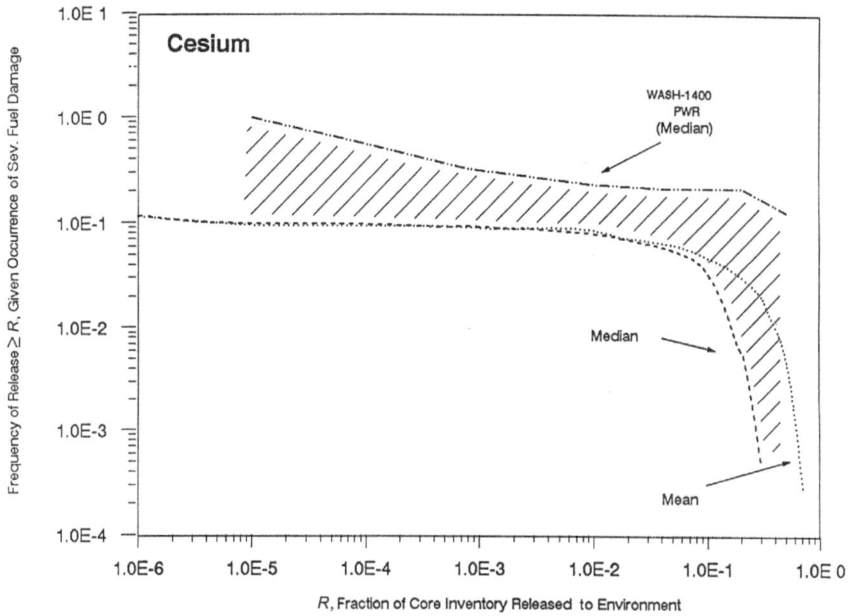

Figure 10.18 Comparison of cesium release fractions calculated for the Surry plant in NUREG-1150 with WASH-1400 release fractions. *Source:* [Kou90].

for the two BWR plants, Peach Bottom and Grand Gulf, in Fig. 10.16 are generally attributed to the differences in BOP designs for the plants.

5. All five LWRs studied in NUREG-1150 meet the NRC safety goals for individuals: 5×10^{-7} early fatalities per reactor-year and 2×10^{-6} latent cancer fatalities per reactor-year.

6. Expert opinions, together with fault trees, have been used to estimate the probabilities and modes of component failures. A formal structure for the elicitation of expert opinions was developed.

7. Monte Carlo calculations, especially Latin Hypercube sampling, were used extensively to tally the risk through a complex sequence of event trees and to evaluate the associated uncertainties.

8. Events with negligible core damage frequencies may contribute large risk, e.g., the interfacing system LOCAs represented as the V sequence of accidents.

9. External events, e.g., fire and earthquake, were considered only for the Surry and Peach Bottom plants. Apart from noting the relevant PDFs for the summary PDS groups in Fig. 10.6, we have chosen not to address external events studied in

NUREG-1150. This is to a large extent due to large uncertainties inherent in external event analyses, as was exemplified by the July 2007 earthquake that rattled the support structures at the Kashiwazaki-Kariwa Nuclear Power Plant in Japan. The analysis of the magnitude-6.6 earthquake [Nor07] suggests that "the relation between ground accelerations and the loads imposed on buildings is not fully understood."

10. As a general observation regarding PRA studies for nuclear power plants, the primary value of the PRA study should be recognized as a means to discover vulnerabilities in systems and plant operating procedures, as was the case with the Zion CCWS issue during the NUREG-1150 study. The bottom-line PRA results for core damage frequency and radionuclide release rates should be used with due recognition for significant uncertainties involved with them.

11. Subsequent to the release of NUREG-1150, all nuclear power plants in the United States have gone through PRA studies under the Individual Plant Examination (IPE) program of the U.S. Nuclear Regulatory Commission. The studies were limited to Level 1 and Level 2 evaluations, comprising system and containment analyses discussed in Section 10.2.6.

10.3 SIMPLIFIED PRA IN THE STRUCTURE OF NUREG-1150

10.3.1 Description of the Simplified PRA Model

With the basic PRA structure illustrated in Fig. 10.4, we present a simplified PRA study for the Surry plant that could augment the risk integration results summarized in Fig. 10.15 and obtain physical insights into the risk calculations. The method begins with a simplified form of Eqs. (10.16) and (10.17), where we skip the first step involved with the initiating events and start the source term calculation with the summary PDS groups given in the summary report of NUREG-1150, i.e., Fig. 10.6. We simplify further by grouping the AP bins into the early-containment failure (ECF) and late-containment failure (LCF) bins.

Radionuclide release fractions $f(S_i|A_j)$ and fission product inventories Q_i in Eq. (10.15) are obtained from the summary NUREG-1150 report and other sources, without resort to complex numerical calculations. The final step for atmospheric dispersion and dose rate calculations is performed through the Gaussian plume model and the simple health effects model discussed in Sections 8.6 and 8.7.

The simplified PRA model consists of the following steps:

1. With the assumption that ECF events have similar characteristics as containment bypass scenarios, obtain the AP vector representing frequencies for A_1 = ECF and A_2 = LCF,

$$P_{\mathbf{A}}^T = [P(\mathrm{A}_1) \quad P(\mathrm{A}_1)] \,. \tag{10.22}$$

Table 10.4 Equilibrium Mass Inventory in Nine Radionuclide Groups for the Surry Plant

Group	Elements	Total mass (kg)
1	Xe, Kr	273.4
2	I, Br	12.4
3	Cs, Rb	145.7
4	Te, Sb, Se	25.4
5	Sr	47.6
6	Ru, Rh, Pd, Mo, Tc	369.5
7	La, Zr, Nd, Eu, Nb, Pm, Pr, Sm, Y	538.7
8	Ce, Pu, Np	626.0
9	Ba	61.2

Source: [NRC90].

For this purpose, use numerical values given in Fig. 10.6 to obtain the PDS frequency vector for internal events,

$$P_{\mathbf{D}}^{T} = [P(\text{SBO}) \quad P(\text{ATWS}) \quad P(\text{transient}) \quad P(\text{LOCA}) \quad P(\text{bypass})]\,, \tag{10.23}$$

and combine it with the the conditional probability matrix again from Fig. 10.6,

$$\mathbf{P}_{\mathbf{D}\rightarrow\mathbf{A}} = \begin{bmatrix} 0.011 & 0.081 & 0.008 & 0.006 & 1 \\ 0.079 & 0.046 & 0.013 & 0.055 & 0 \end{bmatrix}. \tag{10.24}$$

Equation (10.10) then yields the AP vector

$$P_{\mathbf{A}} = \mathbf{P}_{\mathbf{D}\rightarrow\mathbf{A}} P_{\mathbf{D}} = \begin{bmatrix} 3.87 \\ 2.64 \end{bmatrix} \times 10^{-6}/\text{reactor-year}. \tag{10.25}$$

2. Obtain the RN inventories for ECF events by combining three different inventory data. A radionuclide inventory summary given in Appendix A provides the equilibrium inventory data for a 3560-MWt reactor in units of MCi, while Table 10.4 lists the equilibrium mass inventory for the 2500-MWt Surry plant in the nine RN groups. Finally, Table 10.5 provides the mass inventory of RNs for the Surry plant at about $T = 10$ hours into a long-term SBO accident, when the pressure vessel is postulated to fail.

(a) Assume that the equilibrium radioactivity inventory is proportional to the rated power level and obtain the equilibrium radioactivity inventory $Q_i(0), i = 1, \ldots, 9$, in units of MCi for the Surry plant.

(b) Assume that the radioactivity inventory $Q_i T), i = 1, \ldots, 9$, is proportional to the corresponding RN inventory $M_i(t)$, both for $t = 0$ at the beginning of the accident and for $t = T$ at the vessel breach, and that the RN inventory $M_i(T)$ provides a reasonable approximation to the inventory for all ECF events,

$$\frac{Q_i(T)}{Q_i(0)} = \frac{M_i(T)}{M_i(0)}, \; i = 1, \ldots, 9. \tag{10.26}$$

Exclude structural materials Sn, Zr, Fe, Cr, Ni, Ag, Cd, and In from Table 10.5 when computing $M_i(T)$ values to determine $Q_i(T), i = 1, \ldots, 9$, in units of 100 MCi as

$$Q^T(T) = \begin{bmatrix} 0.09 & 0.20 & 0.004 & 1.15 & 1.56 & 4.20 & 5.71 & 15.42 & 1.20 \end{bmatrix} \tag{10.27}$$

for a total inventory of 2.954 BCi at $T = 10$ hours.

(c) Although the LCF events will involve a substantially less RN inventory, use Table 10.5 and $Q_i(T), i = 1, \ldots, 9$, obtained in step (b) as an approximate estimate for these events as well. We will comment on this approximation after we present our simplified PRA study for the Surry plant.

3. Use Figs. 10.10 and 10.19 to obtain the mean RN release fractions $f(S_i|A_j)$ of Eq. (10.15) for A_1 = ECF and A_2 = LCF events and nine source groups represented as a matrix,

$$f^T_{\mathbf{A}\to\mathbf{S}} = \begin{bmatrix} 0.8 & 0.23 & 0.2 & 0.12 & 0.024 & 0.005 & 0.002 & 0.006 & 0.024 \\ 0 & 0.027 & 8\times10^{-4} & 0.002 & 1\times10^{-4} & 5\times10^{-5} & 2\times10^{-5} & 2\times10^{-5} & 1\times10^{-4} \end{bmatrix}. \tag{10.28}$$

Equation (10.15) is then used to combine $f_{\mathbf{A}\to\mathbf{S}}$ with the radioactivity inventory $Q_i(T), i = 1, \ldots, 9$, from step 2 to obtain the inventory of RNs released in units of MCi,

$$P^T_{\mathbf{A}\to\mathbf{S}} = \begin{bmatrix} 7.53 & 4.60 & 0.08 & 13.79 & 3.76 & 1.89 & 1.03 & 8.48 & 2.87 \\ 0 & 0.54 & 3\times10^{-4} & 0.195 & 0.02 & 0.021 & 0.01 & 0.031 & 0.013 \end{bmatrix}. \tag{10.29}$$

Note that Eqs. (10.27) and (10.28) represent the radiological source terms in two groups, corresponding to early and late release times, but explicitly for nine RN groups. This is in contrast to the 30 to 60 source term groups considered in Eq. (10.16) for NUREG-1150.

4. Perform the matrix multiplication of Eq. (10.16) with two AP bins, A_1 and A_2, to obtain the source term vector in units of Ci/reactor-year,

$$P^T_{\mathbf{S}} = \begin{bmatrix} 29.13 & 19.24 & 0.318 & 53.88 & 14.6 & 7.37 & 4.00 & 32.9 & 11.14 \end{bmatrix}, \tag{10.30}$$

and a total radioactivity release rate q = 172.6 Ci/reactor-year due to postulated accidents.

5. Use the Gaussian plume model of Eq. (8.34) with the Pasquill Type F dispersion coefficients based on the recommendation of Regulatory Guide 1.4 [NRC74] to obtain the atmospheric dispersion factor for a ground-level release of RNs and at 5 km from the release:

$$\frac{\chi}{Q} = \frac{1}{\pi u \sigma_y \sigma_z} = 6.4\times10^{-5}\ \frac{\text{s}}{\text{m}^3}. \tag{10.31}$$

Table 10.5 Mass Inventory of the Core Melt at the Time of Vessel Failure for the Surry plant

Element	Mass (kg)	Element	Mass (kg)
Cs	4.6	Cd	75.9
I	0.46	In	494
Xe	9.4	Ce	131
Kr	0.49	Rb	0.55
Te	18.0	Br	0
Ag (FP)	0	Ru	104
Sb	0	Rh	20.9
Ba	60.6	Pd	52.5
Sn	249	Nd	171
Tc	37.1	Eu	8.90
UO_2	79,650	Gd	0
Zr (Struct)	7,480	Nb	2.70
Zr (FP)	81.3	Pm	7.20
Fe	23,200	Pr	50.7
Mo	155	Sm	34.0
Sr	47.6	Y	22.9
Cr	6,370	Np	26.0
Ni	3,540	Pu	469
Mn	0	Se	0
La	62.3	FeO	12,660
Ag (Struct)	2,610	ZrO_2	12,030

Source: [NRC90]

6. Introduce a simplifying assumption that the RN release due to ECF and LCF events consists of 1-MeV gammas and takes place at a constant rate over 8 hours following the containment failure, together with the infinite cloud model of Eq. (8.42), to obtain the radiological dose:

$$\begin{aligned} \text{Dose} &= 0.507\chi\bar{E}_\gamma T = 0.507\bar{E}_\gamma\frac{\chi}{Q}QT = 0.507\bar{E}_\gamma\frac{\chi}{Q}q \\ &= 5.6 \text{ mrem/reactor-year.} \end{aligned} \tag{10.32}$$

Here, we set $\bar{E}_\gamma = 1$ MeV and the exposure time $T = 8$ hours for release rate Q in units of Ci/s due to accidents per reactor-year. Note also in Eq. (10.32) that the choice of the exposure time $T = 8$ hours is completely arbitrary and immaterial, since all we need finally to calculate the dose is the total RN release rate $q = 172.6$ Ci/reactor-year obtained in step 4.

7. Finally, use the health effect model of BEIR-III, used in NUREG-1150, which suggests 2×10^{-4} fatalities/person-rem, rather than 5.7×10^{-4} fatalities/person-rem suggested in the recent BEIR-VII report [NAP05], to obtain a point frequency

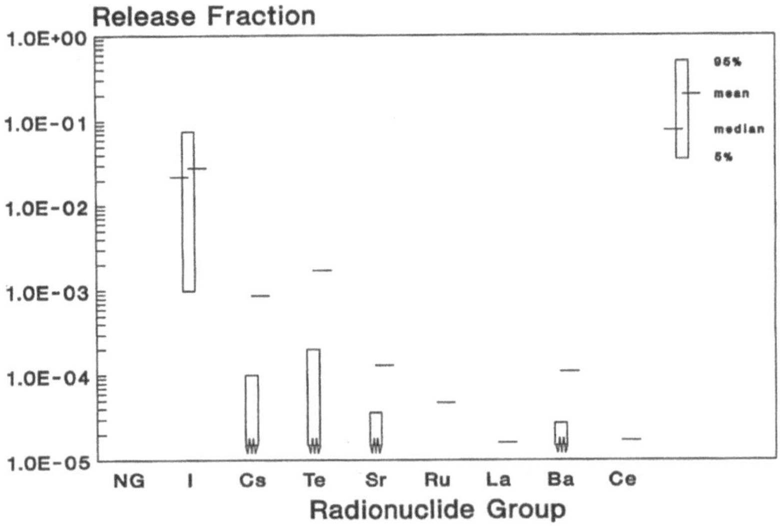

Figure 10.19 Probability density functions for release fraction $f(S_i|A_j)$ in nine radionuclide groups for Surry late-containment failure events. *Source:* [NRC90, Vol. 1].

estimate for early fatalities, which may be approximately considered as a mean value:

$$\text{Mean frequency of early fatality} = 1 \times 10^{-6} \text{ fatalities/reactor-year.} \quad (10.33)$$

This result for the fatality estimate may now be compared with the NUREG-1150 estimate of 2×10^{-6} fatalities/reactor-year for the Surry plant given in Fig. 10.13.

10.3.2 Parametric Studies and Comments on the Simplified PRA Model

Even without applying the building-wake dispersion correction factor [NRC74] in the range of 1.1 to 3.0 to Eq. (10.31), the agreement between our approximate estimate of Eq. (10.33) and the NUREG-1150 early fatality frequency estimate is within a factor of 2, which must certainly have benefited from the cancellation of effects due to several approximations introduced.

Parametric studies illustrate the usefulness of the simple model. One simplifying assumption inherent in the entire model presented in Section 10.3.1 is the use of the same RN inventory for both the ECF and LCF releases. To understand the effect of this approximation, repeat the same source term calculation for $P(S)$ of Eq. (10.30) with LCF inventory decreased by a factor of 4. The resulting total radioactivity release rate is 170.9 Ci/reactor-year, a decrease of only 1% from the base calculation of 172.6 Ci/reactor-year. This result illustrates simply but succinctly that the overall RN release and hence the risk are dominated by early-containment failure events. This is a valuable insight that could have been gained without the parametric study, certainly with a bit of hindsight, but nonetheless illustrates the usefulness of our seven-step PRA model.

We could also perform another set of parametric studies where we single out just the contributions from the SBO and bypass events alone. This may be accomplished simply by considering the SBO and bypass event frequencies separately in Eq. (10.23) for the PDS vector $P(D)$ and repeating steps 1 through 4 to arrive at new estimates for the source term $P(S)$ in Eq. (10.30). With the simplifying assumption that early fatalities are determined entirely by the total amount of radioactivity released, we arrive at contributions of 9% and 87% to the total early fatalities from the SBO and bypass events, respectively. The relative contributions obtained from our simple analysis compare with ~16% and ~83%, respectively, from Fig. 10.15. Although one could say there are substantial differences between our seven-step PRA estimates and NUREG-1150 results here, we are able to get approximate but valuable insights to major contributors to early fatality estimates.

These two parametric studies illustrate the usefulness of our simplified seven-step PRA model in providing physical insights into the risk estimates. The method may also be useful in performing risk-benefit analyses when certain system modifications or procedure changes are to be evaluated on a preliminary basis. It should of course be recognized that such analyses are possible only after full, detailed PRA studies for a particular plant have been completed and should supplement any parametric studies that can be performed with a full-scope database and PRA software. The simplified method may also provide useful risk comparisons for various NPPs, including the five LWRs studied in NUREG-1150.

References

[AEC57] "Theoretical Possibilities and Consequences of Major Accidents in Large Nuclear Power Plants," U.S. Atomic Energy Commission (1957).

[Bri75] G. A. Briggs, "Plume Rise Prediction," in *Proc. of Workshop: Lectures on Air Pollution and Environmental Analysis,* American Meteorological Society (1975).

[Cha90] D. I. Chanin, H. Jow, J. A. Rollstin, et al., "MELCOR Accident Consequence Code System (MAACS)," NUREG/CR-4691, vols. 1–3, U.S. Nuclear Regulatory Commission (1990).

[Dos89] S. S. Dosanjh, "MELPROG-PWR/MOD1: A Two-Dimensional, Mechanistic Code for Analysis of Reactor Core Melt Progression and Vessel Attack Under Severe Accident Conditions," NUREG/CR-5193, U.S. Nuclear Regulatory Commission (1989).

[Eva85] J. S. Evans, D. W. Moeller, and D. W. Cooper, "Health Effects Model for Nuclear Power Plant Accident Consequence Analysis," NUREG/CR-4214, U.S. Nuclear Regulatory Commission (1985).

[Gie79] J. A. Gieseke, P. Baybutt, H. Jordan, and R. G. Jung, "Fission Product Analysis," NUREG/CR-0697, U.S. Nuclear Regulatory Commission (1979).
[Gie86] J. A. Gieseke et al., "Source Term Code Package: A User's Guide," NUREG/CR-4587, U.S. Nuclear Regulatory Commission (1986).
[Has02] F. E. Haskin, A. L. Camp, S. A. Hodge, and D. A. Powers, "Perspectives on Reactor Safety," NUREG/CR-6042, rev. 2, U.S. Nuclear Regulatory Commission (2002).
[Ima84] R. L. Iman and M. J. Shortencarier, "A Fortran 77 Program and User's Guide for the Generation of Latin Hypercube and Random Samples for Use with Computer Models," NUREG/CR-3624, U.S. Nuclear Regulatory Commission (1984).
[Int77] International Commission on Radiological Protection, "Recommendations of ICRP," Publication 26, *Annals of ICRP* **1**, no. 3 (1977).
[Int78] International Commission on Radiological Protection, "Limits for Intakes of Radionuclides by Workers," Publication 30, *Annals of ICRP* **2**, nos. 3 and 4 (1978).
[Jow93] H. N. Jow, W. B. Murfin and J. D. Johnson, "XSOR Codes Users Manual," NUREG/CR-5360, U.S. Nuclear Regulatory Commission (1993).
[Koc81] D. C. Kocher, "Dose Rate Conversion Factors for External Exposure to Photons and Electrons," NUREG/CR-1918, U. S. Nuclear Regulatory Commission (1981).
[Kou90] H. J. C. Kouts, G. Apostolakis, E. H. A. Birkhofer, L. G. Hoegberg, W. E. Kastenberg, L. G. LeSage, N. C. Rasmussen, H. J. Teague, and J. J. Taylor, "Special Committee Review of the Nuclear Regulatory Commission's Severe Accident Risks Report (NUREG-1150)," NUREG-1420, U.S. Nuclear Regulatory Commission (1990).
[Mor93] M. G. Morgan, "Risk Analysis and Management," *Sci. Am.* **269**, 32 (1993).
[NAP05] *Health Risks from Exposure to Low Levels of Ionizing Radiation*, BEIR VII—Phase 2, Biological Effects of Ionizing Radiation Committee, National Academies Press (2005).
[Nor07] D. Normile, "Quake Underscores Shaky Understanding of Ground Forces," *Science* **317**, 438 (2007).
[NRC74] "Assumptions Used for Evaluating the Potential Radiological Consequences of a Loss of Coolant Accident for Pressurized Water Rectors," Regulatory Guide 1.4, U.S. Nuclear Regulatory Commission (1974).
[NRC75] "Reactor Safety Study—An Assessment of Accident Risks in U.S. Commercial Nuclear Power Plants," WASH-1400, U.S. Nuclear Regulatory Commission (1975).
[NRC86] "Safety Goals for the Operation of Nuclear Power Plants," Title 10, *Code of Federal Regulations*, Part 50, Policy Statement, U. S. Nuclear Regulatory Commission (1986).
[NRC90] "Severe Accident Risks: An Assessment for Five U.S. Nuclear Power Plants," NUREG-1150, U.S. Nuclear Regulatory Commission (1990).
[NRC01] "RELAP5/MOD3.3 Code Manual, Volume 1: Code Structure, Systems Models, and Solution Methods," NUREG/CR-5535, U.S. Nuclear Regulatory Commission (2001).

[Sta84] D. S. Stack, "A SETS User's Manual for Accident Sequence Analysis," NUREG/CR-3547, U.S. Nuclear Regulatory Commission (1984).
[Sta02] M. Stamatelatos, "Probabilistic Risk Assessment Procedures Guide for NASA Managers and Practitioners," Version 1.1, Office of Safety of Mission Assurance, National Aeronautics and Space Administration (2002).
[Sum95] R. M. Summers, R. K. Cole, Jr., R. C. Smith, D. S. Stuart, S. L. Thompson, S. A. Hodge, C. R. Hyman, and R. L. Sanders, "MELCOR Computer Code Manuals," NUREG/CR-6119, U.S. Nuclear Regulatory Commission (1995).
[Swa87] A. D. Swain III, "Accident Sequence Evaluation Program—Human Reliability Analysis Procedure," NUREG/CR-4772, U.S. Nuclear Regulatory Commission (1987).
[Was91] K. E. Washington, K. K. Murata, R. G. Gido, F. Gelbard, N. A. Russell, S. C. Billups, D. E. Carroll, R. O. Griffith, and D. L. Y. Louie., "Reference Manual for the CONTAIN 1.1 Code for Containment Severe Accident Analysis," NUREG/CR-5715, U.S. Nuclear Regulatory Commission (1991).

Exercises

10.1 A PRA study for a PWR plant reports the consequences of severe accidents in terms a CCDF of Eq. (10.3),

$$G(x) = \exp(-ax^b), \text{ with } a = 12 \text{ and } b = 0.05,$$

which represents the probability $\{$number of early fatalities per year $> x\}$. (a) Obtain the probability density function corresponding to the above CCDF and (b) calculate the mean number of early fatalities expected per year for the plant.

10.2 Repeat the simplified PRA analysis of Section 10.3.1 with the inventory of Table 10.5 reduced by a factor of 10 for AP bin A_2 in an effort to account for the expected reduction in the radionuclide inventory available for the late-containment leakage. Determine an alternate estimate for the total release rate of radionuclides and discuss the result.

10.3 Repeat the simplified PRA analysis of Section 10.3.1 using an alternate table of equilibrium radioactivity, e.g., Table 5.1-1 of NUREG/CR-6042 [Has02], and compare with the dose estimate of Eq. (10.32).

CHAPTER 11

PASSIVE SAFETY AND ADVANCED NUCLEAR ENERGY SYSTEMS

All of the advanced nuclear energy systems under development feature enhanced safety features as well as system designs that would allow for reduced construction costs and improved operational efficiency. The enhanced safety features are typically represented in terms of higher levels of passive safety, with goals to accomplish (a) self-shutdown capability even with failures in the reactor protection system and (b) long-term cooling capability for the entire plant without reliance on forced circulation of coolant. We begin Section 11.1 with a discussion of passive safety tests performed at the Experimental Breeder Reactor Unit II (EBR-II) in 1986 and the physical basis for the tests. This is followed by specific examples of advanced reactor designs in Sections 11.2 and 11.3. In addition to the description of the specific features of these designs, a few sample techniques that may be applied for system and safety analyses of the advanced reactor designs are also presented.

11.1 PASSIVE SAFETY DEMONSTRATION TESTS AT EBR-II

A set of two safety tests was performed [Pla86] at the EBR-II in April 1986 that succinctly established the feasibility of passive safety in a sodium-cooled fast reactor (SFR) with metallic fuel. The tests also raised the possibility of introducing safety features into other nuclear power plant designs that support the basic principles of passive safety. The EBR-II passive demonstration tests were performed a couple of weeks before the ill-fated Chernobyl accident and provided further impetus for studying passive safety features for the future generation of NPPs. Some of these features are discussed in connection with Generation III+ NPPs in Section 11.2. Passive safety was mostly referred to as inherent safety originally in connection with the 1986 EBR-II tests, but the term *passive safety* is now more commonly used.

We begin with a brief description of the pool-type primary system design of the EBR-II plant in Section 11.1.1, together with a simplified system model. This will illustrate the role of large negative-feedback effects of metallic fuel behind the self-shutdown capability of the EBR-II. The passive safety demonstration tests involving a loss of flow without scram (LOFWS) transient and a loss of heat sink without scram (LOHSWS) event are then described in Section 11.1.2. A simplified fuel channel analysis is presented in Section 11.1.3, followed by a discussion in Section 11.1.4 of the implications of the 1986 EBR-II tests for subsequent NPP designs.

11.1.1 EBR-II Primary System and Simplified Model

The EBR-II operated successfully for 30 years before it was shut down for decommissioning in 1995. It featured a pool-type primary system involving a Na-Na intermediate heat exchanger (IHX) coupled to a steam generator. The reactor generated 20 MWe (62.5 MWt) of power which was fed to the local grid. A schematic of the EBR-II primary system is presented in Fig. 11.1. Note that the primary pumps pick up the sodium coolant from the pool and supply the sodium to the reactor core at the inlet plenum. The sodium discharged from the outlet plenum is circulated by the auxiliary pump into the IHX, where the sodium is returned to the pool.

We develop two simplified energy balance equations that describe the basic features of the EBR-II system dynamics, one for the core and the other for the primary loop. For the core dynamics, we introduce a fuel channel model comprising a cylindrical fuel rod surrounded by a coolant channel using basic energy conservation equations for the fuel and coolant regions coupled through the heat flux at the fuel rod surface. For the primary loop dynamics, we derive a macroscopic energy balance equation representing lumped-parameter models for the core, IHX, and inlet and outlet plena.

11.1.1.1 Lumped-Parameter Fuel Channel Model for the Core For a simplified thermal-hydraulic (TH) analysis of the core, we introduce macroscopic energy balance equations for a fuel channel illustrated in Fig. 11.2. The energy balance for the fuel rod is represented by a time-dependent heat conduction equation for fuel temperature T_f with volumetric heat source S,

$$\rho_f C_f \frac{\partial T_f}{\partial t} = -\nabla \cdot \mathbf{q} + S, \tag{11.1}$$

where ρ_f and C_f are the density and heat capacity of the fuel rod, respectively, and $\mathbf{q}$ is the heat flux at the rod surface in contact with the coolant. Assuming constant

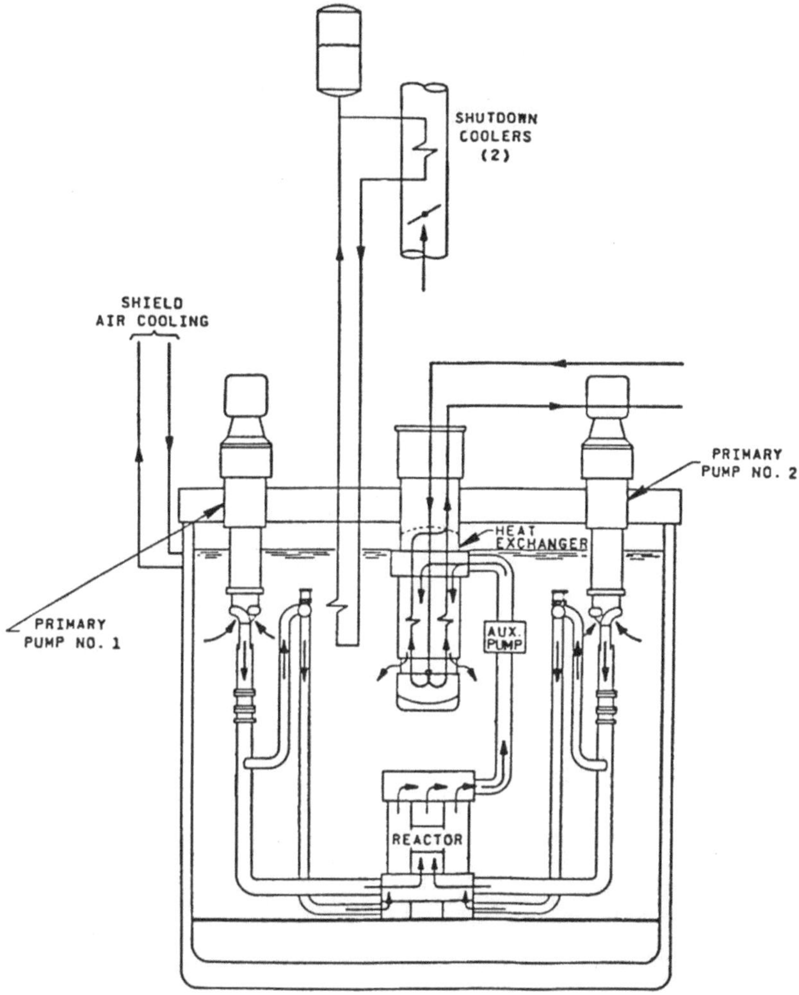

Figure 11.1 Schematic diagram of the EBR-II primary system. *Source:* [Gol87].

heat capacity and integrating Eq. (11.1) over the fuel volume V_f, with total fuel mass M_f, yields

$$M_f C_f \frac{\partial T_f}{\partial t} = -\int_{A_f} \mathbf{q} \cdot \mathbf{n} dA + V_f S = -qMH + V_f S. \tag{11.2}$$

Here, we have used the Gauss divergence theorem to convert the volume integral of $\nabla \cdot \mathbf{q}$ into an integral of the wall heat flux q over the fuel rod surface area $A_f = MH$, where M and H are the wetted perimeter and length of the coolant channel,

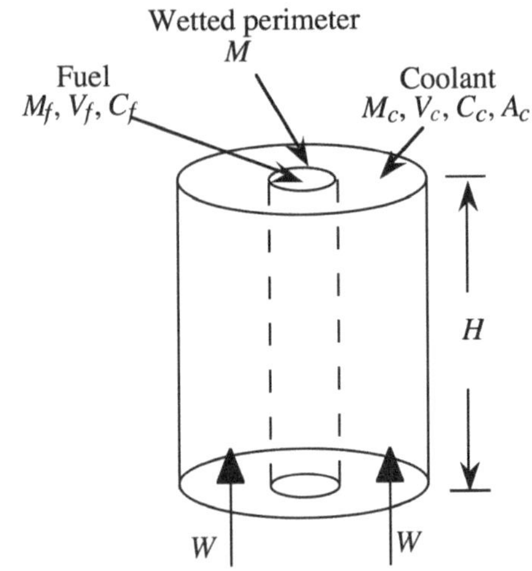

Figure 11.2 Lumped-parameter fuel channel model.

respectively. By invoking Newton's law of cooling with the overall heat transfer coefficient U introduced, the wall heat flux q at the fuel-coolant interface may be written as

$$q = U(T_f - T_c), \tag{11.3}$$

where the channel-average coolant temperature T_c is obtained from the energy balance equation for the coolant channel and T_f is obtained as an arithmetic average of inlet and outlet fuel temperatures, $T_{f,in}$ and $T_{f,out}$, for the fuel rod,

$$T_f = (T_{f,out} + T_{f,in})/2. \tag{11.4}$$

For the coolant channel, we neglect the kinetic energy, potential energy, and viscous heating to write an energy balance in terms of the coolant enthalpy h,

$$\frac{\partial}{\partial t}\rho_c h = -\nabla \cdot (\rho \mathbf{v} h) - \nabla \cdot \mathbf{q} + \frac{\partial p}{\partial t} + S, \tag{11.5}$$

where ρ_c and $\mathbf{v}$ are the coolant density and fluid velocity, respectively. For slow transients that are characteristic of passive safety systems, we may drop the pressure derivative term and neglect the volumetric heat source S representing the direct deposition of gamma energy released in the fission process. In addition, we treat the coolant as an incompressible fluid with heat capacity C_c and one-dimensional, vertical fluid speed v. With these simplifying assumptions, Eq. (11.5) is integrated to yield an energy balance equation in terms of coolant temperature T_c:

$$M_c C_c \frac{\mathrm{d}T_c}{\mathrm{d}t} = -\rho_c v A_c \int_0^H \frac{\partial h}{\partial z}\mathrm{d}z + A_f q = -W C_c \Delta T_c + MHU(T_f - T_c), \tag{11.6}$$

where M_c and A_c are the total coolant mass and cross-sectional area of the coolant channel, respectively. In addition, the coolant mass flow rate $W = \rho_c v A_c$ is introduced, together with the coolant temperature rise across the core height H,

$$\Delta T_c = T_{c,out} - T_{c,in}, \tag{11.7}$$

in terms of the inlet and outlet coolant temperatures $T_{c,in}$ and $T_{c,out}$, respectively, and, similar to Eq. (11.4), the channel-average coolant temperature is obtained as an arithmetic average of inlet and outlet coolant temperatures:

$$T_c = \tfrac{1}{2}(T_{c,out} + T_{c,in}). \tag{11.8}$$

We now propose to use the simplified fuel channel model to represent the core dynamics. Thus, T_f and T_c of Eqs. (11.4) and (11.8) now represent the core-average fuel and coolant temperatures, respectively, while ΔT_c of Eq. (11.7) represents the core-average coolant temperature rise across the core height. Furthermore, noting that the transients resulting from the LOFWS and LOHSWS events are sufficiently slow, we may introduce a quasistatic assumption [Ott88,Pla87,Wad88] for macroscopic energy balance equations (11.2) and (11.6) by setting the time derivatives to zero. Equation (11.2) then simply shows that the total heat flux $MHq(t)$ into the coolant channel is equal to the total heat generation rate $V_f S(t)$ in the core or the total core power. Introducing the relative power $P(t)$ and relative flow rate $F(t)$, the coolant energy balance equation (11.6) reduces to

$$W(0)F(t)C_c \Delta T_c(t) = MHq(0)P(t) \equiv P_T P(t), \tag{11.9}$$

where P_T is the rated total core power. Finally, the time-dependent coolant temperature rise across the core is obtained,

$$\Delta T_c(t) = \frac{MHq(0)}{W(0)C_c}\frac{P(t)}{F(t)} = \Delta T_c(0)\frac{P(t)}{F(t)}. \tag{11.10}$$

This is a simple, intuitive equation that shows that, in slow transients, the coolant temperature rise across the core is proportional to the core power and inversely proportional to the coolant flow rate. Equation (11.10) may now be used to obtain a variation in the coolant temperature rise, given a variation in the power-to-flow ratio:

$$\delta\left[\Delta T_c(t)\right] = \Delta T_c(0)\delta\left[\frac{P(t)}{F(t)}\right] = \Delta T_c(0)\left[\frac{P(t)}{F(t)} - 1\right], \tag{11.11}$$

which yields a relationship connecting variations in the outlet and inlet coolant temperatures:

$$\delta T_{c,out}(t) = \delta T_{c,in}(t) + \Delta T_c(0)\left[\frac{P(t)}{F(t)} - 1\right]. \tag{11.12}$$

11.1.1.2 Primary Loop Dynamics Model In this section, we develop a simplified energy balance equation for the primary loop consisting of the core, IHX, and inlet and outlet plena based on the macroscopic energy balance equations (11.2) and (11.6). We make a simplifying assumption that the coolant sodium inventory in the core and IHX is small compared with the coolant inventory in the inlet and outlet plena. This is a reasonable assumption in view of a large sodium inventory in the pool of the pool-type SFR design illustrated in Fig. 11.1. Thus, in terms of the total heat capacities C_{in} and C_{out} for the inlet and outlet plena, respectively, and relative heat rejection rate P_r from the IHX, we set up a macroscopic energy balance equating the rate of change of the total sodium internal energy to the difference between the core power and IHX heat rejection rate:

$$\frac{\mathrm{d}}{\mathrm{d}t}\left[C_{in}T_{c,in}(t) + C_{out}T_{c,out}(t)\right] = MHq(0)\left[P(t) - P_r\right] = P_T\left[P(t) - P_r\right]. \tag{11.13}$$

The macroscopic energy balance of Eq. (11.13) is schematically illustrated in Fig. 11.3. Introducing the total primary loop sodium inventory $C_p = C_{in} + C_{out}$, together with Eq. (11.12), we rewrite Eq. (11.13) as

$$\frac{\mathrm{d}}{\mathrm{d}t}\left\{C_p\delta T_{c,in}(t) + C_{out}\Delta T_c(0)\left[\frac{P(t)}{F(t)} - 1\right]\right\} = P_T\left[P(t) - P_r\right]. \tag{11.14}$$

Equation (11.14) may be integrated to yield an equation for the inlet plenum temperature variation in terms of the relative power and power-to-flow ratio:

$$C_p\delta T_{c,in}(t) = P_T\int_0^t \left[P(t') - P_r\right]\mathrm{d}t' - C_{out}\Delta T_c(0)\left[\frac{P(t)}{F(t)} - 1\right], \tag{11.15}$$

where the IHX heat rejection rate is assumed to remain constant during the slow transient. Equation (11.15) agrees with an expression that Sevy presented [Sev85] without a derivation.

11.1.1.3 Quasistatic Reactivity Feedback Model The reactivity feedback effects in metal-fueled pool-type SFRs are a bit more complex than the fuel temperature feedback, primarily associated with the Doppler broadening of absorption resonances, and moderator density feedback for LWR cores discussed in Section 8.5.3. The fuel temperature feedback involves primarily thermal expansion and bowing effects of metallic fuel rods, both in the axial and radial directions, as well as the Doppler effect. In addition, the fuel pool as a whole may undergo axial expansion effectively resulting in an increased insertion of control rods into the core, with the upper end of the control rods fixed at the control rod housing outside the pool. The bottom fuel grid plate may also experience thermal expansion and movement. At the same time, heating and thermal expansion of sodium may result in an increase in the reactivity due to the hardening of the neutron flux spectrum. This potential for a positive sodium void coefficient will be discussed in Section 11.3.1. All these thermal feedback effects play a major role in the reactivity feedback and hence passive safety of metal-fueled SFRs.

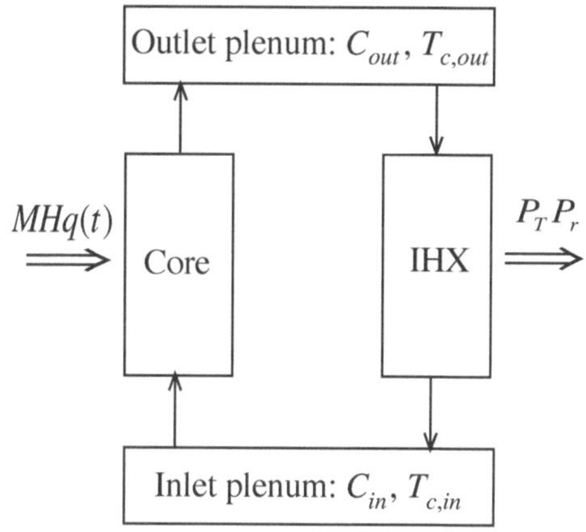

Figure 11.3 Illustration of primary loop energy balance.

Consistent with the macroscopic energy balance of Eqs. (11.2), (11.6), and (11.15), we also assume that the net reactivity variation is negligibly small throughout the transient. Thus, the temperature and flow feedback effects may be combined [Ott88,Pla87,Wad88] to yield

$$\delta K(t) = A[P(t) - 1] + B\left[\frac{P(t)}{F(t)} - 1\right] + C\delta T_{c,in}(t) \simeq 0, \tag{11.16}$$

where

A = fuel temperature coefficient of reactivity

B = flow coefficient of reactivity

C = inlet temperature coefficient of reactivity

The sum of *A* and *B* is essentially the power coefficient of reactivity representing power changes affecting both the fuel and coolant temperature distributions. For the EBR-II, the feedback coefficients *A*, *B*, and *C* are all negative. Substituting the primary loop balance equation (11.15) into Eq. (11.16) results in a combined reactivity balance equation

$$A[P(t) - 1] + B'\left[\frac{P(t)}{F(t)} - 1\right] + \frac{CP_T}{C_p}\int_0^t [P(t') - P_r]\,dt' = 0, \tag{11.17}$$

where $B' = B - (C_{c,out}C/C_p)\Delta T_c(0)$. Equations (11.16) and (11.17) are the two key expressions representing the quasistatic formulation of the core and primary loop dynamics.

For the EBR-II LOFWS event we analyze in Section 11.1.2, the inlet coolant temperature $T_{c,in}$ remains nearly constant during the transient and the associated feedback effect may be neglected. Equation (11.16) may then be solved for the time-dependent power-to-flow ratio

$$\frac{P(t)}{F(t)} = \frac{1 + A/B}{1 + (A/B)F(t)}. \tag{11.18}$$

An asymptotic value of the power-to-flow ratio can be obtained approximately in the limit as $F(t)$ becomes vanishingly small,

$$\lim_{t\to\infty} \frac{P(t)}{F(t)} = 1 + \frac{A}{B} > 1.0, \tag{11.19}$$

which simplifies Eq. (11.12) to

$$\delta T_{c,out}(\infty) = \frac{A}{B}\Delta T_c(0). \tag{11.20}$$

For the LOHSWS event, the heat sink decreases, i.e., $P(\infty) = 0$, but the primary coolant flow remains nearly constant, allowing us to set $F(t) = 1$ in Eq. (11.16), which provides a simple expression for the asymptotic inlet coolant temperature rise

$$\delta T_{c,in}(\infty) = \frac{A+B}{C} = \frac{\text{power coefficient of reactivity}}{\text{inlet temperature coeffcient of reactivity}}. \tag{11.21}$$

In fact, in this transient, the coolant temperature rise ΔT_c decreases as the heat sink is lost and the inlet coolant temperature rise decreases the reactivity, thereby rendering the core to a high-temperature, low-power state. Equation (11.17) is also simplified to

$$(A + B')\,[P(t) - 1] + \frac{CP_T}{C_p}\int_0^t [P(t') - P_r]\,\mathrm{d}t' = 0, \tag{11.22}$$

which can be converted to a differential equation for $P(t)$ and integrated again to yield

$$P(t) - P_r = (1 - P_r)\exp(-t/\tau), \text{ with } \tau = (A + B')\,C_p/CP_T. \tag{11.23}$$

Equation (11.23) shows that in a LOHSWS event the power level would decrease approximately exponentially with time constant τ. Finally, Eq. (11.12) indicates that the asymptotic outlet coolant temperature variation would reach

$$\delta T_{c,out}(\infty) = \delta T_{c,in}(\infty) - \Delta T_c(0) = \frac{A+B}{C} - \Delta T_c(0), \tag{11.24}$$

which simply yields $T_{c,out}(\infty) = T_{c,in}(\infty)$, consistent with the premise of the LOHS $\Delta T_c(\infty) = 0$.

With typical values [Ott88] estimated for the reactivity feedback coefficients A, B, and C and $\Delta T_c(0) = 140$ K for the EBR-II [Fel87], we obtain the asymptotic coolant temperature increases of Eqs. (11.20) and (11.21) in Table 11.1. The quasistatic formulations provide simple but valuable comparisons of the expected coolant temperature rises for the metal- and oxide-fueled SFR configurations. The temperature increases for both the postulated LOFWS and LOHSWS are much smaller for the metal-fueled core than those for the oxide-fueled core, indicating a greater potential for passive safety and relatively mild transients expected in metal-fueled pool-type SFRs.

Table 11.1 Representative Feedback Coefficients and Temperature Rises

Fuel type	$A(\$)$	$B(\$)$	$C(\$/\text{K})$	$\delta T_{c,out}(\infty)$ (K) LOFWS	$\delta T_{c,in}(\infty)$ (K) LOHSWS
Metal	-0.15	-0.30	-0.003	70	150
Oxide	-1.70	-0.40	-0.004	595	525

With the feedback coefficients used in Table 11.1, an estimate [Ott88] $C_p/P_T \simeq$ 1.0 second, and the simplifying assumption that equal inventories of sodium in the upper and lower plena are involved in the primary loop dynamics, i.e., $C_{out} = 0.5C_p$, for a metal-fueled core we estimate the time constant of Eq. (11.23) for the power reduction in a LOHSWS transient,

$$B' = B - 0.5C\Delta T_c(0) = -0.09\ \$ \text{ and } \tau = 80 \text{ s.} \tag{11.25}$$

The corresponding time constant for an oxide-fueled core is 455 seconds, indicating that the transient would be much faster in a metal-fueled core.

11.1.2 Unprotected Loss-of-Flow and Loss-of-Heat-Sink Tests

11.1.2.1 Loss of Flow Without Scram Test The LOFWS test was initiated, in the morning of March 3, 1986, with the reactor operating at its rated power, by turning off the primary and secondary sodium pumps and bypassing the loss of flow (LOF) scram circuit. To be prudent with the demonstration test, the auxiliary sodium pump was kept on a battery power at 3 to 4% of rated flow, although subsequent computer simulations indicated that the auxiliary pump had negligible contributions to the overall outcome of the test. Thus, the test also effectively simulated a SBO event, which would disable both the primary and secondary sodium pumps, followed by a scram failure. In the SFR community, a scram failure event is also called an unprotected transient event, e.g., the abbreviation LOFWS is used synonymously with ULOF.

Figure 11.4 illustrates the transient behavior of the LOFWS test, which was formally referred to as the SHRT-45 test [Gol87,Pla87]. In addition to the test data, results of the simulation of the test with the DSNP code [Sap93] are plotted for the relative or normalized sodium flow rate $F(t)$, relative power $P(t)$, outlet sodium temperature $T_{c,out}(t)$, and reactivity $K(t)$ for the 500-second duration of the test. Note first that the maximum reactivity variation was less than 40 cents, or on the order of 100 pcm, which justifies the quasistatic approximation of Eq. (11.16). The maximum increase in the outlet sodium temperature $T_{c,out}(t)$ was approximately 220 K, in contrast to the inlet sodium temperature rise of less than 40 K during the transient. This justifies the approximation not to explicitly account for inlet sodium temperature variations in Eq. (11.18).

Comparison of the relative flow and power variations in Fig. 11.4 shows that the power variation follows the flow variation but with a time lag. This is understandable

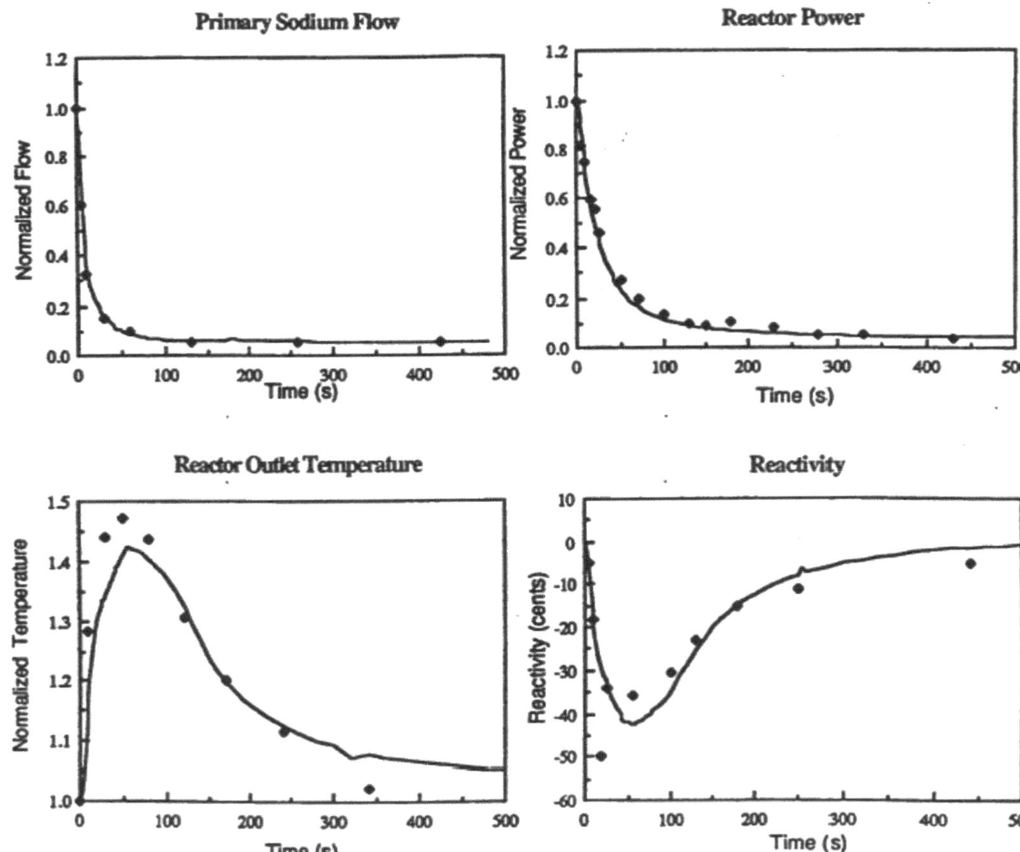

Figure 11.4 Evolution of primary sodium flow, reactor power, outlet sodium temperature, and reactivity during the SHRT-45 test. DSNP simulation results plotted as curves are compared with the test data.

since the primary sodium flow coastdown introduces a negative reactivity insertion, which decreases the power level to a low asymptotic level at the end of the 500-second transient. This self-shutdown behavior of the EBR-II may be also illustrated by noting that the power-to-flow ratio in Eq. (11.18) remains greater than unity, since $F(t) < 1.0$ during the flow coastdown, and asymptotically reaches the ratio given in Eq. (11.19).

The temperature measured in the instrumented fuel assembly XX09 during the SHRT-45 test is plotted [Cha87] in Fig. 11.5, together with computer predictions for the assembly temperature and the cladding temperature of the hottest driver fuel assembly. The SHRT-45 data indicated [Cha87] that the cladding temperature in the hottest fuel assembly exceeded the eutectic temperature of 988 K for the U-Zr metallic fuel and 316 stainless steel clad for approximately 50 seconds. This time period of 50 seconds was estimated to be approximately 2% of the time duration allowed for the clad temperature to exceed the eutectic point without inducing the actual damage due to eutectic formation. The computer prediction was made with a combination of the NATDEMO code [Moh81] for the coupled nuclear-TH plant calculations and the HOTCHAN code [Moh87] for the calculation of individual fuel assembly temperatures. The reactor restarted immediately upon the completion of the LOFWS test and no fuel breach was experienced at the EBR-II.

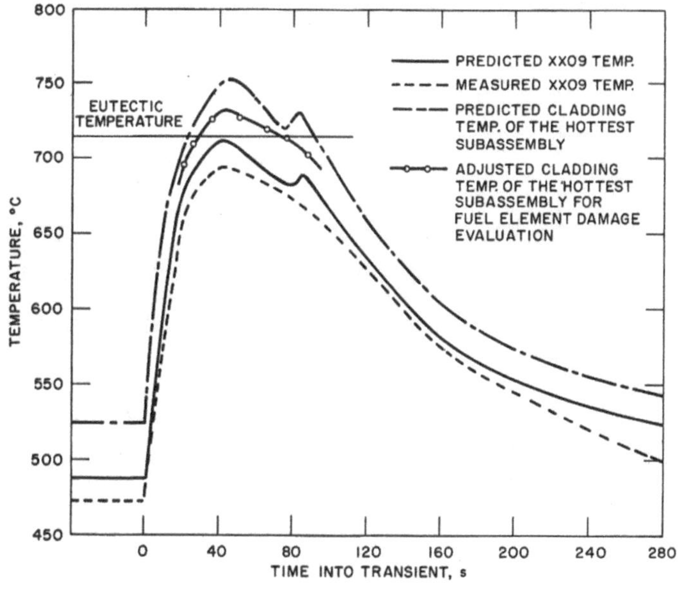

Figure 11.5 Inner region temperature of instrumented fuel assembly XX09 and cladding temperature of the hottest driver assembly during the SHRT-45 transient. *Source:* [Cha87].

11.1.2.2 Loss of Heat Sink Without Scram Test A LOHSWS test was performed at the EBR-II in the afternoon of March 3, 1986, following the LOFWS test in the morning. This second test was initiated by disabling the reactor scram and stopping the sodium flow in the IHX, by physically tripping the secondary sodium pump and reversing the voltage on the electromagnetic pump. This test simulated faults in the secondary heat transfer loop or power conversion system coupled with a scram failure. In this test, the decrease in the heat sink for the IHX resulted in a corresponding decrease in the primary sodium temperature rise $\Delta T_c(t)$, inducing an increase in the inlet primary sodium temperature $T_{c,in}(t)$. This resulted in a negative reactivity insertion and a smooth decrease in the power level over a period of approximately 1200 seconds, as illustrated [Fel87] in Fig. 11.6. The peak increase in $T_{c,in}(t)$, registered at the high-pressure plenum (HPP) inlet, was 40 K, while the outlet primary sodium temperature $T_{c,out}(t)$, measured in the instrumented assembly XX09 inner region, decreased by 95 K. In this unprotected LOHS test, the 40 K increase in $T_{c,in}(t)$ was considerably smaller than an estimate of 150 K in Table 11.1, perhaps due to inaccuracies in our estimate of the feedback coefficients. The time constant τ = 80 seconds estimated in Eq. (11.25) also appears to be short compared with that indicated by Fig. 11.6.

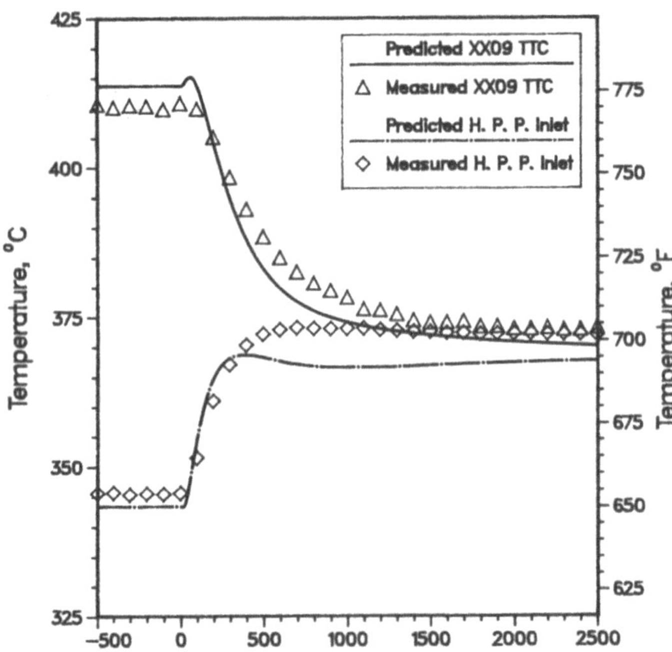

Figure 11.6 Temperatures of HPP inlet and inner region (TTC) of instrumented fuel assembly XX09 during the SHRT-45 transient. *Source:* [Fel87].

The LOFWS and LOHSWS tests eloquently established the self-shutdown capability of metal-fueled SFRs that benefits from low fuel temperatures and large thermal expansions possible in metallic fuel. These tests, together with subsequent transient overpower tests at the EBR-II, essentially demonstrated the possibility of accommodating ATWS events in SFRs without adverse consequences. The actual planning of the passive safety demonstration tests was done in multiple incremental power levels to provide a safe experimental approach and establish the necessary semi-empirical correlations. One key correlation obtained through this process for computer simulations is the amount of sodium pool inventory that is involved in the natural circulation cooling of the core during the tests.

11.1.3 Simplified Fuel Channel Analysis

As a simple demonstration of the macroscopic fuel channel model developed in Section 11.1.1, we present a numerical solution of Eqs. (11.2) and (11.6) for an SFR core subject to a flow coastdown with time constant τ_c that simulates a LOF event:

$$F(t) = 0.2 + \frac{0.8}{1 + t/\tau_c}. \tag{11.26}$$

The flow coastdown causes the reactor power to decrease to 30% of full power with time constant τ_f,

$$P(t) = 0.3 + \frac{0.7}{1 + t/\tau_f}. \tag{11.27}$$

As discussed in the analysis of the EBR-II LOFWS test, assume that the inlet coolant temperature $T_{c,in}$ and fuel temperature $T_{f,in}$ at the inlet of the fuel channel remain constant during the transient. Assume also that thermodynamic properties of fuel and coolant and the overall heat transfer coefficient U characterizing heat transfer from the fuel to the coolant channel remain constant throughout the transient. Table 11.2 presents design parameters estimated from an SFR design [Tho91].

The solution of the macroscopic energy balance equation for an SFR fuel channel is illustrated in Fig. 11.7, together with the flow and power coastdown profiles in Fig. 11.8. We note that the outlet coolant temperature $T_{c,out}$ initially increases but eventually tapers off. This trend is somewhat similar to the outlet coolant temperature plots in Figs. 11.4 and 11.5 for the EBR-II LOFWS test. The difference, $T_f(t) - T_c(t)$, between the core-average fuel and coolant temperatures is approximately proportional to the core power as indicated by Eqs. (11.3) and (11.9). From the plot, we note that $T_f(\infty) - T_c(\infty) \simeq 50$ K, which is equal to $[T_f(0) - T_c(0)]P(\infty)$, with $P(\infty) \simeq 0.32$. Likewise, we can quickly verify that $\Delta T_c(\infty) = 227 \text{ K } = \Delta T_c(0)P(\infty)/F(\infty) = 150 \text{ K} \times 0.32/0.21$, in agreement with Eq. (11.10). One visible limitation of the simplified model for the calculation of average fuel and coolant temperatures with Eqs. (11.2) and (11.6), respectively, appears as the rather unphysical result that the outlet coolant temperature could be higher than the fuel temperature at the channel outlet.

Table 11.2 Design Parameters for a Typical SFR Design

Rated power	470 MWt
Fuel rod diameter	7 mm
Equivalent coolant channel diameter	9 mm
Fuel length	1.35 m
Fuel material	U-10 wt.% Zr alloy
Power density	43.4 kW/kg of fuel
Coolant flow rate per channel	0.1858 kg/s
Flow coastdown time constant τ_c	15 s
Power coastdown time constant τ_f	30 s
Fuel density	15.7 $\mathrm{Mg/m^3}$
Coolant sodium density	0.847 $\mathrm{Mg/m^3}$
Heat capacity of fuel rod	0.22 kJ/kg·K
Heat capacity of coolant	1.27 kJ/kg·K
Overall heat transfer coefficient U	7.949 $\mathrm{kW/m^2{\cdot}K}$
Inlet coolant temperature $T_{c,in}$	600 K
Fuel temperature at channel inlet $T_{f,in}$	750 K
Fuel melting temperature	1487 K
Sodium boiling temperature	1156 K

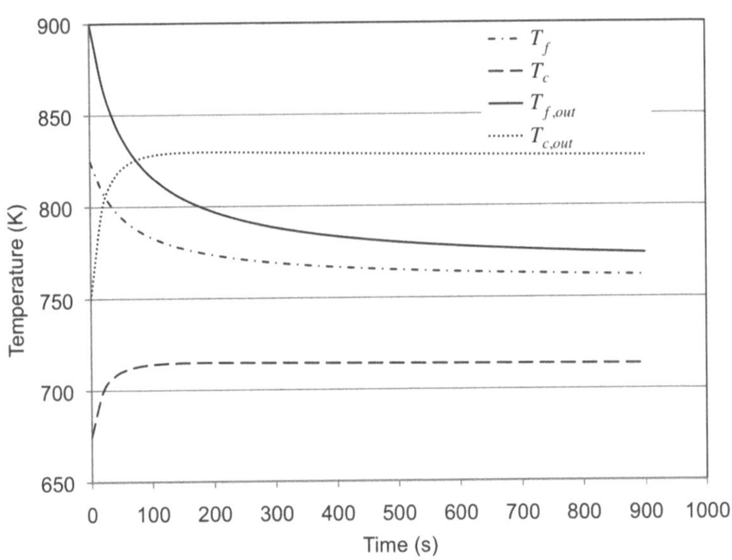

Figure 11.7 Fuel and coolant temperatures following power and flow coastdown.

11.1.4 Implications of EBR-II Passive Safety Demonstration Tests

The successful demonstration of passive safety features of SFRs through the 1986 EBR-II tests was a significant milestone for the nuclear community and motivated the development of passive safety features for other NPP designs during the late

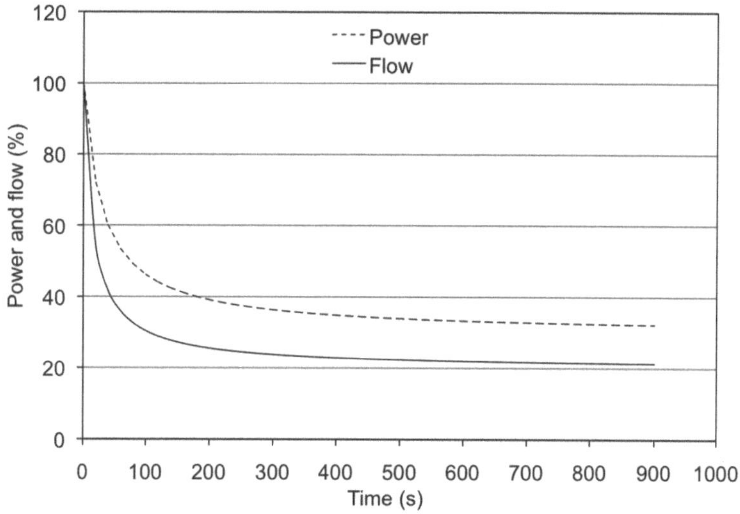

Figure 11.8 Power and flow coastdown curves.

1980s and early 1990s. Much of the effort focused on natural circulation cooling and enhanced depressurization capabilities in LWRs. One particular concept that received considerable attention was the density lock, featuring hot-cold interfaces, for the process inherent ultimate safety (PIUS) reactor design [For89]. The PIUS design could allow for natural circulation cooling of the core in transient events where the overheating of the primary system occurs. This design and other similar concepts led to the AP600 and simplified boiling water reactor (SBWR) designs that featured substantial passive features. These designs evolved into the AP1000 and economic SBWR (ESBWR) designs discussed in Section 11.2.

Together with these efforts, the nuclear industry recognized the need to avoid the costly operator errors in the 1979 TMI-2 accident discussed in Section 9.1. This resulted in the development of the Utilities Requirements Document (URD) [Dev95], eventually released by the Electric Power Research Institute (EPRI). The URD is an industry-sponsored effort that seeks to define the technical basis for advanced LWR designs [Mar93], including the full reliance on passive-safety-grade systems for 72 hours with no need for operator action in responding to postulated design basis accidents. It is intended that all active systems will be nonsafety grade and therefore outside the scope of the stringent regulatory oversight imposed on safety-grade systems, implying high levels of reliability associated with the passive safety systems.

One particular passive safety feature adopted in both Generation III+ designs discussed in Section 11.2 is the ability to depressurize the primary system in time so that reservoirs of coolant inside the containment building may be effectively used to keep the core cooled following postulated DBAs. This feature is particularly evident in the in-containment refueling water storage tank (IRWST) for the AP1000 and the gravity-driven cooling system (GDCS) for the ESBWR covered in Section 11.2.

11.2 SAFETY CHARACTERISTICS OF GENERATION III+ PLANTS

Key safety features of the AP1000 and ESBWR plants are described in Sections 11.2.1 and 11.2.3, respectively, as examples of Generation III+ advanced reactor designs. In addition to descriptions of new or enhanced safety features of the AP1000 design, the effectiveness of the passive systems is illustrated with a discussion of a small-break LOCA in Section 11.2.2. For the second Generation III+ example, an approach for the reliability uncertainty quantification is illustrated in Section 11.2.4 for the analysis of the passive containment cooling system of the 600-MWe SBWR design.

11.2.1 AP1000 Design Features

The AP1000 is a four-loop plant featuring 157 fuel assemblies with an active fuel length of 4.27 m (14 ft) for power output of 1150 MWe (3411 MWt) and evolved from the 600-MWe AP600 design [Wes92], for which the concepts and applicability of passive safety features were developed. The AP1000 design [Wes03] received the design certification from the U.S. Nuclear Regulatory Commission in 2006 and satisfies the URD requirement for full reliance on passive safety systems without operator actions for three days into postulated accidents.

No pumps, fans, diesel generators, chillers, or other rotating machinery are required for the safety systems in normal operating conditions and postulated accidents for the AP1000 design. A few simple valves align passive safety systems when they are automatically actuated, with valves designated as fail safe. They require power to stay in their normal closed position and a loss of power causes them to open into their safety alignment.

The AP1000 design [Bru04,Pau02,Wes92,Wes03] includes the passive safety injection (SI), passive residual heat removal (PRHR), and passive containment cooling system (PCCS). The pressurizer has a volume of 45.3 m^3 (1600 ft^3), which is 30% larger than those normally used in plants of comparable power rating, so that there is no need for PORVs. This in turn eliminates a possible source of reactor coolant system leakage and reduces maintenance tasks. Simplified designs featuring passive systems allow support systems to be nonsafety grade, e.g., the service water system and associated safety cooling tower. Key safety systems for the AP1000 design illustrated in Fig. 11.9 include:

11.2.1.1 Containment System

(a) Containment Structures

The system design maintains the containment peak pressure below the design limit for a double-ended break of a primary or secondary side pipe. For primary system breaks, the design analysis assumes loss of offsite power, together with the failure of one of the valves for cooling water flow of the passive containment cooling system.

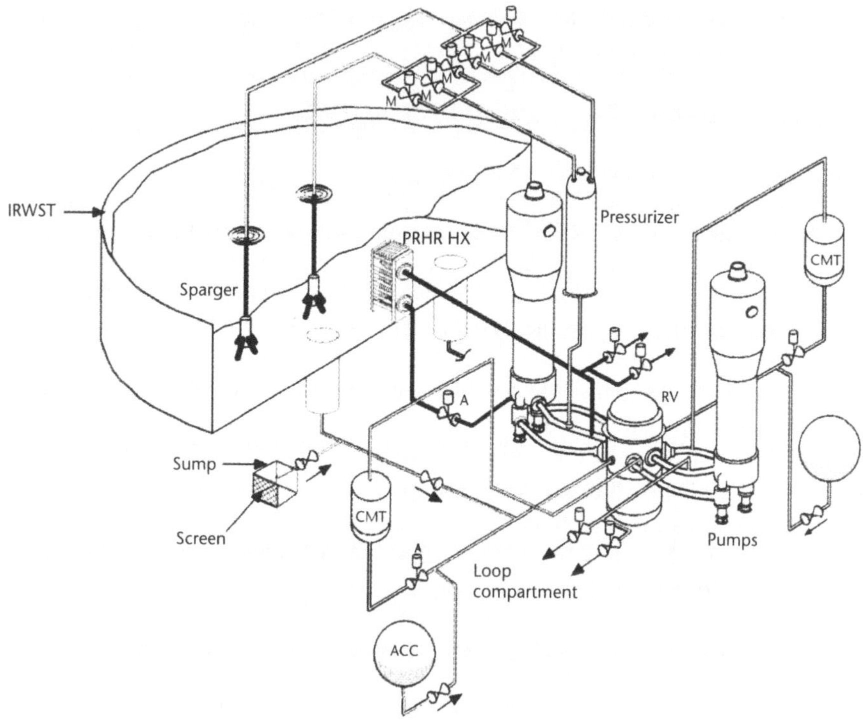

Figure 11.9 Passive core cooling features of AP1000. *Source:* Reprinted with permission from [Pau02]. Copyright © 2002 Progressive Media Group.

(b) Passive Containment Cooling System
This passive system is designed to reduce containment pressure and temperature following a LOCA or main steam line break (MSLB) so that the containment pressure is to remain below the design limit with no operator action required for three days and to fall below half the design limit within 24 hours. The PCCS transfers heat directly from the steel containment vessel to the environment and the system relies on a number of components, including the PCCS gravity-drain water tank, air baffle, air inlet and exhaust, and water distribution system.

(c) Containment Isolation System
The system consists of piping, valves, and actuators to isolate the containment while allowing for the passage of emergency fluids in case of accidents. Two barriers are provided for each isolation system, e.g., a check valve inside the containment and a motor-operated valve outside the containment.

(d) Containment Hydrogen Control System
The system is designed to monitor the hydrogen concentration and maintain it below the flammability limit, utilizing hydrogen recombiners located inside the

containment for a design basis LOCA. Hydrogen igniters are also distributed within the containment to protect the system in case of a core melt accident.

11.2.1.2 Passive Core Cooling System (PXS)

(a) Safety Injection and Depressurization System
The accumulators (ACCs), core makeup tanks (CMTs), in-containment refueling water storage tank (IRWST), and containment sump provide passive injection of coolant during a LOCA so that the PXS operates without the use of active equipment, e.g., pumps and AC power sources, although it requires one-time alignment of valves in the system. The IRWST, normally isolated by check valves, injects coolant into the RCS through four stages of the automatic depressurization system (ADS), which are actuated by the reduction in the system pressure. The valves in the last stage of the ADS are squib-actuated for reliable leak-free delivery of the coolant. In addition, the CMTs automatically provide coolant to the RCS in non-LOCA events.

(b) Passive Residual Heat Removal Heat Exchanger (PRHR HX)
The passive heat exchanger automatically actuates when the secondary heat removal capability is lost due to a steam generator tube rupture (SGTR), loss of feedwater (LOFW), or MSLB event. The IRWST provides the heat sink for the PRHR HX. The IRWST water absorbs decay heat, and the steam eventually generated in the IRWST passes to the containment, condenses on the steel containment vessel, and drains by gravity back into the IRWST.

11.2.1.3 Habitability System The system provides ventilation and passive heat sinks to the main control room and maintains radiation monitoring, fire protection, and emergency lighting.

11.2.1.4 Fission Product Removal and Control System The system relies on natural fission product removal processes, including aerosol removal and pool scrubbing, which are provided by the containment systems. One set of containment air filtration valves is assumed to remain open during a LOCA.

11.2.2 Small-Break LOCA Analysis for AP1000

For the AP1000 design, a pipe rupture involving a total cross-sectional area ≥ 1.0 ft^2 (0.09 m^2) is considered [Wes03] a LBLOCA and classified as a design basis or condition IV event discussed in Section 8.2.1. The AP1000 design features allow the injection of large volumes of water from the CMT and IRWST into the reactor vessel, following the initial blowdown of the RCS inventory through the broken pipe. This passive delivery of coolant provides long-term cooling of the core. In this section, we discuss the sequence of events following a SBLOCA involving a 2-inch (0.05-m) break in the cold leg connected to CMT-1 to highlight the role of passive safety systems in the AP1000 design. The SBLOCA is classified as a condition III event, which may occur infrequently during the life of the plant.

The AP1000 safety system is designed to provide a controlled depressurization of the RCS if the break is greater than the makeup capability of the charging system. This will then allow the injection of large volumes of borated water from the ACCs, CMTs, and IRWST into the RCS, and the decay heat is removed through the PRHR HX, thereby preventing or minimizing core uncovery. The analysis of SBLOCA transients is performed with the NOTRUMP code [Mey85], with features similar to but simpler than the RELAP5 code [NRC01]. While the RELAP5 code offers one-dimensional two-fluid thermal-hydraulic models for LOCA analyses in general, the NOTRUMP code performs one-dimensional nonequilibrium drift-flux calculations based on macroscopic fluid conservation equations.

The flow channels are discretized into a number of fixed control volumes or cells, each consisting of an upper vapor region and a lower mixture region. The exchange of mass and energy between cells and the upper and lower regions of a cell is represented through a variety of flow paths and networks. For each region with volume V and mass inventory M, the continuity equation is written in terms of incoming and outgoing mass flow rates W_{in} and W_{out}, respectively,

$$\frac{\mathrm{d}M}{\mathrm{d}t} = W_{in} - W_{out}, \tag{11.28}$$

where the flow rates include the mass exchanges between the upper and lower regions of the cell. Similarly, the energy conservation equation (11.5) is written for the internal energy inventory U of each region of the cell in terms of the enthalpies h_{in} and h_{out} for the incoming and outgoing flows, respectively,

$$\frac{\mathrm{d}U}{\mathrm{d}t} = W_{in}h_{in} - W_{out}h_{out} + Q - P\frac{\mathrm{d}V}{\mathrm{d}t}, \tag{11.29}$$

where Q represents the sum of heat fluxes or volumetric heat sources and P the single pressure for the entire cell. The momentum conservation is written for W to represent normal pressure drops, including acceleration, gravitational, frictional, and form loss terms. A staggered mesh structure is used with a momentum integral formulation [Mey61] to duly represent flow rates at junctions involving area changes. In addition, a number of specific models are included in the NOTRUMP code to handle thermal-hydraulic phenomena encountered in LOCA events, including (a) the drift flux model [Wal69] to represent vertical countercurrent two-phase flows, (b) the bubble rise model [NRC01] to calculate the bubble escape rate from the lower region of a stratified interior fluid volume, and (c) the critical flow model to represent the fluid flow out of a broken pipe, together with the appropriate equations of state and various empirical correlations.

For a discussion of the SBLOCA transient, Fig. 11.10 recasts Fig. 11.9 to highlight the connections between key passive safety systems, in particular, the ACCs, CMTs, and IRWST. The simulated RCS pressure transient following the break at $t = 0$ is plotted in Fig. 11.11 corresponding to the following sequence of events:

- Break opens at $t = 0$.
- Reactor trip signal is delivered at $t = 54.7$ seconds.
- Turbine stop valves close at $t = 60.7$ seconds.

- "S" (safety injection) signal is received at $t = 61.9$ seconds.
- Main feed isolation valves begin to close at $t = 63.9$ seconds.
- Reactor coolant pumps start to coast down at $t = 67.9$ seconds.
- ADS stage 1 actuates at $t = 1334.1$ seconds.
- ADS stage 2 actuates at $t = 1404.1$ seconds.
- ACC injection starts at $t = 1405$ seconds.
- ADS stage 3 actuates at $t = 1524.1$ seconds.
- ACC empties at $t = 1940.2$ seconds.
- ADS stage 4 actuates at $t = 2418.6$ seconds.
- CMTs become empty $t = 2895$ seconds.
- IRWST injection initiates at $t = 3280$ seconds.

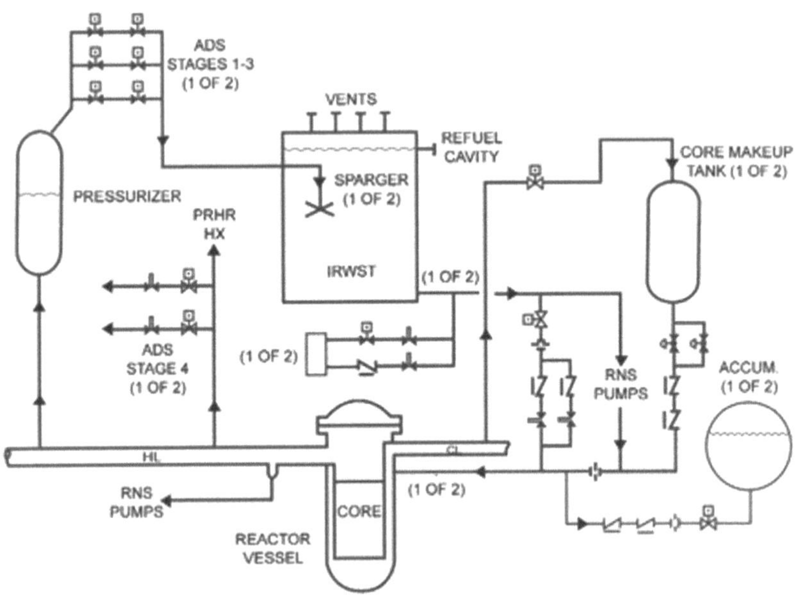

Figure 11.10 Schematic diagram of the AP1000 passive safety system, illustrating the connections between the passive safety systems and non-safety-related normal residual heat removal system (RNS). *Source:* Reprinted with permission from [Wes07]. Copyright © 2007 Westinghouse Electric Company.

Due to the coolant leak through the broken pipe, the RCS coolant inventory and pressurizer (PZR) level continue to decrease, as illustrated in Fig. 11.12, and a reactor trip signal is triggered by a low-PZR pressure at 54.7 seconds into the transient. The reactor trip in turn causes the isolation of steam lines for the steam generators, closing the turbine stop valves at 60.7 seconds. The CMT and PRHR isolation valves begin to inject borated water into the RCS, following the receipt of the "S," or safety injection,

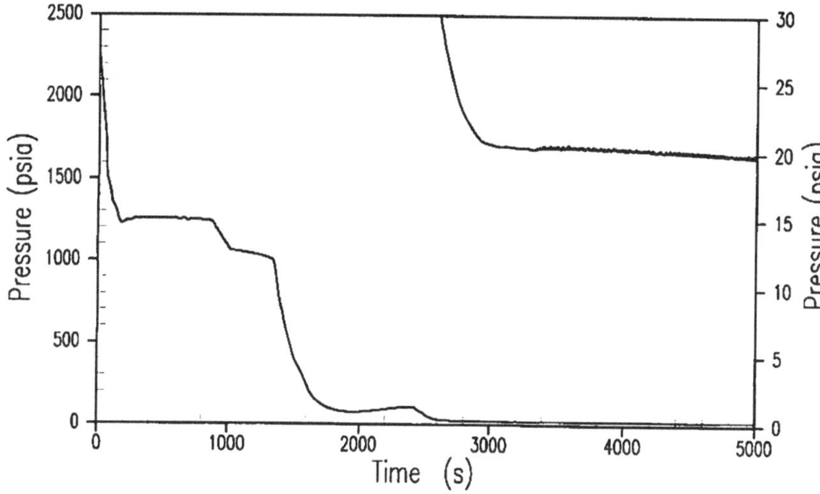

Figure 11.11 AP1000 RCS pressure transient for a SBLOCA event. *Source:* [Wes03].

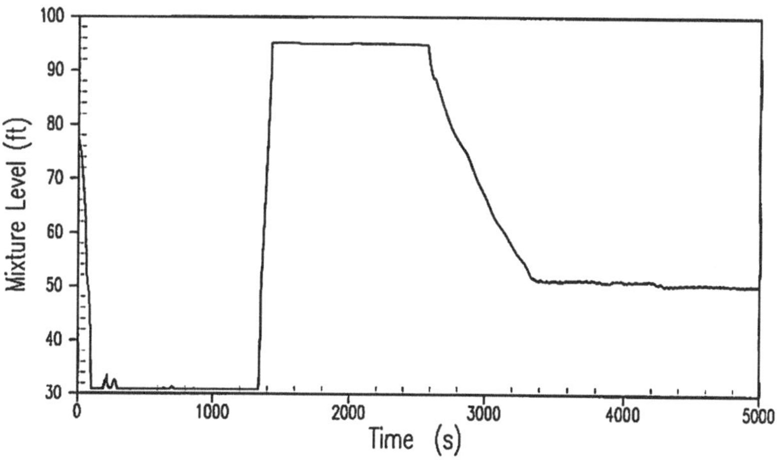

Figure 11.12 AP1000 pressurizer mixture level transient for a SBLOCA event. *Source:* [Wes03].

signal at 61.9 seconds, when the PZR pressure falls below 1700 psia (12 MPa). The reactor coolant pumps trip after the "S" signal with a 6.0-second time delay.

For approximately the next 20 minutes, the mixture level in the downcomer of the reactor pressure vessel (RPV) continues to drop, although the core remains completely covered. When the CMTs drain to the 67.5% level, the ADS stage 1 valves begin to discharge water through an opening at the top of the PZR at 1334.1 seconds, resulting in a rapid restoration of the PZR mixture level shown in Fig. 11.12. Upon the opening of the ADS valves for stages 2 and 3, the increased ADS flow causes a more rapid RCS depressurization, as depicted in Fig. 11.11, allowing the ACC discharge to begin at 1405 seconds. The ACC discharge flow in turn reduces the CMT flow temporarily. The firing of squib-actuated valves in ADS stage 4 initiates when the CMT water level is reduced to 20%, allowing the PZR discharge mixture into the hot legs. Once the downcomer pressure falls below the IRWST injection setpoint, large volumes of IRWST water begin to flow into the RPV at 3280 seconds, retaining the reactor water level at the hot-leg elevation for the remainder of the transient. For the 2-inch SBLOCA, the RPV water level remains at least 1.5 m (5 ft) above the top of the active fuel region throughout the transient and the peak clad temperature occurs at the initiation of the transient. Figure 11.13 also illustrates the effective role of the PRHR HX in decay heat removal during the interval of 600 to 1800 seconds, when the ACC injection is made possible with the actuation of the ADS.

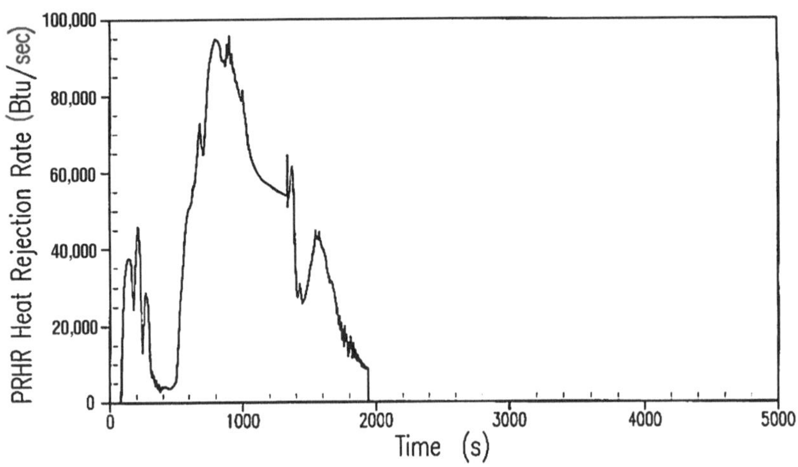

Figure 11.13 AP1000 PRHR heat flux variation for a SBLOCA event. *Source:* [Wes03].

The AP1000 passive safety system makes an effective combined use of (a) the ACCs that provide high flow rates for ~9 minutes, (b) the CMTs that provide relatively high flow rates for ~45 minutes, and (c) the IRWST that provides low flow rates for a longer period of time. As a key passive safety component, the ADSs allow a rapid RPV depressurization and thereby facilitate the replenishment of the coolant inventory in the RPV through the three passive sources of water in SBLOCA events. Compared with the current generation of PWRs, the AP1000 simulation shows a similar RPV pressure decrease during the blowdown period but indicates a significantly faster pressure drop with the actuation of ADSs. In contrast to the AP1000 transient behavior, a typical PWR plant would indicate only a gradual, continuous decrease in the RPV pressure. This reflects a key difference in the engineered safety features because the charging and safety injection pumps, not the ACCs as in the AP1000 system, serve as the primary injection source for a SBLOCA until the residual heat removal system can provide the necessary heat removal.

11.2.3 Economic Simplified Boiling Water Reactor

The Generation III+ BWR system evolved from the current generation of BWRs that require external recirculation pumps into the advanced BWR (ABWR) design that features internal recirculation pumps, thereby significantly reducing the consequences of LOCAs. The ABWR subsequently evolved into the 600-MWe SBWR design, which eliminated the recirculation pumps altogether, relying entirely on natural circulation cooling for normal and emergency operations. The ESBWR increases the power rating to 1550 MWe (4500 MWt), making the design economically competitive with other energy systems. The ESBWR design was docketed in December 2005 for review by the NRC and was approved in 2014.

Natural circulation cooling of the core is achieved through the installation of a tall chimney and associated increase in the RPV height, combined with a decrease in the active fuel length from the conventional 3.7 m (12 ft) to 3.0 m. The increase in power output is obtained by increasing the number of fuel assemblies from 800 and 872 for the BWR/6 and SBWR designs, respectively, to 1132 for the ESBWR. A large inventory of water and steam in the RPV, combined with passive safety features, eliminates safety-grade pumps and AC power for postulated accidents. The ESBWR safety-grade system consists of the emergency core cooling system and PCCS. The ECCS comprises the ADS and GDCS, while the PCCS relies on isolation and passive containment cooling condensers. The ESBWR design allows the rejection of a full load subject to a turbine trip without the need to shut down the reactor. This allows a quick recovery to power production in the event of secondary-system malfunctions. A noteworthy feature of the ESBWR is the basemat internal melt arrest and coolability (BiMAC) core catcher installed below the RPV to protect the plant in accidents resulting in containment failures.

A schematic diagram of the ESBWR plant is given in Fig. 11.14 [Col07], with key safety features [GEH07,GEN92] described below.

11.2.3.1 Containment System

(a) Pressure Suppression Containment
The containment system prevents the release to the environment of fission products, steam, and water released in DBAs, including the large-break LOCAs, and consists of the drywell and suppression chambers, with the connecting vent systems.

(b) Passive Containment Cooling system
As illustrated in Fig. 11.15, the PCCS serves as one of two key safety-grade systems for the ESBWR, and is designed to remove for 3 days the decay heat

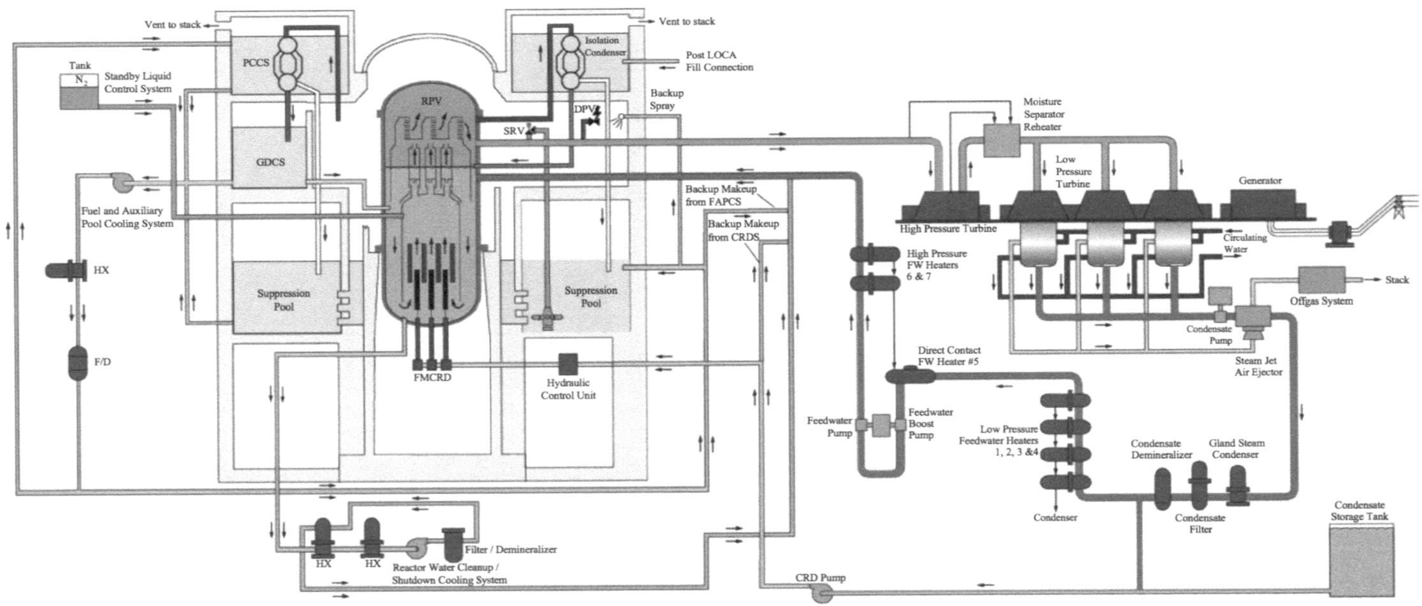

Figure 11.14 Schematic diagram of the ESBWR plant. *Source:* Reprinted with permission from [Col07]. Copyright © 2007 GE Hitachi Nuclear Energy.

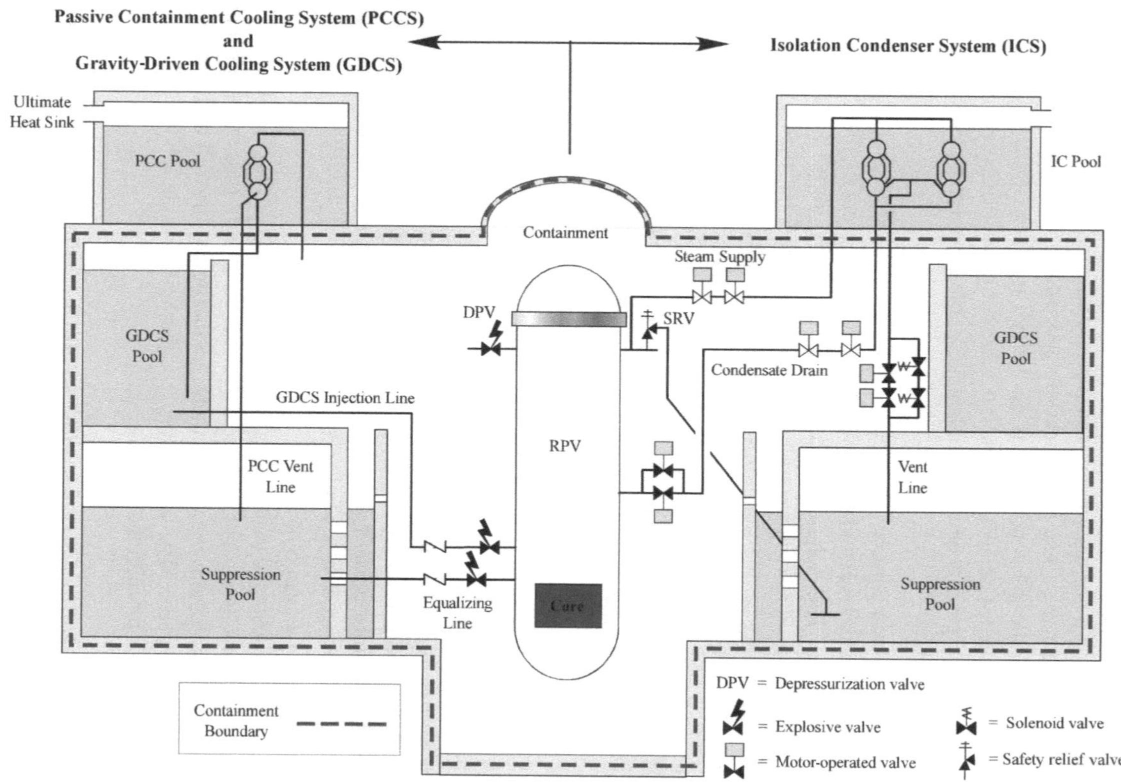

Figure 11.15 ESBWR passive safety systems. *Source:* Reprinted with permission from [Col07]. Copyright © 2007 GE Hitachi Nuclear Energy.

rejected to the containment in a LOCA, thus meeting the URD requirement. Located within pools above and outside the drywell are the isolation and passive containment cooling (PCC) condensers. Both types of condensers or condensing heat exchangers are located in elevated water pools outside the RPV that are vented to the atmosphere.

During medium- and large-break LOCAs, the steam released into the containment is picked up via a normally open inlet piping and condensed on the tube side of the PCC condensers. The latent heat of vaporization is then deposited into the PCC pools, while the condensate is discharged into the GDCS pool located in the upper drywell. Connected to the lower header of the PCC condenser is a vent line that discharges any entrained noncondensable gases below the surface of the suppression pool.

The isolation condensers are intended for all events that challenge the RPV pressure or water level and share with the PCC condensers the PCC pools in the upper drywell. The isolation condensers are connected to the RPV in a closed loop, while the PCC condensers have open flow paths that include a normally open inlet piping for the steam flow as discussed above.

(c) Secondary Containment
This refers to the safety envelope that provides a barrier to fission products released from the containment and houses the entire safety-related equipment, except that associated with the main steam tunnel and isolation condenser and PCC pools. Normally kept at a negative pressure relative to the environment, the safety envelope is designed to isolate automatically upon detection of high drywell pressure, low reactor water level, or high radiation levels in ventilation exhaust.

(d) Containment Isolation System
The system provides the isolation of the containment and prevents the release of radionuclides to the environment through isolation valves, piping, and controls. The motor-operated valves remain in their last position upon failure, while pneumatic-operated valves fail in a closed position.

(e) Flammability Control System
The system is designed to mitigate the buildup of combustible gases from a 100% metal-water reaction in a LOCA and comprises glow-plug-type hydrogen igniters distributed throughout the containment, including the drywell and suppression chamber air space. The glow plug is a thermal ignition device activated by electric current through the tip.

11.2.3.2 Emergency Core Cooling System As the second key component of the safety-grade systems for the ESBWR, the ECCS provides protection against LOCAs for 3 days without operator action through a combined action of the GDCS and ADS. The GDCS provides passive coolant flow to the annulus region of the reactor through dedicated nozzles from pools within the drywell above the core. This ensures that the water level is maintained at least 1 m above the core in all postulated accidents.

Through a combined use of safety relief valves (SRVs) and depressurization valves (DPVs), the ADS discharges steam to the suppression pool (SP) and depressurizes the RPV so that the GDCS can provide coolant to the RPV. The SRV blowdown is directed through lines that are connected to spargers located in the bottom of the suppression pool. These valves are traditional dual-action valves, mounted vertically, so that they open either on primary system overpressure, against a retaining spring, or by a pneumatic power assist. The DPVs are squib-actuated non-reclosing valves mounted horizontally and discharge directly into the drywell airspace and the DPV discharge flow eventually collects in the suppression pool. The squib valves offer high-reliability leak-free capability, with an explosive pushing a piston that shears a sealing cap for the valve. A squib-activated valve provides a larger flow area than standard SRVs. The GDCS flow provides long-term post-LOCA cooling capability and if necessary may drain to the lower drywell in a core melt accident, when the molten fuel reaches the drywell cavity floor.

11.2.3.3 Control Room Habitability System The system provides a safe operating environment, including missile protection, radiation shielding, radiation monitoring, ventilation, and fire protection. The sealed emergency operating area (SEOA) includes all instrumentation and controls necessary for safe shutdown of the plant after a DBA. The emergency breathing air system is a redundant safety-grade system which supplies stored, compressed air to the SEOA for breathing and pressurization to minimize leakage.

11.2.3.4 Fission Product Removal System Included in the system are containment sprays in the upper drywell and suppression chamber, but the sprays are not safety grade, i.e., the operability of the sprays is not assumed in DBA evaluations. When the safety envelope (secondary containment) is isolated, it can be serviced by the reactor building heating, ventilation and air conditioning (HVAC) system with the help of containment filters. Fission products released to the RPV or to the drywell will be entrained in the pool water as they pass through the suppression pool.

11.2.4 Reliability Quantification of SBWR Passive Safety Containment

In this section we present a study to quantify the uncertainty associated with the reliability calculated for the SBWR passive safety containment system. This will illustrate a general approach to obtain an uncertainty estimate for complex system simulations and at the same time provide additional understanding of the passive safety characteristics of an advanced reactor design. Although the study was performed for the 600-MWe SBWR design, the characteristics for the PCCS are essentially the same as those for the 1.55-GWe ESBWR design.

The PCCS reliability quantification [Van96,Van97,Cui01] is based on transient PCCS performance calculations with the CONTAIN code [Mur90] for a postulated MSLB accident. The unreliability or failure probability of the PCCS was determined through a limit surface (LS) delineating the limiting containment pressure, which is obtained through a sequence of CONTAIN cases. To minimize the number of complex CONTAIN cases required to construct the LS, a query learning algorithm was developed that can provide, via a genetic algorithm [Gol89], optimal sets of CONTAIN input parameters for sequential executions of the CONTAIN code. The LS for the containment pressure is characterized by five key parameters affecting the PCCS performance. Hence, each sequential CONTAIN case represents a variation from the previous case in these five parameters. The need to develop optimization algorithms may be readily understood by recognizing that five intervals each for five variables would require 5^5, or 3125, CONTAIN cases. The LS was actually determined with 130 CONTAIN cases and constructed through both an artificial neural network [Hay94] and an alternating conditional expectation algorithm [Bre85,Kim97].

11.2.4.1 Key Features of the CONTAIN Code The CONTAIN code [Mur90, Was91] solves a set of lumped-parameter fluid conservation equations together with the equation of state for a mixture of gases and liquids to predict containment behavior during reactor accidents. The system volume is divided into control volumes or cells, and each cell comprises an upper atmosphere region and, if necessary, a lower pool region. The exchange of mass, energy, and momentum between cells as well as between the upper and lower regions of a cell may be represented through a variety of flow paths and energy transfer processes. The code cannot represent detailed in-vessel phenomena but may accept time-dependent primary system mass and energy flow rates from other system codes, e.g., the RELAP5 code [NRC01], as a boundary condition.

The basic representation of the CONTAIN control volumes may be illustrated conveniently by the mass conservation equations in terms of mass inventory $M^u_{k,i}$ for component k in the upper atmospheric region of cell i,

$$\frac{\mathrm{d}M^u_{k,i}}{\mathrm{d}t} = \sum_j \left(f^u_{k,j} W^u_{j\to i} - f^u_{k,i} W^u_{i\to j}\right) + W^u_{k,i,source} - W^u_{k,i,sink}, \tag{11.30}$$

where $f^u_{k,j} = M^u_{k,j} / \sum_n M^u_{n,j}$ is the fraction of the upper region mass in component k for cell j and $W^u_{j\to i}$ the total upper region mass flow rate from cell j to cell i so that $f^u_{k,j} W^u_{j\to i}$ represents the upper region flow rate for component k from cell j to cell i. The last two terms in Eq. (11.30) represent the effective flow rates of component k into and out of cell i, respectively, due to evaporation, condensation, or direct flows, e.g., the suppression chamber vent flow out of the cell. The lower pool comprises only liquid water and hence the continuity equation is written simply in terms of the mass M^p_i, flow rate $W^p_{j\to i}$ between cells, and direct flow rates into and out of cell i,

$$\frac{\mathrm{d}M^p_i}{\mathrm{d}t} = \sum_j \left(W^p_{j\to i} - W^p_{i\to j}\right) + W^p_{i,source} - W^p_{i,sink}. \tag{11.31}$$

The energy conservation equations for each cell are solved for the total energy inventories of the mixture and water for the upper and lower regions, respectively, without separating out contributions from each component. Contributions from convection, conduction, gravitational forces, and direct flows are represented in the energy conservation equations. Because containment analyses typically involve low flow rates, however, kinetic energy terms are neglected. The momentum conservation equation is written for a single pressure for each cell to represent normal pressure drops, including acceleration, gravitational, frictional, and form loss terms. For a cell consisting of multiple flow segments, the time derivative of the mass velocity is integrated in space over the entire flow path in the cell in a momentum integral approach [Mey61], similar to the NOTRUMP model [Mey85] discussed in Section 11.2.2. The equation of state is written separately for the upper and lower regions, with all gases represented via an ideal gas approximation.

11.2.4.2 The CONTAIN Model for the PCCS Response to an MSLB A simplified CONTAIN model of Fig. 11.16 illustrates key features of the ESBWR engineered safety features described in Section 11.2.3. The nodalization scheme [Van96] focuses on the transient response of the PCCS and ECCS, including the GDCS, SP, DPV, and SRV. Among the simplifications in the 8-cell, 11-path CONTAIN model illustrated in Fig. 11.16 is the absence of isolation condensers, because a previous study indicated that the condensing exchangers play a minor role in an MSLB accident. Figure 11.16 also illustrates flow paths 9 and 10 for the PCCS noncondensable lines and SP vent, respectively, that are represented only in some portions of the CONTAIN runs. Note also that cell 3 for the RPV serves as a repository for the GDCS flow, while blowdown flow rates and other in-vessel phenomena are obtained from a separate RELAP5 [Fle92] calculation.

Cell 1 Upper drywell (DW) head and annulus DW:

Included in this cell are the upper DW head region directly above the RPV and the annular region between the RPV and the reactor shield wall, with the hemispherical steel DW head modeled as an external boundary structure. The shield wall physically separates cell 1 from cell 2 but flow path 1 represents a gas flow path located between the shield and top DW slab.

Cell 2 Upper DW and central DW:

The upper DW region housing the main steam lines, feedwater lines, and GDCS pools are included in this cell, together with the central DW region between the suppression chamber (SC) and annular region of cell 1. In addition to the connection to cell 1 discussed above, cell 2 connects to (a) cell 4, the lower DW via path 2, representing vertical vents, (b) cell 3, the RPV, via path 3, which includes both the MSLB and all DPV flow areas, (c) cell 6 via path 4, representing the PCCS inlet lines, (d) cell 5, via path 8, representing vacuum breakers, (e) cell 5, via path 10, representing the SBWR horizontal vent system, and (f) cell 3, via path 11, representing the liquid injection line from the GDCS pool to the RPV. The vacuum breakers not explicitly illustrated in Fig. 11.16 are included in the CONTAIN model to allow for the flow from the SC to DW when the SC pressure exceeds the DW pressure. Flow path 7, also connecting to cell 5, models leakage flows between the DW and SC, postulated to occur in the vacuum breakers.

Cell 3 Reactor pressure vessel:

A pool region is specified in this cell to model the RPV inventory, with the flow rates during the RPV blowdown obtained from a detailed RELAP5 run as discussed

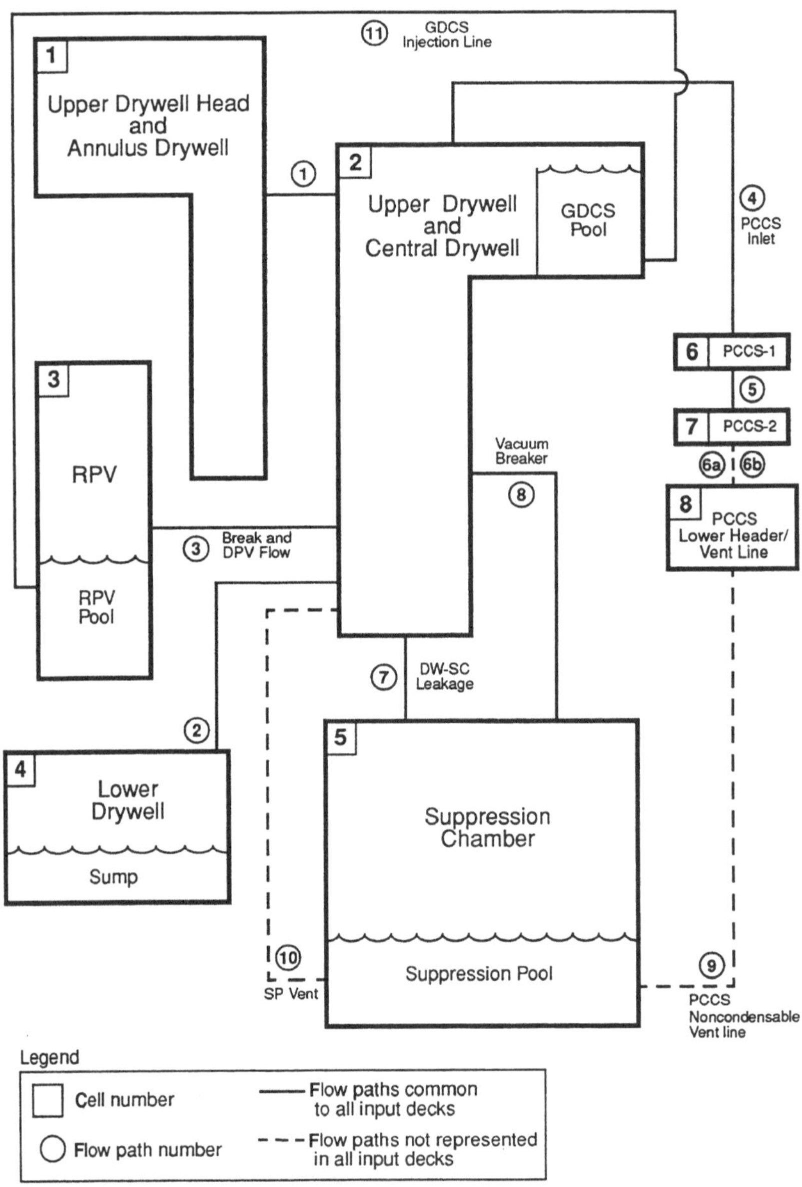

Figure 11.16 CONTAIN 8-cell nodalization for the SBWR passive containment. *Source:* [Van96].

earlier. Flow path 3 models the MSLB and DPV flow areas, while path 11 provides the RPV makeup flow via the GDCS injection line.

Cell 4 Lower DW:

This cell models gas flow from cell 2 via path 2 representing vertical vents and the containment sump that collects condensate runoff from all structures during the accident.

Cell 5 Suppression chamber:

In addition to the vacuum breaker flows represented via paths 7 and 8 discussed for cell 2, the horizontal vent system provides via path 10 the RPV blowdown flow to the suppression pool modeled in this cell during the initial phase of the accident. Path 9, representing the noncondensable vent line, is included in the CONTAIN runs following the reactor depressurization.

Cell 6 Upper PCCS heat exchanger (PCCS-1):

The upper portion of PCCS heat exchangers is modeled in this cell and is connected to cell 2 via flow path 4, representing the normally open PCCS inlet lines. Runoff from the tube surfaces in this cell is directed to cell 7 via path 5.

Cell 7 Lower PCCS heat exchanger (PCCS-2):

This cell represents the lower PCCS heat exchangers and may connect via path 6a to noncondensable vent lines modeled in cell 8 and eventually connect via path 9 to the SP. Via path 6b without the noncondensable vent lines modeled, cell 7 may also provide gas flow to the lower header of the heat exchangers during the initial phase of the accident.

Cell 8 PCCS heat exchanger lower header and noncondensable vent line:

This cell serves as a repository for noncondensable gas, which passes through the PCCS heat exchanger during the initial phase, and later as a conduit for the vent line.

11.2.4.3 Main Steam Line Break Sequence The CONTAIN cases simulate the response of the PCCS following a postulated double-ended guillotine break of one main steam line inside the containment, following the successful reactor scram and closure of the main steam isolation valves. Heat removal by active non-safety-grade systems, including the reactor water clean-up system in conjunction with either the control rod drive, low-pressure coolant injection, or condensate system is, however, assumed unavailable to test the effectiveness of the SBWR passive safety features.

With the MSLB at $t = 0$, key events follow the sequence:

- Reactor scram initiated on high DW pressure at $t = 1$ second
- MSIV closure initiated on low RPV pressure at $t = 1$ to 5 seconds
- RPV water level reaching the ADS actuation point at $t = 570$ seconds
- ADS actuation following the elapse of 10-second ADS timer at $t = 580$ seconds
- Sequential actuation of SRVs and DPSs during $t = 580$ to 725 seconds
- Firing of short-term GDCS squib valves at $t = 730$ seconds
- Firing of equalizing-line squib valves due to continuing decrease in RPV water level at $t > 2380$ seconds

The CONTAIN cases were run to simulate the system response for 72 hours following the MSLB, representing the URD considered in Section 11.1.4. The objective of the reliability quantification study was to determine the probability that the design pressure limit $p_{limit} = 0.483$ MPa for the containment will not be exceeded without operator intervention for 72 hours following a postulated MSLB accident. For this purpose, an LS separating the regions of system success and failure, defined via the design pressure limit, was constructed in terms of five system variables $\mathbf{x} = \{x_1, \ldots, x_5\}$:

x_1 = DW-SC leak area
x_2 = PCCS tube heat transfer coefficient
x_3 = PCCS inlet line flow coefficient
x_4 = GDCS line flow coefficient
x_5 = GDCS line check valve reverse flow fraction

These variables were identified as most relevant to the successful PCCS performance following an MSLB accident.

11.2.4.4 Incremental Query Learning Algorithm for LS Generation To minimize the number of CONTAIN cases required to generate a sufficiently accurate LS in a five-dimensional space spanned by $\{x_1, \ldots, x_5\}$, a genetic algorithm (GA) [Gol89] was implemented to sequentially obtain sets of the five variables for CONTAIN input decks in an optimal progression. The five-dimensional LS for the maximum pressure is represented by a set of training examples for an artificial neural network (ANN) [Hay94], and the training set is incrementally expanded through a series of GA queries to optimally locate new examples in untested regions of the LS.

For mapping a nonlinear relationship between input and output variables for a complex system, an ANN links input and output nodes via suitable activation and threshold functions emulating human cognition and learning processes. In the training step for the network, an improved representation of the activation and threshold functions is attained through a back-propagation process that strives to minimize the difference between the actual and simulated output variables. An ANN architecture involving two hidden layers with seven and six nodes each were used in mapping the LS.

A GA emulates biological evolution processes by constructing bit strings that encode a fitness function and uses stochastic algorithms to manipulate bit positions of the strings via crossover and mutation operators. Through multiple generations involving parent and progeny strings, a string with the highest fitness emerges, which then provides the desired optimal fitness function and hence the optimal set of input variables for the next CONTAIN run. At each iteration, the current LS representation and training set are used to formulate a multiobjective fitness function [Van97]

$$J(\mathbf{x}) = \frac{f_b(\mathbf{x}) f_d^2(\mathbf{x})}{f_\rho(\mathbf{x})}, \tag{11.32}$$

which is chosen to reward low density, balance, and nearness to the LS. As illustrated in Fig. 11.17 for an idealized two-dimensional surface, the distance factor $f_d(\mathbf{x})$ takes the form of a Gaussian distribution centered on the current LS estimate and penalizes potential query points away from the surface. The density factor $f_\rho(\mathbf{x})$ assigns a low importance to the sites with sparse neighboring points, with a threshold distance R_0, while the balance factor $f_b(\mathbf{x})$ preserves the global balance of the training set. Thus,

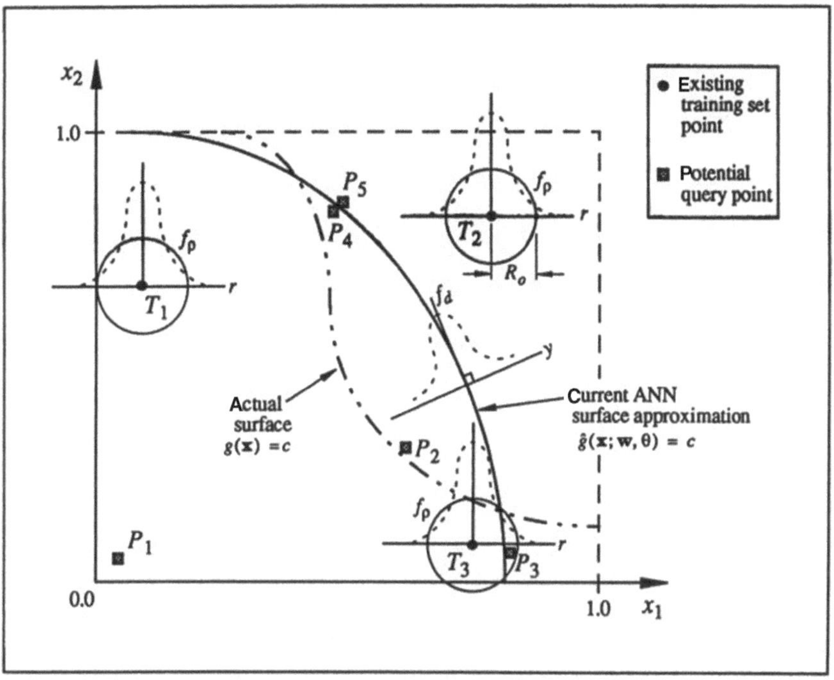

Figure 11.17 Search for the next training set point via a fitness function $J(\mathbf{x})$. *Source:* [Van96].

$f_d(\mathbf{x})$ precludes points P_1 and P_2 because they are far from the current LS estimate, although they may be close to the actual surface yet to be discovered. Point P_3 is precluded by $f_\rho(\mathbf{x})$ because it lies within R_0 of an existing point T_3, and finally $f_b(\mathbf{x})$ favors P_5 over P_4 because P_5 adds to the global balance of the training set. The training of the ANN began with a set of 35 random CONTAIN runs and 95 points were added sequentially to the ANN training set via the GA for a total of 130 points. A three-dimensional projection of the converged ANN representation of the five-dimensional LS is plotted in Fig. 11.18 in terms of normalized system variables.

Once a converged LS was obtained, a Monte Carlo sampling of the PDF $f(\mathbf{x})$ for system variables $\mathbf{x}$ was performed to obtain the limiting containment pressure $p(\mathbf{x})$ and calculate the probability of the system failing to maintain pressure within $p_{limit} = 0.483$ MPa, i.e., the unreliability F of the PCCS:

$$F = \int H[p(\mathbf{x}) - p_{limit}]\, f(\mathbf{x})\mathrm{d}\mathbf{x}, \tag{11.33}$$

in terms of the Heaviside step function H. With $f(\mathbf{x})$ derived from the SBWR standard safety analysis report (SSAR) [GEN92] aided by engineering judgment, a mean unreliability $F = 6.2 \times 10^{-4}$/demand was obtained reflecting nominal degradations in all five system variables. This shows consistency with an SSAR

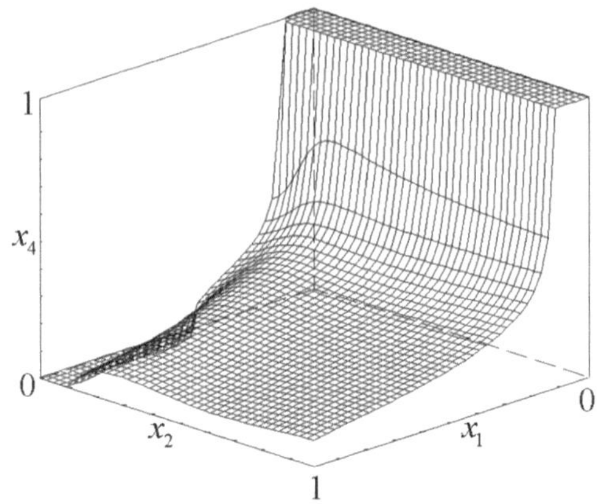

Figure 11.18 Three-dimensional projection of the five-dimensional LS. *Source:* [Van96].

estimate of $F = 3 \times 10^{-4}$/demand for the PCCS failure solely reflecting degradation of the DW-SC vacuum breaker, i.e., x_1 = DW-SC leakage. Additional parametric studies representing different levels of system degradations, together with statistical estimates of uncertainties in the reliability estimates, are reported [Van96,Van97].

With a relatively small number of CONTAIN runs delineating the LS obtained via the ANN-GA incremental learning algorithm, the reliability of the PCCS subject to multiple concurrent system degradations was determined, together with the quantification of uncertainties in the unreliability estimates. For the reliability quantification of systems described by complex nonlinear transient models, a response surface representing general system evolutions may be constructed in a manner similar to the limit surface representing a limiting system parameter considered in this section. Alternate nonlinear mapping tools, including the alternating conditional expectation (ACE) algorithms [Bre85,Kim97], may also be used to represent limit surfaces. For the incremental learning algorithms, in lieu of the combined GA-ANN approach discussed in this section, a Bayesian inference method [Lor04] could be formulated in a structure similar to the Bayesian recursive relationship of Section 13.3.1. These alternate approaches may be fruitful areas for future studies of the reliability quantification of passive safety systems.

11.3 GENERATION IV NUCLEAR POWER PLANTS

Among the six Generation IV concepts selected [Gif02] following the evaluation of more than 100 concepts submitted, the U.S. Department of Energy has supported further studies primarily on two designs, the SFR and very high temperature reactor (VHTR). Only a brief review of the two concepts is presented in this section, together with a discussion of additional safety and design issues for the SFR concept. A systematic process to evaluate safety issues for nuclear systems is presented with the VHTR concept as an example.

11.3.1 Sodium-Cooled Fast Reactor

In addition to the superb passive safety features demonstrated in the 1986 EBR-II tests as discussed in Section 11.1, the SFR offers significant advantages over LWRs both as a transmuter of legacy used nuclear fuel and as a breeder of fissile material. The first advantage derives from the fact that at neutron energies around 100 to 200 keV all of the transuranics, including Np, Am, and Cm, function reasonably well as nuclear fuel, although not as efficiently as fissile Pu isotopes ^{239}Pu and ^{241}Pu. The SFR offers a significant potential as a breeder because the number η of fission neutrons released per absorption in a ^{239}Pu nucleus is 2.6, compared with $\eta = 2.1$ for ^{235}U, in the fast spectrum. The SFR could function equally well with a ^{232}Th-^{233}U cycle because $\eta = 2.3$ for ^{233}U in the fast spectrum.

Figure 11.19 is a schematic diagram of a pool-type SFR design [Gif02] that includes a secondary loop featuring an IHX coupled to a steam generator and turbo-generator facilities. The pool-type design eliminates the possibility of LOCAs due to primary pipe breaks, in addition to a large heat sink provided by the sodium pool as discussed in Section 11.1. We now present two additional points regarding the safety of SFRs.

11.3.1.1 Sodium Void Coefficient of Reactivity The March 1986 EBR-II tests, followed by similar tests involving unprotected transient overpower (UTOP) events, demonstrated that the pool-type SFRs possess superb safety characteristics. In addition, metallic fuel SFR designs benefit from a large thermal conductivity of fuel rods resulting in lower fuel temperatures, and hence lower energy contents, compared with oxide fuel designs. Metallic fuel rods also allow a larger thermal expansion than oxide fuel rods, which essentially forms the basis for the self-shutdown capability demonstrated in the EBR-II passive safety tests. There remains, however, the possibility of a positive void coefficient of reactivity in Pu-fueled SFR cores, due to a particular spectral hardening effect present in fast spectrum reactors.

In SFRs typically fueled with ^{239}Pu and cooled with liquid sodium, coolant voiding could increase the reactivity. This is primarily due to a peculiar behavior of ^{239}Pu cross sections around 100 to 200 keV, which is typically the mean energy of neutrons for these reactors. Around this energy, the capture-to-fission cross section ratio $\alpha = \sigma_c/\sigma_f$ decreases as the neutron energy increases, as illustrated in the ENDF/B-V plot [McL88] of Fig. 11.20. Thus, if sodium voiding were to take place and harden the flux spectrum, the parameter $\eta = \nu/(1+\alpha)$, representing the number of neutrons released per absorption in fuel, would increase, resulting in an increase in k_∞. This tendency for a positive VCR is partly mitigated by an increase in diffusion constant D which would increase the neutron leakage. The net effect of the sodium voiding is determined primarily by these two competing phenomena so that in most viable SFR designs the VCR is positive near the core center where the leakage effect is small but tends to become negative as the periphery of the core and the blanket regions are

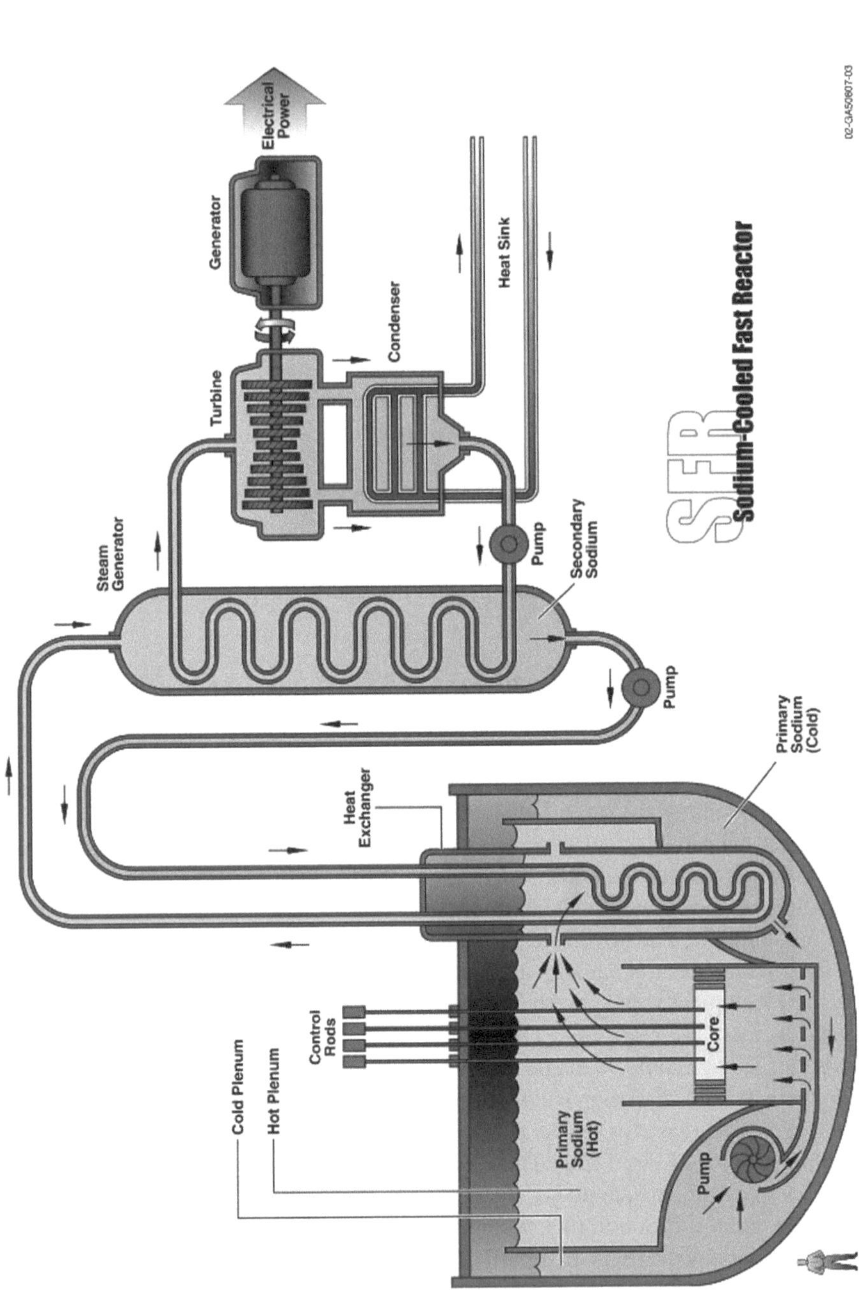

Figure 11.19 Pool-type SFR coupled to an IHX and steam generator. *Source:* [Gif02].

voided. A pancake-shaped core illustrated in Fig. 11.19 would enhance the neutron leakage and minimize the magnitude of the positive VCR. In addition, a small core height would reduce the requirement for the pumping power associated with the liquid metal coolant. In any case, the net power coefficient of reactivity invariably would be sufficiently large negative in SFR designs studied. This behavior of the sodium VCR, however, has long been a concern in the development of this type of reactor. It should of course be recognized that the actual reactivity effects of sodium voiding in the SFR would involve a number of other phenomena, including those associated with neutron slowing down, besides the two primary effects discussed here.

11.3.1.2 Undercooling and Reactivity-Induced Transient Events Just as power level variations affect the reactivity of a reactor through thermal-hydraulic feedback, so do the reactivity coefficients affect the transient behavior of the reactor. This was evident in the Chernobyl accident discussed in Chapter 9. Reactivity coefficients also played a key role in the EBR-II passive safety tests studied in Section 11.1. To illustrate the point further, we discuss two types of transient events for metal-fueled SFR designs, which call for a self-shutdown capability of the reactor, even in the case of a scram failure.

For the unprotected loss of flow (ULOF) event, the resulting transient in power is sufficiently slow so that we will again assume a quasistatic neutronic behavior, i.e., that the net reactivity remains vanishingly small during the transient. Furthermore, the power transient primarily raises the fuel temperature, while the sodium coolant temperature is determined largely by the flow coastdown rate. This allows us to represent the reactivity balance in terms of a power coefficient of reactivity α_p decoupled from a coolant coefficient of reactivity α_c:

$$\delta K = \frac{\partial \ln k}{\partial T_f}\frac{\partial T_f}{\partial P}\delta P + \frac{\partial \ln k}{\partial T_c}\delta T_c \equiv \alpha_p \delta P + \alpha_c \delta T_c \simeq 0. \tag{11.34}$$

Since both α_p and α_c are negative, an undercooling event with $\delta T_c > 0$ can be terminated at a low power level corresponding to $\delta P < 0$, even in the case of a scram failure. To minimize the terminal power level, i.e., to have the largest possible reduction in power, however, we desire to make the power coefficient α_p as small negative as feasible.

This objective to reduce the magnitude of the negative power coefficient of reactivity is certainly contrary to the general concept behind the inherent safety of nuclear reactors. In fact, if we consider a reactivity-induced transient initiated by the insertion of positive reactivity δK_{ex}, we may again use a quasistatic approximation to obtain a reactivity balance:

$$\delta K = \delta K_{ex} + \alpha_p \delta P \simeq 0. \tag{11.35}$$

Here, to minimize the power increase δP, it is clearly desirable to maximize the magnitude of the negative power coefficient α_p. This simple example illustrates rather succinctly that the passive safety of nuclear power plants requires a careful balance between a number of conflicting objectives. This then is merely one of the

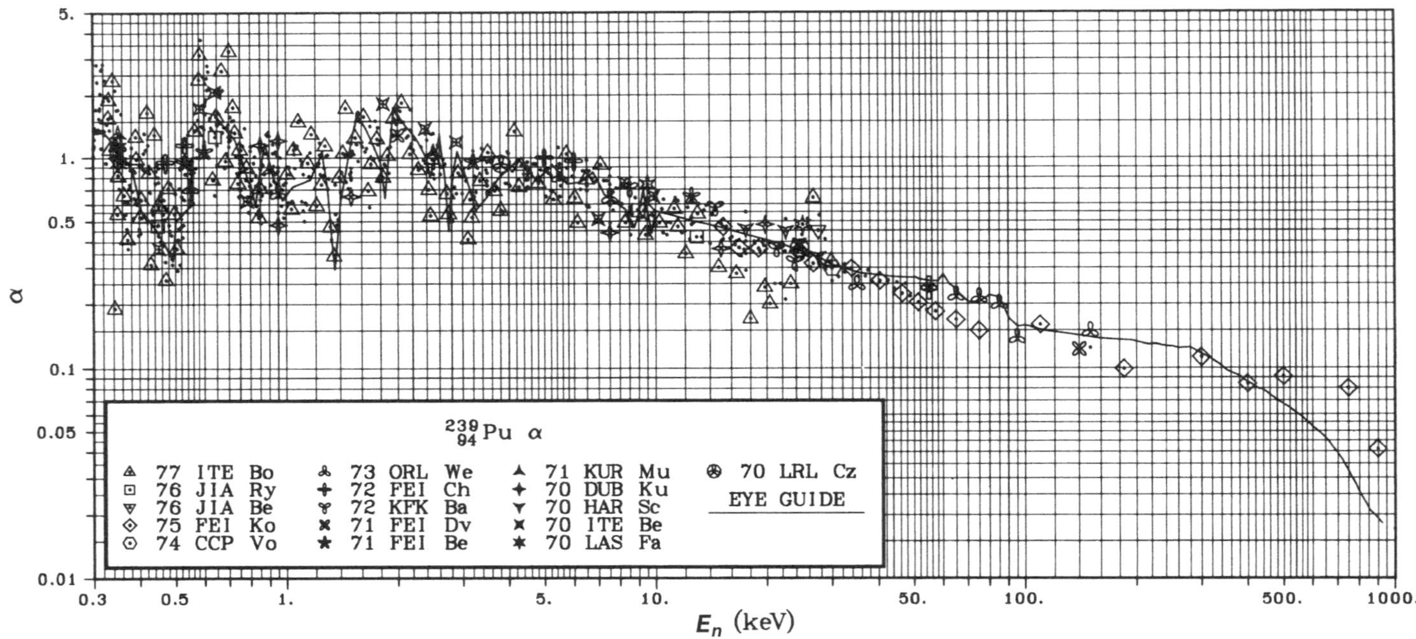

Figure 11.20 Capture-to-fission cross section ratio for ^{239}Pu. *Source:* Reprinted with permission from [McL88]. Copyright © 1988 Elsevier Ltd.

many challenges that lie ahead for nuclear engineers in the further development of Generation IV nuclear energy systems. Some tradeoff studies suggested by Eqs. (11.34) and (11.35) for SFR designs have been presented in [Wad88].

11.3.2 Hypothetical Core Disruptive Accidents for Fast Reactors

Associated with the potential for a positive VCR for SFRs discussed in Section 11.3.1 is the possibility that a SFR core that has suffered melting and collapsed into a compact geometry could undergo a secondary criticality and release a significant amount of energy due to runaway fission reactions, somewhat reminiscent of the superprompt critical transient of the ill-fated Chernobyl accident. This recriticality possibility exists in SFRs voided of sodium because in fast spectrum reactors the fissile enrichment, e.g., ^{239}Pu content in a Pu-fueled core, would be significantly higher than that for LWRs due to generally smaller reaction cross sections for all nuclides, including Pu isotopes, for fast neutrons than for thermal neutrons. Thus, a molten SFR core, devoid of all coolant and collapsed into a more compact geometry, could become supercritical, while such secondary criticality is not a physical possibility for a molten LWR core with a low fissile enrichment. Furthermore, if the sodium VCR is positive, any reactivity-induced accident that results in overheating of the core and voiding of sodium in the core could increase the reactivity in an autocatalytic manner.

The resulting supercritical transient could eventually result in significant disruption of the core, or even a violent disassembly of the core such as that experienced in the Chernobyl accident. This sequence of events is known as the hypothetical core disruptive accident (CDA) and the potential release of fission energy in a supercritical CDA has been studied over the years, starting with simple, bounding calculations known as the Bethe-Tait (B-T) model [Bet56] and its variants [Nic62,Lee72].

Largely for historical interests as well as for a general understanding of the phenomena involved, a simplified version of the B-T model is presented here. The model is based on the assumption that a large segment of a molten core would drop by gravity to the remaining core volume, thereby inserting a large positive reactivity, and the resulting superprompt critical transient is eventually terminated by a disassembly of the consolidated mixture of fuel and structural materials.

The model consists of the point kinetics equation for transient reactor power calculations coupled to a set of equations that governs the motion of the molten material and a first-order perturbation equation that yields the reactivity changes due to the fuel motion. The fuel motion is represented by a combination of (a) the equation of continuity, (b) the equation of motion, and (c) a threshold-type equation of state. For fast transients of the type considered, a simplifying assumption is made that the energy produced in the transient is deposited entirely in the fuel material, thereby obviating the need for heat transfer equations. Furthermore, fuel temperature feedback effects on the reactivity, including the Doppler effect, are ignored.

A key feature of the B-T model is the equation of state

$$p(r,t) = (\gamma - 1)\rho(r,t)\left[E(r,t) - Q^*\right], \tag{11.36}$$

which provides pressure $p(r,t)$ in a spherical core at position r and time t for a molten fuel-structure mixture of density $\rho(r,t)$ and energy density $E(r,t)$ in terms of a threshold energy density Q^* and the ratio γ of heat capacities. The threshold equation (11.36) suggests that there is no reactivity feedback due to material motion until a sufficient amount of energy is generated to exceed Q^*. Therefore, the transient calculation is partitioned into two phases: the initiating phase, where a ramp insertion of positive reactivity is made to drive the core into a superprompt critical configuration until $E(r,t) = Q^*$, followed by the disassembly phase, during which a significant movement of the molten mixture takes place.

11.3.2.1 Initiating phase The power rise $n(t)$ due to the insertion of reactivity at constant rate m ($/s) is calculated by the point kinetics equation with an infinite delayed approximation for delayed neutron precursors,

$$\frac{dn(t)}{dt} = \frac{K(t)-\beta}{\Lambda}n(t) + \frac{n(0)}{\Lambda} \text{ with } K(t) = mt, \tag{11.37}$$

so that the power n_{pc} at prompt criticality, i.e., $K(t) = \beta$, is determined approximately [Ash79] as

$$n_{pc} = n\left(\frac{1}{m}\right) \simeq n(0)\beta\sqrt{\frac{\pi}{2m\Lambda}}. \tag{11.38}$$

For power $n(t)$ beyond the superprompt criticality, delayed neutron effects may be ignored to yield

$$n(\tau) = n_{pc}\exp\left(\frac{m}{2\Lambda}\tau^2\right), \tag{11.39}$$

where $\tau = t - 1/m$ is the time measured from the prompt criticality. The energy density $E(r,\tau)$ is then calculated as the time integral of power $Q(\tau)$, modulated by a normalized flux distribution $\phi(r)$ for a spherical reflected core with radius R and buckling B^2,

$$E(r,\tau) = Q(\tau)\phi(r), \tag{11.40}$$

where, for algebraic convenience, the flux may be approximated by a Taylor series

$$\phi(r) = \frac{1}{Br}\sin Br \simeq 1 - q\frac{r^2}{R^2} \text{ with } q = \frac{B^2R^2}{6}. \tag{11.41}$$

Since the energy released in a superprompt critical transient is very much larger than the initial energy of the molten core at the prompt criticality, the magnitude $Q(\tau)$ of the fission energy density may be estimated by

$$Q(\tau) = \int_0^\tau n(t)dt \text{ with } Q(0) \simeq 0. \tag{11.42}$$

Integrating Eq. (11.42) by parts, following the substitution of Eq. (11.39), yields an expression for time τ^* at which the peak energy density reaches the threshold energy density Q^*,

$$Q^* = n_{pc}\frac{\Lambda}{m\tau^*}\exp\left[\frac{m}{2\Lambda}\left(\tau^*\right)^2\right]. \tag{11.43}$$

Equation (11.43) is rewritten as

$$\frac{m}{\Lambda}\left(\frac{Q^*}{n_{pc}}\right)^2\frac{m}{\Lambda}\left(\tau^*\right)^2 = \exp\left[\frac{m}{\Lambda}\left(\tau^*\right)^2\right] \tag{11.44}$$

and expanded as

$$\ln\left[\frac{m}{\Lambda}\left(\frac{Q^*}{n_{pc}}\right)^2\right] + \ln\left[\frac{m}{\Lambda}\left(\tau^*\right)^2\right] = \frac{m}{\Lambda}\left(\tau^*\right)^2. \tag{11.45}$$

With $m\left(\tau^*\right)^2/\Lambda \gg 1$, the term $\ln\left[m\left(\tau^*\right)^2/\Lambda\right]$ may be neglected compared with $m\left(\tau^*\right)^2/\Lambda$ in Eq. (11.45) to yield an explicit, albeit approximate, expression for τ^*,

$$\left(\tau^*\right)^2 \simeq \frac{\Lambda}{m}\ln\left[\frac{m}{\Lambda}\left(\frac{Q^*}{n_{pc}}\right)^2\right]. \tag{11.46}$$

11.3.2.2 Disassembly Phase Once the threshold energy density Q^* is reached, the pressure rises rapidly and the disassembly phase begins, while an exponential rise in energy density continues for $\tau \geq \tau^*$,

$$Q(\tau) = Q^*\exp\left[\frac{K_m}{\Lambda}(\tau - \tau^*)\right], \tag{11.47}$$

without further insertion of positive reactivity, i.e., with $K_m = \beta + m\tau^* \simeq m\tau^*$. The transient is terminated when the disassembly reactivity feedback $K_d(\tau)$ cancels K_m, with $K_d(\tau)$ obtained through first-order perturbation theory applied [Dud76] to the one-group diffusion equation

$$K_d(\tau) = \frac{\delta k}{k} = \frac{\int_V \left[\left\{\delta\nu\Sigma_f(r) - \delta\Sigma_a(r)\right\}\phi^2(r) - \delta D(r)\nabla\phi(r)\cdot\nabla\phi(r)\right]\mathrm{d}\mathbf{r}}{\int_V \nu\Sigma_f\phi^2(r)\mathrm{d}\mathbf{r}}, \tag{11.48}$$

where the integrals are carried out over volume V of the spherical core and the time dependence in the perturbed terms is suppressed. With microscopic cross sections assumed unperturbed, the perturbed parameters are given by fractional changes in the density $\rho(r)$ of the mixture associated with the material movement

$$\delta\ln\Sigma_a = \delta\Sigma_a/\Sigma_a = \delta\ln\nu\Sigma_f = -\delta\ln D = \delta\ln\rho. \tag{11.49}$$

With Eqs. (11.41) and (11.49) substituted, Eq. (11.48) can be written as

$$K_d(\tau) = \int_V \delta\rho(r,\tau)w(r)\mathrm{d}\mathbf{r}, \tag{11.50}$$

where the material worth function $w(r)$ is obtained up to terms of order r^2,

$$w(r) = w(0)\left(1 - \frac{4q}{3}\frac{r^2}{R^2}\right) \text{ with } w(0) = \frac{2q}{3\,(1 - 6q/5 + 3q^2/7)\,\rho\nu\Sigma_f\Sigma_a V R^2}. \tag{11.51}$$

We note that the effect on $K_d(\tau)$ due to the increase in the overall core volume should be included, but the effect is usually small and is ignored in Eqs. (11.48) and (11.50).

The density change $\delta\rho(r,\tau)$ can be calculated from the equation of continuity written in terms of the material displacement $\mathbf{u}(r,\tau)$,

$$K_d(\tau) = -\int_V \nabla \cdot [\rho(r,\tau)\mathbf{u}(r,\tau)]w(r)\mathrm{d}\mathbf{r}, \tag{11.52}$$

which can be rewritten via Green's theorem as

$$K_d(\tau) = -\int_V \rho(r,\tau)\mathbf{u}(r,\tau) \cdot \nabla w(r)\mathrm{d}\mathbf{r}. \tag{11.53}$$

With the approximation that the density change is small, Eq. (11.53) is differentiated twice and combined with the equation of motion,

$$\rho(r,\tau)\frac{\mathrm{d}^2\mathbf{u}(r,\tau)}{\mathrm{d}\tau^2} = -\nabla p(r,\tau), \tag{11.54}$$

to yield a relationship between the pressure buildup $p(r,\tau)$ and the disassembly reactivity:

$$\frac{\mathrm{d}^2 K_d(\tau)}{\mathrm{d}\tau^2} = \int_V \nabla p(r,\tau)\cdot\nabla w(r)\mathrm{d}\mathbf{r}. \tag{11.55}$$

Another application of Green's theorem and the substitution of Eqs. (11.36) and (11.51) allow the evaluation of the volume integral in Eq. (11.55) in terms of the energy density $Q(\tau)$ of Eq. (11.47),

$$\begin{aligned}\frac{\mathrm{d}^2 K_d(\tau)}{\mathrm{d}\tau^2} &= \int_V p(r,\tau)\nabla^2 w(r)\mathrm{d}r \\ &= -\frac{5}{2}\frac{K_m}{x}\left(\frac{K_m}{\Lambda}\right)^2\left[\left(1 - \frac{3}{5}q\right)\frac{Q(\tau)}{Q^*} - 1\right],\end{aligned} \tag{11.56}$$

where

$$\frac{1}{x} = \frac{64\pi q(\gamma - 1)\rho Q^* w(0) R\Lambda^2}{15K_m^3}. \tag{11.57}$$

For a large energy release $Q_f = Q(\tau_f) \gg Q^*$ at the end of the disassembly phase, Eq. (11.56) may be integrated twice with τ_f defined as the time when the disassembly reactivity $K_d(\tau_f)$ cancels the maximum reactivity K_m attained through the collapse and compacting of the molten core during the initiating phase,

$$K_d(\tau_f) = -K_m \simeq -\frac{5}{2}\frac{K_m}{x}\left(1 - \frac{3}{5}q\right)\frac{Q_f}{Q^*}. \tag{11.58}$$

An estimate for the total energy release in the postulated accident is finally obtained as

$$Q_f \simeq \frac{2}{5}\frac{xQ^*}{1-3q/5} = \frac{3K_m^3}{32\pi q(1-3q/5)(\gamma-1)\rho w(0)R\Lambda^2}, \tag{11.59}$$

or, equivalently, with Eq. (11.46) substituted for K_m,

$$Q_f \simeq \frac{3\left[m \ln\left\{m(Q^*)^2/\Lambda n_{pc}^2\right\}\right]^{3/2}}{32\pi q(1-3q/5)(\gamma-1)\rho w(0)R\Lambda^{1/2}}. \tag{11.60}$$

This shows there is a strong dependence of the final energy release on the maximum reactivity K_m inserted or equivalently on the ramp rate m of reactivity insertion.

The potential energy release in a CDA may be estimated via the basic structure outlined in this section, but with fewer approximations introduced [Nic62,Lee72]. The improvements in the model could include a more realistic equation of state than the threshold-type equation (11.36), multigroup formulation for the disassembly reactivity, cylindrical geometry, zoned fuel loading, and a more realistic equation of motion allowing for pressure wave propagation. Additional improvements could include Doppler feedback and a numerical integration of the governing equations bypassing various approximations introduced to derive the analytic solution of Eq. (11.60). To provide the order of magnitude for some of the parameters introduced in deriving the equation for the final energy release, the results from a numerical calculation [Lee72] for a small SFR core are presented.

Example 11.1 A 400-MWt reflected reactor has a core height of 0.9 m, with a height-to-diameter ratio of 0.78 and a (Pu-U)O_2 fuel inventory of 3.2 Mg. For a CDA initiated by a postulated ramp reactivity insertion of 100 $/s into an initially molten core at 3030 K with a power level of 2.5 kWt, an energy release on the order of 9.0 GJ is calculated with the two-dimensional MARS code [Hir67] with the Doppler feedback ignored. The energy release estimate is reduced by a factor of 2 with a Doppler coefficient $\mathrm{d}k/\mathrm{d}T = -0.005/T$. Figure 11.21 illustrates, through the plots of $\mathrm{d}K_d(\tau)/\mathrm{d}\tau$ and $K_d(\tau)$, the trends and time scales involved with various reactivity components during the disassembly phase of the postulated accident. The disassembly phase begins after a maximum reactivity insertion $K_m \simeq \$1.8$ and the entire transient is terminated in ~20 ms. The transient is much faster and the postulated energy release smaller than the pulse duration of ~ 4 s and energy release estimate of 200 to 300 GJ, respectively, for the ill-fated Chernobyl accident discussed in Section 9.3.3. The reactivity plots of Fig. 11.21 show that the reactivity effect due to the core boundary expansion, ignored in Eqs. (11.48) and (11.50), is indeed usually small and that the disassembly takes place over a period of ~1 ms. ◇

Over the past three decades, a number of complex coupled nuclear-thermal-hydraulic-structural computer codes have been developed, together with significant experimental validation efforts [Fuk09], to perform more realistic simulations of CDAs. One good example is the SIMMER-III code [Kon99], which features a two-dimensional, three-velocity-field, multiphase, multicomponent, Eulerian fluid dynamics formulation coupled with a space-dependent reactor kinetics model. Modern CDA simulation codes typically represent the accident scenario [Fau02] in multiple

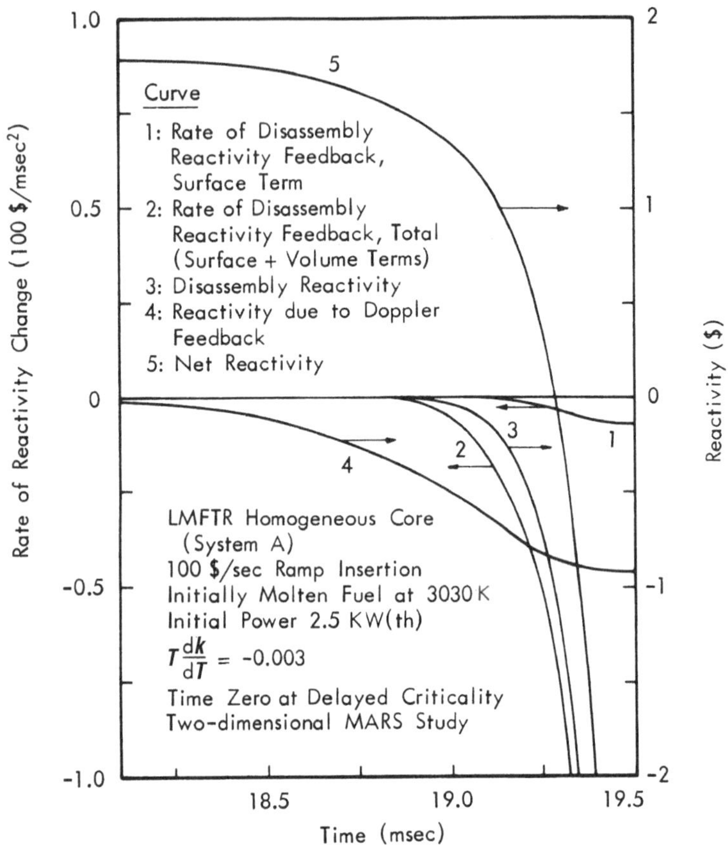

Figure 11.21 Reactivity behavior during a disassembly transient in a homogeneous cylindrical SFR core. *Source:* [Lee72].

phases: (a) initiating phase involving fuel pin disruption, (b) transition or core meltout phase resulting in subassembly disruption, pool formation and fuel escape, and (c) postaccident material relocation or heat removal phase. Emphasis is placed away from fuel-coolant thermal interactions to potential energetic recriticality events in the transition phase, as well as the in-vessel retention of molten corium.

Significant efforts have also been made in recent years to incorporate various design features [Wig09], including the use of metallic fuel with enhanced passive safety characteristics discussed in Section 11.1, and to reduce the CDA probability as a beyond design basis accident (BDBA) for SFRs. Because it is difficult, however, to accurately model the transition phase and to ensure the prevention of core disruption and recriticality, there is a continued concern regarding the CDA possibility. This is manifest in the fuel assembly with inner duct structure (FAIDUS) concept that was featured as part of the Gen-IV SFR design from the Japan Nuclear Cycle Development Institute. In this concept [Fau02,Nag08], ducts will be constructed into fuel assemblies so that molten corium would be channeled gradually into a core catcher, thereby eliminating the possibility of a rapid insertion of positive reactivity due to a large mass of molten fuel collapsing into a supercritical assembly.

11.3.3 VHTR and Phenomena Identification and Ranking Table

The VHTR design was selected as one of the six concept groups in the Generation IV Roadmap and has been under development as the next generation nuclear plant (NGNP) in the United States. The primary incentive for the VHTR development is its potential to produce He at a temperature of up to 1273 K, which may then be used to produce hydrogen by dissociating water. The VHTR design offers additional safety measures associated with multiple pyrolytic carbon coatings in the 1-mm-diameter tristructural-isotropic (TRISO) particles that form the basic building block for the core. Figure 11.22 illustrates the dual-purpose VHTR plant that produces hydrogen as well as electricity, while Fig. 11.23 shows the structure of fuel pin cells or compacts, which are packed with TRISO particles. The compacts are then loaded into hexagonal graphite fuel assemblies stacked up in the reactor vessel. In the alternate pebble bed reactor (PBR) design, the TRISO particles are packed into graphite spheres with a diameter of 60 mm and are then loaded and circulated in the vessel. In addition, some VHTR designs have proposed siting the reactor vessel underground for added security of the plant.

The key distinguishing features [Bal08] of the VHTR and PBR designs include: (a) TRISO particles that offer the capability to contain fission products in operating and postulated accident conditions, (b) the primary system cooled by inert He gas, (c) the graphite-moderated core with a low power density and large heat capacity, (d) the negative fuel and moderator temperature coefficients of reactivity, (e) the reactor cavity cooling system utilizing natural convection cooling, (e) a confinement-style reactor building structure, and (f) an IHX. The IHX would be required for coupling the high-temperature, high-pressure primary He system to the process heat loop.

A phenomena identification and ranking table (PIRT) is a structured process to perform expert assessment to support decision making. The process is grouped into three major activities involving a total of nine distinct steps illustrated in Fig. 11.24. The first activity group entails defining the issues, objectives, and hardware and scenarios for the PIRT and specifying the appropriate evaluation criteria. The second group identifies and compiles the current knowledge base and plausible phenomena that should be addressed. The third group of activities finally ranks the importance of the phenomena analyzed and assesses the knowledge base for the phenomena and provides the overall documentation.

The NRC sponsored a PIRT study [Bal08] to assess the analytical tools and relevant research activities that will be required for a timely review of an NGNP license application. With this common issue for step 1, five different expert panels were formed to cover five different areas, including the accidents and thermal fluids (ACTH) area, and associated objectives. The ACTH panel identified in step 2 the objectives to identify safety-significant phenomena for operating and postulated

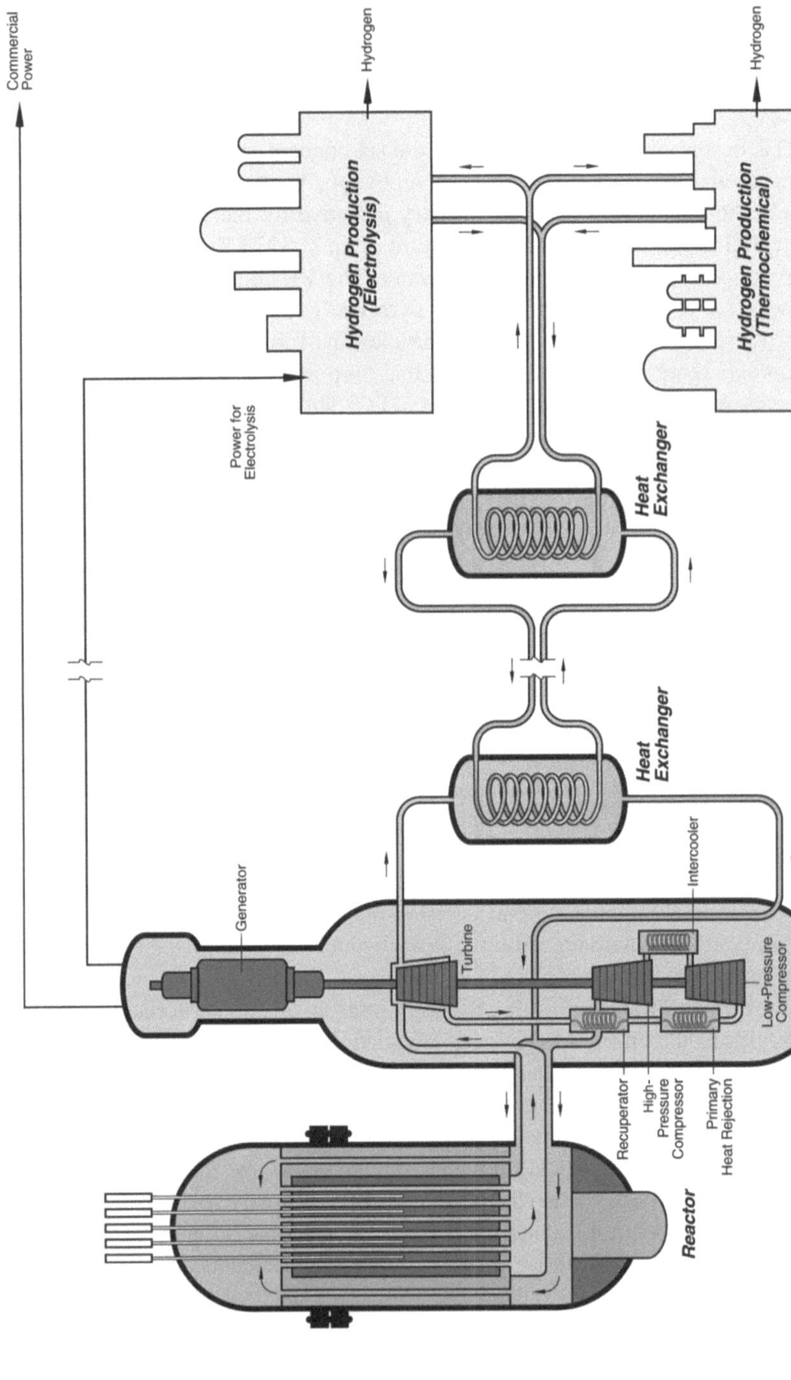

Figure 11.22 Schematic diagram of the VHTR. *Source:* [Gif02].

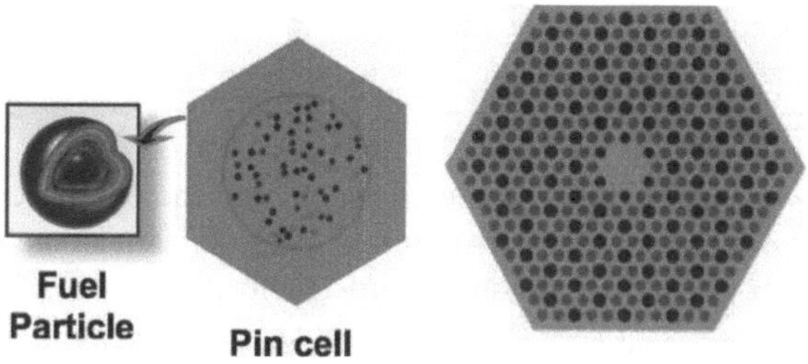

Figure 11.23 TRISO particle, pin cell, and prismatic fuel assembly for the VHTR. *Source:* [Gif02].

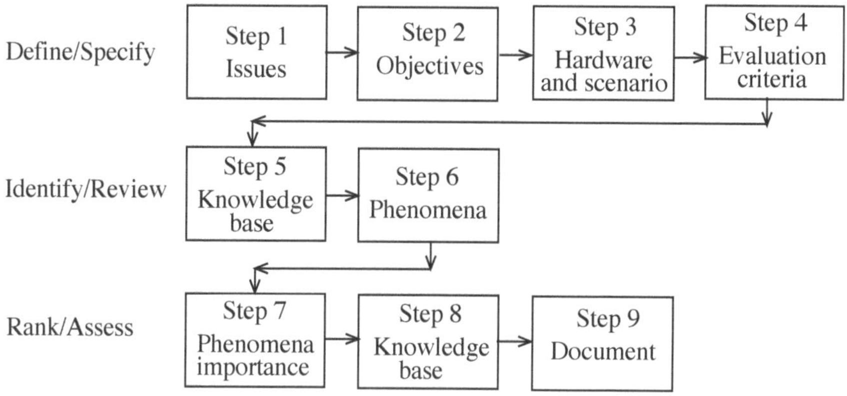

Figure 11.24 Nine PIRT steps grouped into three activities.

accident conditions and evaluated in step 3 a set of phenomena associated with reactor systems including passive cooling of the reactor core, reactor pressure vessel, and reactor cavity cooling system via appropriate combinations of radiation, convection, and conduction. The panel considered five events as safety significant:

(a) Pressurized loss of forced circulation (LOFC) accident
(b) Depressurized LOFC accident
(c) Depressurized LOFC accident followed by air ingress
(d) Reactivity-induced transients, including anticipated transients without scram
(e) Events related to the reactor-to-process heat coupling

The above events were reviewed in three classes according to the event frequencies as (a) anticipated transients, (b) DBAs, and (c) BDBAs, with expected frequencies of 0.01/plant-year, 10^{-4} to 10^{-2}/plant-year, and 5×10^{-7} to 10^{-4}/plant-year, respectively. The classification is in line with the general classification system for the current generation of nuclear plants discussed in Section 8.2.1. The ACTH panel eventually decided not to consider steam-water ingress events as credible for the VHTR or PBR design studied, which does not include a steam generator in the primary loop.

The PIRT panels identified the most significant phenomena, with the importance ranked high and the corresponding knowledge base ranked low or medium, in the five topical areas:

(a) Accidents and thermal fluids: heat transport and reactor physics phenomena that impact the primary system temperatures, and postulated air ingress accidents that could possibly result in substantial core damage.
(b) Fission product transport and dose: radiological source term, transport phenomena during an air ingress accident, and transport of fission products into the confinement building and the environment.
(c) High-temperature materials: high-temperature stability of key components, thermal aging and degradation, and heavy-section properties of the reactor vessel.
(d) Graphite: irradiation effects on material properties, consistency of graphite quality, and graphite dust that could impact the source term.
(e) Process heat for hydrogen cogeneration: external threat to the nuclear plant due to the release of oxygen gases from the hydrogen plant.

A variant of the PIRT process was used in the selection of the six concept groups for the Generation IV Roadmap [Gif02] from 100+ advanced reactor concepts submitted to the U. S. Department of Energy. The process was also used in Design Control Documents for various advanced reactors, including the AP1000 design, for which the NRC issued the final design certification in 2006.

References

[Ash79] M. Ash, *Nuclear Reactor Kinetics*, 2nd ed., McGraw-Hill (1979).

[Bal08] S. J. Ball and S. E. Fisher, "Next Generation Nuclear Plant Phenomena Identification and Ranking Tables (PIRTs)," NUREG/CR-6944, vol. 1, U.S. Nuclear Regulatory Commission (2008).

[Bet56] H. A. Bethe and J. H. Tait, "An Estimate of the Order of Magnitude of the Explosion When the Core of a Fast Reactor Collapses," UKAEA-RHM (56)/113, United Kingdom Atomic Energy Authority (1956).

[Bre85] L. Breiman and J. H. Friedman, "Estimating Optimal Transformations for Multiple Regression and Correlation," *J. Am. Stat. Assoc.* **80**, 955 (1985).

[Bru04] H. J. Bruschi, "The Westinghouse AP1000—Final Design Approved," *Nucl. News*, 30 (November 2004).

[Cha87] L. K. Chang, J. F. Koenig, and D. L. Porter, "Whole-Core Damage Analysis of EBR-II Driver Fuel Elements Following SHRT Program," *Nucl. Eng. Design* **101**, 67 (1987).

[Col07] M. Colby, "Economic Simplified Boiling Water Reactor (ESBWR) Core Engineering," colloquium presentation, University of Michigan (2007).

[Cui01] Z. Cui, J. C. Lee, J. J. Vandenkieboom, and R. W. Youngblood, "Unreliability Quantification of a Containment Cooling System through ACE and ANN Algorithms," *Trans. Am. Nucl. Soc.* **85**, 178 (2001).

[Dev95] J. C. Devine, Jr., W. Layman, D. E. W. Leaver, and J. Santucci, "The Passive ALWR Approach to Assuring Containment Integrity," *Nucl. Eng. Design* **157**, 469 (1995).

[Dud76] J. J. Duderstadt and L. J. Hamilton, *Nuclear Reactor Analysis,* Wiley (1976).

[Fau02] H. K. Fauske, K. Koyama, and S. Kubo, "Assessment of the FBR Core Disruptive Accident (CDA): The Role and Application of General Behavior Principles (GBPs)," *J. Nucl. Sci. Technol.* **39**, 615 (2002).

[Fel87] E. E. Feldman, D. Mohr, L. K. Chang, H. P. Planchon, E. M. Dean, and P. R. Betten, "EBR-II Unprotected Loss-of-Heat-Sink Predictions and Preliminary Test Results," *Nucl. Eng. Design* **101**, 57 (1987).

[Fle92] C. D. Fletcher and R. R. Schultz, "RELAP5/MOD3 Code Manual," NUREG/CR-5535, U. S. Nuclear Regulatory Commission (1992).

[For89] C. W. Forsberg, D. L. Moses, E. B. Lewis, R. Gibson, R. Pearson, W. J. Reich, G. A. Murphy, R. H. Staunton, and W. E. Kohn, "Proposed and Existing Passive and Inherent Safety-Related Structures, Systems and Components for Advanced Light Water Reactors," ORNL-6554, Oak Ridge National Laboratory (1989).

[Fuk09] Y. Fukano, K. Kawada, I. Sato, A. E. Wright, D. J. Kilsdonk, R. W. Aeschlimann, and T. H. Bauer, "CAIE Experiments on the Flow and Freezing of Metal Fuel and Cladding Melts (1): Test Conditions and Overview of the Results," in *Proc. Int. Conf. Fast Reactors and Related Fuel Cycles, FR09*, IAEA-CN-176/03-11P (2009).

[GEH07] "ESBWR Design Control Document 26A6642AT," GE-Hitachi Nuclear Energy (2007).

[GEN92] "SBWR Standard Safety Analysis Report," no. 25A5113, rev. A, GE Nuclear Energy (1992).

[Gif02] "Generation IV Nuclear Energy Systems," gif.inel.gov/roadmap (2002).

[Gol87] G. H. Golden, H. P. Planchon, J. I. Sackett, and R. M. Singer, "Evolution of Thermal-Hydraulics Testing in EBR-II," *Nucl. Eng. Design* **101**, 3 (1987).

[Gol89] D. E. Goldberg, *Genetic Algorithms in Search, Optimization, and Machine Learning*, Addison-Wesley (1989).

[Hay94] S. Haykin, *Neural Networks: A Comprehensive Foundation*, Macmillan College Publishing (1994).

[Hir67] N. Hirakawa, "MARS, A Two-Dimensional Excursion Code," APDA-198, Atomic Power Development Associates, Inc. (1967).

[Kim97] H. G. Kim and J. C. Lee, "Development of Generalized Critical Heat Flux Correlation Through the Alternating Conditional Expectation Algorithm," *Nucl. Sci. Eng.* **127**, 300 (1997).

[Kon99] S. Kondo, Y. Tobita, K. Morita, D. J. Brear, K. Kamiyama, H. Yamano, S. Fujita, M. Maschek, E. A. Fischer, E. Kiefhaber, G. Buckel, E. Hesselschwerdt, M. Flad, P. Costa, and S. Pigny, "Current Status and Validation of the SIMMER-III LMFR Safety Analysis Code," in *Proc. Int. Conf. Nucl. Eng., ICONE-7* (1999).

[Lee72] J. C. Lee and T. H. Pigford, "Explosive Disassembly of Fast Reactors," *Nucl. Sci. Eng.* **48**, 28 (1972).

[Lor04] T. J. Loredo, "Bayesian Adaptive Exploration," *Proc. Am. Inst. Phys. Conf. Bayesian Inference and Maximum Entropy Methods in Science and Engineering* **707**, 330 (2004).

[Mar93] T. U. Marston, W. H. Layman, and G. Bockhold Jr., "Utility Requirements for Safety in the Passive Advanced Light-Water Reactor," *Nucl. Safety* **34**, 85 (1993).

[McL88] V. McLane, D. L. Dunford, and P. F. Rose, *Neutron Cross Sections,* vol. 2, *Neutron Cross Section Curves,* 753, Academic Press (1988).

[Mey61] J. E. Meyer, "Hydrodynamic Models for the Treatment of Reactor Thermal Transients," *Nucl. Sci. Eng.* **10**, 269 (1961).

[Mey85] P. E. Meyer, "NOTRUMP—A Nodal Transient Small-Break and General Network Code," WCAP-10080-A, Westinghouse Electric Corporation (1985).

[Moh81] D. Mohr and E. E. Feldman, "A Dynamic Simulation of the EBR-II Plant During Natural Convection with the NATDEMO Code," in *Decay Heat Removal and Natural Convection in Fast Breeder Reactor*, A. K. Agrawal and G. P. Guppy, eds., 207, Hemisphere Publishing (1981).

[Moh87] D. Mohr, L. K. Chang, P. R. Betten, E. E. Feldman, and H. P. Planchon, "Validation of the HOTCHAN Code for Analyzing EBR-II Driver Following Loss of Flow Without Scram," in *Second Proc. ASME-JSME Thermal Engineering Joint Conf.*, CONF-870304-2 (1987).

[Mur90] K. K. Murata. D. E. Carroll, K. E. Washington, F. Gelbard, G. D. Valdez, D. C. Williams, and K. D. Bergeron, "User's Manual for CONTAIN 1.1: A Computer Code for Severe Nuclear Reactor Accident Containment Analysis," NUREG/CR-5026, rev. 1.11, U.S. Nuclear Regulatory Commission (1990).

[Nag08] M. Nagamura, T. Ogawa, S. Ohki, T. Mizuno, and S. Kubo, "Development of Advanced Loop-Type Fast Reactor Design in Japan (6): Minor Actinide Containing Oxide Fuel Core Design Study for the JSFR," in *Proc. Int. Cong. Advances in Nuclear Power Plants, ICAPP'08*-8082 (2008).

[Nic62] R. B. Nicholson, "Methods for Determining the Energy Release in Hypothetical Reactor Meltdown Accidents," APDA-150, Atomic Power Development Associates, Inc. (1962); see also *Nucl. Sci. Eng.* **18**, 207 (1964).

[NRC01] "RELAP5/MOD3.3 Code Manual, Volume 1: Code Structure, Systems Models, and Solution Methods," NUREG/CR-5535, rev. 1, U.S. Nuclear Regulatory Commission (2001).

[Ott88] K. O. Ott, "Inherent Shutdown Capabilities of Metal-Fueled Liquid-Metal-Cooled Reactors During Unscrammed Loss-of-Flow and Loss-of-Heat-Sink Incidents," *Nucl. Sci. Eng.* **99**, 13 (1988).

[Pau02] C. K. Paulson, "AP1000: Set to Compete," *Nucl. Eng. Int.* **47**, 20 (2002).

[Pla86] H. P. Planchon, R. M. Singer, D. Mohr, E. E. Feldman, L. K. Chang, and P. R. Betten, "The Experimental Breeder Reactor II Inherent Shutdown and Heat Removal Tests—Results and Analysis," *Nucl. Eng. Design* **91**, 287 (1986).

[Pla87] H. P. Planchon, J. I. Sackett, G. H. Golden, and R. H. Sevy, "Implications of the EBR-II Inherent Safety Demonstration Test," *Nucl. Eng. Design* **101**, 75 (1987).

[Sap93] D. Saphier, "The Simulation Language of DSNP: Dynamic Simulator for Nuclear Power Plants," ANL-CT-77-20, rev. 3.5, Argonne National Laboratory (1993).

[Sev85] R. H. Sevy, Argonne National Laboratory, Private Communication (Oct. 1985).

[Tho91] M. L. Thompson, C. L. Cockey and T. Wu, "Actinide Recycle Enhancement," GEFR-00898, GE Nuclear Energy (1991).

[Van96] J. J. Vandenkieboom, "Reliability Quantification of Advanced Reactor Passive Safety Systems," PhD Thesis, University of Michigan (1996).
[Van97] J. J. Vandenkieboom, R. W. Youngblood, J. C. Lee, and W. Kerr, "Reliability Quantification of Advanced Reactor Passive Safety Systems," *Trans. Am. Nucl. Soc.* **76**, 296 (1997).
[Wad88] D. C. Wade and Y. I. Chang, "The Integral Fast Reactor Concept: Physics of Operation and Safety," *Nucl. Sci. Eng.* **100**, 507 (1988).
[Wal69] G. B. Wallis, *One-Dimensional Two-Phase Flow*, McGraw-Hill (1969).
[Was91] K.E. Washington, K.K. Murata, R.G. Gido, F. Gelbard, N.A. Russell, S.C. Billups, D.E. Carroll, R.O. Griffith, and D.L.Y. Louie, "Reference Manual for the CONTAIN 1.1 Code for Containment Severe Accident Analysis," NUREG/CR-5715, U.S. Nuclear Regulatory Commission (1991).
[Wes92] "AP600 Standard Safety Analysis Report," DE-AC03-90SF18495, Westinghouse Electric Corporation (1992).
[Wes03] "AP1000 Design Control Document," APP-GW-GL-700, rev. 3, Westinghouse Electric Company (2003).
[Wes07] "AP1000 Simple, Safe, Innovative," www.AP1000.westinghousenuclear.com (2007).
[Wig09] R. A. Wigeland and J. E. Calahan, "Mitigation of Sodium-Cooled Fast Reactor Severe Accident Consequences Using Inherent Safety Principles," in *Proc. Int. Conf. Fast Reactors and Related Fuel Cycles, FR09*, IAEA-CN-176/03-02 (2009).

Exercises

11.1 Numerically integrate the fuel channel models of Eqs. (11.2) and (11.6) using the SFR parameters given in Table 11.2 and verify the plots given in Figs. 11.7 and 11.8. You may consider the Crank-Nicolson algorithm for the integration.
11.2 Perform parametric studies with the computer program developed in Exercise 11.1 varying the time constants for the flow and power coastdown and discuss the results.
11.3 Compare the results of the quasistatic formulation for LOFWS and LOHWS events with published computer simulation results for the EBR-II or other SFR designs and suggest possible improvements in the quasistatic formulation.
11.4 Starting from the point kinetics equation with an infinite delayed approximation for delayed neutron precursors, Eq. (11.37), obtain the power n_{pc} at prompt criticality given in Eq. (11.38).
11.5 Starting from Eqs. (11.41) and (11.48), derive the material worth function of Eq. (11.51).
11.6 Indicate approximations made in deriving Eqs. (11.59) and (11.60) from Eq. (11.55).
11.7 For a reflected critical core with uniform composition, the following parameters are given:

$$\beta = 0.003, \qquad V = 6 \text{ m}^3, \qquad q = 0.6,$$
$$\Sigma_a = 2.652 \times 10^{-3} \text{ cm}^{-1}, \; \nu\Sigma_f = 3.788 \times 10^{-3} \text{ cm}^{-1}, D = 1.463 \text{ cm}.$$

(a) Evaluate the material worth $w(0)$ in Eq. (11.51) at the center of the core for material density ρ.
(b) For a fractional density increase $\delta\rho/\rho = 0.05$, perform the integral of Eq. (11.50) to determine the reactivity change.
11.8 A power escalation maneuver of a PWR core may be described by the lumped-parameter fuel channel model of Eqs. (11.2) and (11.6). In this maneuver involving a moderator temperature programming, the average coolant temperature $T_c(t)$ and core flow rate $W(t)$ remain constant. Following a 10% step increase in core power, derive expressions for the average fuel temperature $T_f(t)$ and the coolant temperature rise $\Delta T_c(t)$ across the core.

CHAPTER 12

RISK-INFORMED REGULATIONS AND RELIABILITY-CENTERED MAINTENANCE

There has been continuing interest in applying PRA studies for nuclear power plant licensing and regulations, starting certainly from WASH-1400, the landmark PRA study on NPPs. Despite some initial skepticism about the validity of the results, general usefulness of the Reactor Safety Study was validated in some sense by the unfortunate TMI-2 accident of 1979, discussed in Chapter 9. This is because one of the key results of WASH-1400 was that a large contribution to the NPP risk might be due to small-break LOCAs of the type that initiated the TMI-2 accident, rather than design basis accidents, e.g., large-break LOCAs. This was discussed in more detail in Section 10.1. With the completion of the NUREG-1150 PRA studies for three PWRs and two BWRs in 1990, the NRC has continued to explore ways to use quantitative risk measures to supplement deterministic risk estimates in regulations and licensing of NPPs. An important part of the recent effort in this direction is to employ risk and reliability calculations in performing various maintenance activities in a systematic way. In this chapter, we begin with a brief history of the steps taken by the U.S. nuclear industry in the use of PRAs and specific examples adopted by the NRC to implement risk-informed regulations (RIRs) in Section 12.1. This is followed by a discussion of reliability-centered maintenance approaches in Section 12.2.

12.1 RISK MEASURES FOR NUCLEAR PLANT REGULATIONS

Following the TMI-2 accident, while the NUREG-1150 study was underway, in 1986 the NRC released a Policy Statement on Safety Goals [NRC86], following a period of trial use of a draft statement. As discussed in Section 8.2.3, this is the first key step that the NRC took to formally suggest the use of quantitative safety goals and plant performance guidelines to supplement deterministic guidelines, in particular, the General Design Criteria in 10 CFR 50, Appendix A [NRC71].

Together with the NUREG-1150 study, the NRC requested [NRC88] NPP licensees to perform an individual plant examination for severe accident vulnerabilities for each plant, with the examination restricted to level 1 and limited level 2 PRAs, excluding thereby any offsite consequence analyses. This was followed by the NRC request to perform individual plant examinations for external events (IPEEE) in 1991 [NRC91]. With the completion of NUREG-1150, together with IPEs and IPEEEs, significant interest emerged in making increased uses of PRA methods in NPP regulations that led to an in-depth review of NRC staff uses of PRA in 1994 [NRC94] and the release of a PRA policy statement in 1995 [NRC95]. The policy statement encouraged the use of PRA and associated analyses, e.g., sensitivity and uncertainty analyses, to reduce unnecessary conservatism, especially in support of the Backfit Rule of 10 CFR 50.109 [NRC07a] introduced in the aftermath of the TMI-2 accident. A number of issues were raised, however, in actual applications [Has02] of the PRA policy statement, including (1) the treatment of uncertainties in PRA results and (2) the approach to represent events not fully considered in limited IPE/IPEEE studies, e.g., low-power/shutdown events or external events. Therefore, a decision was made to change the terminology for activities implementing the PRA policy statement from *risk-based* to *risk-informed* regulations.

12.1.1 Principles of Risk-Informed Regulations and Licensing

One recent example of the NRC uses of quantitative risk information is in the reactor oversight process (ROP) [Has02]. The objective of the ROP is to move regulations close to a performance basis rather than a prescriptive, compliance basis, with a full use of quantitative risk information. The process consists of six elements:

1. Reactor safety cornerstones covering reactor safety, radiation safety, and materials safeguards
2. Performance indicators in four color-coded bands, green, white, yellow, and red
3. Baseline inspections
4. Significance determination process
5. NRC action matrix
6. Licensee's corrective action program

With performance objectives defined by the three cornerstones for reactor safety (element 1), the NRC specifies the performance indicators (element 2) in terms of (a) unplanned and risk-significant scrams and transients, (b) availability of mitigating systems, (c) integrity of barriers to the releases of radioactivity, (d) emergency preparedness, (e) limitation of public and occupational radiation exposure, and (f) nuclear materials safeguards. In defining the thresholds for the green (licensee response) and white (increased regulatory response) bands, the NRC staff made an attempt [Has02] to use the nominal core damage frequencies (CDFs) of 10^{-5}/reactor-year and 10^{-4}/reactor-year, respectively, with just the monitored event. Unscheduled scram frequencies are suggested as one of several performance indicators for the yellow (required regulatory response) and red (unacceptable performance) bands, with >6 and >25 such scrams per 7000 critical hours, respectively, as the thresholds.

The NRC also uses the baseline inspection program (element 3) to monitor the performance of licensees in all safety cornerstones considered in element 1. The inspections eventually lead to the determination of the significance of the inspection findings (element 4), classified in color-coded bands, similar to the performance indicator bands, with a red finding signaling an event of high safety significance. The inspection findings represent the degree of degradation of safety cornerstones (element 1) and are used as input to determine escalating responses stipulated in the NRC action matrix (element 5). The licensee responses required by the action matrix range from routine senior resident inspector interactions to NRC meetings with senior licensee management and are finally resolved through the licensee's corrective action program (element 6). The ROP has essentially transformed [Has02] routine inspections at NPPs into inspections of the licensee's corrective action program and the licensee's ability to identify and resolve safety issues.

The NRC implemented the 1995 PRA policy statement through the release of a series of chapters in the Standard Review Plan (SRP) [NRC98a] and Regulatory Guides (RGs) [NRC98b,NRC98c,NRC98d,NRC98e,NRC02]. The SRP chapters describe how the NRC staff should review license applications, while the RGs provide detailed guidelines on methods that the licensees may use in their applications to the NRC. The 2007 version [NRC07b] of the SRP for light water reactors offers 19 chapters and includes a discussion of issues related to the severe accidents, PRA, human factors engineering, and combined construction and operating license (COL). The SRP describes the scope of reviews, acceptance criteria, review procedures, and evaluation findings documented as the Safety Evaluation Report. A key NRC document describing the principles of RIRs and licensing procedures is RG 1.174 [NRC02], which was discussed briefly in connection with the Davis-Besse incident in Section 9.6.

The principles and guidelines in RG 1.174 are suggested to make an efficient use of calculated risks as part of the information available in decision-making processes. As discussed earlier in the section, the approach is supposed to be not simply risk based but rather risk informed. In fact, the NRC proposes five principles in applying the basic defense-in-depth philosophy to license changes and other regulatory decisions, as illustrated in Fig. 12.1. The five safety principles leading to integrated decision making in RG 1.174 suggest that:

1. Any proposed licensing change meets current regulations.
2. The change is consistent with the defense-in-depth philosophy.
3. The change maintains sufficient safety margins.

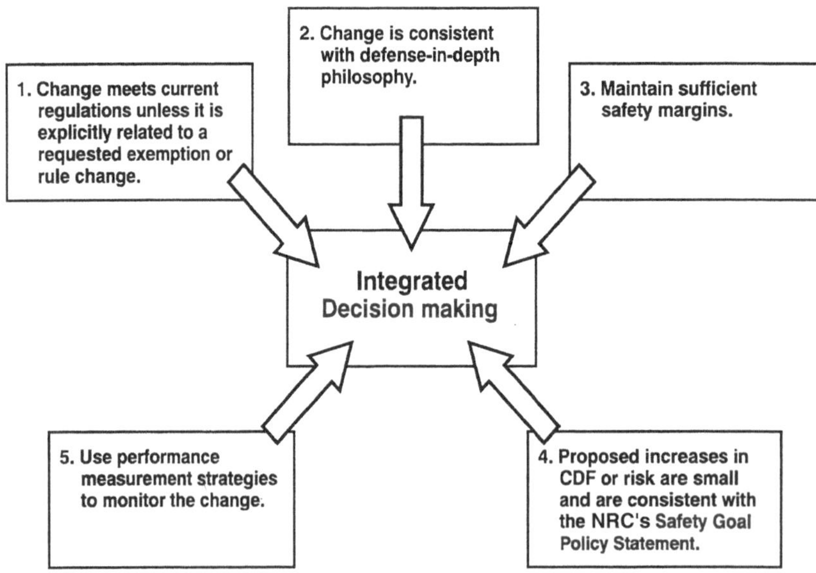

Figure 12.1 Principles of risk-informed integrated decision-making process. *Source:* [NRC02].

4. Any increase in the core damage frequency or risk resulting from the change, if any, is small and is consistent with the Safety Goal Policy Statement [NRC86].
5. Use performance measurement strategies to monitor the impact of the proposed change.

To satisfy the five safety principles in the integrated decision-making process, RG 1.174 also provides step-by-step guidelines:

1. Define the proposed change: Identify (a) plant licensing bases, including the final safety analysis report, technical specifications, and licensing conditions, and (b) structures, systems, and components (SSCs), procedures, and activities that may be affected by the proposed changes, and (c) available engineering studies, codes, operational experience, PRA findings, and other information relevant to the proposed changes.
2. Perform engineering analysis: Evaluate the proposed change with regard to (a) the defense-in-depth principles articulated in the General Design Criteria, 10 CFR 50 Appendix A [NRC71], (b) safety margins that should be maintained, and (c) changes in the quantitative risk in terms of both the CDF and large early release frequency for the plant.
3. Define an implementation and monitoring program: Develop a program that can adequately track the performance of SSCs that could be impacted by the proposed change.

One particular quantitative measure proposed in step 2 is in direct support of principle 4 and was illustrated in Fig. 9.19 in terms of incremental CDF and incremental LERF. The NRC staff apparently tried [Lee04] to apply the five RG 1.174 principles in the 2001 decision-making process that allowed a delayed shutdown of the Davis-Besse plant. The February 2003 NRC region III report [Dye03] documenting the significance of the Davis-Besse incident, in hindsight, illustrates that the NRC staff had difficulty in duly applying the five principles and that the five RG 1.174 principles were not met.

12.1.2 Uncertainties in Risk-Informed Decision Making

One important issue in greater utilization of RIRs is the need to account for uncertainties in PRAs. Detailed guidance on representing uncertainties in risk-informed decision-making processes is presented in a recent NRC report [Dro09]. With the understanding that aleatory uncertainties should be handled through the stochastic methods discussed in Section 6.7.2, detailed approaches are suggested for representing three types of epistemic uncertainties associated with (a) parameters, (b) models, and (c) incompleteness in PRA models.

Parameter uncertainties are related to the statistical representation of probabilities and frequencies of basic events and processes making up the PRA model. In propagating uncertainties using sampling techniques of the type discussed in Section 6.7.2, it is important to consider the state-of-knowledge correlation or epistemic correlation (EC) for events that are correlated. This point may be illustrated by considering two components in an active-parallel configuration with failure probabilities x_1 and x_2 represented by probability density functions $p(x_1)$ and $p(x_2)$, respectively. If the components are uncorrelated, then Eqs. (2.50) and (4.8) yield an expectation value for failure probability $z = x_1 x_2$ of the two-component system

$$E(z) = E(x_1 x_2) = \int_0^\infty x_1 p(x_1) \mathrm{d}x_1 \int_0^\infty x_2 p(x_2) \mathrm{d}x_2 = \langle x_1 \rangle \langle x_2 \rangle \,, \qquad (12.1)$$

which simplifies to $E(z) = \langle x \rangle^2$ for the case of identical components $x = x_1 = x_2$. In contrast, if the two components are correlated, then Eq. (2.52) provides the corresponding expectation value for the system, $E(z) = \langle x \rangle^2 + V(x)$. Thus, the uncertainty in the system failure probability would be underestimated if ECs are not properly represented for correlated components. Care should be taken in modeling basic events with the same parameters or the same state of knowledge.

Modeling uncertainties include those associated with the logic structure or the choice of models for the fault and event trees utilized in the PRA. Effort should be made to perform sensitive studies and evaluate important measures discussed in Section 6.7.3 so that potentially significant contributors to the relevant risk measures are not neglected. Incompleteness uncertainties relate to basic limitations of the PRA model due to unknown phenomena or events not represented. There is, in principle, no systematic way to account for this type of uncertainties. It is, however, suggested [Dro09] that bounding or conservative calculations be performed to gain some measure of the effects of potential unknown events on the risk measures. Finally, all three types of uncertainties should be considered in applying the calculated risk measures in the risk-informed decision-making process illustrated in Fig. 12.1, with due care given in applying the quantitative guidelines of Fig. 9.19.

12.1.3 Other Initiatives in Risk-Informed Regulations

Despite the need to apply the risk-informed decision-making principles with due care, there is no question that quantitative risk measures obtained through PRA studies should play an important role in NPP regulations and licensing. Together with RG 1.174, several other RGs describe more specific applications: (1) RG 1.175 [NRC98b] on in-service testing, (2) RG 1.176 [NRC98c] on graded quality assurance, (3) RG 1.177 [NRC98d] on technical specifications, (4) RG 1.178 [NRC98e] on in-service inspection of piping, and (5) RG 1.160 [NRC97] on maintenance. Among the industry reports that support the risk-informed regulatory activities are (a) PSA application guidelines [Tru95] proposed by the Electric Power Research Institute (EPRI) and (b) industry-sponsored guidelines NUMARC 93-01 [NEI96] for monitoring maintenance activities.

For many of the activities covered by the RGs and industry guidelines, such as maintenance activities, it may become necessary to use an expert panel to determine the risk significance of components together with risk importance measures discussed in Section 6.7. The task takes on a significant importance when it is realized that a typical PRA will address no more than about 2000 SSCs out of a total of 24,000 SSCs subject to the maintenance rule [Has02]. A detailed discussion on the use of risk importance measures is given in [Wal01].

Among recent NRC initiatives that support the RIR activities is the recharacterization of SSCs [Has02], illustrated in Fig. 12.2. Box 1 of Fig. 12.2 contains safety-related SSCs that a risk-informed evaluation concludes are safety significant (RISC-1) and hence would remain consistent with the current classification. Nonsafety SSCs that are risk significant are grouped in box 2 as RISC-2 SSCs, e.g., emergency diesel generators and auxiliary feedwater pumps. Box 3 (RISC-3) contains safety-related SSCs that may not pose significant risk, which may then be subject to less stringent regulatory oversight. Box 4 shows SSCs that have little safety significance and would remain in that category. Applications of risk-informed decision-making principles to NPP regulations and licensing are also discussed in various publications [Kad07,Kel05,Bor01,Wal01].

12.2 RELIABILITY-CENTERED MAINTENANCE

An important part of the RIR activities for NPPs is the consideration of the risk and reliability associated with the performance and maintenance of key SSCs, as discussed in Section 12.1.3. With this perspective in mind, this section focuses on the maintenance of nuclear systems within the context of the reliability, availability, and maintainability of the systems, coupled to the safety (S) implications of the maintenance activities, thus leading to the RAMS structure. Since any RAM or RAMS

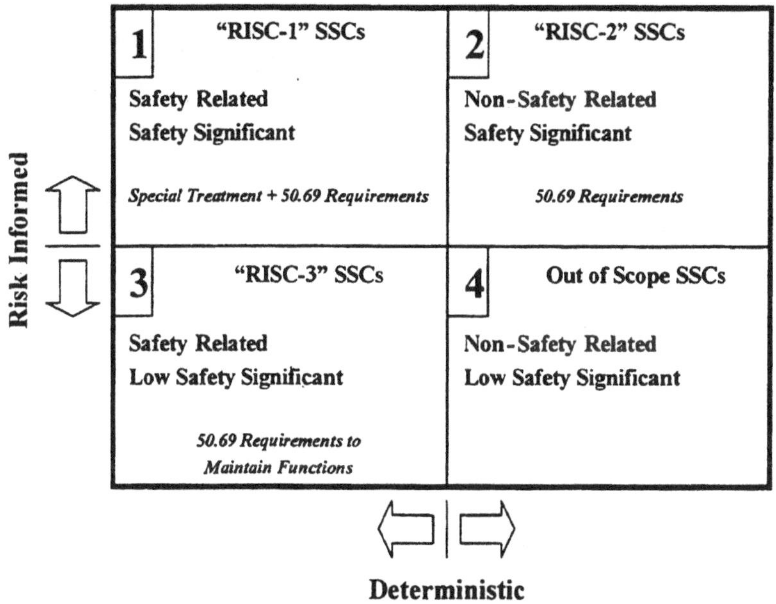

Figure 12.2 Proposed categorization of safety-related SSCs. *Source:* [Has02].

program typically includes reliability-centered maintenance (RCM) activities, general considerations for such activities are presented, together with simple illustrative examples in Section 12.2.1. The implementation of RCM programs for NPP systems is discussed in Section 12.2.2.

Maintenance activities are usually classified as either preventive or corrective activities as discussed in Section 1.5. Preventive maintenance (PM) represents planned maintenance that is performed when the equipment is functioning properly to avoid failures during subsequent operation. Corrective maintenance (CM) is carried out when an item has failed to restore the equipment to functionality or to switch in a standby equipment to restore the system.

12.2.1 Optimization Strategy for Preventive Maintenance

Two simple examples to optimize PM strategies are presented in this section, leading to detailed considerations required for the implementation of realistic RCM programs in Section 12.2.2. The first example illustrates basic concepts behind the minimization of the cost associated with PM strategies in general, while the second discusses how the time interval for the inspection and testing of a system could be selected to maximize the availability of the system.

For operating interval T and an exponential failure probability with constant hazard rate λ, Eq. (2.94) provides the fraction of time a device is unavailable as the time average of the cumulative distribution function $F(t)$ for the failure event

$$\langle F(t)\rangle_T = T^{-1}\int_0^T F(t)\mathrm{d}t \simeq \lambda T/2. \tag{12.2}$$

For repair or replacement interval T for the system, the mean time between repair (MTBR) may likewise be obtained as

$$\begin{aligned}\tau_r(T) &= T - \{\text{time interval for the system in repair state}\}\\ &= T - \textstyle\int_0^T F(t)\mathrm{d}t = \int_0^T R(t)\mathrm{d}t,\end{aligned} \tag{12.3}$$

which yields the familiar result of Eq. (2.92) for the mean time to failure (MTTF)

$$\lim_{T\to\infty} \tau_r(T) = \tau = 1/\lambda = \mathrm{MTTF}. \tag{12.4}$$

The total maintenance cost should include the cost C_p associated with a PM task and the cost C_c associated with a CM task, where C_p is expected to be generally less than C_c. This is because a CM task would involve more substantial work and often entail a longer system outage. The total cost per replacement or repair period T may be written [Rau04] as a sum of C_p and C_c weighted by the CDF $F(t)$ of Eq. (12.2),

$$C(T) = C_p + C_c F(T), \tag{12.5}$$

where, for simplicity, the PM cost is assumed independent of time. The total cost may also be written as a product of the mean total cost per unit time $\langle C(T)\rangle$ and MTBR, so that

$$\langle C(T)\rangle = \frac{C_p + C_c F(T)}{\tau_r(T)}. \tag{12.6}$$

In the limit as $T \to \infty$, Eq. (12.4) simplifies Eq. (12.6) to

$$\langle C(\infty)\rangle = \frac{C_p + C_c F(\infty)}{\tau} = \frac{C_p + C_c}{\tau}, \tag{12.7}$$

which represents the maintenance cost for the case when no PM is performed during operation. With this understanding, a ratio of Eq. (12.7) to Eq. (12.6) may be taken as a measure of the PM effectiveness (PME),

$$\begin{aligned}\mathrm{PME} &= \frac{\langle C(\infty)\rangle}{\langle C(T)\rangle} = \frac{C_p + C_c}{C_p + C_c F(T)}\frac{\tau_r(T)}{\tau}\\ &= \frac{1+r}{1+rF(T)}\frac{\tau_r(T)}{\tau} \text{ with } r = \frac{C_c}{C_p},\end{aligned} \tag{12.8}$$

so that a PM policy should be considered effective if PME > 1.0. In fact, given the ratio r of the CM-to-PM cost, a search can be made for the optimum PM interval T [Rau04]. For the example considered, a Weibull distribution of Eq. (2.124) is used to represent the age-related increase in the failure rate for several different values of the cost ratio r. A limiting case for the PME of Eq. (12.8) may also be considered for a time-independent hazard rate λ, where Eq. (12.8) may be reduced to

$$\text{PME} = \frac{\langle C(\infty)\rangle}{\langle C(T)\rangle} = \frac{(C_p + C_c)F(T)}{C_p + C_c F(T)}. \tag{12.9}$$

In this case, because $F(T) \leq 1.0$, PME ≤ 1.0, which simply states that, if the failure rate remains constant during operation, then a preventive maintenance or replacement of the system is not justified.

Another example for the PM strategy optimization involves maximizing the system availability. We consider a testing and inspection interval T, allowing duly for the finite time required for testing and for repair or maintenance. In this simplified illustration, we determine the total fraction of time the system is unavailable by combining the time the system is in a failed state, Eq. (12.2), with the testing time τ_t and repair time τ_m:

$$\langle \bar{A} \rangle = \frac{\lambda T}{2} + \frac{\tau_t}{T} + \lambda \tau_m. \tag{12.10}$$

By setting the derivative of Eq. (12.10) with respect to T to zero, we obtain an expression for the optimal test interval

$$T_{opt} = \sqrt{2\tau_t/\lambda}, \tag{12.11}$$

which yields the minimum unavailability

$$\langle \bar{A} \rangle_{\min} = \sqrt{2\lambda\tau_t} + \lambda\tau_m. \tag{12.12}$$

This simple analysis indicates the need to find a reasonable testing interval, since a short testing/inspection interval T, compared with the time required for testing, would result in an increased wear and tear of the system. Furthermore, it is necessary to account for the possibility of imperfect repair and undetected test-induced failures.

12.2.2 Reliability-Centered Maintenance Framework

The practice for NPP maintenance has to balance the reliability, cost, and safe performance of the system, which requires a shift toward RCM. Reducing the maintenance cost without sacrificing the system reliability and availability of risk-important SSC depends a lot on selecting the right PM strategies and maintenance intervals. That is why a program such as RCM is very useful in developing a cost-effective maintenance program since it provides a full framework to optimize the maintenance tasks in a systematic way.

There have been different versions of the RCM program depending on the applications [Now78,Ber05], such as maintenance steering group (MSG)-3, RCM2, streamlined RCM (SRCM), and reliability-centered asset management (RCAM). MSG-3 is used in the aviation industry and RCM2 provides a separate treatment of environmental aspects of failures. SRCM was developed by the EPRI to reduce the resources and steps required to carry out a traditional RCM program while RCAM was developed as a result of the experience from application studies for the electrical distribution system [Ber05]. Although there are various versions of the RCM program, the principles and concepts of traditional RCM have remained essentially unchanged. The principles from the first definition [Smi93] of RCM are:

1. Preservation of system function
2. Identification of failure modes
3. Prioritizing of function needs
4. Selection of applicable and effective maintenance tasks

Figure 12.3 illustrates the main procedures and logic steps for developing RCAM plans as described by Bertling [Ber05]. The flow model in Fig. 12.3 is divided into three main stages:

Stage 1. *System reliability analysis*: Define the system and evaluate critical components for system reliability.
Stage 2. *Evaluation of PM and component behavior*: Analyze the components in detail and, using the necessary input data, define a quantitative relationship between system reliability and PM measures.
Stage 3. *System reliability and cost/benefit analysis*: Evaluate the cost for different PM strategies and methods by using the information gained on the effect of PM on system reliability.

The main challenge for all RCM programs is to develop meaningful relationships between the PM activities and system reliability in stage 2. This is particularly challenging for nuclear systems subject to significant radiation-induced degradations. In this regard, the ability to perform online diagnostics or surveillance of degradations and aging of various SSCs in nuclear systems will become increasingly important, especially as the extension of the operating licenses of the current fleet of NPPs to a total operating period of 80 years is considered actively. Reliability calculations for systems undergoing stages of degradation may require multistate semi-Markov models [Mar05], which account explicitly for the time required for the maintenance activities, often expressed as a sojourn time. The dynamic event tree algorithms discussed in Chapter 13, together with efficient, accurate fault tree algorithms, including the binary decision algorithm of Section 7.5, may also find increased applications in online surveillance of SSCs undergoing degradations in nuclear systems.

12.2.3 Cost-Benefit Considerations

Stage 3 of Fig. 12.3 represents the cost-benefit analysis step, where PM methods and strategies are compared to select a cost-effective PM strategy. What is needed to make the optimal choice is a balance equation representing all cost elements and an optimization algorithm for various maintenance strategy costs. An optimization algorithm can be selected once a formula for a maintenance strategy cost can be quantified. Hence, the main challenge here is to account for all costs for each maintenance strategy in a mathematical form that can then be optimized.

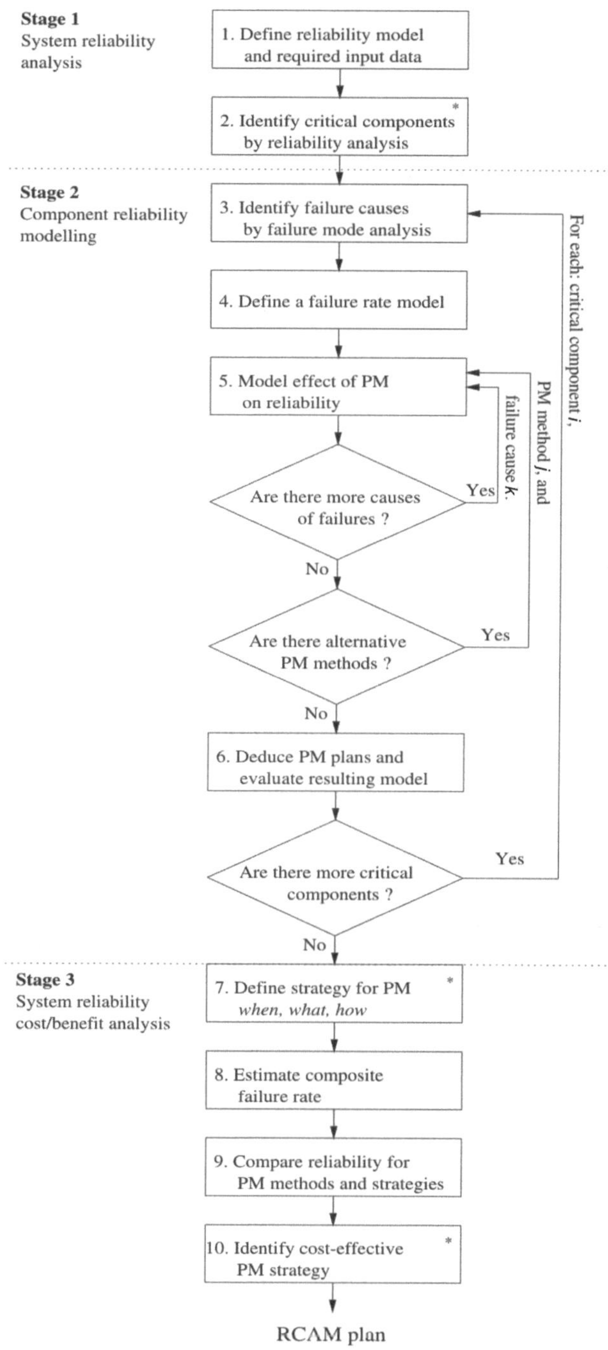

Figure 12.3 Logic for RCAM method. *Source:* Reprinted with permission from [Ber05]. Copyright © 2005 The Institute of Electrical and Electronics Engineers.

When a cost-benefit analysis is considered, it usually entails balancing the elements of risk, cost, and loss. In the context of selecting maintenance policies for nuclear systems, the element of risk is the level of reliability that is desired to avoid the failure to generate electricity. Costs include operating and maintenance costs of the system. The EPRI has developed a computerized cost-benefit analysis module (CBAM) that analyzes maintenance cost at the component, system, unit, and plant levels [EPR99]. The formula used in the CBAM software is a generalization of Eq. (12.5),

$$C(T) = C_p^* + C_c^* F(T) + C_{nr}, \tag{12.13}$$

where C_p^* and C_c^* are the effective PM and CM costs, respectively, including the replacement power costs associated with the maintenance activities, and the additional term C_{nr} accounts for the one-time, nonrecurring cost of implementing the optimized maintenance program distributed over the remaining life of the equipment. The element of loss includes the loss of revenue due to unexpected failures between scheduled outages, replacement power costs, and damage of the reputation or credibility of the company as a result of the outage or accident.

Conducting a cost-benefit analysis for all three elements in a consolidated formulation is challenging because it is hard to balance the risk, cost, and loss, represented in different units. A common approach to address this difficulty involves expressing all of these three attributes in equivalent monetary terms and optimizing the combined cost of the maintenance strategy proposed. In a somewhat alternate approach, Hadavi [Had09] used a value theory to define the "value" as "function/resources," "worth/cost," or "satisfaction of needs/resources" for incorporating the three competing optimization criteria, risk *R*, cost *C* , and loss *L*, into a single evaluation function *E* to optimize maintenance scheduling,

$$E = w_R R + w_C C + w_L L, \tag{12.14}$$

in terms of three weighting factors $w_R, w_C,$ and w_L. With a judicious selection of the weighting factors, Hadavi provided a cost-benefit analysis of PM strategies for the auxiliary feedwater system of a PWR.

In an alternate approach combining the RAMS framework explicitly with the cost analysis, Martorell et al. [Mar05] modified the fractional unavailability of Eq. (12.10) to explicitly account for both PM and CM times and incorporated it into a generalized form of the cost function of Eq. (12.13). In a PM study for a system of emergency diesel generators, with an annual PM schedule with realistic testing and maintenance times, a multiobjective optimization algorithm was developed to minimize the total cost. The optimization algorithm sought to maintain the reliability and availability of the system within the constraints of the technical specifications for the system. Various NRC Regulatory Guides [NRC97,NRC00] and publications [Sam95] related to the management of risk associated with maintenance activities were also considered in formulating the optimization problem.

References

[Ber05] L. Bertling, R. Allan, and R. Eriksson, "A Reliability-Centered Asset Maintenance Method for Assessing the Impact of Maintenance in Power Distribution Systems," *IEEE Trans. Power Sys.* **20**, 75 (2005).

[Bor01] E. Borgonovo and G. E. Apostolakis, "A New Importance Measure for Risk-Informed Decision Making," *Reliab. Eng. Sys. Safety* **72**, 193 (2001).

[Dro09] M. Drouin, G. Parry, J. Lehner, G. Martinez-Guridi, J. LaChance, and T. Wheeler, "Guidance on the Treatment of Uncertainties Associated with PRAs in Risk-Informed Decision Making," NUREG-1855, vol. 1, U.S. Nuclear Regulatory Commission (2009).

[Dye03] J. E. Dyer, "Davis-Besse Control Rod Drive Mechanism Penetration Cracking and Reactor Pressure Vessel Head Degradation Preliminary Significance Assessment," Report No. 50-346/2002-08(DRS), U.S. Nuclear Regulatory Commission (2003).

[EPR99] "Cost Benefit Analysis for Maintenance Optimization," TR-107902, Electric Power Research Institute (1999).

[Had09] S. M. H. Hadavi, "A Heuristic Model for Risk and Cost Impacts of Plant Outage Maintenance Schedule," *Ann. Nucl. Energy* **36**, 974 (2009).

[Has02] F. E. Haskin, A. L. Camp, S. A. Hodge, and D. A. Powers, "Perspectives on Reactor Safety," NUREG/CR-6042, rev. 2, U.S. Nuclear Regulatory Commission (2002).

[Kad07] A. C. Kadak and T. Matsuo, "The Nuclear Industry's Transition to Risk-Informed Regulation and Operation in the United States," *Reliab. Eng. Sys. Safety* **92**, 609 (2007).

[Kel05] W. Keller and M. Modarres, "A Historical Overview of Probabilistic Risk Assessment Development and Its Use in the Nuclear Power Industry; A Tribute to the Late Professor Norman Carl Rasmussen," *Reliab. Eng. Sys. Safety* **89**, 271 (2005).

[Lee04] J. C. Lee, T. H. Pigford, and G. S. Was, "Report of the Committee to Review the NRC's Oversight of the Davis-Besse Nuclear Power Station," Appendix II, GAO-04-415, U.S. General Accounting Office (2004).

[Mar05] S. Martorell, J. F. Villanueva, S. Carlos, Y. Nebot, A. Sanchez, J. L. Pitarch, and V. Serradell, "RAMS+C Informed Decision-Making with Application to Multi-Objective Optimization of Technical Specifications and Maintenance Using Genetic Algorithms," *Reliab. Eng., Sys. Safety* **87**, 65 (2005).

[NEI96] "Industry Guideline of Monitoring the Effectiveness of Maintenance at Nuclear Power Plants," NUMARC 93-01, rev. 2, Nuclear Energy Institute (1996).

[Now78] F. S. Nowlan and H. F. Heap, "Reliability-Centered Maintenance," A066-579, U.S. Department of Commerce (1978).

[NRC71] "General Design Criteria for Nuclear Power Plants," Title 10, *Code of Federal Regulations*, Part 50, Appendix A, U.S. Nuclear Regulatory Commission (1971).

[NRC86] "Safety goals for the Operation of Nuclear Power Plants," Title 10, *Code of Federal Regulations*, Part 50, Policy Statement, U.S. Nuclear Regulatory Commission (1986).

[NRC88] "Individual Plant Examination for Severe Accident Vulnerabilities," 10 CFR 50.54(f), Generic Letter 88-20, U.S. Nuclear Regulatory Commission (1988).

[NRC91] "Individual Plant Examination of External Events (IPEEE) for Severe Accident Vulnerabilities," 10 CFR 50.54(f), Generic Letter 88-20, Supplement 4, U.S. Nuclear Regulatory Commission (1991).

[NRC94] "A Review of NRC Staff Uses of Probabilistic Risk Assessment," NUREG-1489, U.S. Nuclear Regulatory Commission (1994).

[NRC95] "Use of Probabilistic Risk Assessment Methods in Nuclear Regulatory Activities: Final Policy Statement," *Federal Register*, 60FR42622, U.S. Nuclear Regulatory Commission (1995).

[NRC97] "Monitoring the Effectiveness of Maintenance at Nuclear Power Plant," Regulatory Guide 1.160, U.S. Nuclear Regulatory Commission (1997).

[NRC98a] "Use of Probabilistic Risk Assessment in Plant-Specific Risk-Informed Decisionmaking: General Guidance," Standard Review Plan, NUREG-0800, Chapter 19, U.S. Nuclear Regulatory Commission (1998).

[NRC98b] "An Approach for Plant-Specific Risk-Informed Decisionmaking: In-Service Testing," Regulatory Guide 1.175, U.S. Nuclear Regulatory Commission (1998).

[NRC98c] "An Approach for Plant-Specific Risk-Informed Decision-Making: Graded Quality Assurance," Regulatory Guide 1.176, U.S. Nuclear Regulatory Commission (1998).

[NRC98d] "An Approach for Plant-Specific Risk-Informed Decisionmaking: Technical Specifications," Regulatory Guide 1.177, U. S. Nuclear Regulatory Commission (1998).

[NRC98e] "An Approach for Plant-Specific Risk-Informed Decisionmaking: In-Service Inspection of Piping," Regulatory Guide 1.178, U.S. Nuclear Regulatory Commission (1998).

[NRC00] "Assessing and Managing Risk Before Maintenance Activities at Nuclear Power Plants," Regulatory Guide 1.182, U.S. Nuclear Regulatory Commission (2000).

[NRC02] "An Approach for Using Probabilistic Risk Assessment in Risk-Informed Decisions on Plant-Specific Changes to the Licensing Basis," Regulatory Guide 1.174, rev. 2, U.S. Nuclear Regulatory Commission (2002).

[NRC07a] "Backfitting," 10 CFR 50.109, U.S. Nuclear Regulatory Commission (2007).

[NRC07b] "Standard Review Plan for the Review of Safety Analysis Reports for Nuclear Power Plants: LWR Edition," NUREG-0800, U.S. Nuclear Regulatory Commission (2007).

[Rau04] M. Rausand and A. Hoyland, *System Reliability Theory—Models, Statistical Methods, and Applications*, Wiley (2004).

[Sam95] P. K. Samanta, I. S. Kim, T. Mankamo, and W. E. Veseley, "Handbook of Methods for Risk-Based Analyses of Technical Specifications," NUREG/CR-6141, U.S. Nuclear Regulatory Commission (1995).

[Smi93] A. M. Smith, *Reliability-Centered Maintenance*, McGraw-Hill (1993).

[Tru95] D. True, K. Fleming, G. Parry, B. Putney, and J-P. Sursock, "PSA Applications Guide," TR-105396, Electric Power Research Institute (1995).
[Wal01] I. B. Wall, J. J. Haugh, and D. H. Worlege, "Recent Applications of PSA for Managing Nuclear Power Plant Safety," *Prog. Nucl. Energy* **39**, 367 (2001).

Exercises

12.1 Two hardware modifications costing $1million each are suggested for the Surry Unit 1 plant. Modification A is expected to reduce the probability of ATWS by a factor of 3, while modification B is expected to reduce the LOCA probability by a factor of 3. Based on the event tree analysis of Section 10.3.1 and risk-informed regulation guidelines, determine which modification should be given the first priority and discuss the result.
12.2 An operating procedure change is suggested for the Surry Unit 1 plant which is expected to halve the LOCA probability but double the ATWS probability. It is suggested to determine the merit of the proposed modification in terms of the early release of iodine and bromine isotopes only. Based on the simplified event tree analysis of Section 10.3.1, would you accept the procedural change proposed?

CHAPTER 13

DYNAMIC EVENT TREE ANALYSIS

Basic approaches for PRA of complex engineered systems through combined use of FT and ET structures were discussed in Chapter 6, followed by a summary of key PRA tools available for the risk assessment of nuclear systems in Chapter 7. Two major PRA studies for nuclear power plants were reviewed and discussed in Chapter 10, highlighting the complexities involved in such studies, with a large number of transient and accident sequences that need to be analyzed. Figure 10.4 illustrates the four large blocks of ETs that form the basis for the NUREG-1150 PRA studies for five LWR plants. One key structure implemented to manage the computational requirements for the PRA studies entails grouping transients and accidents of similar nature with representative time evolutions and probability density functions for the aggregated groups.

In this chapter we now introduce techniques that accurately represent, under a dynamic event tree (DET) structure, detailed evolutions of transient events without restrictions placed on preselected groups of events. This chapter covers recent developments for PRAs of complex engineered systems so, to be consistent with that literature where vectors are not denoted by boldface fonts, the symbols in this chapter for vectors also will not be in boldface.

13.1 BASIC FEATURES OF DYNAMIC EVENT TREE ANALYSIS

Together with the massive effort undertaken to reassess the safety of NPPs in the aftermath of the TMI-2 accident of 1979, the need to represent the evolution of transients and accidents realistically, accounting for the actuation of engineered safety features and actions of human operators, was readily recognized. This recognition led to the development of probabilistic reliability techniques as summarized in a review paper [Lab00]. One earlier paper [Ame81], in particular, introduced the dynamical logical analytical methodology (DYLAM) with the objective of accounting accurately for the timing of system failures and operator actions in accident evolutions exemplified by the TMI-2 accident. We begin with a brief review of the DYLAM concept and illustrate the basic features of DET approaches in this section.

Among the limitations of the classic PRA techniques recognized [Ame81,Aco93, Lab00] for NPP risk assessment are:

1. Branches in the ET structure typically represent binary on-off or success-failure events and thereby are not able to meaningfully account for partial or degraded operation of systems.
2. Various transients or accident scenarios are grouped into a few representative ETs with pre-selected branching times having aggregated branching probabilities.
3. Each accident scenario is treated as a set of hardware failures and operator errors, making it difficult to explicitly represent the likelihood of either hardware failures or operator errors

In an effort to remedy these limitations or concerns related to traditional ET methods, the DET approaches allow the formation of ET branches at times selected via logical rules as the actual transient evolves for each transient scenario, with explicit representations of operator actions or hardware degradations at each branch point. With detailed representations of individual transients, the DET methods also could account more accurately for dependent system failures [Aco93] and dynamic interactions between system states and component states [Ame81].

To illustrate the additional details that could be represented via the DET methods, Amendola and Reina [Ame81] considered a loss of flow accident in a sodium-cooled fast reactor which is initiated by a pump coastdown. In a simplified SFR model, the reactor is represented by 11 state variables, representing (1) each of two coolant channels for the core with outlet temperature sensors, (2) shutdown logic signal, (3) channel outlet temperature controller, (4) flow rate sensor, (5) scram actuator, (6) flow rate controller, (7) core reactivity, (8) reactivity sensor, (9) reactivity controller, and (10) pumps. The core flow rate is postulated to decrease with a time constant of 5.71 seconds, together with the failure of the flow rate sensor and channel 1 temperature sensor. The resulting transient is illustrated in Fig. 13.1 for seven key system variables for $\sim$5 seconds into the postulated accident, with other possible system evolutions indicated with dashed curves. An accounting for the probabilities of system evolutions for each scenario yields the probability for a terminal state involving the occurrence of fuel melting in the postulated accident simulations. The system evolutions illustrated in Fig. 13.1 highlight the importance of representing the

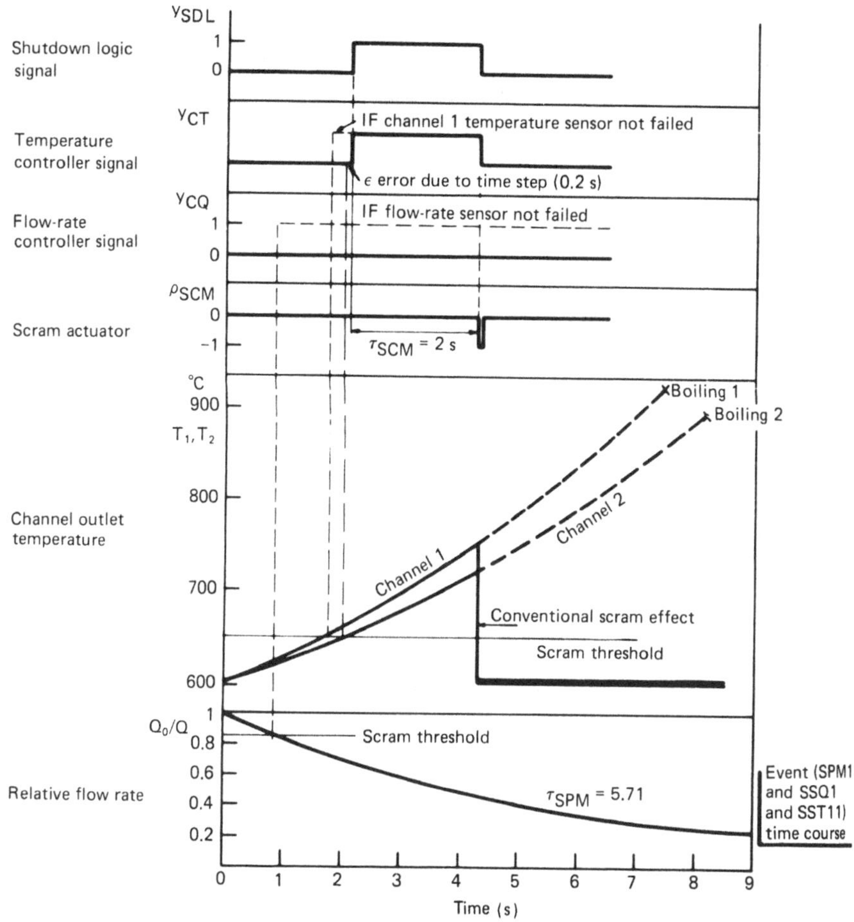

Figure 13.1 Evolution of seven system variables in a postulated LOF event. *Source:* Reprinted with permission from [Ame81]. Copyright ©1981 American Nuclear Society, La Grange Park, Illinois.

correct timing and associated probabilities for system and sensor failures, which is not surprising to any system analyst.

The importance of correctly representing operator actions in NPP risk assessment is illustrated [Aco93] in a generalization of the DYLAM concept, designated as the dynamic event tree analysis method (DETAM), for the analysis of a postulated steam generator tube rupture event in a PWR. The SGTR event was recognized in the NUREG-1150 PRA discussion of Chapter 10 as a major contributor to the V sequence of events leading to containment bypass. In the DETAM analysis of the SGTR event, accompanied by a failure of the emergency feedwater system

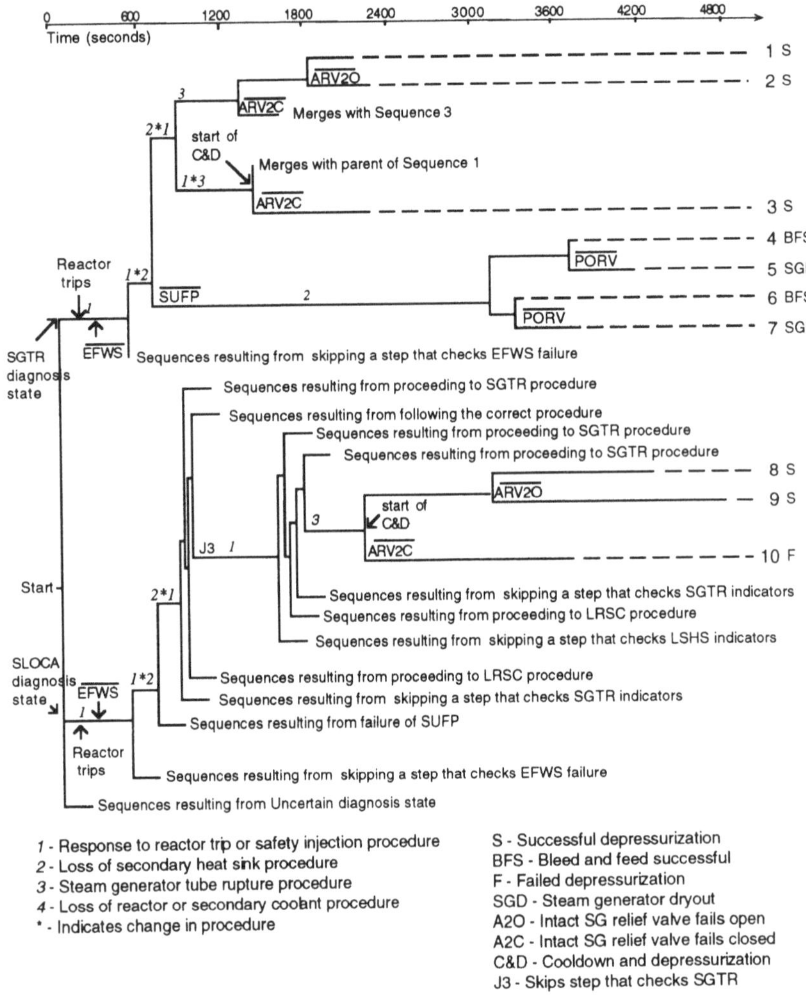

Figure 13.2 Partial dynamic event tree for a SGTR event with EFWS failure. *Source:* [Aco93].

(EFWS), special focus was given to a detailed representation of operator actions via DETs so that the likelihood of operator errors is explicitly represented with the appropriate branching times specified throughout the management of the event. In this approach, the system state is explicitly represented as a function of the physical or hardware state, process or component state, and operator action state, allowing for explicit delineation of the dependences between failure events. The sequences of operator actions and resulting scenarios are illustrated in Fig. 13.2, where five scenarios resulting in successful depressurization (S state) and one scenario leading to failed depressurization (F state) are indicated together with four others requiring various operator actions. In the figure, for example, $\overline{\text{EFWS}}$ represents the failure of EFWS.

One feature that can be readily recognized in the part of the DET highlighted for the $\overline{\text{EFWS}}$ in Fig. 13.2 is the multiple branches with irregular branch times representing various operator actions and accident management procedures. The SGTR event accompanied by the EFWS failure requires the representation of 128 possible hardware states, associated with seven binary system states, plus the failed EFWS, as well as 324 possible operator planning states and 2304 operator diagnosis states, for a total of 9.6×10^7 distinct states for each time step. Application of a number of simplifying assumptions and realistic cutoff frequencies, however, resulted in a relatively manageable ET structure involving only 52 scenarios illustrated in Fig. 13.2. Nonetheless, the total number of possible scenarios increases geometrically as the transient time increases. This is the nature of a DET analysis that currently limits its application as a tool to augment, rather than replace, conventional ET algorithms. This issue is discussed further in Sections 13.3 and 13.4 and in [Aco93,Lab00].

13.2 CONTINUOUS EVENT TREE FORMULATION

With the strengths and limitations of DETs discussed in Section 13.1 through a review of two early publications, we now present the theoretical foundation for the probabilistic reliability analysis that Devooght and Smidts introduced [Dev92]. Although the formulation was originally presented to represent continuous event trees (CETs) for dynamical systems without statistical fluctuations, we extend [Aum06] the formulation here to account for fluctuations in the system state vector $x(t)$ as well as in the measurements $y(t)$ from which $x(t)$ is obtained. The generalized formulation will be used in Sections 13.3 and 13.4 and simplifies to the CET formulation when $x(t)$ is directly measurable and noise free, as would be the case for probabilistic dynamic system analyses.

13.2.1 Derivation of the Stochastic Balance Equation

Consider a physical system represented by system state vector $x(t)$, e.g., system power and coolant temperature, at time t corresponding to component state vector $c(t)$, e.g., steam valve or pump speed, subject to white Gaussian noise vector $w_x(t)$ with covariance Q_x:

$$\dot{x} = f(x, c) + w_x(t) \quad \text{with } \langle w_x(t) \rangle = 0, \big\langle w_x(t) w_x^T(t') \big\rangle = Q_x(t - t'), \tag{13.1}$$

where $\dot{x} = \mathrm{d}x/\mathrm{d}t$ and $f(x, c)$ is the functional describing the system evolution. Component state c is generally unobservable, while x is determined indirectly through measurements y, with nonlinear measurement function h that is subject to white Gaussian noise vector v with covariance R:

$$y = h(x) + v(t) \quad \text{with } \langle v(t) \rangle = 0, \big\langle v(t) v^T(t') \big\rangle = R\delta(t - t'). \tag{13.2}$$

We desire to obtain a stochastic balance equation for the PDF $p(x, c, t)$ so that $p(x, c, t)\,\mathrm{d}x\,\mathrm{d}c$ represents the probability that the system is in $(x \sim x + \mathrm{d}x, c \sim c + \mathrm{d}c)$ at time t. For notational convenience, we will also use z to represent the combination of the system state and component state variables:

$$z = (x \;\; c)^T . \tag{13.3}$$

The balance equation will then be obtained [Gar85] as a combination of the master equation representing the system transition probabilities $W(z|z')$ for transitions from state z' to z in a Markov process,

$$\frac{\partial p(z,t)}{\partial t} = \int [W(z|z')p(z',t) - W(z'|z)p(z,t)]\,\mathrm{d}z', \tag{13.4}$$

and the Fokker-Planck equation representing all possible system transitions subject to fluctuations or undergoing a diffusion process,

$$\begin{aligned} \frac{\partial}{\partial t}p(x,c,t) &= -\sum_j \frac{\partial}{\partial x_j}[\eta_j(x,c)p(x,c,t)] \\ &+ \frac{1}{2}\sum_{j,k}\frac{\partial^2}{\partial x_j\,\partial x_k}[\sigma_{jk}(x,c)p(x,c,t)] . \end{aligned} \tag{13.5}$$

Here η and σ are the two first moments of the probability of system transition, from state x at time t to state x' at time $t + \Delta t$, for a given component state c:

$$\begin{aligned} \eta_j(x,c) &= \lim_{\Delta t \to 0}\frac{1}{\Delta t}\int_{-\infty}^{\infty}(x'_j - x_j)p(x',c,t+\Delta t|x,c,t)\mathrm{d}x' \\ &= \lim_{\Delta t \to 0}\left\langle \frac{\Delta x_j(t)}{\Delta t}\right\rangle, \end{aligned} \tag{13.6}$$

$$\sigma_{jk}(x,c) = \lim_{\Delta t \to 0}\left\langle \frac{\Delta x_j(t)\Delta x_k(t)}{\Delta t}\right\rangle . \tag{13.7}$$

The master equation (13.4) may be considered a continuum generalization of the Markov equation (5.10), with the matrices $W(c|c', x)$ and $W(c'|c, x)$ representing probabilities per unit time of *discrete* component state transitions $c' \to c$ and $c \to c'$, respectively, for a given system state x. Finally, after introducing the total probability per unit time of leaving component state c as

$$\Gamma(c,x) = \int W(c'|c,x)\mathrm{d}c', \tag{13.8}$$

and combining Eqs. (13.4) and (13.5), we obtain the desired *stochastic balance equation*

$$\begin{aligned} \frac{\partial}{\partial t}p(x,c,t) &= -\sum_j \frac{\partial}{\partial x_j}[\eta_j(x,c)p(x,c,t)] + \frac{1}{2}\sum_{j,k}\frac{\partial^2}{\partial x_j \partial x_k}[\sigma_{jk}(x,c)p(x,c,t)] \\ &+ \int_{c'} W(c|c',x)p(x,c',t)\mathrm{d}c' - \Gamma(c,x)p(x,c,t). \end{aligned} \tag{13.9}$$

Equation (13.9) is usually known [Gar85] as the *differential Chapman-Kolmogorov equation* and forms the basis for our discussion on the probabilistic reliability formulation for the balance of the chapter.

For the purpose of a CET formulation for dynamic system analysis involving significant system transitions, we may introduce an approximation that the system evolves deterministically for a constant component state. This approximation, together with Eq. (13.6) written as $\eta_j(x, c) = \dot{x}_j = f_j(x, c)$, corresponding to the jth component of vector x, allows us to drop the diffusive term in Eq. (13.9),

$$\begin{aligned}\frac{\partial}{\partial t}p(x,c,t) &= -\sum_j \frac{\partial}{\partial x_j}\left[\dot{x}_j p(x,c,t)\right] + \int_{c'} W(c|c',x)p(x,c',t)\mathrm{d}c' \\ &\quad - \Gamma(c,x)p(x,c,t). \end{aligned} \tag{13.10}$$

The similarity between this form of the stochastic balance equation and the one-group time-dependent neutron transport equation is recognized and discussed in [Dev92], with the component state vector c and $\dot{x} = f(x, c)$ playing the roles of the unit directional vector Ω and the velocity vector v in the angular neutron flux $\phi(x, \Omega, t)$, respectively. Thus, the solution to Eq. (13.10) may be conveniently obtained by numerical techniques similar to those developed for the solution of the neutron transport equation. One key difference, however, between the two integrodifferential equations is that, while the vector v may lie anywhere in the velocity space of neutrons for the transport equation, $\dot{x} = f(x, c)$ governs and restricts the trajectory of the system states in the stochastic balance equation. Nonetheless, the solution for the distribution of the probability fluid via Eq. (13.10) may be conveniently obtained via an integral form of the equation, which will be derived in Section 13.2.2.

13.2.2 Integral Form of the Stochastic Balance Equation

Conversion of Eq. (13.10) into an integral form of the balance equation for the joint PDF $p(x, c, t)$ was obtained through a series of algebraic manipulations [Dev92], but a somewhat simpler approach is taken here for the same task. The first step involves integrating Eq. (13.10) over the space of x:

$$\frac{\partial}{\partial t}p(c,t) + \langle\Gamma(c)\rangle\, p(c,t) = \int_{c'} \langle W(c|c')\rangle\, p(c',t)\mathrm{d}c' \equiv Q(c,t), \tag{13.11}$$

where weighted-average transition probabilities are defined as

$$\langle\Gamma(c)\rangle = \int_x \Gamma(x,c)p(x,c,t)\mathrm{d}x/p(c,t) \ \text{ with } p(c,t) = \int_x p(x,c,t)\mathrm{d}x, \tag{13.12}$$

$$\langle W(c|c')\rangle = \int_x W(c|\,c',x)p(x,c',t)\mathrm{d}x/p(c',t). \tag{13.13}$$

In deriving Eq. (13.11), the Gauss divergence theorem is used, with the outward unit normal vector $\widehat{n}$, to obtain

$$\int_{V_x} \nabla\cdot f(x,c)p(x,c,t)\mathrm{d}V = \int_{S_x} \widehat{n}\cdot f(x,c)p(x,c,t)\mathrm{d}S, \tag{13.14}$$

which will vanish for a sufficiently large surface S_x corresponding to a large state space volume V_x. This represents the fact that the system cannot evolve to infinity in finite time [Dev92].

Recognizing that the state space evolutions are limited to a set of feasible trajectories, we consider some feasible trajectory x^*, for which we may set

$$p(x) = \delta(x - x^*). \tag{13.15}$$

Invoking the product axiom for probabilities of Eq. (2.4), together with Eq. (13.15), allows us to write

$$p(c,t) = \int_x p(x,c,t)\mathrm{d}x = \int_x p(c,t\,|x\,)\,p(x)\mathrm{d}x = p\,(c,t\,|x^*\,)\,, \tag{13.16}$$

which then recasts Eqs. (13.12) and (13.13) as

$$\langle\Gamma(c)\rangle = \frac{\int_x \Gamma(x,c)p\,(c,t\,|x\,)\,\delta(x-x^*)\mathrm{d}x}{p\,(c,t\,|x^*\,)} = \Gamma(x^*,c), \tag{13.17}$$

$$\langle W(c|c')\rangle = W(c|c',x^*). \tag{13.18}$$

Integrating Eq. (13.11), with Eqs. (13.16) through (13.18) substituted, yields

$$\begin{aligned} p\,(c,t\,|x^*\,) &= \exp\left\{-\int_0^t \Gamma\left[x^*(s),c\right]\mathrm{d}s\right\} \\ &\times \left\{\int_0^t Q(c,\tau)\exp\left[\int_0^\tau \Gamma\left[x^*(s),c\right]\mathrm{d}s\right]\mathrm{d}\tau + p(c,0\,|u)\right\}, \end{aligned} \tag{13.19}$$

where $Q(c,\tau) = \int_{c'} W(c|c',x^*)p(c',\tau\,|x^*\,)\mathrm{d}c'$ and the initial condition is $u = x^*(0)$.

We formally write the solution to the system equation (13.1), without the noise term, as

$$x(t) = g(u,c,t),\ u = x(0), \tag{13.20}$$

so that

$$p(x\,|u\,) = \delta\left[x - g(u,c,t)\right]. \tag{13.21}$$

Similarly, with $x(t) = g(v,c,t-\tau)$ so that $v = x(\tau)$, we write

$$p(x\,|v\,) = \delta\left[x - g(v,c,t-\tau)\right],\ 0 < \tau \le t. \tag{13.22}$$

Again using a form of the product axiom for probabilities, Eq. (2.4), or a general relationship for joint PDFs, sometimes known as the Chapman-Kolmogorov equation [Gar85], we may construct the joint PDF for a combined state (x,c) at time t:

$$p(x,c,t) = \int \mathrm{d}v\, p(x,c,t\,|v)p(v) = \int \mathrm{d}v\, p(c,t\,|x)p(x\,|v)p(v)\,. \tag{13.23}$$

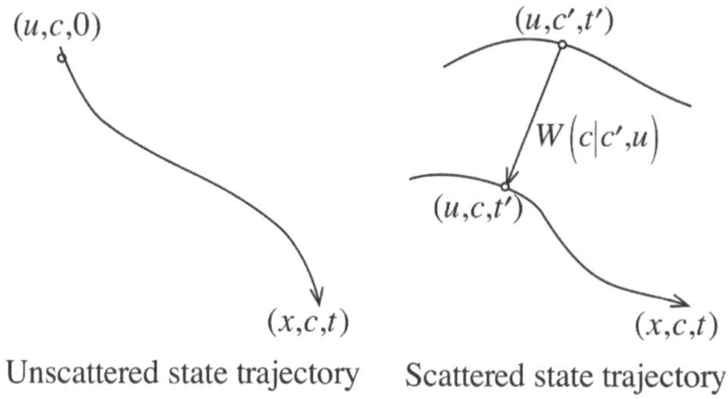

Figure 13.3 Two types of state trajectories represented in Eq. (13.24). *Source:* [Aum94].

Substituting Eqs. (13.19), (13.21), and (13.22) into Eq. (13.23), we finally obtain the desired integral form of the stochastic balance equation:

$$\begin{aligned} p(x,c,t) &= \int \mathrm{d}u\, p(u,c,0)\delta\left[x-g(u,c,t)\right]\exp\left\{-\int_0^t \mathrm{d}s\,\Gamma\left[g(u,c,s)\right]\right\} \\ &\quad + \int_0^t \mathrm{d}t' \int \mathrm{d}u \int \mathrm{d}c' p(u,c',t')\delta\left[x-g(u,c,t-t')\right] \\ &\quad \times \exp\left\{-\int_{t'}^t \mathrm{d}s\,\Gamma\left[g(u,c,s-t')\right]\right\} W(c|c',u), \end{aligned} \tag{13.24}$$

where a notational simplification $\Gamma\left[g(u,c,s)\right] = \Gamma\left[c, g(u,c,s)\right]$ is made.

With the component transition probabilities W and Γ defined in Eqs. (13.4) and (13.8), we note that the first term of the solution for $p(x,c,t)$ represents the system evolution from the initial state $u = x(0)$ to $x(t) = g(u,c,t)$ without undergoing any component transition over the time interval $[0,t]$. In the analogy of neutron transport, this term may be considered an "unscattered" trajectory, as illustrated in Fig. 13.3. Likewise, the second term characterizes the system evolution that involves a component transition from c' to c during the interval $[0,t']$ and remaining in c over the remaining interval $[t',t]$, represented by the "scattered" trajectory in Fig. 13.3.

13.2.3 Numerical Solution of the Stochastic Balance Equation

It is much more convenient to solve the integral form of the stochastic balance equation (13.24) for $p(x,c,t)$ than the differential form in Eq. (13.10), typically via Monte Carlo techniques [Smi92]. In this approach, the time t for component transition may be sampled by selecting a random number ξ_1 uniformly distributed over [0,1],

$$\xi_1 = \exp\left\{-\int_0^t \Gamma\left[g(u,c,s)\right]\mathrm{d}s\right\} = \exp\left\{-\int_0^t \Gamma\left[c,x(s)\right]\mathrm{d}s\right\}, \tag{13.25}$$

coupled to the solution of Eq. (13.1) for the system state $x(t)$. To determine a transition from component state c'_n to a new state c_n for component n, another random number ξ_2 uniformly distributed over $[0, 1]$ yields

$$\xi_2 = \frac{\int_0^{c_n} W(\alpha\,|c'_n, x)\,\mathrm{d}\alpha}{\Gamma(c'_n, x)}, \tag{13.26}$$

in a manner similar to the selection of scattering angles in particle transport algorithms.

Alternatively [Aum94,Aum06,Buc08], Monte Carlo techniques may be used to search for various trajectories involving component transitions with discrete time steps. With a judicious selection of time steps, we may introduce a simple approximation that the component transitions or failures take place over a short time interval $\tau = t - t'$, which allows us to assume that only one transition takes place deterministically at the beginning of the interval and that $\tau\Gamma(c, s) \ll 1$. With this approximation, the exponential term for the "scattered" trajectory may be set to unity. Equation (13.24) may then actually be evaluated for the case with N component states as

$$p\,(x, c) = \prod_{n=1}^{N} p(x' \to x, c'_n \to c_n), \tag{13.27}$$

where

$$p(x' \to x, c'_n \to c_n)$$
$$= \begin{cases} \int \mathrm{d}x' p(x', c'_n, t')\delta\left[x - g(x', c, t - t')\right] \exp\left[-\Gamma(c'_n, x)\tau\right] & \text{(a)}, \\ \tau \int \mathrm{d}x' \int \mathrm{d}c'_n p(x', c'_n, t')\delta\left[x - g(x', c, t - t')\right] W(c'_n \to c_n|x) & \text{(b)}, \end{cases} \tag{13.28}$$

with (a) no transition and (b) one transition. In Eq. (13.28), we have set $W(c|c', x') \simeq W(c|c', x)$ and $\Gamma(c, x') \simeq \Gamma(c, x)$ after recognizing that the dependence of the component transition probabilities on the system state is relatively weak [Aum06]. As illustrated in Fig. 13.4, Eq. (13.28) allows us [Aum06] to obtain $p(x, c)$, for each time step, as the probability of the top event of a simple fault tree comprising an AND gate with N basic events each with the no-transition or transition probability of Eq. (13.28). It should be noted that the difference in the computational requirements between Eqs. (13.24) and (13.27) is small for Monte Carlo sampling of point values of transition probabilities but could become significant when the transition probabilities are represented as probability density functions.

13.3 CELL-TO-CELL MAPPING FOR PARAMETER ESTIMATION

A Bayesian formulation for parameter estimation that has been developed for DET analysis is the cell-to-cell mapping (CCM) technique presented in a number of publications [Ald87,Din97,Wan04,Buc08]. The CCM technique is based on a discretized

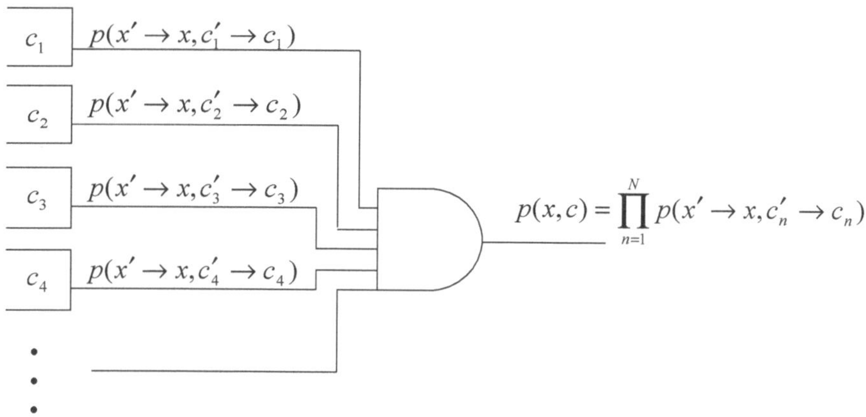

Figure 13.4 Fault tree representation of component transition probabilities. *Source:* [Aum06].

representation of the combined state $z = (x\ c)^T$ introduced in Eq. (13.3) and may be used for system simulation and parameter estimation as well as for dynamic system reliability analysis. We begin in this section with a general derivation of the method that could be adapted and simplified for various specific applications.

13.3.1 Derivation of the Bayesian Recursive Relationship

Before moving into the task of deriving the Bayesian relationship, we need to augment the system equation (13.1) to explicitly consider the component state that could undergo random, discrete transitions and stay constant between these transitions, as discussed in connection with Eqs. (13.27) and (13.28). Thus, with a white Gaussian noise $w_c(t)$ with covariance Q_c assumed for the component state, Eq. (13.1) is generalized to

$$\dot{z}(t) \equiv \begin{pmatrix} \dot{x} \\ \dot{c} \end{pmatrix} = \begin{pmatrix} f(x,c) \\ 0 \end{pmatrix} + \begin{pmatrix} w_x(t) \\ w_c(t) \end{pmatrix} = \begin{pmatrix} f(z) \\ 0 \end{pmatrix} + w(t), \qquad (13.29)$$

where

$$\langle w(t) w^T(t') \rangle = \begin{pmatrix} Q_x & 0 \\ 0 & Q_c \end{pmatrix} \delta(t-t') \equiv Q\delta(t-t'). \qquad (13.30)$$

Now consider a time-discrete form of Eqs. (13.2) and (13.29) at the kth time step t_k, where a collection of measurement vectors $Y_{k-1} = \{y_1, y_2, \ldots, y_{k-1}\}$ is available. Starting from the conditional PDF $p\left(z_{k-1} \left| Y_{k-1}\right.\right)$, i.e., the distribution for the joint state (x_{k-1}, c_{k-1}) given Y_{k-1}, we desire to obtain a conditional PDF $p(z_k|Y_k)$ providing a new system estimate z_k subject to a new observation y_k added to the existing observation base Y_{k-1}.

Writing $Y_k = \{y_k, Y_{k-1}\}$ and applying the Bayes rule of Eq. (3.22) suggest that

$$p(z_k|Y_k) = p(z_k|y_k, Y_{k-1}) = \frac{p(y_k|z_k)\,p(z_k|Y_{k-1})}{\displaystyle\int p(y_k|z_k)\,p(z_k|Y_{k-1})\,\mathrm{d}z_k}, \tag{13.31}$$

where $p(z_k|Y_{k-1})$ is the prior distribution for the new state estimate z_k before the new measurement y_k and $p(y_k|z_k)$ is the likelihood function that provides the relationship between z_k and y_k given new observation y_k. The prior distribution is obtained in the *predictor* part of the recursive algorithm based on the initial PDF $p(z_{k-1}|Y_{k-1})$ at t_{k-1} with

$$p(z_k|Y_{k-1}) = \int p(z_k|z_{k-1})\,p(z_{k-1}|Y_{k-1})\mathrm{d}z_{k-1}, \tag{13.32}$$

where system equation (13.29) is solved to yield the transition probability $p(z_k|z_{k-1})$ over the time step $[t_{k-1}, t_k]$. Substitution of Eq. (13.32) into Eq. (13.31), representing the *corrector* step of the algorithm, provides a combined expression for the updated PDF, fully reflecting the new state simulation z_k and new observation y_k at time t_k:

$$p(z_k|Y_k) = \frac{\displaystyle\int p(y_k|z_k)\,p(z_k|z_{k-1})\,p(z_{k-1}|Y_{k-1})\mathrm{d}z_{k-1}}{\displaystyle\int \mathrm{d}z_k \int \mathrm{d}z_{k-1} p(y_k|z_k)\,p(z_k|z_{k-1})\,p(z_{k-1}|Y_{k-1})}. \tag{13.33}$$

The recursive Bayesian algorithm of Eq. (13.33) may be derived much more formally [Wan04] with due accounting given for the fact that the new observation y_k is related to z_k and w_k but not to Y_{k-1}.

Figure 13.5 illustrates the flow of information associated with the system simulations and measurements in the recursive Bayesian structure of Eq. (13.33), where ovals represent PDFs and boxes show the simulation and observation activities. It is clear from the flow of information that the prior distribution of Eq. (13.32) is constructed before the new observation is made and that the prior and likelihood distributions are combined to yield the updated PDF $p(z_k|Y_k)$. The updated PDF replaces the initial PDF $p(z_{k-1}|Y_{k-1})$ for the next time step in a recursive manner.

Additional remarks will now clarify the two PDFs $p(z_k|z_{k-1})$ and $p(y_k|z_k)$ that form the basis for the recursive algorithm of Eq. (13.33). Because the uncertainties or fluctuations in the system equation (13.29) are assumed to be Gaussian distributions and because the system equation is integrated in a discretized form to yield a nominal estimate $\hat{z}_k$ at time t_k, the PDF associated with the system simulation is given by a Gaussian N with mean $\hat{z}_k$ and variance Q,

$$\begin{aligned} p(z_k|z_{k-1}) &= N(\hat{z}_k, Q) \\ &= \frac{1}{(2\pi)^{m/2}|Q|}\exp\left[-\frac{(z_k-\hat{z}_k)^T Q^{-1}(z_k-\hat{z}_k)}{2}\right] \end{aligned} \tag{13.34}$$

for an m-dimensional system vector z. Similarly, the measurement PDF may be represented by another Gaussian,

$$p(y_k|z_k) = N[h(z_k), R]. \tag{13.35}$$

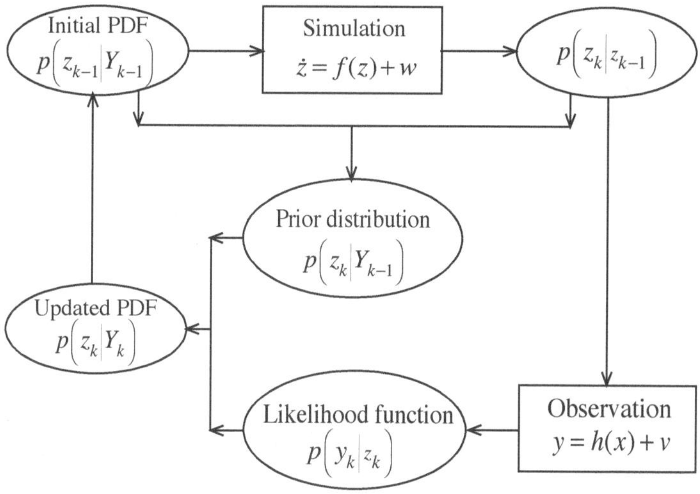

Figure 13.5 Bayesian framework for dynamic reliability analysis.

These two specifications for the simulation and measurement PDFs now complete all the information necessary for the Bayesian framework of Eq. (13.33). For the idealized cases where the simulation and measurement errors are negligibly small, Eqs. (13.34) and (13.35) simplify to

$$p(z_k\,|z_{k-1}) = \delta(z_k - \hat{z}_k), \qquad p\,(y_k\,|z_k\,) = \delta\,[y_k - h(z_k)]\,. \tag{13.36}$$

For system transient simulations without measurement or equivalently if system states are directly monitored [Din97], the measurement PDF $p(y_k\,|z_k)$ may be simply set to unity in Eq. (13.33). Discretization of Eq. (13.33) is considered here for this simple case [Din97], with more general cases left for further reading [Wan04]. For this purpose, partition the phase volumes covering the combined system variable $z = (x\ c)^T$ into cells,

$$\begin{aligned} V_{k-1} &= \{\Delta V(x_{k-1}); \Delta V(c_{k-1})\} \\ &= \{\Delta V_{i'}, i' = 1, \ldots, I; \Delta V_{j'}, j' = 1, \ldots, J\} \text{ for } t = t_{k-1}, \\ V_k &= \{\Delta V(x_k); \Delta V(c_k)\} \\ &= \{\Delta V_i, i = 1, \ldots, I; \Delta V_j, j = 1, \ldots, J\} \text{ for } t = t_k, \end{aligned} \tag{13.37}$$

and define the discretized PDF

$$p(i,j) = \int_{\Delta V_i} dx_k \int_{\Delta V_j} dc_k\, p(x_k, c_k). \tag{13.38}$$

With similar definitions for other probabilities, Eq. (13.33) may then be discretized to yield the *cell-to-cell mapping* (CCM) algorithm

$$p(i,j) = \sum_{j'=1}^{J} \sum_{i'=1}^{I} g(i' \to i, j') \cdot h(i', j' \to j) \cdot p(i', j'), \tag{13.39}$$

where the matrix element $g(i' \to i, j')$ represents the transition probability $p(x_k \mid x_{k-1}, c_{k-1})$ for a system state evolving from cell $\Delta V_{i'}$ to cell ΔV_i with the component state remaining in cell $\Delta V_{j'}$, and the matrix element $h(i', j' \to j)$ represents likewise the component state transition probability $p(c_k \mid c_{k-1}, x_{k-1})$. Thus, if a cell is not reached in a time step selected, the transition probability into the cell is set to zero. Furthermore, the cells are usually partitioned to allow component transitions to take place across the cell boundaries [Lab00]. This form of the CCM algorithm for dynamic system reliability analysis is equivalent to a discretized form of the CET algorithm represented by Eq. (13.27) and (13.28) and illustrated in Fig. 13.4, where state space evolutions with and without component transitions are duly represented. Thus, for statistically independent component transitions or failures, $h(i', j' \to j)$ is evaluated as a product of individual component transition or no transition probabilities. The timestep size is to be selected large enough so that some system states will experience transition across the cells but small enough so that not more than one component state transition takes place in one time step [Buc08].

13.3.2 CCM Technique for Dynamic Event Tree Construction

The use of the CCM technique in dynamic system reliability was illustrated [Ald87, Buc08] for a simple level control problem involving a water tank. The tank is represented by one state variable x, which is the water level in the tank. The tank is connected to two supply lines (components c_1 and c_2) and one outflow line (component c_3). Each component comprises a separate valve and level sensor, as illustrated in Fig. 13.6.

The purpose of the control exercise is to maintain water level x within the low and high setpoints, $(lsp = -1 \text{ m}) \le x \le (hsp = 1 \text{ m})$, and avoid two failure modes: (a) dryout for $x < (L = -3 \text{ m})$ and (b) overflow for $x > (H = 3 \text{ m})$. Each component can be in one of four states: on, off, failed on, and failed off. The three components are controlled to achieve the objective:

$$\begin{aligned}
&1.\ lsp \le x \le hsp \Rightarrow c_1 = \text{on},\ c_2 = \text{ off},\ c_3 = \text{on},\\
&2.\ x < lsp \Rightarrow c_1 = \text{ on},\ c_2 = \text{ on},\ c_3 = \text{ off},\\
&3.\ x > hsp \Rightarrow c_1 = \text{ off},\ c_2 = \text{ off},\ c_3 = \text{ off}.
\end{aligned} \tag{13.40}$$

The system equation represents the rate of water-level changes proportional to the net sum of inflow and outflow rates,

$$\frac{dx}{dt} = W(c_1 + c_2 - c_3), \tag{13.41}$$

where $W = 0.01$ m/minute and the component states $c_n = 1,\ n = 1, 2, 3$, if the unit is on or failed on and $c_n = 0,\ n = 1, 2, 3$, otherwise. Failures of the three components are statistically independent of one another and the failure rates are specified as point values independent of x. The water level is discretized into nine equal intervals.

Starting from a normal operating state $-0.33\,\text{m} \le x \le 0.33$ m, with component state $\{c_1, c_2, c_3 \,|\text{on, off, on}\}$, stochastic search algorithms are used to find all possible system transitions that could result in three possible outcomes {overflow, dryout,

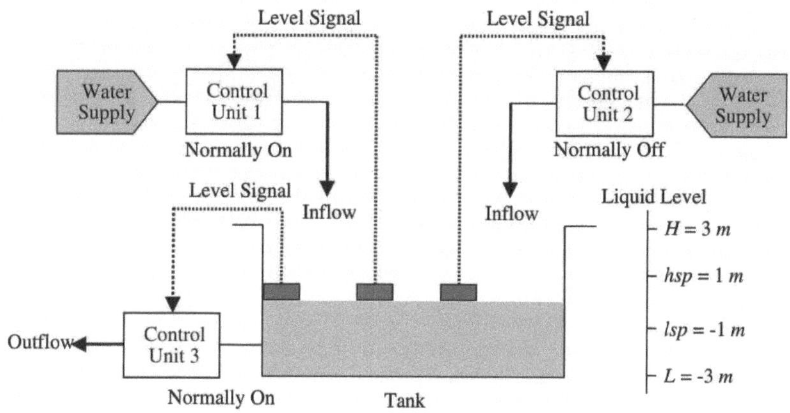

Figure 13.6 Water tank with a level control system. *Source:* [Buc08].

Table 13.1 Time Evolution of One Possible Dryout Scenario

Time	Unit 1 (in)	Unit 2 (in)	Unit 3 (out)	Level (x)
−8	OK (on)	OK (off)	OK (on)	$-0.33 \leqslant x \leqslant 0.33$
−7	Failed OFF	OK (off)	OK (on)	$-0.33 \leqslant x \leqslant 0.33$
−6	Failed OFF	OK (off)	Failed ON	$-1.0 \leqslant x \leqslant -0.67$
−5	Failed OFF	OK (on)	Failed ON	$-1.33 \leqslant x \leqslant -1.0$
−4	Failed OFF	Failed OFF	Failed ON	$-1.33 \leqslant x \leqslant -1.0$
−3	Failed OFF	Failed OFF	Failed ON	$-2.0 \leqslant x \leqslant -1.67$
−2	Failed OFF	Failed OFF	Failed ON	$-2.33 \leqslant x \leqslant -2.0$
−1	Failed OFF	Failed OFF	Failed ON	$-3.0 \leqslant x \leqslant -2.67$
0	Failed OFF	Failed OFF	Failed ON	$x \leqslant -3.0$ (DRY)

Source: [Buc08].

and no failure} over 8 time steps or 4 hours of operation. As summarized in Fig. 13.7, 508, 64, and 169 trajectories are determined for the overflow, dryout, and no-failure states, respectively, for a total of 741 trajectories. One possible scenario, out of 64 possible dryout scenarios, resulting in the component state $\{c_1, c_2, c_3 \,|\text{off, off, on}\}$ is summarized in Table 13.1, where the time is presented backward in units of 30-minute time steps with the initial state at $t = -8$. The probabilities of various end states at $t = 0$ are calculated via the CCM algorithm of Eq. (13.39) and presented in a normalized form in Fig. 13.7. For this idealized test problem with statistically independent, constant values of component failure rates, Eq. (13.39) would take a simple form, with $h(i', j' \to j)$ determined as a product of individual component failure rates. A number of trajectories resulting in very low probabilities are discarded in the search process.

```
Initial State: 4        on/off/on       -0.33 <= x <=  0.33 (NO FAILURE)
State    Paths    Probability    Configuration    Level
---------------------------------------------------------------------------
98       7        0.1168194473   ON/ON/on         x > 3.00 (OVERFLOW)
164      85       0.0976659975   ON/OFF/OFF       x > 3.00 (OVERFLOW)
186      124      0.0877072660   OFF/ON/OFF       x > 3.00 (OVERFLOW)
175      11       0.0745854679   off/ON/OFF       x > 3.00 (OVERFLOW)
296      85       0.0665902764   ON/ON/ON         x > 3.00 (OVERFLOW)
131      7        0.0546680051   ON/off/OFF       x > 3.00 (OVERFLOW)
197      189      0.0019635399   ON/ON/OFF        x > 3.00 (OVERFLOW)
---------------------------------------------------------------------------
Total:   508      0.5000000000
State    Paths    Probability    Configuration    Level
---------------------------------------------------------------------------
251      64       0.1250000000   OFF/OFF/ON       x < -3.00 (DRYOUT)
---------------------------------------------------------------------------
Total:   64       0.1250000000
State    Paths    Probability    Configuration    Level
---------------------------------------------------------------------------
257      13       0.1250000000   ON/OFF/ON        -0.33 <= x <=  0.33 (NO FAILURE)
281      16       0.0688316559   OFF/ON/ON         1.00 <= x <=  1.67 (NO FAILURE)
145      18       0.0589711047   OFF/OFF/OFF      -1.67 <= x <= -1.00 (NO FAILURE)
149      16       0.0503159982   OFF/OFF/OFF       1.00 <= x <=  1.67 (NO FAILURE)
277      18       0.0278641030   OFF/ON/ON        -1.67 <= x <= -1.00 (NO FAILURE)
278      19       0.0274108370   OFF/ON/ON        -1.00 <= x <= -0.33 (NO FAILURE)
146      19       0.0150594141   OFF/OFF/OFF      -1.00 <= x <= -0.33 (NO FAILURE)
279      13       0.0005438601   OFF/ON/ON        -0.33 <= x <=  0.33 (NO FAILURE)
147      13       0.0003979661   OFF/OFF/OFF      -0.33 <= x <=  0.33 (NO FAILURE)
280      12       0.0003495440   OFF/ON/ON         0.33 <= x <=  1.00 (NO FAILURE)
148      12       0.0002555170   OFF/OFF/OFF       0.33 <= x <=  1.00 (NO FAILURE)
---------------------------------------------------------------------------
Total:   169      0.3750000000
```

Figure 13.7 Summary of 741 trajectories identified for the water tank control problem. *Source:* [Buc08].

Each of the 741 trajectories represents one end state of a DET. The large number of end states making up the DET can be readily appreciated once it is recognized that at the first time step $t = -7$ there are 27 possible distinct branches corresponding to three different states {operational, failed on, failed off} for each of the three components for the water tank. In fact, the Monte Carlo search algorithms [Buc08] include procedures to group states with common system configurations at each step to manage the multitude of trajectories and eventually determine the 741 distinct trajectories.

Each of the 741 trajectories can now be represented in terms of a time sequence of AND gates in the structure of a dynamic fault tree (DFT) illustrated in Fig. 13.8. Each of the nine bottom-level AND gates represents the configuration of the three components $\{c_1, c_2, c_3 \,|\mathrm{U1, U2, U3}\}$ and system state x at each time step, starting from the initial state at $t = -8$ and ending in the tank dryout state at $t = 0$. The second-level AND gate, comprising all nine bottom-level AND gates, then represents one dryout scenario corresponding to the trajectory summarized in Table 13.1. The top event for the dryout FT in turn is structured as an OR gate consisting of 64 two-level AND gates representing all possible trajectories culminating in the one dryout end state after eight 30-minute time steps. Finally, there will be 7 overflow OR gates and 11 no-failure OR gates corresponding to the two other outcomes summarized in Fig. 13.7. Each of the 18 top-event OR gates consists of a differing number of

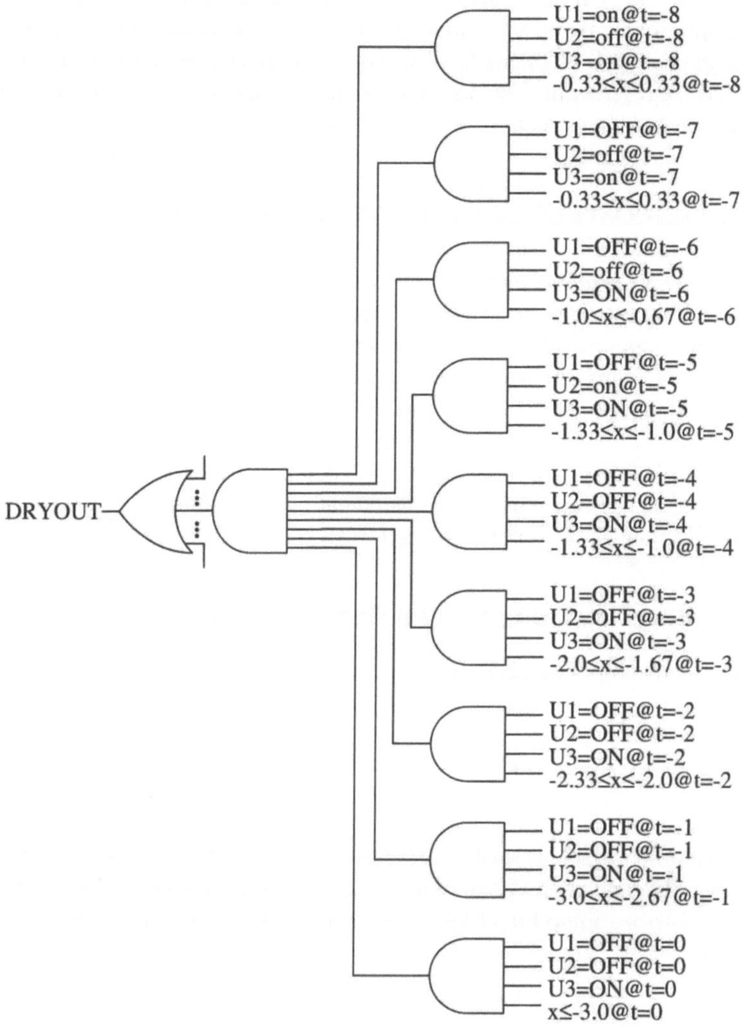

Figure 13.8 Three-level dynamic FT representing the dryout end state. Only 1 out of 64 two-level AND gates is explicitly shown. *Source:* [Buc08].

two-level AND gates corresponding to the number of trajectories for each distinct end state at $t = 0$.

The DET, even for this simple test problem, is quite complicated compared with standard ET/FT structure but can address, in principle, various concerns about conventional ET/FT techniques raised in Section 13.1. Furthermore, once the DET is generated, the structure could be used to represent different initial system states. Implementation of the DETs, once generated, into standard PRA codes is also illustrated

[Buc08] with the SAPHIRE code [NRC08] discussed in Chapter 7 for a detailed representation of an auxiliary feedwater system for a PWR plant. For this task, instead of entering individual FTs graphically, it becomes much more convenient to construct an ASCII file representing the multitude of the FTs and import the file into SAPHIRE through the MAR-D structure.

13.4 DIAGNOSIS OF COMPONENT DEGRADATIONS

The CET formulation of Section 13.2 was applied [Aum06] for probabilistic diagnosis of multiple-component degradations via an alternate use of the Bayes rule for the system model of Eq. (13.29). In this application, the objective is to identify the transitions or degradations of components c when deviations from expected system observations y are detected. System transition probabilities are then used to determine the likelihood of different component transitions that could account for the discrepancy in the system observations. We begin with a basic formulation in Section 13.4.1 and discuss the numerical implementation of a balance-of-plant (BOP) model for a BWR plant as an example in Section 13.4.2.

13.4.1 Bayesian Framework for Component Diagnostics

The product rule of Eq. (2.4) provides the conditional PDF for component state c given observation y at time t:

$$p(c|y) \equiv p(c,t|y) = \frac{p(c,y)}{p(y)} = \frac{\int p(x,c,y)\mathrm{d}x}{p(y)} = \frac{\int p(y|x,c)p(x,c)\mathrm{d}x}{p(y)}, \tag{13.42}$$

where the time variable is suppressed for notational convenience. Observation y does not explicitly depend on c, according to Eq. (13.2), which allows the substitution of $p(y|x,c) = p(y|x)$ into Eq. (13.42), together with the Bayes rule of Eq. (3.22),

$$p(c|y) = \frac{\int p(y|x)p(x,c)\mathrm{d}x}{p(y)} = \int \frac{p(x|y)p(x,c)\mathrm{d}x}{p(x)}. \tag{13.43}$$

Because $p(x,t) = \int p(x,c,t)\mathrm{d}c$, the evaluation of Eq. (13.43) requires two PDFs, $p(x,c,t)$ and $p(x,t|y)$. For this purpose, we make two separate uses of the stochastic balance equation (13.9) with different simplifying assumptions: (a) Eq. (13.10) to represent deterministic system state evolutions subject to component state transitions and (b) the Fokker-Planck equation (13.5) to account for uncertain state evolutions $p(x,t|y)$ at constant component states given observation y. We have already discussed how Eq. (13.10) may be solved via the FT structure of Eqs. (13.27) and (13.28).

Although Eq. (13.5) may be directly solved together with the system equation (13.29) and measurement equation (13.2), it is much more convenient to use an efficient minimum variance estimation algorithm, the Kalman filter [Jaz70], which is derived in Appendix D. Thus, Eqs. (13.29) and (13.2) are recast in terms of the state transition matrix $\Phi(k|k-1)$ and measurement matrix $M(k)$ at time step k,

$$z(k) = \Phi(k|k-1)z(k-1) + w(k), \tag{13.44}$$

$$y(k) = M(k)z(k) + v(k), \quad M(k) = (M_x(k)\ M_c(k)). \tag{13.45}$$

The system estimate and covariance predicted before the measurement $y(k)$ at time step k are given in terms of the optimal estimate $\hat{z}(k-1)$ and covariance $P(k-1) = \left\langle (z-\hat{z})(z-\hat{z})^T \right\rangle$ at time step $k-1$:

$$\hat{z}^-(k) = \Phi(k|k-1)\hat{z}(k-1), \tag{13.46}$$

$$P^-(k) = \Phi(k|k-1)P(k-1)\Phi^T(k|k-1) + Q(k), \tag{13.47}$$

where $Q(k)$ is defined in Eq. (13.30) for time step k. With a new measurement $y(k)$ obtained, the optimal estimate and covariance are updated as

$$\hat{z}(k) = \hat{z}^-(k) + K(k)\left[y(k) - M(k)\hat{z}^-(k)\right] \equiv \hat{z}^-(k) + K(k)\xi(k), \tag{13.48}$$

$$P(k) = [I - K(k)M(k)]\,P(k-1). \tag{13.49}$$

Here I is the identity matrix and the Kalman gain matrix

$$K(k) = P^-(k)M^T(k)[M(k)P^-(k)M^T(k) + R(k)]^{-1}, \tag{13.50}$$

with the Gaussian covariance matrix $R(k)$, minimizes the covariance $P(k)$ representing

$$P(k) = \left\langle [z(k) - \hat{z}(k)][z(k) - \hat{z}(k)]^T \right\rangle. \tag{13.51}$$

We note that the augmented state transition matrix

$$\Phi(k|k-1) = \begin{pmatrix} \phi(k|k-1) & 0 \\ 0 & I \end{pmatrix} \tag{13.52}$$

consists of transition matrix $\phi(k|k-1)$ for system state x evolving in time and an identity matrix for the constant component state. The key step in Kalman filtering consists of Eq. (13.48), which corrects the prior estimate $\hat{z}^-(k)$ using the measurement residual

$$\xi(k) = y(k) - M(k)\hat{z}^-(k), \tag{13.53}$$

thereby accounting for the difference between the latest measurement and prediction. In diagnostic applications of Kalman filtering, an unexpectedly large value of $\xi(k)$ would be indicative of component degradations.

To utilize the measurement residual for component degradation monitoring, Eq. (13.53) is used to determine the expectation of ξ^2,

$$\langle \xi^2 \rangle = MP^-(k)M^T + R, \tag{13.54}$$

and the covariance matrix and measurement matrix for system and component states are separated out as

$$P(k-1) = \begin{pmatrix} P_x & 0 \\ 0 & P_c \end{pmatrix}, \quad M = (M_x \; M_c), \tag{13.55}$$

where the time index k is suppressed in the measurement matrix M. With $P_c = 0$ at time step $k-1$ and $\hat{c}(k-1) = c =$ constant, Eq. (13.47) yields

$$P^-(k) = \begin{pmatrix} \phi P_x \phi^T + Q_x & 0 \\ 0 & Q_c \end{pmatrix}, \tag{13.56}$$

which gives the expected covariance of the measurement residual before recognizing any component transition

$$T = M_x \phi P_x \phi^T M_x^T + M_x Q_x M_x^T + R, \tag{13.57}$$

and a finite Q_c may be introduced to account for modeling errors associated with the component transition indicated in Eq. (13.56).

If a significant difference is observed between the actual ξ^2 and expected covariance T, then a modeling deficit is assumed, i.e., a component state transition has occurred. Since neither the type nor the magnitude of the fault is known, it is expedient to consider the impact of a single component fault at a time. Thus, with the assumption that individual component transitions are uncorrelated, a diagonal noise covariance Q_c is introduced in Eq. (13.56) such that

$$M_c Q_c M_c^T = \xi^2 - T, \tag{13.58}$$

and the Kalman filter is reinitialized [Jaz70]. In the process, Jacobian matrices connecting observation y to state variable x and component state c are used [Aum06] to perturb the component state c for each hypothesis to be tested. The adaptive Kalman filtering is allowed to proceed through the normal recursive process for each hypothesized component transition until the adapted system evolution indicates a converged PDF for the component state c. If the new component state c is sufficiently different from the initial state c' before the modeling deficit is detected, the conditional PDF required for Eq. (13.43) may be obtained as a Gaussian

$$p(x|\,y) \equiv p(x,t|\,y) = N\left[\hat{x}(k), Q_x(k)\right]. \tag{13.59}$$

The component state transition probability $p_j(x,c) = p_j(x,c,t)$ for hypothesis j is then evaluated via Eqs. (13.27) and (13.28) and substituted into Eq. (13.43), together with Eq. (13.59), to determine $p_j(c\,|y)$, which is finally integrated to yield the likelihood p_j for the hypothesis

$$p_j = \int \mathrm{d}c\, p_j(c|y) = \int \mathrm{d}x \int \mathrm{d}c \frac{p(x|y)p_j(x,c)}{p(x)} = \int \mathrm{d}x\, \frac{p(x|y)p_j(x)}{p(x)}. \tag{13.60}$$

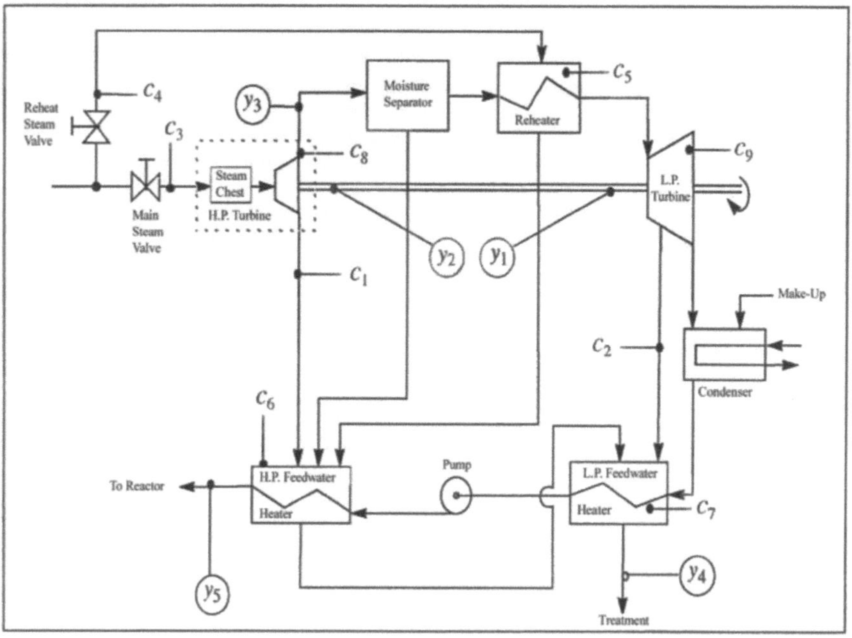

Figure 13.9 Schematic diagram for the Big Rock Point BOP. *Source:* [Aum06].

13.4.2 Implementation of the Probabilistic Diagnostic Algorithm

The probabilistic framework of Section 13.4.1 for multicomponent degradation diagnosis was implemented and tested [Aum06] with a BOP model [Sha77] for the Big Rock Point BWR plant. The BOP is represented via 11 system state variables, including (a) steam enthalpy and density at the high-pressure (HP) turbine feed, (b) steam flow rates out of the HP and low-pressure (LP) turbines, (c) steam enthalpy and density out of the reheater, (d) reheat steam flow and feedwater enthalpy in and out of the HP feedwater heater (FWH), and (e) reheat steam flow rate to the HP and LP FWH. Nine component variables are highlighted in Fig. 13.9 and include (a) HP and LP steam bleed flow rates, (b) effective flow areas for the main steam and reheat steam valves, (c) heat transfer coefficients for the reheater, HP FWH, and LP FWH, and (d) efficiency for the HP and LP turbines. Also highlighted in Fig. 13.9 are five variables monitored for system diagnostics, including (a) torques on LP and HP turbine shafts, (b) HP turbine exhaust pressure, (c) feedwater flow rate out of the LP FWH, and (d) feedwater temperature out of the HP FWH.

At $t = 50$ seconds into nominal operation, binary system faults involving a 5% increase from nominal in the effective flow area c_3 of the main steam valve and a 10% decrease in the LP turbine efficiency c_9 are modeled. Two of five simulated system observations, y_4 representing the LP FWH flow rate and y_5 the HP FWH exit temperature, both with a zero-mean white Gaussian noise of 1% superimposed, are presented in Figs. 13.10 and 13.11, respectively. Note that the effects of the binary component faults on both noisy observed signals are barely noticeable.

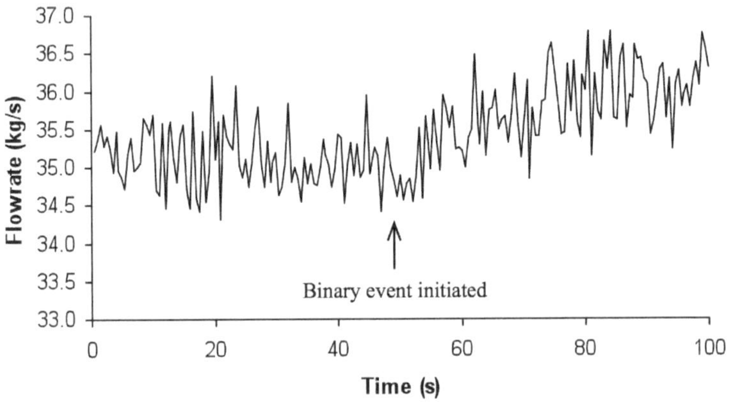

Figure 13.10 System observation y_4, LP feedwater heater flow rate. *Source:* [Aum06].

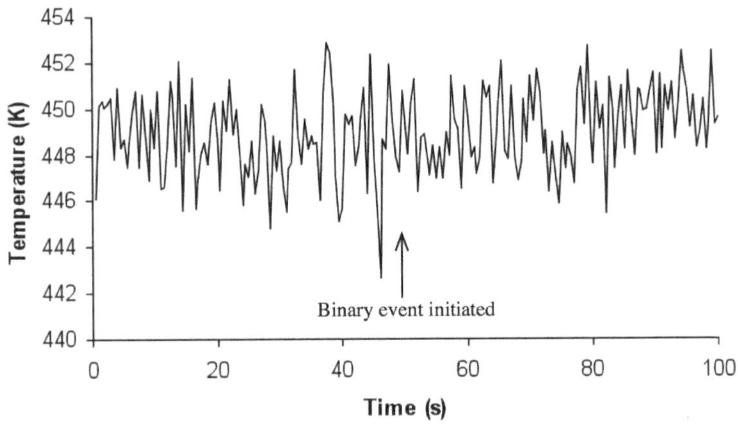

Figure 13.11 System observation y_5, HP fedwater heater exit temperature. *Source:* [Aum06].

For the identification of components faulted, the Bayesian diagnostic algorithm of Section 13.4.1 was applied together with discretized transition matrices $W(c_n \,|c_n', x)$ for the nine component states. Each transition matrix is represented by (11×11)

lognormal frequencies covering both the degradation and enhancement of the component performance from nominal operation and simple environmental multipliers represent the dependence of $W(c_n \,|c'_n, x)$ on the state variable x. The diagnostic algorithm comprising the adaptive Kalman filter was run with a 5-second time step.

Once the onset of the faults was detected between 50 and 55 seconds via a χ^2 test [Aum06] for the measurement residual ξ of Eq. (13.53), a bank of adaptive Kalman filters was run to identify the likely component transitions. Allowing for each of the $N = 9$ components to be either in a faulted or normal operating state requires a bank of $J = 2^N = 512$ adaptive Kalman filters in general. By limiting the number of simultaneous faults to three, however, the required number of fault/no-fault combinations was reduced from $J = 512$ to $J = 130$ for the execution of the adaptive Kalman filter for the next 45 seconds. Representative component state evolutions for two hypotheses for components c_3 and c_9 are presented in Figs. 13.12 and 13.13, respectively. Note that the adaptive Kalman filtering is able to provide an effective convergence of the component states to the correct faulted states, despite the varying rates of evolution. Out of the 130 hypotheses simulated, only 8 are statistically identified as significantly unique and sufficiently different from the nominal case. For the eight unique hypotheses, the transition PDFs $W\left(c_n | \, c'_n, x\right)$ are evaluated via a Latin hypercube sampling (LHS) routine [Ima84] to efficiently determine the component transition probability $p_j(x, c), j = 1, \ldots, 8$, of Eq. (13.27). Together with the PDF $p\left(x | \, y\right)$ determined from the Kalman filter, the likelihood p_j for each of the eight hypotheses is calculated via Eq. (13.60) and summarized in Table 13.2, together with the root mean square (RMS) uncertainty σ_{RSS}.

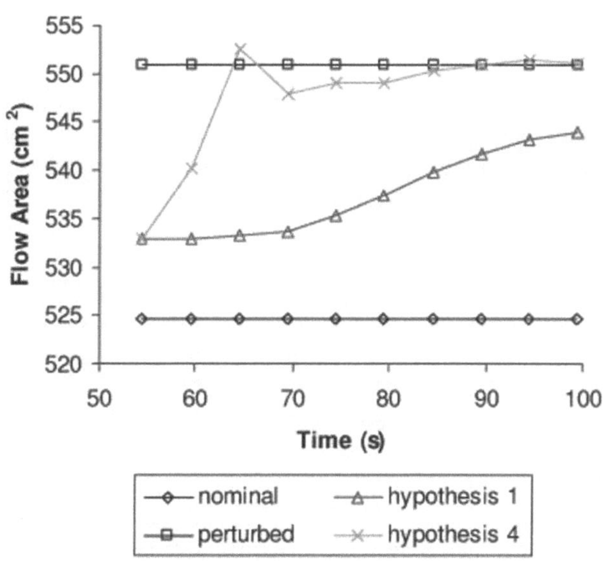

Figure 13.12 Evolution of component c_3, main steam valve flow area, obtained through Kalman filter for two hypotheses. *Source:* [Aum06].

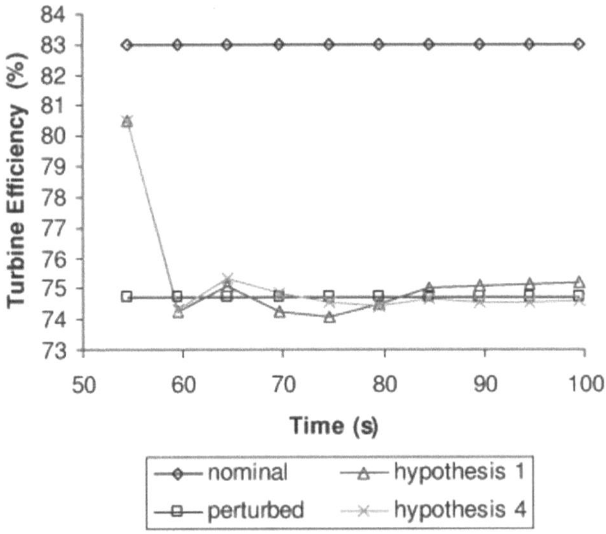

Figure 13.13 Evolution of component c_9, LP turbine efficiency, obtained through Kalman filter for two hypotheses. *Source:* [Aum06].

Note from Table 13.2 that the correct binary fault event involving $\Delta c_3 = 5\%$ and $\Delta c_9 = -10\%$ is identified in two feasible hypotheses, $j = 1, 4$. The magnitudes of the faults identified [Aum06] are also quite acceptable: $\Delta c_3 = 3.7\%$ and $\Delta c_9 = -9.4\%$ for $j = 1$ and $\Delta c_3 = 5.1\%$ and $\Delta c_9 = -10.1\%$ for $j = 4$, as illustrated in Figs. 13.12 and 13.13. Thus, the probabilistic diagnostic algorithm appears effective in identifying the correct fault scenario from noisy indirect observations involving small-magnitude faults. This is achieved through full utilization of statistical signatures extracted from the nominal and perturbed states hypothesized, not merely relying on the difference between the mean values of the nominal and perturbed component PDFs.

The probabilistic diagnostic algorithm is able to efficiently identify both the types and magnitudes of the faults through the use of Kalman filtering, which represents a continuum component state with a Gaussian of Eq. (13.59) involving only two parameters, mean and variance. This is to be contrasted with discretized diagnostic algorithms, e.g., involving the CCM technique of Section 13.3, which will in general require a number of discrete intervals to characterize a continuum distribution. Nonetheless, a multistep DET structure will be required for long-term monitoring of dynamic systems through Eq. (13.60), because $p_j\,(c|\,y)$ will have to serve as the branching probability at each branch point. This may require pruning some less likely scenarios as well as condensing some scenarios, as was noted also in the discussion [Buc08] for the CCM-based DET analysis presented in Section 13.3.2.

These are some of the challenges for the probabilistic diagnostic formulation before it can effectively utilize the system simulation, plant observation, and component reliability data in a synergistic manner. Furthermore, the reliability databases

Table 13.2 Attributes of Feasible Component Hypotheses

Hypothesis j	Probability, p_j (%)	Components Possibly Faulted	Hypothesis Uncertainty, σ_{RSS} (%)
1[a]	80.65 ± 3.40	3,9	1.68
2	$<10^{-12}$	3,5,9	17.34
3	$5.3 \times 10^{-8} \pm 5.9 \times 10^{-15}$	3,4,9	6.38
4[a]	19.34 ± 3.31	3,9	1.71
5	$6.0 \times 10^{-11} \pm 4.1 \times 10^{-22}$	3,5,9	11.78
6	$2.0 \times 10^{-6} \pm 7.5 \times 10^{-14}$	1,3,9	3.91
7	$1.2 \times 10^{-8} \pm 1.1 \times 10^{-17}$	1,2,9	4.98
8	$5.3 \times 10^{-9} \pm 4.8 \times 10^{-19}$	1,2,9	15.46

[a]Correct fault sequence.

Source: [Aum06]

for various system components, even for common systems for power plants and electric grids [IEE07], provide sparse probabilistic information for the evaluation of transition probabilities $W(c_n | c'_n, x)$.

References

[Aco93] C. Acosta and N. Siu, "Dynamic Event Trees in Accident Sequence Analysis: Application to Steam Generator Tube Rupture," *Reliab. Eng. Sys. Safety.* **41**, 135 (1993).

[Ald87] T. Aldemir, "Computer-Assisted Markov Failure Modeling of Process Control Systems," *IEEE Trans. Reliab.* **R-36**, 133 (1987).

[Ame81] A. Amendola and G. Reina, "Event Sequences and Consequences Spectrum: A Methodology of Probabilistic Transient Analysis," *Nucl. Sci. Eng.* **77**, 297 (1981).

[Aum94] S. E. Aumeier, "Probabilistic Techniques for System Diagnostics and Surveillance," PhD Dissertation, Department of Nuclear Engineering, University of Michigan (1994).

[Aum06] S. E. Aumeier, B. Alpay, J. C. Lee, and A. Z. Akcasu, "Probabilistic Techniques of Diagnosis of Multiple Component Degradations," *Nucl. Sci. Eng.* **153**, 101 (2006).

[Buc08] P. Bucci, J. Kirschenbaum, L. A. Mangan, T. Aldemir, C. Smith, and T. Wood, "Construction of Event-Tree/Fault-Tree Models from a Markov Approach to Dynamic System Reliability," *Reliab. Eng. Sys. Safety* **93**, 1616 (2008).

[Dev92] J. Devooght and C. Smidts, "Probabilistic Reactor Dynamics—I: The Theory of Continuous Event Trees," *Nucl. Sci. Eng.* **111**, 229 (1992).

[Din97] L. Dinca and T. Aldemir, "Parameter Estimation Toward Fault Diagnosis in Nonlinear Systems Using a Markov Model of System Dynamics," *Nucl. Sci. Eng.* **127**, 199 (1997).

[Gar85] C. W. Gardiner, *Handbook of Stochastic Methods*, 2nd ed., Springer Verlag (1985).

[Ima84] R. L. Iman and M. J. Shortencarier, "A Fortran-77 Program and User's Guide for the Generation of Latin Hypercube and Random Samples for Use with Computer Models," NUREG-CR-3624, U.S. Nuclear Regulatory Commission (1984).

[Jaz70] A. H. Jazwinski, *Stochastic Processes and Filtering Theory*, Academic Press (1970).

[Lab00] P. E. Labeau, C. Smidts, and S, Swaminathan, "Dynamic Reliability: Towards an Integrated Platform for Probabilistic Risk Assessment," *Reliab. Eng. Sys. Safety* **68**, 219 (2000).

[NRC08] "Systems Analysis Program for Hands-On Integrated Reliability Evaluations (SAPHIRE), Technical Reference," NUREG/CR-6952, vol. 2, U.S. Nuclear Regulatory Commission (2008).

[Sha77] P. V. G. Shankar, "Simulation Model of a Nuclear Reactor Turbine," *Nucl. Eng. Des.* **44**, 269 (1977).

[Smi92] C. Smidts and J. Devooght, "Probabilistic Reactor Dynamics—II: A Monte Carlo Study of a Fast Reactor Transient," *Nucl. Sci. Eng.* **111**, 241 (1992).

[Wan04] P. Wang and T. Aldemir, "Some Improvements in State/Parameter Estimation Using the Cell-to-Cell Mapping Technique," *Nucl. Sci. Eng.* **147**, 1 (2004).

Exercises

13.1 Prove that the Kalman gain matrix of Eq. (13.50) minimizes the covariance of Eq. (13.49), rewritten as Eq. (D.19).

13.2 Establish the equivalence between Eqs. (D.19) and (D.21).

13.3 Starting from Eq. (13.53), derive the expectation of the measurement residual squared given in Eq. (13.54).

13.4 For a system modeled by state vector x and measurement vector y, assume that the state is completely measurable, i.e., $x = y$. Furthermore, if the covariance matrix P^- for state x before measurements at time step t_k is approximated by a diagonal matrix whose elements are uniformly equal to σ_x^2 and the measurement covariance matrix R is represented by a diagonal matrix of elements σ_y^2, obtain a simplified expression for the Kalman gain matrix K of Eq. (13.50) for the system at $t = t_k$. Provide a physical interpretation of the result. Show also that the optimal posterior system estimate x^+ given the measurement y represents the variance-weighted average of measurement y and prior system estimate x^-.

APPENDIX A

REACTOR RADIOLOGICAL SOURCES

This appendix supplements the discussion in Chapters 1, 8, and 10 by providing more background on the radioactivity in nuclear reactors and the health effects of radiation exposure.

A.1 FISSION PRODUCT INVENTORY AND DECAY HEAT

The fission products (FPs) and transuranic (TRU) elements produced in an operating reactor each exhibit significantly different chemical activity, volatility, and half-life, as indicated by the seven groups of radionuclides in Table A.1. In the comprehensive PRA study for five LWR plants [NRC90], NUREG-1150, as discussed extensively in Chapter 10, the fission products are classified in nine groups: noble gases, I, Cs, Te, Sr, Ru, La, Ce, and Ba.

Further groupings of the radionuclides yield some understanding of their risk significance:

- Noble gases Xe and Kr are chemically inactive or inert and not of significant inhalation concern in risk and safety assessment of postulated nuclear accidents.

Table A.1 Activity of Radionuclides at a 3560-MWt Reactor

Group	Radionuclide	Radioactive Inventory (10^6 Ci)	Half-Life (day)
A	*Noble gases*		
	Krypton-85	0.60	3,950
	Krypton-85m	26	0.183
	Krypton-87	51	0.0528
	Krypton-88	73	0.117
	Xenon-133	183	5.28
	Xenon-135	37	0.384
B	*Iodines*		
	Iodine-131	91	8.05
	Iodine-132	129	0.0958
	Iodine-133	183	0.875
	Iodine-134	204	0.0366
	Iodine-135	161	0.280
C	*Alkali metals*		
	Rubidium-86	0.028	18.7
	Cesium-134	8.1	750
	Cesium-136	3.2	13.0
	Cesium-137	5.1	11,000
D	*Tellurium-antimony*		
	Tellurium-127	6.3	0.391
	Tellurium-127m	1.2	109
	Tellurium-129	33	0.048
	Tellurium-129m	5.7	34.0
	Tellurium-131m	14	1.25
	Tellurium-132	129	3.25
	Antimony-127	6.6	3.88
	Antimony-129	35	0.179
E	*Alkaline earths*		
	Strontium-89	101	52.1
	Strontium-90	4.0	11,030
	Strontium-91	118	0.403
	Barium-140	172	12.8

- Volatile species comprising halogens I and Br, alkali metals Cs and Rb, and tellurium group Te, Se, and Sb pose significant risk potential, because they typically have relatively short half-lives and are easily dispersed in the environment. Cesium and iodine isotopes alone make up approximately an inventory of 200 kg and produce approximately 1×10^9 Ci of radioactivity. Tellurium furthermore decays into Cs, providing a continued source of radioactivity. The main concern with the halogens in a postulated accident is that, if present as

Table A.1 Activity of Radionuclides at a 3560-MWt Reactor (*continued*)

Group	Radionuclide	Radioactive Inventory (10^6 Ci)	Half-Life (day)
F	*Noble metals and cobalt*		
	Cobalt-58	0.84	71.0
	Cobalt-60	0.31	1,920
	Molybdenum-99	172	2.8
	Technetium-99m	151	0.25
	Ruthenium-103	118	39.5
	Ruthenium-105	77	0.185
	Ruthenium-106	27	366
	Rhodium-105	53	1.50
G	*Rare earths, refractory oxides, and transuranics*		
	Yittrium-90	4.2	2.67
	Yittrium-91	129	59.0
	Zirconium-95	161	65.2
	Zirconium-97	161	0.71
	Niobium-95	161	35.0
	Lanthanium-140	172	1.67
	Cerium-141	161	32.3
	Cerium-143	140	1.38
	Cerium-144	91	284
	Praseodymium-143	140	13.7
	Neodymium-147	65	11.1
	Neptunium-239	1800	2.35
	Plutonium-238	0.061	32,500
	Plutonium-239	0.023	8.9×10^6
	Plutonium-240	0.023	2.4×10^6
	Plutonium-241	3.7	5,350
	Americium-241	0.0018	1.5×10^5
	Curium-242	0.54	163
	Curium-244	0.025	6,630

Source: Reprinted with permission from F. J. Rahn, A. G. Adamantiades, J. E. Kenton, and C. Baum, *A Guide to Nuclear Power Technology*. Copyright © 1984 John Wiley & Sons, Inc.

elemental I_2, radioactive iodine can be absorbed and could affect the health of the thyroid gland. It should be noted, however, that I and Cs more likely will be present as CsI and CsOH in the reducing environment resulting from Zr-H_2O reactions accompanied by core melt in postulated LWR accidents.

- Among the nonvolatile species are (a) alkaline earth elements Ba and Sr, (b) noble metals Ru, Rh, Pd, Mo, and Tc, (c) rare earths or lanthanides Y, La, Ce, Pr,

Nd, Pm, Sm, and Eu, (d) refractory oxides Zr and Nb, and (e) transuranics Np, Pu, Am and Cm. Note that ^{239}Np alone presents an equilibrium radioactivity of 1.8×10^9 Ci in Table A.1.

As a concluding observation regarding the radioactivity inventory in a nuclear power plant, note that the FP decays account for 6 to 7% of the equilibrium heat generation and that the FP decay heat decreases slowly over days to hundreds of years. Decay heat calculations are usually performed via the American Nuclear Society 5.1 standard [ANS94] discussed in Section 8.3.2. The significant amount of residual heat continually generated via the radioactive decay of FPs, which requires post-shutdown heat removal systems in an NPP, is an important distinguishing feature of risk and safety assessment of NPPs, together with a large inventory of radioactivity discussed in connection with Table A.1.

A.2 HEALTH EFFECTS OF RADIATION EXPOSURE

Several billion curies of radioactivity that is present in an operating NPP clearly requires the radionuclides be contained as best as possible even in the worst possible accident scenarios postulated. To gain an understanding of the potential biological hazards associated with even a small fraction of the radionuclides released, it should be noted that the radiation exposure rate at 1.0 m from an unshielded 1.0 Ci ^{60}Co source is approximately 1.0 rem/hr, or 10 mSv/hr.

With this perspective, a brief review of the health effects of radiation exposure is presented in three categories: (a) acute effects, (b) latent effects, and (c) genetic effects.

Acute health effects primarily due to a single dose of radiation exposure generally are considered detectable with a threshold of 50 rem. An acute dose of 500 rem has been traditionally termed lethal dose (LD) 50, indicating a 50% fatality among those receiving the dose over a period of weeks. Furthermore, doses of 5000 rem may result in fatality in days, while doses in excess of 15,000 rem could result in fatality in a matter of hours. In the aftermath of the 1986 accident at Chernobyl, it was revealed, however, that with proper and prompt medical attention, those receiving large radiation exposures well above the LD50 may survive.

Latent effects of radiation exposure typically refer to the incidence of cancer of various types and have been the subject of studies by the Biological Effects of Ionizing Radiation (BEIR) Committee of the U.S. National Academy of Sciences for the past several decades. At an interval of every decade or so, the BEIR committee has released a comprehensive report summarizing the most current understanding of the health effects of low-level exposure to ionizing radiation. The BEIR-VII report [NAP05] presents the most recent findings of the BEIR committee as of 2005 but is not quite free from the controversies that accompanied previous issues of the BEIR reports [Rah84].

The BEIR-VII report notes that the worldwide population-average background radiation exposure is 240 mrem/year, which includes 120 mrem/year of radon exposure. The total population-average background exposure for the United States is somewhat higher, at 300 mrem/year, due in part to higher average radon levels. A breakdown of the total annual exposure indicates that 82% of the exposure comes from natural background and 18% from man-made radiation, of which medical applications account for 79% and nuclear fuel cycle 1%. Radiation exposure contributes to the annual death rate from cancer, which in 2003 was somewhat in excess of 20% of the 2.5 million deaths in the United States [Klu06].

With these statistics regarding the background radiation exposure level of about 240 to 300 mrem/year and a cancer death rate of 20%, we note the main conclusion of the BEIR-VII report that suggests a cancer death rate of 5.7×10^{-4} additional cancer deaths/person-rem above background based on a linear no-threshold (LNT) theory for solid cancers (5.1×10^{-4}) and linear-quadratic model for leukemia (0.61×10^{-4}). BEIR-VII further recommends that the estimate for cancer death rate be reduced by a dose rate effectiveness factor (DREF) of 1.5 for low dose rates.

The latent radiation effects are to be considered on a statistical average basis, not for individual exposures separately, as illustrated for two simple examples. (a) Continuous lifetime exposure at 0.1 rem/yr for a population of 10^6 persons: (0.1 rem/yr $\times$ 70 years $\times 10^6$ persons = 7.0×10^6 person-rem) $\times$ (5.7×10^{-4} deaths/person-rem) = 4.0×10^3 additional deaths/10^6 persons = 2% increase in cancer death rate. (b) Single 10-rem exposure for 10^6 persons: 10^7 rem $\times$ 5.7×10^{-4} deaths/person-rem = 5.7×10^3 additional deaths/10^6 persons = 3% increase in cancer death rate.

For many years controversies have surrounded the correlation of consequences with very low doses of radiation. Such a correlation is important because not only can the uncertainties in nuclear risk analyses be reduced, but the communication of the meaning of the risks of low radiation doses to the public can be improved.

The latest BEIR-VII recommendation for the latent effects of low-level ionizing radiation can be compared with earlier recommendations:

- BEIR-III (1980): 2×10^{-4} additional deaths/person-rem above background.
- BEIR-V (1990): 8×10^{-4} additional deaths/person-rem above background.

In addition, the BEIR-V report estimates the risk to an embryo as 4×10^{-3} *additional* cancer deaths/person-rem, which is five times larger than the population average effect. This historical comparison of BEIR committee recommendations shows that the bottom-line BEIR-VII recommendation is somewhat of an average of the two previous recommendations. BEIR-VII still retains, however, the basic premise that the health effects of low-level radiation exposure should be estimated based on the LNT model. This is to be contrasted with the threshold model, where low levels of radiation exposure are believed to have no deleterious effects, as well as with hormesis theory, which suggests beneficial effects of radiation exposure at low levels. The BEIR-VII recommendation appears to reflect significant uncertainties associated with extrapolating the health effects associated with large exposures, including those of the Japanese bomb victims.

To conclude the discussion on the health effects of radiation exposure, we simply note that there appears to be no definitive answer at this point regarding the long-term genetic effects of low-level radiation exposure. This reflects various epidemiological studies, including those on radiation workers at Sellafield (Windscale) in the United Kingdom. The 1957 accident at a graphite-moderated reactor at Windscale resulted in a fire that released 10 Mg of uranium and a significant amount of radionuclides into the surrounding areas. Epidemiological studies of the offspring of the Windscale workers have yet to reveal definitive genetic effects of the radiation exposure.

Based on the 1991 International Committee on Radiological Protection (ICRP) Guidelines on Exposure Limits, the U.S. Nuclear Regulatory Commission also revised regulations regarding radiation protection limits in 10 CFR 20 [NRC07]:

- Public exposure: 100 mrem/yr
- Occupational exposure: 5 rem/yr adult, 0.5 rem/yr minor, 0.5 rem/gestation period for a pregnant woman

Together with these radiation protection limits, the principle of ALARA (as low as reasonably achievable) is used as a part of the NRC licensing and regulation process to minimize radiation exposure in nuclear facilities.

References

[ANS94] "American National Standard for Decay Heat Power in Light Water Reactors," ANSI/ANS-5.1-1994, American Nuclear Society (1994).

[Klu06] J. Kluger, "Why We Worry About the Things We Shouldn't and Ignore the Things We Should," *Time*, 66 (December 4, 2006).

[NAP05] "Health Risks from Exposure to Low Levels of Ionizing Radiation, BEIR VII-Phase 2," Biological Effects of Ionizing Radiation Committee, National Academies Press (2005).

[NRC90] "Severe Accident Risks: An Assessment for Five U.S. Nuclear Power Plants," NUREG-1150, U.S. Nuclear Regulatory Commission (1990).

[NRC07] "Standards for Protection Against Radiation," Title 10, *Code of Federal Regulations*, Part 20, U.S. Nuclear Regulatory Commission (2007).

[Rah84] F. J. Rahn, A. G. Adamantiades, J. E. Kenton, and C. Baum, *A Guide to Nuclear Power Technology*, Wiley (1984).

APPENDIX B

SOME SPECIAL MATHEMATICAL FUNCTIONS

In this appendix some of the special functions introduced in reliability and risk investigations are introduced. Such functions can arise, as in Chapter 2, when computing a failure probability by integrating a failure probability distribution or differentiating a failure probability to obtain a failure probability density. Here we also will see how one of the special functions is related to the standard χ^2-distribution that arises in analyses involving the normal distribution.

B.1 GAMMA FUNCTION

The incomplete gamma function is a two-parameter function that is needed to evaluate the failure probability of Eq. (2.111). It is defined by

$$\gamma(x+1,z)=\int_0^z y^x \exp(-y)\mathrm{d}y. \tag{B.1}$$

The recursion relation for the incomplete gamma function is

$$\gamma(x+1, z) = x\gamma(x, z) - z^x \exp(-z). \tag{B.2}$$

As $z \to \infty$, the incomplete gamma function becomes the gamma function defined by

$$\Gamma(x+1) = \int_0^\infty y^x \exp(-y)\mathrm{d}y, \quad x \neq -n,\ n = 0, 1, 2, \ldots, \tag{B.3}$$

which is tabulated in standard references on mathematical functions [Abr64]. It obeys a recursion relation of the form

$$\Gamma(x+1) = x\Gamma(x). \tag{B.4}$$

For the special case of an integer r,

$$\Gamma(r+1) = r!, \tag{B.5}$$

while another special result is

$$\Gamma(0.5) = \pi^{1/2}. \tag{B.6}$$

Figure B.1 illustrates the behavior of $\Gamma(x)$.

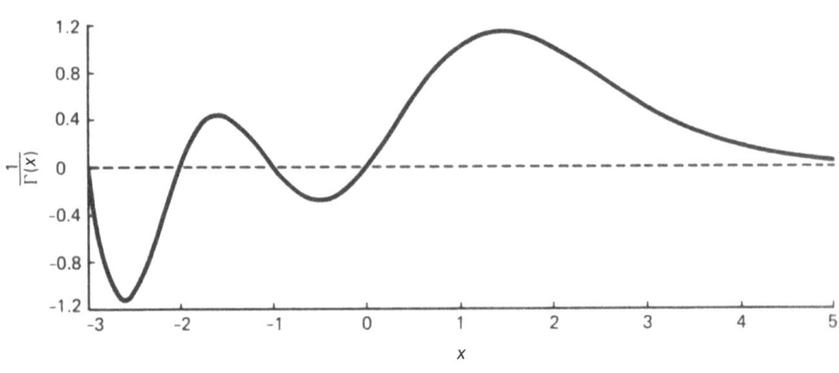

Figure B.1 Reciprocal of the gamma function. *Source:* [Jah60].

A function related to the incomplete gamma function is the exponential integral function defined by [Abr64]

$$E_n(z) = \int_1^\infty x^{-n} \exp(-zx)\mathrm{d}x = z^{n-1} \int_z^\infty y^{-n} \exp(-y)\mathrm{d}y, \quad n = 1, 2, \ldots, \tag{B.7}$$

because

$$E_n(z) = z^{n-1}[\Gamma(1-n) - \gamma(1-n, z)]. \tag{B.8}$$

B.2 ERROR FUNCTION

The error function

$$\begin{aligned}\text{erf } z &= \frac{2}{\sqrt{\pi}}\int_0^z \exp(-u^2)\mathrm{d}u \\ &= \gamma(0.5, z^2)/\Gamma(0.5) \end{aligned} \tag{B.9}$$

arises in the failure probability of Eq. (2.117) for the lognormal distribution. Tables of the error function can be found by using the cumulative probability for the chi-square distribution $P(\chi^2|r)$ for integer r, $r \geq 1$, that is defined by

$$P(\chi^2|r) = [2^{r/2}\Gamma(r/2)]^{-1}\int_0^{\chi^2} t^{r/2-1}\exp(-t/2)\mathrm{d}t. \tag{B.10}$$

This is because erf$z = P(2z^2|1)$.

References

[Abr64] M. Abramowitz and I. Stegun, eds., *Handbook of Mathematical Functions with Formulas, Graphs, and Mathematical Tables*, U.S. Government Printing Office, Washington, D.C. (1964); reprinted by Dover (1970).

[Jah60] E. Jahnke, F. Emde, and F. Lösch, *Tables of Higher Functions*, McGraw-Hill (1960).

APPENDIX C

SOME FAILURE RATE DATA

Table C.1 is an example of failure rate data discussed in Section 6.3 that is available from the Institute of Electrical and Electronics Engineers.

Table C.1 Summary of Failure Rate and Average and Median Downtime per Failure for All Electrical Equipment Surveyed

Equipment	Equipment subclass	Failure rate (failures per unit-year)	Actual hours of downtime per failure	
			Industry average	Median plant average
Transformers	Liquid filled—All	0.0062	356.1	—
	300 kVA to10 000 kVA	0.0059	297.4	—
	10 000+ kVA	0.0153	1178.5	—
Rectifier transformers	Liquid filled 300 kVA to10 000 kVA	0.0153	1664.0	—
Motors > 200 hp	Induction			
	0 to 1000 V	0.0824	42.5	15.0
	1001 V to 5000 V	0.0714	75.1	12.0
	Synchronous 1001 V to 5000 V	0.0762	78.9	16.0
Circuit breakers	Fixed (including molded case)	0.0052	5.8	4.0
	0 to 600 V—All sizes	0.0042	4.7	4.0
	0 to 600 A	0.0035	2.2	1.0
	Above 600 A	0.0096	9.6	8.0
	Above 600 V	0.0176	10.6	3.8
	Metal-clad drawout type—All	0.0030	129.0	7.6
	0 to 600 V—All sizes	0.0027	147.0	4.0
	0 to 600 A	0.0023	3.2	1.0
	Above 600 A	0.0030	232.0	5.0
	Above 600 V	0.0036	109.0	168.0
Motor starters	Contact type: 0 to 600 V	0.0139	65.1	24.5
	Contact type: 601 V to 15 000V	0.0153	284.0	16.0
Generators	Continuous service Steam turbine driven		32.7	—
	Emergency and standby units Reciprocating engine driven Rate per hour in use (0.00536) Failures per start attempt (0.0135)	0.1691	478.0	—
Disconnect switches	Enclosed	0.006100	1.6	2.8

Table C.1 Summary of Failure Rate and Average and Median Downtime per Failure for All Electrical Equipment Surveyed (*continued*)

Equipment	Equipment subclass	Failure rate (failures per unit-year)	Actual hours of downtime per failure	
			Industry average	Median plant average
Switchgear bus—Indoor and outdoor	Insulated: 601 V to 15000 V	0.001129	261.0	28.0
	Bare: 0 to 600 V	0.000802	550.0	27.0
	Bare: Above 600 V	0.001917	17.3	36.0
Bus duct Indoor and outdoor (unit = 1 circuit ft)	All voltages	0.000125	128.0	9.5
Open wire (unit = 1000 circuit ft)	15 000 V	0.01890	42.5	4.0
	Above 15 000 V	0.00750	17.5	12.0
Cable—All types of insulation (unit = 1000 circuit ft)	Aboveground and aerial			
	0 to 600 V	0.001410	457.0	10.5
	601 V to 15 000 V—All	0.014100	40.4	6.9
	In trays aboveground	0.009230	8.9	8.0
	In conduit aboveground	0.049180	140.0	47.5
	Aerial cable	0.014370	31.6	5.3
	Belowground and direct burial			
	0 to 600 V	0.003880	15.0	24.0
	601 V to 15 000 V—All	0.006170	95.5	35.0
	In duct or conduit	0.006130	96.8	35.0
	Above 15000 V	0.003360	16.0	16.0
Cable (unit = 1000 circuit ft)	601 V to 15 000 V			
	Thermoplastic	0.00387	44.5	10.0
	Thermosetting	0.00889	168.0	26.0
	Paper-insulated lead-covered	0.00912	48.9	26.8
	Other	0.01832	16.1	28.5
Cable joints—All types of insulation	601 V to 15 000 V In duct or conduit below-ground	0.000864	36.1	31.2
Cable joints	601 V to 15 000 V			
	Thermoplastic	0.000754	15.8	8.0
	Paper-insulated lead-covered	0.001037	31.4	28.0

Table C.1 Summary of Failure Rate and Average and Median Downtime per Failure for All Electrical Equipment Surveyed (*continued*) [IEE07].

Equipment	Equipment subclass	Failure rate (failures per unit-year)	Actual hours of downtime per failure	
			Industry average	Median plant average
Cable terminations all types of insulation	Aboveground and aerial			
	0 to 600V	0.000127	3.8	4.0
	601 V to 15 000 V—All	0.000879	198.0	11.1
	Aerial cable	0.001848	48.5	11.3
	In trays aboveground	0.000333	8.0	9.0
	In duct or conduit below-ground			
	601 V to 15 000 V	0.000303	25.0	23.4
Cable terminations	601 V to 15 000 V			
	Thermoplastic	0.004192	10.6	11.5
	Thermosetting	0.000307	451.0	11.3
	Paper-insulated lead-covered	0.000781	68.8	29.2
Miscellaneous	Inverters	1.254000	107.0	185.0
	Rectifiers	0.038000	39.0	52.2

Source: Reprinted with permission from "IEEE Recommended Practice for the Design of Reliable Industrial and Commercial Power Systems," IEEE Std 493-2007.

APPENDIX D

LINEAR KALMAN FILTER ALGORITHM

This appendix presents a brief derivation of the Kalman filter, which plays an important role in the Bayesian framework for component diagnostics discussed in Section 13.4.1. The Kalman filter is a minimum-variance parameter estimation algorithm that generates an optimal estimate of system state vector $x(t)$ given observation vector $y(t)$, duly accounting for modeling uncertainties for $x(t)$ and statistical fluctuations in $y(t)$. The optimal estimate $\widehat{x}(t)$ is obtained so that the covariance of the system estimation is minimized.

Consider a dynamical system represented by $x(t)$ subject to white Gaussian noise vector $w(t)$ with covariance Q,

$$\frac{\mathrm{d}x(t)}{\mathrm{d}t} = F(t)x(t) + w(t); \langle w(t) \rangle = 0, \quad \left\langle w(t)w^T(t') \right\rangle = Q\delta(t-t'), \tag{D.1}$$

where $x(t)$ is determined indirectly through observation $y(t)$ subject to white Gaussian noise vector $v(t)$ with covariance R,

$$y(t) = M(t)x(t) + v(t); \langle v(t) \rangle = 0, \quad \left\langle v(t)v^T(t') \right\rangle = R\delta(t-t'). \tag{D.2}$$

The optimal system estimate $\widehat{x}(t)$ may be considered a statistical expectation of the true or exact system state $x(t)$ given observation $y(t)$,

$$\widehat{x}(t) = \langle x(t) \,| y(t) \rangle \,, \tag{D.3}$$

such that the covariance matrix

$$P(t) = \left\langle [x(t) - \widehat{x}(t)] \, [x(t) - \widehat{x}(t)]^T \right\rangle \tag{D.4}$$

is minimized. Although continuum formulations of the Kalman filter are possible [Jaz70], a discretized form of the filter is derived here for various practical applications in mind. For this purpose, the state transition matrix is defined over the time interval $[t_{k-1}, t_k]$,

$$\Phi(k|k-1) = \exp\left[\int_{t_{k-1}}^{t_k} F(t)\mathrm{d}t\right], \tag{D.5}$$

so that Eq. (D.1) may be written in a discretized form

$$x(k) = \Phi(k|k-1)x(k-1) + w(k) = \Phi x(k-1) + w \tag{D.6}$$

together with the measurement equation (D.2) similarly discretized,

$$y(k) = M(k)x(k) + v(k) = Mx(k) + v. \tag{D.7}$$

For notational convenience, the explicit timestep indices are suppressed in the last expression for each of the discretized equations (D.6) and (D.7). The covariance matrix of Eq. (D.4) may be written for time steps $k-1$ and k as

$$P(k-1) = \left\langle [x(k-1) - \widehat{x}(k-1)] \, [x(k-1) - \widehat{x}(k-1)]^T \right\rangle, \tag{D.8}$$

$$P(k) = \left\langle [x(k) - \widehat{x}(k)] \, [x(k) - \widehat{x}(k)]^T \right\rangle. \tag{D.9}$$

The Kalman filter is formulated in a two-step recursive structure beginning with a *prior estimate* at time step k before the measurement is taken:

$$\widehat{x}^-(k) \equiv \widehat{x}(k|k-1) = \Phi\widehat{x}(k-1). \tag{D.10}$$

Here the superscripted estimate $\widehat{x}^-(k)$ is used synonymously with the conditional estimate $\widehat{x}(k\,|k-1)$ to indicate that the estimation is an initial prediction based on the optimal estimate $\widehat{x}(k-1)$ of the previous time step $k-1$. Equations (D.6) and (D.10) yield a prior estimate of the covariance with $w = w(k)$,

$$\begin{aligned} P^-(k) &\equiv P(k|k-1) = \langle [x(k) - \widehat{x}^-(k)][x(k) - \widehat{x}^-(k)]^T \rangle \\ &= \left\langle \{\Phi\,[x(k-1) - \widehat{x}(k-1)] + w\} \,\{\Phi\,[x(k-1) - \widehat{x}(k-1)] + w\}^T \right\rangle \\ &= \Phi \left\langle [x(k-1) - \widehat{x}(k-1)] \, [x(k-1) - \widehat{x}(k-1)]^T \right\rangle \Phi^T + \langle ww^T \rangle . \end{aligned} \tag{D.11}$$

In the last estimation step, the term involving the cross product of the estimation error $[x(k-1) - \widehat{x}(k-1)]$ in step $k-1$ and the modeling error $w = w(k)$ in step k is dropped because the two errors are independent of one another. Equation (D.11) may be simplified by using Eq. (D.8) and a discretized form of covariance Q in Eq. (D.1) to give

$$P^-(k) = \Phi P(k-1)\Phi^T + Q(k) \equiv \Phi P(k-1)\Phi^T + Q. \tag{D.12}$$

In the *correction step* of the Kalman filter, after a new measurement is taken at step k, the objective is to add to the prior estimate of Eq. (D.10) a term proportional to the measurement residual,

$$\xi(k) = y(k) - M\widehat{x}^-(k), \tag{D.13}$$

so that the resulting *posterior estimate*

$$\widehat{x}(k) \equiv \widehat{x}^+(k) = \widehat{x}^-(k) + K\left[y(k) - M\widehat{x}^-(k)\right] \tag{D.14}$$

minimizes the estimation error

$$\varepsilon(k) = x(k) - \widehat{x}(k) \tag{D.15}$$

or equivalently the covariance $P(k)$ of Eq. (D.9). Thus, the key remaining task is to derive an expression for the proportionality constant K introduced in Eq. (D.14). Before inserting $\varepsilon(k)$ of Eq. (D.15) into Eq. (D.9), however, an alternate form of measurement residual $\xi(k)$ is obtained via Eqs. (D.6) and (D.7),

$$\xi(k) = Mx(k) + v(k) - M\widehat{x}^-(k) = M\left[\Phi x(k-1) + w\right] + v(k) - M\widehat{x}^-(k),$$

which can be rewritten with Eq. (D.15) at time step $k-1$ and Eq. (D.10),

$$\begin{aligned}\xi(k) &= M\left[\Phi\left\{\varepsilon(k-1) + \widehat{x}(k-1)\right\} + w\right] + v - M\widehat{x}^-(k)\\ &= M\left[\Phi\varepsilon(k-1) + w\right] + v.\end{aligned} \tag{D.16}$$

With Eqs. (D.10) and (D.16) substituted into Eq. (D.14) and use of Eq. (D.6), a more useful form of Eq. (D.15) is obtained:

$$\begin{aligned}\varepsilon(k) &= \Phi\left[x(k-1) - \widehat{x}(k-1)\right] + w - K\xi(k)\\ &= \Phi\varepsilon(k-1) + w - KM\left[\Phi\varepsilon(k-1) + w\right] - Kv\\ &= (I - KM)\left[\Phi\varepsilon(k-1) + w\right] - Kv.\end{aligned} \tag{D.17}$$

After substituting Eq. (D.17) into Eq. (D.9) and invoking the measurement covariance matrix R of Eq. (D.2), the posterior estimate of the covariance matrix becomes

$$\begin{aligned}P(k) &= P^+(k) = \left\langle \varepsilon(k)\varepsilon^T(k)\right\rangle\\ &= (I - KM)\left[\Phi\left\langle\varepsilon(k-1)\varepsilon^T(k-1)\right\rangle\Phi^T + Q\right](I - KM)^T\\ &\quad + KRK^T,\end{aligned} \tag{D.18}$$

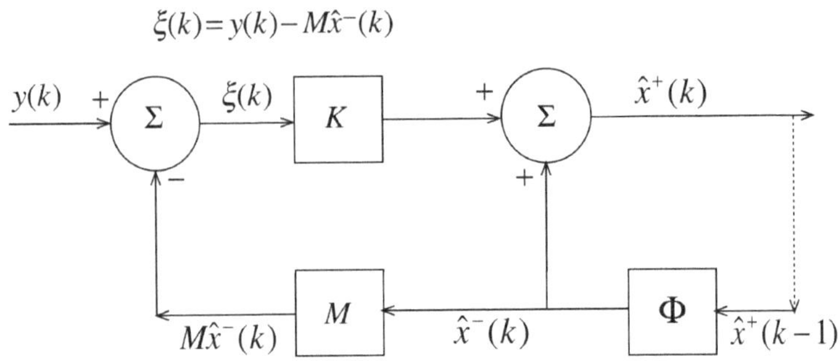

Figure D.1 Flow of information for the Kalman filter.

which can be simplified, via Eq. (D.12), to

$$P(k) = P^{+}(k) = (I - KM)P^{-}(k)(I - KM)^{T} + KRK^{T}. \tag{D.19}$$

Minimization of the posterior covariance matrix $P(k)$ may be accomplished by taking a derivative of the trace of $P(k)$ with respect to K and setting it to zero,

$$-2(I - KM)P^{-}(k)M^{T} + 2KR = 0,$$

which may be rearranged as

$$P^{-}(k)M^{T} - KMP^{-}(k)M^{T} = KR$$

and finally solved to give the *Kalman gain matrix* at time step k,

$$K(k) = P^{-}(k)M^{T}\left[MP^{-}(k)M^{T} + R\right]^{-1}. \tag{D.20}$$

Use of Eq. (D.20) also yields an alternate, simpler form of the posterior covariance matrix,

$$P(k) = P^{+}(k) = (I - KM)P^{-}(k). \tag{D.21}$$

In summary, the discretized linear Kalman filter algorithm can be recursively applied through the following steps:

(i) Obtain prior estimates before the measurement via Eqs. (D.10) and (D.12).

(ii) Update the prior estimates into the posterior estimates via Eqs. (D.14) and (D.19) or (D.21), together with the Kalman gain matrix of Eq. (D.20).

The flow of information for the Kalman filter algorithm is illustrated in Fig. D.1.

When the system equation (D.1) or the measurement equation (D.2) is nonlinear, then the equations may be successively linearized as the system evolves in time. This approach is known as the extended Kalman filter. More recently, an unscented Kalman filter algorithm [Vos04] has been developed that allows for direct uses of nonlinear equations, albeit at additional computational costs.

References

[Jaz70] A. H. Jazwinski, *Stochastic Processes and Filtering Theory*, Academic Press (1970).

[Vos04] H. U. Voss, J. Timmer, and J. Kurths, "Nonlinear Dynamical System Identification from Uncertain and Indirect Measurements," *Int. J. Bifurcation and Chaos* **14**, 1904 (2004).

ANSWERS TO SELECTED EXERCISES

Chapter 2

2.1 $H = BC + AD$
2.2 $H = \overline{A}B$
2.3 $H = C + \overline{A}\,\overline{B} + AB$
2.4 $H = AB$
2.5 $\sum_{n=1}^{4} P(E_n)$
2.6 $\sum_{n=1}^{4} P(E_n) - \sum_{n=1}^{3} \sum_{m=n+1}^{4} P(E_n E_m) + \sum_{n=1}^{2} \sum_{m=1}^{3} \sum_{j=1}^{4} P(E_n E_m E_j) - P(E_1 E_2 E_3 E_4)$
2.7 (a) 0.0846 (b) 0.016
2.8 (a) 0.09879 (b) 0.09478
2.9 0.159
2.10 (a) 0.1353 (b) 0.2706 (c) 0.2706 (d) 0.382×10^{-4}
2.11 (a) 0.1947 (b) 0.2356
2.12 (a) 0.143
2.13 0.0140
2.14 (a) 0.8187 (b) 0.8187 (c) 0.8187
2.15 (a) 0.031 (b) 0.99902
2.16 (a) $t/(t+2)$ (b) $2/(t+2)^2$
2.17 (a) 0.0304 (b) 911.3 yr

2.18 (a) 177 days (b) 1770 days
2.19 (a) $1/\Gamma(a+1)$ (b) $m = \sigma^2 = a + 1$
2.20 (a) $\exp(-\lambda t), t < t_1$, and $\exp[-\lambda t - k(t-t_1)^2/2], t \geq t_1$ (b) $\lambda \exp(-\lambda t)$, $t < t_1$, and $[\lambda + k(t-t_1)] \exp[-\lambda t - k(t-t_1)^2/2]$, $t \geq t_1$ (c) $\lambda^{-1}[1 - \exp(-\lambda t_1)] + \sqrt{\pi/4k} \exp[-\lambda t_1 + \lambda^2/4k][1 - \text{erf}(\lambda/2\sqrt{k})]$
2.21 (a) 34.45 hr (b) 14.75 hr
2.22 (a) $0, 0 \leq t_1$, and $1-(t/t_1)^{-at_2^2}, t_1 \leq t < t_2$, and $1-(t_2/t_1)^{-at_2^2} \exp[-at_2(t-t_2)], t > t_2$
(b) 951.7 hr c) 232.84 hr
2.23 (a) $1 - \exp(-at^3/3), 0 \leq t \leq t_1$, and $1 - (t/t_1)^{-at_1^3} \exp(-at_1^3/3), t \geq t_1$
(b) 441.3 hr (c) $(9a)^{-1/3}\gamma(1/3, at_1^3/3) + t_1(at_1^3 - 1)^{-1} \exp(-at_1^3/3)$
2.24 (a) $1 - \exp(-at^2/2), t \leq t_1$, and $1 - \exp\{(at_1/b)(1 - bt_1/2 - \exp[b(t-t_1)]\}$, $t > t_1$
(b) 205.2 hr
2.25 (a) $1 - \exp(-\lambda t), 0 \leq t < t_1$, and $1 - \exp[\lambda t_1 - 2\lambda(t_1 t)^{1/2}], t \geq t_1$
(b) $\lambda^{-1}[1 + \exp(-\lambda t_1)/(2\lambda t_1)]$
2.26 (a) $a = \lambda t_1^{1/2}$, $b = \lambda(t_2 - t_1)^{-2}$ (b) $\exp[-2\lambda(tt_1)^{1/2}]$, $0 \leq t \leq t_1$, and $\exp[-\lambda(t + t_1)], t_1 \leq t \leq t_2$, and $\exp\{-(\lambda/3)[4t_1 + 2t_2 + (t-t_1)^3/(t_2-t_1)^2]\}$, $t \geq t_2$
2.27 (a) $1, 0 \leq t < t_1$; $(t/t_1)^{-\lambda t_1}, t_1 \leq t < t_2$; $\{(t_2/t_1)\exp[(t^2 - t_2^2)/2t_2^2]\}^{-\lambda t_1}$, $t \geq t_2$
(b) 1.376×10^5 hr
2.28 (a) $k = (a^2 - b^2)/b$ (b) $\exp(-at)[(a/b)\sinh bt + \cosh bt]$ (c) $2a/(a^2 - b^2)$
2.29 (a) $\beta\lambda^{\alpha/\beta}/\Gamma(\alpha/\beta)$ (b) $\gamma(\alpha/\beta, \lambda t^\beta)/\Gamma(\alpha/\beta)$ (c) $\Gamma[(\alpha+1)/\beta]/\lambda^{1/\beta}\Gamma(\alpha/\beta)$
2.30 (a) $\beta/[t_c\Gamma(1/\alpha)]$ (b) $t_c\Gamma[(1/\alpha) + (1/\beta)]/\Gamma(1/\alpha)$
2.31 (a) $2a$ (b) $1 - 2E_3(at)$ (c) $2E_3(at)$ (d) $aE_2(at)/E_3(at)$ (e) $2/3a$
2.32 (a) $2\beta^{(\alpha+2)/2}/\Gamma[(\alpha+2)/2]$ (b) $\Gamma[(\alpha+3)/2]/\beta^{1/2}\Gamma[(\alpha+2)/2]$
2.33 (a) 0.81 (b) 0.039
2.34 (a) 0.729 (b) 0.493 (c) 0.368 (d) 20
2.35 (a) 0.995 (b) 88,950 hr
2.36 (a) Type I (maximum values) (b) 8.7 m

Chapter 3

3.1 (a) 0.7264 (b) 3.3333E-07
3.2 (a) 0.824 ± 0.006 cm (b) 0.824 ± 0.008 cm
3.3 $\widehat{\alpha} \approx 1.4$ $\widehat{\beta} \approx 1.85 \times 10^5$ sec
3.4 $\widehat{\alpha} \approx 2.3$ $\widehat{\beta} \approx 1.5 \times 10^6$ cycles
3.5 $\widehat{\alpha} \approx 1.95$ $\widehat{\beta} \approx 1.87 \times 10^5$ sec
3.6 1733
3.7 0.0868
3.8 $P(A_1|B) = 0.333$ $P(A_2|B) = 0.667$
3.9 A: 0.045; B: 0.136; C: 0.818
3.10 3.10 0.037

3.11 0.333
3.12 0.5
3.13 (a) M: 0.022; W: 0.688; E: 0.290 (b) 0.0308 (c) 4.36×10^{-6}
(d) 4.39×10^{-6}
3.14 (a) 0.058, 0.280, 0.328, 0.334 (b) 9.37×10^{-3}, 0.136, 0.424, 0.431
3.15 (a) 96 (b) 0.0454/demand (c) 0.0494/demand
3.17 (a) 99.2% (b) Reject
3.18 Reject

Chapter 4

4.1 (a) $R^N(2-R)^N$ (b) $R^N(2-R^N)$
4.2 $\exp[-(\lambda_1+\lambda_2)t][2-\exp(-\lambda_2 t)]$
4.3 $\exp[-(\lambda_1+\lambda_2)t][3-3\exp(-\lambda_2 t)+\exp(-2\lambda_2 t)]$
4.4 $\exp[-(\lambda_1+2\lambda_2)t][3-2\exp(-\lambda_2 t)]$
4.5 (a) 0.504 (b) 0.954
4.6 (a) $R_1 = R^2(8-16R+14R^2-6R^3+R^4)$ $R_2 = R^2(4-2R-4R^2+4R^3-R^4)$
4.7 (a) 0.241 (b) 0.754
4.8 (a) 0.9995 (b) 0.9950
4.9 0.99090432
4.10 0.931
4.11 0.537
4.12 (a) $12\exp(-2\lambda t)-28\exp(-3\lambda t)+27\exp(-4\lambda t)-12\exp(-5\lambda t)+2\exp(-6\lambda t)$
(b) $27/20\lambda$
4.13 (b) 1.25×10^6 hr
4.14 (b) For (1), (2), and (3): $11/6\lambda,\ 5/6\lambda,\ 3/\lambda$
4.15 $2\exp(-\lambda_1 t)-\exp(-2\lambda_2 t)+1.98\lambda_1\{\exp(-\lambda_2 t)-\exp[-(\lambda_1+\lambda_2^*)t]\}/(\lambda_1+\lambda_2^*-\lambda_2)-1.98\lambda_1\{\exp(-\lambda_2 t)-\exp[-(2\lambda_1+\lambda_2^*)t]\}/(2\lambda_1+\lambda_2^*-\lambda_2)$
4.16 (a) 0.99998 (b) 0.998
4.17 (b) $\lambda_1^{-1}+\lambda_2^{-1}$ (c) $\mathrm{MTTF}_{\mathrm{standby}} > \mathrm{MTTF}_{\mathrm{parallel}} > \mathrm{MTTF}_{\mathrm{series}}$
4.18 $[\lambda_1\lambda_2\exp(-\lambda t)]/[(\lambda_1-\lambda)(\lambda_2-\lambda)]+[\lambda\lambda_2\exp(-\lambda_1 t)]/[(\lambda-\lambda_1)(\lambda_2-\lambda_1)]+[\lambda\lambda_1\exp(-\lambda_2 t)]/[(\lambda-\lambda_2)(\lambda_1-\lambda_2)]$
4.19 (a) $R_1(t) = 3\exp(-2\lambda_1 t)-2\exp(-3\lambda_1 t)$ $R_2(t) = 2\exp(-\lambda_2 t)-\exp(-2\lambda_2 t)$
$R_s = 0.992$ (b) $R_{\mathrm{sys}} = 3\exp(-2\lambda_1 t)-2\exp(-3\lambda_1 t)-6\lambda_1 R_s\{[\exp(-2\lambda_1 t)-\exp(-2\lambda_2 t)]/[2(\lambda_1-\lambda_2)]-2[\exp(-2\lambda_1 t)-\exp(-\lambda_2 t)]/[2\lambda_1-\lambda_2]+2[\exp(-3\lambda_1 t)-\exp(-\lambda_2 t)]/[3\lambda_1-\lambda_2]-[\exp(-3\lambda_1 t)-\exp(-2\lambda_2 t)]/[3\lambda_1-2\lambda_2]\}$
4.20 (a) $R_3(R_1+R_4-R_1R_4)+(1-R_3)(1-R_4)R_1R_2+(1-R_3)R_4(R_2+R_5-R_2R_5)$ with $R_n = \exp(-\lambda_n t)$ (b) $1/\lambda$
4.22 (c) $R, 2R-R^2, 3R^2-2R^3$

Chapter 5

5.1 (a) 5.15×10^4 hr (b) 1.5×10^3 hr
5.2 (b) $\sum_{n=1}^{4}(1/\lambda_n)$

5.3 (b) $\mu_1\mu_2/s_1s_2 + (s_1+\mu_1)(s_1+\mu_2)\exp(s_1t)/[s_1(s_1-s_2)] + (s_2+\mu_1)(s_2+\mu_2)\exp(s_2t)/[s_2(s_2-s_1)]$ for $s_{1,2} = \{-(\lambda_1+\lambda_2+\mu_1+\mu_2) \pm [(\lambda_1+\lambda_2+\mu_1+\mu_2)^2 - 4\mu_1\mu_2 - 4\lambda_1\mu_2 - 4\lambda_2\mu_1]^{1/2}\}/2$ (c) $\mu_1\mu_2/(\mu_1\mu_2+\lambda_1\mu_2+\lambda_2\mu_1)$
(d) $(\lambda_1+\lambda_2)^{-1}$
5.4 (b) $-[s_2\exp(s_1t) - s_1\exp(s_2t)]/(s_1-s_2)$ for $s_{1,2} = -[(5\lambda+\mu) \pm (\lambda^2 + 10\lambda\mu + \mu^2)^{1/2}]/2$ (c) $(5\lambda+\mu)/6\lambda^2$
5.5 (b) $1 - 3(\lambda t)^2$ (c) $n\mu(3\lambda+\mu)/[n\mu(3\lambda+\mu)+6\lambda^2]$
5.6 (b) $[2\mu(2\lambda+\lambda^*)+\mu^2]/[2(\lambda+\mu)(2\lambda+\lambda^*)+\mu^2]$ (c) $[4\lambda+\lambda^*+\mu]/[2\lambda(2\lambda+\lambda^*)]$
5.7 (b) $[4\lambda\mu + 2\mu^2 + 2\mu\lambda_c - \lambda\lambda_c]/[4\lambda\mu + 2\mu^2 + 2\mu\lambda_c + \lambda\lambda_c + 2\lambda^2]$ (c) $[3\lambda + \mu + \lambda_c]/[2\lambda^2 + \lambda\lambda_c]$
5.8 (b) $2(\lambda\mu + \mu^2 + \mu\lambda_c)/(2\lambda\mu + 2\mu^2 + \lambda^2 + \lambda\lambda_c + 3\mu\lambda_(c)$ c) $(2\lambda+\mu)/(\lambda^2 + \lambda\lambda_c + \lambda_c\mu)$
5.9 In Example 5.7 let $\mu \to \sigma$ and $\tau \to \mu$
5.10 There are 5 system states
5.11 There are 4 system states
5.12 There are 7 system states
5.13 There are 5 system states

Chapter 6

6.3 (c) $\approx 10^{-8}$ (f) 4×10^{-6}
6.4 (c) 2×10^{-9}
6.7 (b) 3.6×10^{-2} (c) 3.04×10^{-4}
6.8 (b) 3.22×10^{-2} (c) Tank rupture
6.9 1.8×10^{-4}/year
6.10 0.02/year
6.11 1.07×10^{-7}/year
6.12 0.078/year
6.13 (a) 0.031/year (b) 0.021/year
6.14 1.2×10^{-7}/year

Chapter 7

7.1 (a) 3.921×10^{-6}/demand
7.2 8.11×10^{-7}/year

Chapter 8

8.1 (a) 23.6 sec (c) 7%
8.2 (c) 6.5×10^{-3} Way-Wigner, 4.9×10^{-3} ANS 5.1 at $t = 10^5$ sec
8.3 (a) 0.19 GWt (b) 0.23
8.4 −35 pcm/K
8.5 960 ppm

8.6 1.1 rad/s
8.7 6.4×10^{-5} s/m^3
8.8 4.4 kCi
8.9 (a) 8.8 mCi/m^3 (b) 75 nCi/m^3 (c) 1.3×10^5
8.10 (a) 3.4×10^{24} atoms (b) 7.08 kCi
8.11 0.97 nCi/m^3
8.12 0.11 mrem
8.13 (a) 1.35×10^{-6} s/m^3 (c) 0.11 rem
8.14 (a) 0.29 μCi/m^3 (b) 285
8.15 0.6 kW/assembly
8.16 (a) 0.013 μCi/m^3 (b) 373

Chapter 9

9.1 (a) 2.05×10^{-4}/demand (b) 0.01/demand
9.3 9.2×10^{-5}/reactor·year
9.5 4.84×10^{-4}/demand
9.6 (a) 0.029 (c) 2
9.7 0.73 kg/m^2

Chapter 10

10.1 (b) 6.3×10^{-4}/year

Chapter 11

11.7 (a) $1.4 \times 10^{-5}/\rho$ (b) \$2.18
11.8 $T_f(t) - T_c = [T_f(0) - T_c][1 - 0.1 \exp(-At)]$ with $A = MHU/M_f C_f$

Chapter 12

12.1 A
12.2 No

Chapter 13

13.4 $x^+ = (\sigma_y^2 x^- + \sigma_x^2 y)/(\sigma_y^2 + \sigma_x^2)$

INDEX AND ACRONYMS